Radar for Meteorological and Atmospheric Observations

Shoichiro Fukao • Kyosuke Hamazu

Radar for Meteorological and Atmospheric Observations

Consulted by Richard J. Doviak

Springer

Shoichiro Fukao
Professor Emeritus
Kyoto University, Kyoto, Japan

Kyosuke Hamazu
Mitsubishi Electric Corporation and
Mitsubishi Electric Tokki Systems
Corporation
Iga, Japan

Consulted By:
Richard J. Doviak
National Severe Storms Laboratory, NOAA
Affiliated Professor
The School of Meteorology and
the Department of Electrical
and Computer Engineering
The University of Oklahoma

ISBN 978-4-431-54333-6 ISBN 978-4-431-54334-3 (eBook)
DOI 10.1007/978-4-431-54334-3
Springer Tokyo Heidelberg New York Dordrecht London

Library of Congress Control Number: 2013943813

Printed on acid-free paper

Springer is part of Springer Science+Business Media (www.springer.com)

Foreword

During the past several decades an appreciable amount of research and development has been focused on the use of remote sensing techniques to better our understanding of weather and the atmosphere. Radar has the obvious advantage of providing observations with temporal and/or spatial continuity which is leading to improved forecasts of weather.

Observations and interpretation of Doppler and polarimetric weather radar data, combined with in situ observations, have led to giant leaps in our understanding of the dynamics and microphysics of weather systems. Complementary to weather radar observations are those obtained with typically longer wavelength radars (i.e., wavelengths from meters to centimeters versus centimeters to millimeters used to observe precipitation and clouds), observing the precipitation-free atmosphere. The echoing mechanism at these longer wavelengths is typically Bragg scatter from refractive index perturbations caused by turbulent mixing, or reflection from sharp gradients in refractive index. These long-wavelength and super-powerful radars, referred to as atmospheric radars, have mapped the vertical structure of reflectivity and radial winds in the clear atmosphere from below a kilometer to well above 100 km, whereas meteorological radars map the reflectivity and radial velocities of precipitation and cloud particles on horizontal surfaces at various heights in the troposphere. Weather and cloud radar research has attracted the attention of meteorologists whereas atmospheric radar research has primarily attracted the attention of atmospheric physicists.

The authors have done a remarkable job of combing the results of research in these two disciplines to provide readers with a comprehensive overview of the outstanding observations that have been made with radar used as a remote sensor of weather and atmospheric phenomena. This book has a generous amount of figures that display many of the remote sensing facilities to give the reader a quick appreciation for the variety of atmospheric and meteorological radar types around the world, many of which are unique and interesting. Furthermore, liberal reference to publications provides readers a vast reservoir for further pursuit of their preferred topics of interest. In addition this book presents the fundamentals of remote sensing so that students and professors, with a minimal background in physics and electromagnetic theory, and engineers in the field can better understand the potential and limitations of radar in observing weather and the atmosphere while learning about the various instruments and techniques used in remote sensing. The authors plan to maintain a Website where comments from readers can be addressed and where supplements to the book can be found; this will help to keep the book current and up-to-date.

Norman, OK Richard J. Doviak

Preface

With the application of radar to observations of the atmosphere, various weather phenomena and winds in the clear atmosphere can be monitored and mapped in real time. Great progress in understanding weather and the dynamics of the atmosphere has been made using radar, which brings new observational discoveries and promotes further understanding of our environment.

Remote sensing with radar has been developed in the interdisciplinary domains of physical science and engineering. In the past, advances in weather and the atmospheric sciences have developed independently because the respective engineering efforts and scientific studies were conducted within relatively separate communities. However, the scientific and technical bases for atmospheric observations with radar can be treated in common. We worked in academia (Fukao) and industry (Hamazu) and have collaborated to develop various types of weather and atmospheric radars. Routine discussion with our colleagues convinced us that understanding of weather and atmospheric radars can be deepened if they are described comprehensively and systematically in one volume using common approaches whenever possible.

This book is written for scientists, engineers, students, and other interested meteorological and atmospheric personnel. In this book, we try to bridge the gap in our understanding of weather and atmospheric radar. The book consists of two parts. The first half, Chaps. 1–7, mainly discusses the theoretical bases of weather and atmospheric radar, and the last half, Chaps. 8–12, describes actual systems and observations with these radars. This interdisciplinary book was first published in Japanese by the Kyoto University Press in 2005. In the English version, all chapters including those dealing with recent developments contain more in-depth coverage than does the original.

Kyoto, Japan — Shoichiro Fukao
Iga, Japan — Kyosuke Hamazu

Acknowledgements

We are indebted to many people in bringing this book to publication in this form. First and foremost, we wish to express our deepest gratitude to Dr. Richard J. Doviak of the National Severe Storms Laboratory of the National Oceanic and Atmospheric Administration (NOAA) and the University of Oklahoma who has long been a close friend of one of the authors (Fukao) and has been looked upon as mentor for the other (Hamazu). All chapters have been reviewed and edited by him. We were able to complete the work because of his continuous suggestions and stimulating encouragement.

We would like to express our deep gratitude to our colleagues, Drs. Toru Sato, Toshitaka Tsuda, Mamoru Yamamoto, and Hiroyuki Hashiguchi of Kyoto University; Dr. Takuji Nakamura of the National Institute of Polar Research, Japan; Dr. Manabu D. Yamanaka of the Japan Agency for Marine-Earth Science and Technology (JAMSTEC) and Kobe University; and Drs. Hiroaki Miyasita, Atsushi Okamura, Toshio Wakayama, and Shoji Matsuda of Mitsubishi Electric Corporation for various suggestions through discussions. Many of them provided us with the original figures that are included in the book. It is a great pleasure for us to acknowledge the valuable advice and suggestions of Emeritus Professor Hisanao Ogura of Kyoto University; Dr. Hubert Luce of the Université de Toulon et du Var; Drs. Yasushi Fujiyoshi and Takeshi Horinouchi of Hokkaido University; Dr. Toshio Iguchi of the National Institute of Information and Communication Technology (NICT), Japan; Dr. Hiroshi Uyeda of Nagoya University; Dr. Masahiro Ishihara of Kyoto University; Dr. Masayuki Maki of Kagoshima University; Dr. Koyuru Iwanami of the National Research Institute for Earth Science and Disaster Prevision (NIED), Japan; and Drs. Ahoro Adachi and Hiroshi Yamauchi of the Meteorological Research Institute, Japan Meteorological Agency. We would like to express our special thanks to Dr. Yoshiaki Shibagaki of Osaka Electro-Communication University; Seiji Kawamura of the NICT; Drs. Masayuki Yamamoto, Tomohiko Mitani, Junichi Furumoto, and Tatsuhiro Yokoyama of Kyoto University; and Drs. Nobukyuki Kawano and Akihisa Uematsu of the Japan Aerospace Exploration Agency (JAXA). In writing this book, we received warm encouragement from emeritus professors Susumu Kato and Isamu Hirota of Kyoto University.

Finally, we are especially grateful for the support of our respective spouses, Keiko Fukao and Masuni Hamazu, during the writing of this book.

Kyoto, Japan Shoichiro Fukao
Iga, Japan Kyosuke Hamazu

Contents

List of Symbols

a	Attenuation rate $[\mathrm{m}^{-1}]$, mean radius of the Earth (6370 km), semi-major axis diameter of spheroid rain drop
a_e	Effective Earth radius
a_T	Temperature lapse rate
A	Attenuation coefficient $[\mathrm{dB\ km}^{-1}]$, physical antenna aperture
$\boldsymbol{A}$	Vector potential
A_e	Effective antenna aperture
b	Semi-minor axis diameter of spheroid rain drop
B	Frequency bandwidth of the receiver
$\boldsymbol{B}$	Magnetic flux density
B_f	Filter bandwidth
B_n	Noise bandwidth
c	Speed of light (in vacuum) $[\mathrm{m\ s}^{-1}]$
c_a	Sound velocity
$\boldsymbol{c}_\mathrm{a}$	Apparent sound velocity
$\boldsymbol{c}_\mathrm{s}$	True sound velocity
C_n^2	Refractive index structure constant $[\mathrm{m}^{-2/3}]$
C_p	Specific heat capacity at constant pressure ($\simeq 1004$) $[\mathrm{J\ K}^{-1}\ \mathrm{kg}^{-1}]$
d	Distance between successive element antennas
D	Detectability of radar signal, diameter of raindrop, wind direction,
$\boldsymbol{D}$	Electric flux density
D_0	Median volume diameter
D_a	Antenna diameter, distance of two separated antennas
D_m	Mass weighted mean drop diameter
D_r	Dynamic range of A/D conversion
D_rmax	Maximum dynamic range
e	Partial pressure of water vapor [hPa]
E	Total energy of a receiver input signal, withstand voltage $[\mathrm{V\ mm}^{-1}]$
$\boldsymbol{E}$	Electric field strength
$\boldsymbol{E}_0$	Incident electric field
E_a	Array factor

$\boldsymbol{E}_s$	Scattered electric field
f	Radar frequency (transmitted frequency) [Hz]
f_0	Carrier frequency [Hz]
f_c	Frequency of coherent oscillator (COHO)
f_d	Doppler frequency (Doppler shift)
f_{dmax}	Maximum measurable Doppler frequency
f_i	Inertial frequency
f_N	Nyquist frequency
f_p	Pulse repetition frequency
f_s	Frequency of stabilized local oscillator (STALO), sampling frequency
F	Noise figure
F_r	Froude Number
g	Antenna gain at the direction of the maximum radiation pattern (main lobe) in linear unit, radiation pattern of the element antenna (or element pattern), gravitational acceleration
g_{at}	Transmission gain of the RASS
g_D	Directivity of antenna
G	Antenna gain in decibel
h	Altitude (height from sea level), beam height, mountain height
$\boldsymbol{H}$	Magnetic field strength
H_1	Scale height (7.3 km)
$\mathbf{i}_i$	Unit vector along the radar beam direction
I	Electric current, in-phase component of the complex signal
I_a	Acoustic intensity $[\mathrm{W\,m^{-2}}]$
j	Imaginary unit ($j^2 = -1$)
$\boldsymbol{J}$	Electric current density
k	Boltzmann constant ($= 1.38 \times 10^{-23}$ J K^{-1}), radar wave number ($= \omega\sqrt{\varepsilon\mu} = 2\pi/\lambda$)
k_a	Imaginary part of the complex refractive index, wave number of acoustic wave
$\boldsymbol{k}_s$	Scattering vector wave number
K	Thermodynamic temperature measured in kelvins, vertical eddy diffusivity
K_{DP}	Specific differential phase $[\mathrm{deg\,km^{-1}}]$
l	Autocorrelation time lag, length of short dipole (differential antenna), loss value in a true number
$\boldsymbol{l}$	Separation of the scatterer from the volume center
l_0	Inner scale of turbulence
l_K	Kolmogoroff microscale
L	Loss value in decibel
L_B	Maximum scale of eddy in the inertial subrange (or buoyancy lengthscale)
Ldr	Linear depolarization ratio in linear unit
LDR	Linear depolarization ratio in decibel
m	Complex refractive index of drop (or particle), modified refractive index, vertical wavenumber
m_n	The nth moment of drop size distribution

M	Mean molecular weight of the atmosphere, number of DFT or FFT points, number of signal samples along sample time axis (total number of samples), refractive modulus,
M_B	Total number of points of periodogram (FFT points)
M_{coh}	Number of coherent integration
M_I	Number of independent samples
M_{inc}	Number of incoherent integration
M_n	Refractive index gradient (=dn/dz)
M_s	Total number of actual signal samples
M_v	Total water vapor content [kg mm^{-3}]
n	Refractive index
n_r	Real part of the complex refractive index
N	Bit length, Brunt Väisälä frequency, number of element antenna, number of raindrops, number of range samples, Nyquist number
N_0	Parameter of drop size distribution (intercept parameter)
$N(D)$	Drop size distribution (DSD)
N_e	Density of free electron [m^{-3}]
N_T	Total number of raindrops
p	Atmospheric pressure [hPa]
P	Breakdown power, total electric power
$\boldsymbol{P}$	Dielectric polarization
P_a	Transmitted power from sound wave source
P_{ar}	Received power backscattered from sound wave surface
P_r	Received signal power
P_s	Scattered power
P_t	Transmitted power, peak transmitted power
$\boldsymbol{P}_V$	Dipole moment
q	Humidity mixing ratio [kg kg^{-1}]
q_e	Linear density of meteor trail [m^{-1}]
Q	Quadrature phase component of the complex signal
r	Distance between the radar and the scatterer, range
r_a	Maximum observable range
r_R	Distance between bistatic scatterer and receiver
r_T	Distance between transmitter and bistatic scatterer
R	Gas constant, rainfall rate,
R_d	Transmitter's duty cycle
R_f	Flux Richardson number
R_i	Richardson number
R_R	Radiation resistance of short dipole
R_{sp}	Specific constant of drying air ($= 287$ J K^{-1} kg^{-1})
s_f	Frequency stability
$\boldsymbol{s}$	Backscattering matrix of the linear polarization wave
S	Power density, signal power
$\boldsymbol{S}$	Complex Poynting vector
S_i	Incident power density

S_N Power spectral density of noise
S_s Scattered power density
S_S Power spectral density of signal
S_w Vertical shear [s^{-1}]
SNR Signal-to-noise ratio
t Time
T Atmospheric temperature [K], noise temperature [K], period of gravity wave, pulse repetition time (PRT) [s], time period
T_0 Room temperature (290 K)
T_c Correlation time
T_e Equivalent input noise temperature
T_i Input noise temperature, independent sample time
T_s Sample time interval (sampling interval), sky noise temperature
T_{sys} System noise temperature
T_v Temperature of moist atmosphere
T_W Window width
u East-west (zonal) wind
$\overline{u}$ Mean zonal wind
u' Horizontal IGW perturbation from turbulence
U Horizontal wind speed
v Phase velocity of electromagnetic wave
v' Fluctuation component of wind perpendicular to the direction of wave travel
$\boldsymbol{v}$ Wind vector (v_x, v_y, v_z)
v_d Doppler velocity
$\overline{v}_d$ Mean Doppler velocity
v_h Horizontal wind velocity
v_N Nyquist velocity (Nyquist limit)
v_r Radial velocity
V Radar resolution volume
V_6 Resolution volume circumscribed by the 6 dB contour of radar parameters
V_D Volume of raindrop
w Vertical wind velocity (or vertical component of wind velocity; v_z)
w' Vertical IGW perturbation from turbulence
w_T Terminal velocity of precipitation (fall speed)
W Cloud water content (or water content in unit volume) [$g\ m^{-3}$]
W_B Bandwidth of the signal
z Altitude, height from sea level [km]
Z Radar reflectivity factor
Z_{dr} Differential reflectivity in linear unit
Z_{DR} Differential reflectivity in decibel
Z_e Equivalent radar reflectivity factor
Z_i Radar reflectivity factor for ice particles
α Azimuth angle of the baseline formed between two antennas in SDI
β Bistatic angle

γ Specific heat ratio of ideal gas ($\simeq$ 1.4 for dry air)
Γ Dry adiabatic lapse rate ($= g/C_\mathrm{p} \simeq 9.80$) [K km^{-1}]
δ Differential scattering phase, direction of horizontal wind, phase difference between successive element antennas
Δ Resolution of the A/D converter
ε Turbulent energy dissipation rate
ε Permittivity [F m^{-1}]
ε_0 Permittivity in vacuum [F m^{-1}]
ζ Axis ratio b/a, where a is the semi-major axis diameter and b the semi-minor axis diameter of a flat raindrop
η Radar reflectivity
η_1 Efficiency of antenna
η_a Antenna aperture efficiency
η_i Intrinsic impedance (or wave impedance) ($= \sqrt{\mu/\varepsilon}$)
θ Zenith angle of radar beam
θ_1 One-way beamwidth between half-power points (or beam width)
θ_e Elevation angle of radar beam
ϑ_B One way half-power beamwidth in the E-plane [rad]
Θ Potential temperature
κ Wave number for Bragg scattering
$\boldsymbol{\kappa}$ Wave number vector for Bragg scattering
κ_a Wave number vector for acoustic wave
κ_b Wave number corresponding to the Bragg scale
κ_B Wave number corresponding to buoyancy lengthscale ($= 2\pi/L_\mathrm{B}$)
λ Radar wavelength [m]
Λ Parameter of drop size distribution (or slope parameter)
Λ_s Structure wavelength of perturbations within inertial subrange
μ Permeability [H m^{-1}]
μ_0 Permeability in vacuum [H m^{-1}]
ν Kinematic viscosity (dynamic viscosity divided by the fluid density)
ρ Electric charge density [C m^{-1}], radar cross section [m^2]
$|\rho|^2$ Partial reflection coefficient
ρ_a Atmospheric density [kg m^{-3}]
ρ_hv Correlation coefficient between horizontally and vertically polarized waves
ρ_v Water vapor density [g m^{-3}]
ρ_w Density of precipitation particles [g m^{-3}](= 10^6 for water)
σ Electric conductivity [S m^{-1}]
σ_a Absorption cross section
σ_b Backscattering cross section
σ_f Doppler frequency spectrum width [Hz]
σ_s Scattering cross section
σ_t Extinction (or attenuation) cross section
σ_v Doppler velocity spectrum width [m s^{-1}]
σ_{vn} Doppler velocity spectrum width normalized with the Nyquist width
τ Transmitted pules width [s], time lag

τ_i	Independent sample time
τ_c	Correlation time
ϕ	Angular distance from the beam axis in the H-plane
ϕ_h	Phase delay per unit distance (one way) for horizontally polarized wave [rad]
ϕ_v	Phase delay per unit distance (one way) for vertically polarized wave [rad]
Φ_{DP}	Differential phase in two-way ($\Phi_{DP} = \Phi_{hh} - \Phi_{hh}$) [deg]
Φ_{hh}	Phase shift in round trip between radar and scatterer for horizontally polarized wave [deg]
Φ_{vv}	Phase shift in round trip between radar and scatterer for vertically polarized wave [deg]
φ	Phase of received echo signal, zenith angle in the H-plane based on radar beam axis
φ_B	One way half-power beamwidth in the H-plane [rad]
χ	Angle between the direction of polarization of the incident electric field and the direction of scattering vector (= $\pi/2$ for backscattering
Ψ	Differential phase of measured signals between horizontally and vertically polarized waves [deg], scalar potential
ω	Angular frequency [rad s^{-1}]
ω_d	Doppler angular frequency
ω_i	Intrinsic frequency
Ω	Angular velocity of the Earth's rotation ($= 7.292 \times 10^{-5}\ s^{-1}$)

List of Abbreviations

A/D	Analog to digital
AFWS	Air Force Weather Service
AGC	Automatic gain control
AGL	Above ground level
AMeDAS	Automated Meteorological Rata Acquisition System
AMS	American Meteorological Society
ARM	Atmospheric Research Measurement program
ATC	Air traffic control
ATSR	Alternate transmission and simultaneous reception
BL	Boundary layer
BLR	Boundary layer radar
CAP	Cooperative Agency Profiler
CAT	Clear air turbulence
CCIR	International Radio Consultative Committee
CDL	Coherent Doppler lidar
CIRA	Committee on Space Research (COSPA) International Reference Atmosphere
COCO	Coaxial-collinear
COHO	Coherent oscillator
COST	European Cooperation in Science and Technology
CRI	Coherent radar imaging
CST	Central Standard Time
CSU	Colorado State University
DBS	Doppler beam swinging
DFT	Discrete Fourier transform
DIF	Decimation-in-frequency
DIT	Decimation-in-time
DOA	Direction of arrival
DPR	Dual-frequency Precipitation Radar
DRAW	Doppler Radar for Airport Weather
DSD	Drop size distribution

EAR	Equatorial Atmospheric Radar
ECCD	Electromagnetically coupled coaxial dipole
EIK	Extended interaction amplifier
EST	Eastern Standard Time
FAA	Federal Aviation Administration
FCA	Full correlation analysis
FDI	Frequency domain interferometry
FET	Field effect transistor
FFT	First Fourier transform
FII	Frequency domain interferometric imaging
FIR	Finite impulse response
FMCW	Frequency-modulated continuous waves
FRP	Fiber-reinforced plastic
FSA	Full spectral analysis
FWHM	Full width at half maximum
GMAP	Gaussian model adaptive processing
GMS	Geostationary meteorological satellite
GMT	Greenwich mean time
GPM	Global Precipitation Measurement
GPS	Global positioning system
GTS	Global Telecommunication System
HEMT	High electric mobility transistor
HS	Hail signal
HVPS	High-Volume Particle Spectrometer
I	In-phase
IDFT	Inverse discrete Fourier transform
IF	Intermediate frequency
IFFT	Inverse fast Fourier transform
IGW	Inertia-gravity wave
IIR	Infinite impulse response
IR	Infrared radiation
IS	Incoherent scatter
ITU	International Telecommunication Union
JAFNA	Joint Air Force and NASA
JAXA	Japan Aerospace Exploration Agency
JMA	Japan Meteorological Agency
JST	Japan Standard Time
KH	Kelvin–Helmholtz
KIX	Kansai International Airport
LAN	Local area network
LDR	Linear depolarization ratio
LEO	Low Earth orbit
LHC	Left-hand circular
LLJ	Low-level jet
LNA	Low noise amplifier

LO	Local frequency
LT	Local time
LTR	Lower Troposphere Radar
M-P	Marshall–Palmer
MEM	Maximum entropy method
MESFET	Metal-semiconductor FET
ML	Multi-lag
MLIT	Ministry of Land, Infrastructure, Transport and Tourism
MLM	Maximum likelihood method
MMIC	Monolithic microwave integrated circuit
MOPA	Master oscillator and power amplifier
MP	Multi-parameter
MPIfR	Max-Planck-Institut für Radioastronomie
MPM	Millimeter-wavelength propagation model
MRI	Meteorological Research institute
MSM	Mesoscale numerical model
MST	Mesospheric-stratospheric-tropospheric
MU	Middle and Upper atmosphere
MUSIC	Multiple signal classification
NASA	National Aeronautics and Space Administration
NCAR	National Center for Atmospheric Research
NIED	National Research Institute for Earth Science and Disaster Prevention
NOAA	National Oceanic and Atmospheric Administration
NPN	NOAA Profiler Network
NSSL	National Severe Storms Laboratory
NWS	National Weather Service
ORDA	Open radar data acquisition
OTH	Over the horizon
PA	Power-aperture
PANSY	Program of the Antarctic Syowa MST/IS Radar
PBL	Planetary boundary layer
PBS	Post beam steering
PHS	Personal Handy-phone System
POS	Positioning
PPI	Plan position indicator
PR	Precipitation radar
PRF	Pulse repetition frequency
PRT	Pulse repetition time
PSS	Post static steering
PUP	Principal user processor
Q	Quadrature-phase
RASS	Radio acoustic sounding system
RCS	Radar cross section
RDA	Radar data acquisition

rf	Radio frequency
RHC	Right-hand circular
RHI	Range height indicator
RIM	Range imaging
ROPS	Radar Observation data Processing System
RPG	Radar product generator
RPM	Rotation per minute
RX	Receiver
SA	Spaced antenna
SAD	Spaced antenna drift
SCSI	Small Computer System Interface
SDI	Spatial domain interferometry
SI	Le Système International
SNR	Signal-to-noise ratio
SPBS	Sequential post beam steering
SSPA	Solid state power amplifier
ST	Stratospheric-tropospheric
STALO	Stabilized local oscillator
STC	Sensitivity time control
STSR	Simultaneous transmission and simultaneous reception
SVD	Singular value decomposition
T	Tropospheric
TAI	Temps Atomique International
TC	Tropical cyclone
TDWR	Terminal Doppler Weather Radar
TEP	Turbulent eddy profiler
TOGA-COARE	Tropical Ocean Global Atmosphere–Coupled Ocean Atmosphere Research Experiment
TPPN	Trans-Pacific Profiler Network
TR	Transmitter/receiver
TRMM	Tropical Rainfall Measurement Mission
TWT	Traveling wave tube
TX	Transmitter
UHF	Ultrahigh frequency
UTC	Coordinated Universal Time
VAD	Velocity azimuth display
VCP	Volume coverage pattern
VHF	Very high frequency
VIL	Vertical integrated liquid
VVP	Volume velocity processing
WCB	Warm conveyor belt
WCRP	World Climate Research Program
WINDAS	Wind Profiler Data Acquisition System
WRC	World Telecommunication Conference
WMO	World Meteorological Organization

Chapter 1
Introduction

1.1 Principle of Radar

A variety of weather and atmospheric phenomena occur and change every moment in the Earth's atmosphere. This book presents the techniques and sciences of remote sensing various phenomena with radar. Remote sensing is a technique that indirectly measures target without touching it directly in a distant place. Radar is an abbreviation for "RAdio Detection And Ranging", which is an electronic system that generates electromagnetic waves in the transmitter, radiates them into space via antenna, receives the scattered signal returning from the target, and measures the position, movement of the target, etc. Usually, the same antenna is used for transmission of the electromagnetic wave and reception of the return signal. The target position is obtained according to the direction where the scattered signal returns to the antenna, and to the distance calculated by the lapse of time that the electromagnetic waves make in the round-trip between radar and target.

As for the targets that scatter electromagnetic waves, various types of scatterers are known, e.g., isolated objectives such as aircrafts and ships, minute distributed particles such as precipitation and clouds, and perturbations of radio refractive index due to atmospheric turbulence. In this book, the properties of scatterers such as precipitations, clouds, and fogs associated with weather, and refractive index perturbations caused by atmospheric turbulence are presented. The former is mainly observed with meteorological radar (or weather radar), and the latter with atmospheric radar. The conceptual diagrams of meteorological radar and atmospheric radar are shown in Fig. 1.1a and b, respectively. The atmospheric radars typically make observations overhead (i.e., at high elevation angles), whereas meteorological radars typically scan the atmosphere at relatively low elevation angles. Furthermore meteorological radars typically use parabolic reflector antennas whereas atmospheric radars use phased array antennas. Although the frequencies adopted for meteorological and atmospheric radars are different due to the difference of scattering mechanisms of the targets, many aspects of the basic configuration

S. Fukao and K. Hamazu, *Radar for Meteorological and Atmospheric Observations*,
DOI 10.1007/978-4-431-54334-3_1,

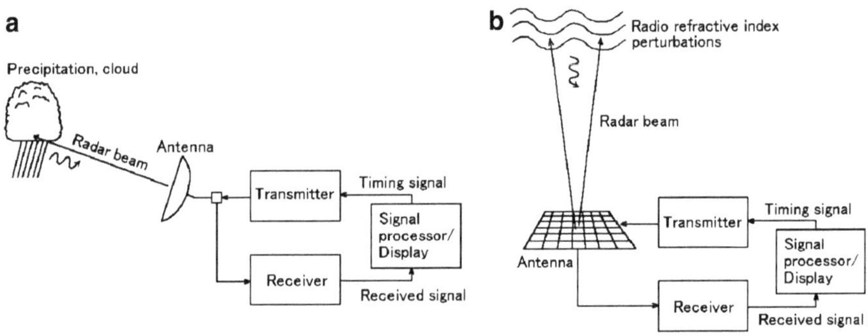

Fig. 1.1 Conceptual diagrams of (**a**) meteorological radar and (**b**) atmospheric radar

and algorithms of signal processing of these radars are common. Therefore, the common components of both radars are stated as uniformly as possible, while the uncommon ones are dealt with in the individual chapters.

The upper atmosphere above the 100 km altitude is the ionosphere where the atmosphere is partially ionized. In the ionosphere, electromagnetic waves are scattered by free electrons, and so the scattering mechanism in this region is not the same as those in other parts of the atmosphere (Gordon 1958). The scattering is very close to incoherent scatter (IS) but extremely weakly affected via Coulomb force by ions. The radar which utilizes IS from the ionosphere is called incoherent scatter radar or IS radar (e.g., Evans 1969). The IS radar is beyond the scope of this book, and we briefly state in Sect. 10.2.1 that some large-scale atmospheric radars have the capability of IS radar.

1.2 History of Meteorological and Atmospheric Radars

The atmosphere has been studied using radar since the 1920s. A basic model of present pulsed meteorological radar in the microwave band was first put in practice for precipitation observations in early 1940s.

Meanwhile, prior to the practical implementation of the meteorological radars, the scattering mechanism of electromagnetic wave had already been theoretically clarified by L. Rayleigh and G. Mie. Rayleigh showed that the magnitude of the backscattering intensity due to precipitation particles in the atmosphere is proportional to the 6th power of the diameter of the scatterer, and inverse proportional to the 4th power of the wavelength of the electromagnetic wave (e.g., Gunn and East 1954; Battan 1973, p. 38).

Moreover, Mie showed that in the area where the diameter of the scatterer is about 1/10 or larger the wavelength for which Rayleigh's scattering theory is not applied, more rigorous scattering theory is necessary (e.g., Gunn and East 1954). These theories are known as Mie scattering theorem and Rayleigh scattering theorem, and will be discussed in Sects. 3.3.3 and 3.3.4, respectively.

Ryde (1946) theoretically estimated the reflection intensity and the attenuation of microwave due to precipitation and cloud particles in the atmosphere based on these theories during 1941 and 1946. His theories became the basics of quantitative observation of weather with radar. When the 1st Conference on Radar Meteorology sponsored by the American Meteorological Society (AMS) was held in the Massachusetts Institute of Technology (MIT) in March 1947, quantitative observations of precipitation had already been advanced around the United States. The relation between radar reflectivity factor and precipitation intensity and the drop size distributions of precipitation found by Marshall and Palmer (1948) became the beginning of the radar meteorology research.

On the other hand, the research on the communication over the horizon (OTH) using the radio wave propagation through the troposphere became active in the 1950s. Understanding about clear air turbulence (CAT) and the scattering mechanism in the troposphere have been rapidly advanced through this research, promoting researches regarding atmospheric remote sensing (Booker and Gordon 1950).

The origin of atmospheric radar is the IS radar at Jicamarca, Peru, and many atmospheric radars had been developed and constructed in the 1970s to 1980s. Woodman and Guillén (1974) showed, for the first time, that the wind and the turbulence in the mesosphere and the stratosphere can be observed using the IS radar at Jicamarca. Atmospheric radars are called as mesospheric–stratospheric–tropospheric (MST) radar, stratospheric–tropospheric (ST) radar, tropospheric (T) radar, or boundary layer (BL) radar (or BLR), according to the observable region for the radars. At the early stage, MST and ST radars that use the frequency band of 50 MHz to 1 GHz were primarily developed for research. Small-scale radars such as BLR that utilize the microwave of 1.3 GHz was rapidly put to practical use soon later.

1.3 Radar Frequency Bands and Usage

For transmission, various frequencies of the wide range from several MHz to around 100 GHz are utilized to detect atmospheric scatterers. The principle of both electromagnetic wave propagation and operation of radar does not change by the operational frequency and is common for various frequencies, although actual radar components greatly vary, depending on the frequency.

In principle, longer wavelengths, i.e., lower transmitted frequencies are suitable for detection of refractive index perturbations. Shorter wavelengths, i.e., higher transmitted frequencies are suitable for detection of minute scatterers. Figure 1.2 shows the operational frequencies of actual radars and their adjoining frequency bands. The radar frequency bands with the main usages are shown in Table 1.1. In passage of the radar development, the individual radar frequency bands have been traditionally named by letters as “L”, “S”, and so on as shown in Table 1.1. However, the letter expression is generally obscure, and inconvenient to show the concrete frequency. Therefore, in this book, we will express the operational frequency by “1.3-GHz” for instance, and not by “L-band.”

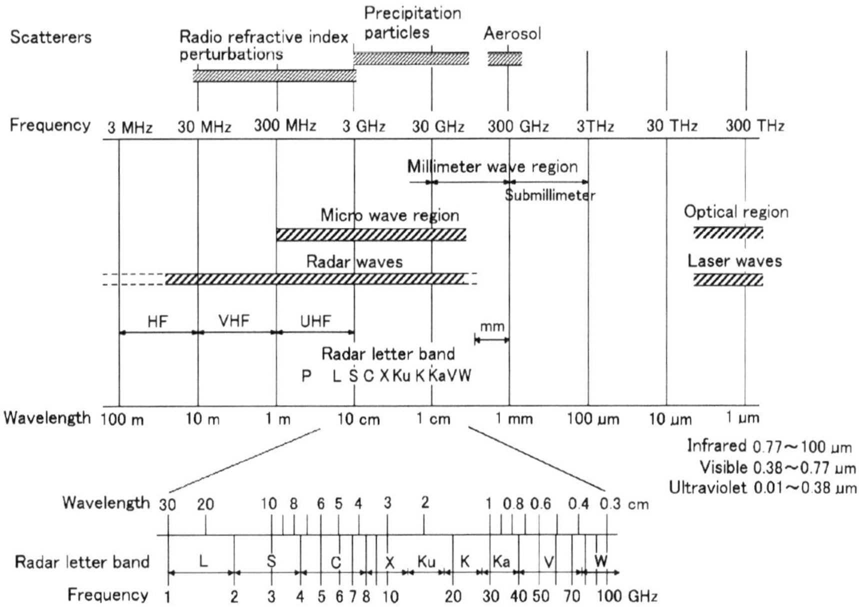

Fig. 1.2 Operational frequency band of various radars and their adjoining frequency bands

The scattering characteristics of electromagnetic wave are closely dependent upon its wavelength. It means that each scattering mechanism has a best combination with a specific wavelength band in the remote sensing of precipitation particles.[1] The diameters of objective particles are between several μm and several mm. Thus, electromagnetic waves of wavelengths of several mm (millimeter wave) to several cm (micro wave) are adopted for these targets. As will be discussed in Sect. 6.3.3, electromagnetic waves of wavelength of less than several cm are attenuated due to precipitation along the propagation path. Thus, it is indispensable to evaluate the influence of the attenuation quantitatively.

In the remote sensing of the atmosphere, refractive index perturbations generated from atmospheric turbulence and waves are the main source of scatterers. The refractive index field has perturbations over a spectrum of spatial scales (i.e., structure wavelengths of perturbations Λ_S), and in general large scale refractive index perturbations contain more intense perturbations. In the radar observation of the atmosphere, only scattering from a specific spatial scale corresponding to half the radar wavelength is detected (Brag scatter; Chap. 3). Therefore, using radar wavelength at twice the largest scales of perturbations, we can more easily observe the backscatter from refractive index perturbations. Practically, wavelengths up

[1]Here, the term "precipitation particles" include both precipitation particles such as raindrop and hailstone, and non-precipitation particles such as cloud and fog particles, and they are not distinguished strictly otherwise mentioned hereafter.

Table 1.1 Radar frequency bands and applications

Letter band	Frequency range	ITU[a] regulation	Applications
HF	3–30 MHz		OTH, atmosphere observation
VHF	30–300 MHz	138–144 MHz	Ultra long range surveillance, atmosphere observation
		216–225 MHz	Ditto
UHF[b]	300–3,000 MHz	420–450 MHz	Ditto
		890–942 MHz	Ditto
L	1–2 GHz	1.215–1.4 GHz	Long range surveillance, air traffic control (ATC), atmosphere observation
S	2–4 GHz	2.3–2.5 GHz	Middle range surveillance, ATC,
		2.7–3.7 GHz	Long range weather observation
C	4–8 GHz	5.25–5.95 GHz	Long range tracking, weather observation
X	8–12 GHz	8.5–10.68 GHz	Short range tracking, weather observation
Ku	12–18 GHz	13.4–14.0 GHz	High resolution satellite altimeter
		15.7–17.7 GHz	Ditto
K	18–27 GHz	24.05–24.25 GHz	Airport surface surveillance
Ka	27–40 GHz	33.4–36 GHz	Short range tracking, weather observation
V	40–75 GHz	59–64 GHz	Remote sensing
W	75–110 GHz		Remote sensing
mm[c]	110–300 GHz		Remote sensing

[a] International Telecommunication Union

[b] 300–1,000 MHz (Skolnik 1990, p1.14). 230 MHz–1 GHz band is often called P-band

[c] Radars of this frequency band have not been achieved yet

to several meters are used for the atmospheric observations in consideration of the reasonable physical size of the antenna aperture and the outer scale (i.e., the largest scales) of turbulence. As will be discussed in Sect. 3.4.2, scattering from perturbations of refractive index due to atmospheric turbulence is dominant at the frequencies lower than 2–3 GHz, and that from precipitation particles are dominant at higher frequencies. Thus, atmospheric radars generally adopt frequencies of several tens MHz to around 1 GHz, and at highest, 3 GHz. Meanwhile, meteorological radars generally adopt frequencies higher than 2–3 GHz. At the frequencies lower than 3 GHz, the attenuation of electromagnetic waves due to precipitation and atmosphere along the propagation path is small and almost negligible.

Among the atmospheric observation techniques, there is one that utilized small floating dust (aerosol) as small as μm $(10^{-9}\,\mathrm{m})$ as the scatterer. In this case, the laser beam of wavelength of the same order as the size of scatterer is adopted. A laser radar is called "lidar" (LIght Detection And Ranging), and is used to observe the air temperature inversion layer, the water vapor content, the atmospheric density, and so on. In recent years, a coherent Doppler lidar (CDL), which is possible to obtain Doppler velocities of aerosol, has become to practical use.

Chapter 2
Electromagnetic Waves

2.1 Characteristics of Electromagnetic Waves

Electric and magnetic fields propagate through space in the form of electromagnetic waves (or radio waves). Various characteristics of the electromagnetic waves can be derived from Maxwell's equations, which will be discussed for thc simplest case in the present chapter. In this chapter, we assume that the waves are propagating in a smoothly varying medium, and that there are no small scale (i.e., of the order of a wavelength) perturbations of refractive index either.

2.1.1 Basic Equations

Maxwell's Equations

Three basic differential equations concerned with electric and magnetic fields, which include Faraday's law, Ampère–Maxwell's law, and Gauss's law, are called by a general name Maxwell's equations. According to Maxwell's equations, electric field strength $\boldsymbol{E}$ and magnetic field strength $\boldsymbol{H}$ are expressed using magnetic flux density $\boldsymbol{B}$, electric current density $\boldsymbol{J}$, and electric flux density $\boldsymbol{D}$ as

$$\nabla \times \boldsymbol{E} = -\frac{\partial}{\partial t}\boldsymbol{B}, \qquad \text{Faraday's law} \qquad (2.1)$$

$$\nabla \times \boldsymbol{H} = \frac{\partial}{\partial t}\boldsymbol{D} + \boldsymbol{J}, \qquad \text{Ampère–Maxwell law} \qquad (2.2)$$

$$\nabla \cdot \boldsymbol{D} = \rho, \qquad \text{Gauss's law for electric flux density} \qquad (2.3)$$

$$\nabla \cdot \boldsymbol{B} = 0. \qquad \text{Gauss's law for magnetic flux density} \qquad (2.4)$$

S. Fukao and K. Hamazu, *Radar for Meteorological and Atmospheric Observations*, DOI 10.1007/978-4-431-54334-3_2, © Springer Japan 2014

The vectors $\boldsymbol{E}$, $\boldsymbol{H}$, $\boldsymbol{J}$, $\boldsymbol{D}$, and $\boldsymbol{B}$ depend on position $(x,\ y,\ z)$ and time (t). In the International System of Unit (SI unit; le Système International d'Unité),[1] the A [ampere] is defined as the electric unit, and other quantities are derived as V [volt, $\mathrm{m^2\,kg\,s^{-3}\,A}$], Wb [weber, $\mathrm{m^2 kg\, s^{-2}\,A^{-1}}$], and C [coulomb, s A]. Applying these quantities, the units of $\boldsymbol{E}, \boldsymbol{H}, \boldsymbol{J}, \boldsymbol{D}$ are derived as $[\mathrm{V\,m^{-1}}]$, $[\mathrm{m^{-1}\,A}]$, $[\mathrm{m^{-2}\,A}]$, and $[\mathrm{C\,m^{-2}}]$, respectively. The symbol ρ denotes electric charge density in the unit of $[\mathrm{C\,m^{-3}}]$, and symbols $\nabla\times$ and $\nabla\cdot$ in vector analysis indicate the differential operator for rotation[2] and divergence,[3] respectively.

In general, $\boldsymbol{D}$ and $\boldsymbol{J}$ are related by a linear function to $\boldsymbol{E}$, and $\boldsymbol{B}$ to $\boldsymbol{H}$. Their relations are expressed by the constitutive equations (2.5)–(2.6) and Ohm's law (2.7) as

$$\boldsymbol{D} = \varepsilon \boldsymbol{E}, \tag{2.5}$$

$$\boldsymbol{B} = \mu \boldsymbol{H}, \tag{2.6}$$

$$\boldsymbol{J} = \sigma \boldsymbol{E}, \tag{2.7}$$

where the permittivity ε $[\mathrm{F\,m^{-1}}]$, the permeability μ $[\mathrm{H\,m^{-1}}]$, and the electric conductivity σ $[\mathrm{S\,m^{-1}}]$ are medium constants. They are scalar quantities for the isotropic[4] medium, and tensor quantities for the aerotropic medium such as ferromagnet and crystalline body. In the present book, they are treated as scalar quantities unless mentioned otherwise.

Wave Equation

Each of Maxwell's equations (2.1) and (2.2) includes both electric and magnetic fields. If they are solved as the simultaneous equations, the equation which contains only electric or magnetic field as a variable is derivable. First, the equation which includes only the electric field is obtained. Differentiating (2.2) with respect to time

---

[1] The MKSA unit system whose basic units are length m [meter], mass [kg], time [s(second)], and electric current A [ampere].

[2] In the orthogonal coordinate system $(x,\ y,\ z)$, the rotation of $\boldsymbol{E}$ is a vector quantity that is expressed by

$$\nabla \times \boldsymbol{E} = \mathrm{rot}\boldsymbol{E} = \left(\frac{\partial E_z}{\partial y} - \frac{\partial E_y}{\partial z}\right)\mathrm{i}_x + \left(\frac{\partial E_x}{\partial z} - \frac{\partial E_z}{\partial x}\right)\mathrm{i}_y + \left(\frac{\partial E_y}{\partial x} - \frac{\partial E_x}{\partial y}\right)\mathrm{i}_z,$$

where i_x, i_y, and i_z are unit vectors in the x, y, and z directions, respectively.

[3] The divergence of $\boldsymbol{D}$ is a scalar quantity that is expressed by

$$\nabla\cdot\boldsymbol{D} = \mathrm{div}\boldsymbol{D} = \frac{\partial D_x}{\partial x} + \frac{\partial D_y}{\partial y} + \frac{\partial D_z}{\partial z}.$$

[4] If the permittivity and the permeability of the medium are independent of the direction of the radiation, the medium is called isotropic.

t and replacing the right-hand side parameters using (2.5) and (2.7),

$$\begin{aligned}\nabla \times \frac{\partial}{\partial t}\boldsymbol{H} &= \frac{\partial^2}{\partial t^2}\boldsymbol{D} + \frac{\partial}{\partial t}\boldsymbol{J} \\ &= \varepsilon\frac{\partial^2}{\partial t^2}\boldsymbol{E} + \sigma\frac{\partial}{\partial t}\boldsymbol{E}. \end{aligned} \tag{2.8}$$

Applying the rotation operator to (2.1), and substituting (2.6) and (2.8) into it,

$$\nabla \times (\nabla \times \boldsymbol{E}) = -\varepsilon\mu\frac{\partial^2}{\partial t^2}\boldsymbol{E} - \mu\sigma\frac{\partial}{\partial t}\boldsymbol{E} \tag{2.9}$$

is obtained. Replacing the left-hand side of (2.9) with an identical equation of the vector operation,

$$\nabla \times (\nabla \times \boldsymbol{E}) = \nabla(\nabla \cdot \boldsymbol{E}) - \nabla^2\boldsymbol{E}, \tag{2.10}$$

where ∇ and ∇^2 are gradient[5] and Laplacian,[6] respectively, and substituting $\nabla(\nabla \cdot \boldsymbol{E}) = \nabla\rho/\varepsilon$, which is obtained from (2.3) and (2.5) under the assumption that permittivity is spatially uniform, into (2.10), the equation for electric field

$$\nabla^2\boldsymbol{E} - \varepsilon\mu\frac{\partial^2}{\partial t^2}\boldsymbol{E} - \mu\sigma\frac{\partial}{\partial t}\boldsymbol{E} = \frac{1}{\varepsilon}\nabla\rho \tag{2.11}$$

is derived.

Next, applying the similar operation to magnetic field,

$$\nabla^2\boldsymbol{H} - \varepsilon\mu\frac{\partial^2}{\partial t^2}\boldsymbol{H} - \mu\sigma\frac{\partial}{\partial t}\boldsymbol{H} = 0 \tag{2.12}$$

[5]In the orthogonal coordinate system (x, y, z), the gradient of a scalar function $\varphi(x,y,z)$ is a vector quantity given by

$$\nabla\varphi = \frac{\partial\varphi}{\partial x}\mathrm{i}_x + \frac{\partial\varphi}{\partial y}\mathrm{i}_y + \frac{\partial\varphi}{\partial z}\mathrm{i}_z.$$

[6]For a scalar function $\varphi(x,y,z)$, $\nabla^2\varphi$ is a scalar quantity given by

$$\nabla^2\varphi = \mathrm{div}(\mathrm{grad}\ \varphi) = \frac{\partial^2\varphi}{\partial x^2} + \frac{\partial^2\varphi}{\partial y^2} + \frac{\partial^2\varphi}{\partial z^2}.$$

is obtained. Equations (2.11) and (2.12), which are called vector wave equations, are the general equations that describe wave propagation in a linear[7] homogeneous[8] medium which is isotropic and nondispersive.[9] Suppose a uniform and lossless dielectric medium ($\sigma = 0$) which has an infinite extent. If the medium has no wave source such as electric charge, $\rho = 0$. The present book concerns the medium which satisfies these conditions. Thus (2.11) and (2.12) are simplified as

$$\nabla^2 \boldsymbol{E} - \varepsilon\mu \frac{\partial^2}{\partial t^2}\boldsymbol{E} = 0, \tag{2.13}$$

$$\nabla^2 \boldsymbol{H} - \varepsilon\mu \frac{\partial^2}{\partial t^2}\boldsymbol{H} = 0. \tag{2.14}$$

In a statistically homogeneous medium in which the permittivity is fluctuating due to, for example, turbulence, waves are scattered by refractive index perturbations. The wave equations for this problem are not derived in this book, and readers should refer to Tatarskii (1971, Chap. 2) or to Doviak and Zrnić (2006, Chap. 11) for them.

Plane Wave

Electromagnetic wave is called a plane wave if it has electric and magnetic fields within the plane perpendicular to the direction of wave propagation. When the electromagnetic wave spreads radially from wave source, it is locally a plane wave when observed sufficiently far from the wave source. If the plane electromagnetic wave shows sinusoidal oscillation with single angular frequency ω [rad s^{-1}] ($= 2\pi f; f$ is frequency [Hz]), the time factor is expressed as $e^{j\omega t}$, where j is the imaginary unit ($j^2 = -1$). Then, (2.13) and (2.14) lead to

$$\nabla^2 \boldsymbol{E} + k^2 \boldsymbol{E} = 0, \tag{2.15}$$

$$\nabla^2 \boldsymbol{H} + k^2 \boldsymbol{H} = 0, \tag{2.16}$$

respectively, where

$$k = \omega\sqrt{\varepsilon\mu}. \tag{2.17}$$

---

[7] The medium is called linear medium if proportionality holds between $\boldsymbol{J}$ and $\boldsymbol{E}$, $\boldsymbol{B}$ and $\boldsymbol{H}$, and $\boldsymbol{J}$ and $\boldsymbol{E}$, respectively, and nonlinear medium otherwise. In nonlinear medium, the permittivity, the permeability, or the conductivity are the function of $\boldsymbol{E}$ or $\boldsymbol{H}$.

[8] The medium is called homogeneous medium if the permittivity and the permeability are spatially uniform, while inhomogeneous otherwise.

[9] The medium with constant permittivity and permeability irrespective of frequency is called nondispersive, otherwise it is dispersive.

Parameter k is called the (radar) wave number, which is related to radar wavelength λ [m] as $k = 2\pi/\lambda$. The equations in the form of (2.15) and (2.16) are called Helmholtz equations. The solution of (2.15) at the position $\boldsymbol{r}$ is generally given by

$$\boldsymbol{E}(\boldsymbol{r}) = \boldsymbol{E}_1 \mathrm{e}^{-j\boldsymbol{k}\cdot\boldsymbol{r}}, \tag{2.18}$$

where $\boldsymbol{E}_1$ is the vector which does not depend on the position. If $\rho = 0$ and $\sigma = 0$, substituting (2.18) into (2.1) to (2.4),

$$\boldsymbol{k} \times \boldsymbol{E} = \omega\mu \boldsymbol{H}, \tag{2.19}$$

$$\boldsymbol{k} \times \boldsymbol{H} = -\omega\mu \boldsymbol{E}, \tag{2.20}$$

$$\boldsymbol{k} \cdot \boldsymbol{E} = 0, \tag{2.21}$$

$$\boldsymbol{k} \cdot \boldsymbol{H} = 0 \tag{2.22}$$

are obtained. Equations (2.21) and (2.22) mean that both $\boldsymbol{E}$ and $\boldsymbol{H}$ are orthogonal to $\boldsymbol{k}$. Furthermore, (2.19) and (2.20) indicate that the sequence of $\boldsymbol{E}$, $\boldsymbol{H}$, and $\boldsymbol{k}$ are in the sense of rotation to advance a right-handed screw. The vectors $\boldsymbol{E}$ and $\boldsymbol{H}$ are in-phase, and related as

$$\boldsymbol{H} = \frac{1}{\eta_\mathrm{i}} \mathrm{i}_k \times \boldsymbol{E}, \tag{2.23}$$

where i_k is the unit vector in the direction of $\boldsymbol{k}$, and η_i is the intrinsic impedance (or wave impedance) given by

$$\eta_\mathrm{i} = \sqrt{\mu/\varepsilon}. \tag{2.24}$$

Denoting ε and μ in a vacuum as ε_0 and μ_0, respectively, $\varepsilon_0 \simeq 8.8542 \times 10^{-12}$ [F m^{-1}] and $\mu_0 = 4\pi \times 10^{-7} \simeq 1.2566 \times 10^{-6}$ [H m^{-1}]. Therefore, the wave impedance η_0 becomes $\eta_0 = \sqrt{\mu_0/\varepsilon_0}$ ($\simeq 376.7 \simeq 120\pi$) [$\Omega$].

$\boldsymbol{E}_2 \mathrm{e}^{+j\boldsymbol{k}\cdot\boldsymbol{r}}$ also satisfies (2.15), and accordingly the solution of (2.15) is, in general, given by the sum of both values as

$$\boldsymbol{E}(\boldsymbol{r}) = \boldsymbol{E}_1 \mathrm{e}^{-j\boldsymbol{k}\cdot\boldsymbol{r}} + \boldsymbol{E}_2 \mathrm{e}^{+j\boldsymbol{k}\cdot\boldsymbol{r}}. \tag{2.25}$$

Multiplying (2.25) by time factor $\mathrm{e}^{j\omega t}$,

$$\boldsymbol{E}(\boldsymbol{r}, t) = \boldsymbol{E}_1 \mathrm{e}^{j(\omega t - \boldsymbol{k}\cdot\boldsymbol{r})} + \boldsymbol{E}_2 \mathrm{e}^{j(\omega t + \boldsymbol{k}\cdot\boldsymbol{r})} \tag{2.26}$$

is obtained. The above equation means that waves $\boldsymbol{E}_1$ and $\boldsymbol{E}_2$ propagate to the directions $+\boldsymbol{k}$ and $-\boldsymbol{k}$, respectively, with the phase velocity of

$$v = \frac{\omega}{k} = 1/\sqrt{\varepsilon\mu}, \tag{2.27}$$

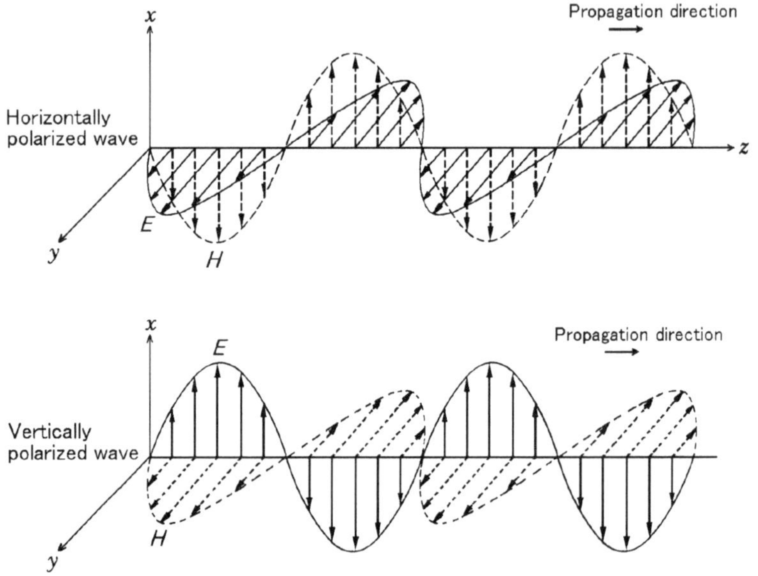

Fig. 2.1 Horizontally (*top*) and vertically (*bottom*) polarized waves. *Solid lines* show electric field and *dotted lines* show magnetic field

where v is the speed of light and equal to c in a vacuum. As a result, it is derived that

$$c = 1/\sqrt{\varepsilon_0 \mu_0} \simeq 2.9979 \times 10^8 \left[\mathrm{m\ s^{-1}}\right]. \tag{2.28}$$

2.1.2 *Polarization*

When a plane electromagnetic wave propagates in the positive z direction, the electric field vector lies in the xy plane perpendicular to the z-axis. The tip of the electric field traces a curve in the xy plane, corresponding to the change of amplitude. If the curve shows a straight line, the plane wave is called linearly polarized, while circularly polarized when the curve shows a circle. In general, the wave is called elliptically polarized when it shows an ellipse.

If the electric field of a linearly polarized plane wave oscillates in the horizontal (y) direction as shown in the upper panel of Fig. 2.1, the wave is called horizontally polarized. On the other hand, if the wave oscillates is in the vertical (x) direction as shown in the bottom panel of the same figure, it is called vertically polarized. The horizontal or vertical polarization is defined with the ground as the reference. The planes where the electric and magnetic field vectors exist are called the E-plane and the H-plane, respectively.

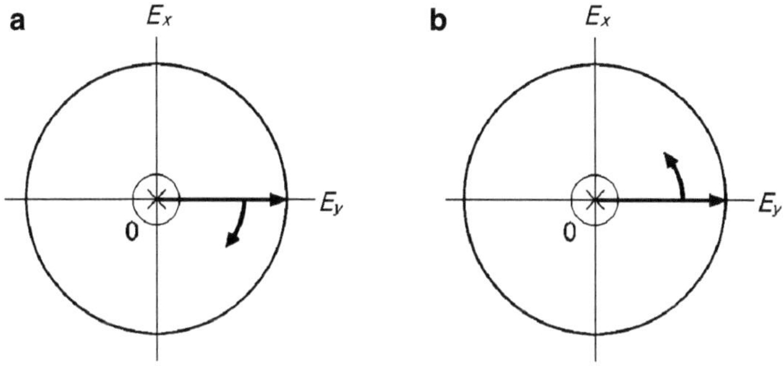

Fig. 2.2 Rotation direction of circularly polarized waves viewed in the direction of propagation. (**a**) Right-hand circular (RHC) and (**b**) Left-hand circular (LHC) polarization

Here, a plane electromagnetic wave propagates in the z direction, and the unit vectors in the x and y axes are denoted i_x and i_y, respectively. When there are two linearly polarized waves in the x direction and in the y direction whose phases are ϕ_x and ϕ_y, respectively, the sum of the two waves becomes

$$\boldsymbol{E}_c = \mathrm{i}_x E_x \mathrm{e}^{[j(\omega t - kz) + j\phi_x]} + \mathrm{i}_y E_y \mathrm{e}^{[j(\omega t - kz) + j\phi_y]}. \tag{2.29}$$

The tip of the above wave, the sum of two orthogonally polarized waves, draws an ellipse and is called elliptically polarized wave. If $\phi_x - \phi_y = n\pi$ $(n = 0, 1, 2, \cdots)$, the wave is linearly polarized. On the other hand, if $E_x = E_y$ and $\phi_x - \phi_y = \pm\pi/2$, the wave is circularly polarized. In the case of circularly and elliptically polarized waves the tip moves either clockwise or counterclockwise. When $\phi_x - \phi_y = \pi/2$, the electric field which is viewed from the tail to the z direction rotates clockwise as time changes. As shown in Fig. 2.2a, it is called right-hand circular (RHC) polarization. On the other hand, if $\phi_x - \phi_y = -\pi/2$, as shown in Panel (b), it rotates counterclockwise, and is called left-hand circular (LHC) polarization. Arbitrary elliptically polarized wave can be considered as the composition of RHC and LHC waves of different magnitude, or that of the horizontally and vertically polarized waves.

The circular polarized wave which is radiated to the scatterers such as raindrops, whose reflectivity is approximately identical for horizontal and vertical polarizations, is backscattered as circular polarized wave but in reversed rotation. On the other hand, the backscattered signal from aircrafts or vessels becomes an elliptically polarized wave. In the present book, only one polarization, horizontal or vertical, is treated for simplicity if not otherwise mentioned. Furthermore we assume that the transmitted and received signals are of the same polarization. If multiple polarizations are treated as in the case of polarimetric radar, it needs to pay attention to these points. The polarimetric radar will be discussed in detail in Sect. 6.4.

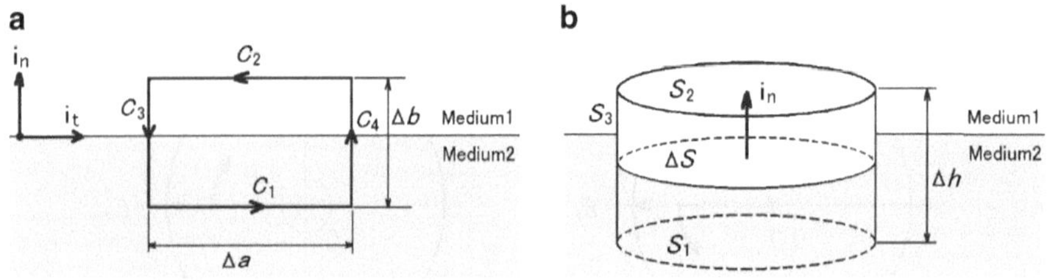

Fig. 2.3 Boundary surface between two mediums. (**a**) Tangential and (**b**) normal components

2.1.3 Reflection and Refraction

Boundary Condition

The boundary condition for propagation of electromagnetic waves at the surface where the medium 1 and the medium 2 contact as shown in Fig. 2.3 is derived below (e.g., Stratton 2007, Chap. 1). First, Faraday's law[10] is applied, as shown in Fig. 2.3a, to the closed curve $C = C_1 + C_2 + C_3 + C_4$ which encircles the minute area of breadthwise Δa and lengthwise Δb in the static magnetic field. When Δb becomes infinitesimally small, contribution from C_3 and C_4 to the line integral can be neglected, and thus

$$\begin{aligned}\lim_{\Delta b \to 0} \oint_C \boldsymbol{E} \cdot \mathrm{d}\boldsymbol{l} &= \int_{C_1} \boldsymbol{E}_1 \cdot \mathrm{i_t} \mathrm{d}l + \int_{C_2} \boldsymbol{E}_2 \cdot (-\mathrm{i_t} \mathrm{d}l) \\ &= \int_{C_1} E_{1\mathrm{t}} \mathrm{d}l - \int_{C_2} E_{2\mathrm{t}} \mathrm{d}l = (E_{1\mathrm{t}} - E_{2\mathrm{t}})\Delta a = 0,\end{aligned} \tag{2.30}$$

where $E_{1\mathrm{t}} = \boldsymbol{E}_1 \cdot \mathrm{i_t}$ and $E_{2\mathrm{t}} = \boldsymbol{E}_2 \cdot \mathrm{i_t}$ are $\mathrm{i_t}$ (tangential) components of $\boldsymbol{E}_1$ and $\boldsymbol{E}_2$, respectively. From (2.30),

$$E_{1\mathrm{t}} = E_{2\mathrm{t}}. \tag{2.31}$$

That is, the tangential components of electric field E_t of both sides becomes equal at the interface of two mediums. Similarly, if surface current does not exist at the interface of the two mediums, the tangential components of magnetic field satisfy

$$H_{1\mathrm{t}} = H_{2\mathrm{t}}, \tag{2.32}$$

where $H_{1\mathrm{t}} = \boldsymbol{H}_1 \cdot \mathrm{i_t}$ and $H_{2\mathrm{t}} = \boldsymbol{H}_2 \cdot \mathrm{i_t}$ are similarly the $\mathrm{i_t}$ components of $\boldsymbol{H}_1$ and $\boldsymbol{H}_2$, respectively.

---

[10]The integral form of (2.1) is expressed as $\oint \boldsymbol{E} \cdot \mathrm{d}\boldsymbol{l} = -\frac{\partial}{\partial t} \oint_S \boldsymbol{B} \cdot \mathrm{d}\boldsymbol{S}$, where $\boldsymbol{S}$ is encircled by an arbitrary closed curve C. Regarding the static magnetic field in a dielectric, $\oint \boldsymbol{E} \cdot \mathrm{d}\boldsymbol{l} = 0$.

Secondly, Gauss's law[11] is applied to the minute volume with cross section ΔS and height Δh which pass through the interface of two mediums as shown in Fig. 2.3b. The surface area of the volume is $S = S_1 + S_2 + S_3$, where S_1, S_2, and S_3 are areas of the bottom, the upper, and the side panels, respectively. In the limit that Δh becomes close to 0, contribution from S_3 to the surface integral can be neglected, and consequently if there is no electric charge at the interface,

$$\begin{aligned}\lim_{\Delta h \to 0} \oint_S \boldsymbol{D} \cdot \mathrm{d}\boldsymbol{S} &= \int_{S_1} \boldsymbol{D}_1 \cdot (-\mathrm{i}_\mathrm{n}) \mathrm{d}S + \int_{S_2} \boldsymbol{D}_2 \mathrm{i}_\mathrm{n} \mathrm{d}S \\ &= -\int_{S_1} D_{1\mathrm{n}} \mathrm{d}S + \int_{S_2} D_{2\mathrm{n}} \mathrm{d}S = (D_{2\mathrm{n}} - D_{1\mathrm{n}}) \Delta S = 0 \end{aligned} \tag{2.33}$$

is obtained, where $D_{1\mathrm{n}} = \boldsymbol{D}_1 \cdot \mathrm{i}_\mathrm{n}$ and $D_{2\mathrm{n}} = \boldsymbol{D}_2 \cdot \mathrm{i}_\mathrm{n}$ are the i_n (normal) components of $\boldsymbol{D}_1$ and $\boldsymbol{D}_2$, respectively. From (2.33),

$$D_{1\mathrm{n}} = D_{2\mathrm{n}}. \tag{2.34}$$

The above equation means that the normal components of electric flux density in the two mediums become equal at the interface. Similarly, the normal components of magnetic flux density become identical at the surface, thus

$$B_{1\mathrm{n}} = B_{2\mathrm{n}}, \tag{2.35}$$

where $B_{1\mathrm{n}} = \boldsymbol{B}_1 \cdot \mathrm{i}_\mathrm{n}$ and $B_{2\mathrm{n}} = \boldsymbol{B}_2 \cdot \mathrm{i}_\mathrm{n}$ are the i_n components of $\boldsymbol{B}_1$ and $\boldsymbol{B}_2$, respectively.

Reflection and Refraction of Plane Wave

The reflection and refraction of a plane wave at $z = 0$ of the interface xy between two mediums 1 and 2 shown in Fig. 2.4 is discussed. The incident wave $\boldsymbol{E}_\mathrm{i}$ in xz-plane propagates from the medium 1 to 2. A part of the incident wave is reflected as $\boldsymbol{E}_\mathrm{r}$ back in the medium 1 and the other is refracted as $\boldsymbol{E}_\mathrm{t}$ into the medium 2. If the angles between z-axis and the traveling directions are θ_i, θ_r, and θ_t, respectively, the incident wave, the reflected wave, and the refracted wave are expressed as $\boldsymbol{E}_\mathrm{i}\mathrm{e}^{-jk_1(x\sin\theta_\mathrm{i}+z\cos\theta_\mathrm{i})}$, $\boldsymbol{E}_\mathrm{r}\mathrm{e}^{-jk_1(x\sin\theta_\mathrm{r}-z\cos\theta_\mathrm{r})}$, and $\boldsymbol{E}_\mathrm{t}\mathrm{e}^{-jk_2(x\sin\theta_\mathrm{t}+z\cos\theta_\mathrm{t})}$, respectively, where k_1 and k_2 are the propagation constants of mediums 1 and 2, respectively. From the boundary condition (2.31), the tangential components of the electric field at the interface between mediums 1 and 2 should be continuous, that is

$$E_\mathrm{i} e^{-jk_1 x \sin\theta_\mathrm{i}} + E_\mathrm{r} e^{-jk_1 x \sin\theta_\mathrm{r}} = E_\mathrm{t} e^{-jk_2 x \sin\theta_\mathrm{t}}. \tag{2.36}$$

---

[11] The integral form of (2.3) is expressed as $\oint_S \boldsymbol{D} \cdot \mathrm{d}\boldsymbol{S} = \oint_V \rho \mathrm{d}\boldsymbol{V} = Q$, where space V is encircled by an arbitrary closed surface, and Q is an electric charge. In the free space with no electric charge inside, $\oint_S \boldsymbol{D} \cdot \mathrm{d}\boldsymbol{S} = 0$.

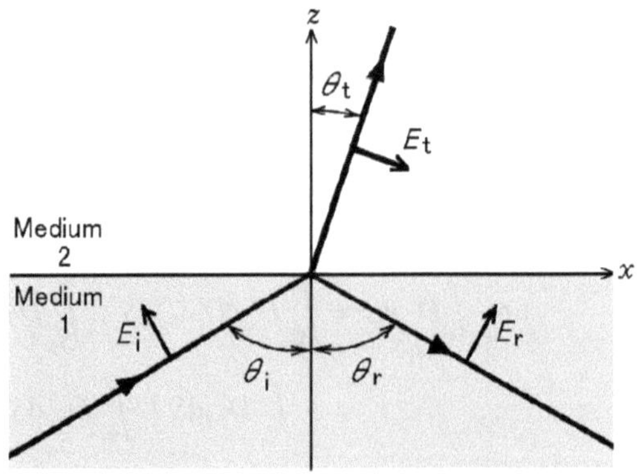

Fig. 2.4 Reflection and refraction of a plane wave at the boundary surface between two mediums

For (2.36) to be applied at any place of the interface, both phase and amplitude of each wave must be identical. As for the phase,

$$k_1 \sin\theta_i = k_1 \sin\theta_r = k_2 \sin\theta_t \tag{2.37}$$

is satisfied. From (2.37),

$$\theta_i = \theta_r, \tag{2.38}$$

$$\frac{\sin\theta_i}{\sin\theta_t} = \frac{k_2}{k_1} \tag{2.39}$$

is derived. The above relation means that the angles of incidence and reflection are identical, and the ratio between their sines is constant regardless of the incident angle. This is one form of Snell's law in the optics.

The refractive index is defined as the ratio of the speed of light in a vacuum to that in a non-vacuum medium by

$$n \equiv \frac{c}{v}, \tag{2.40}$$

which is called as the absolute refractive index. Thus, if the (absolute) refractive indexes of mediums 1 and 2 are n_1 and n_2, respectively, and the phase velocities v_1 and v_2, respectively, (2.39) is modified using (2.27) as follows. Since angular frequency ω does not change as the wave passes across the interface,

$$\frac{\sin\theta_i}{\sin\theta_t} = \frac{k_2}{k_1} = \frac{v_1}{v_2} = \frac{n_2}{n_1} \tag{2.41}$$

is the relative refractive index of medium 2 to medium 1. In particular, if the medium 1 is vacuum and the medium 2 is the atmosphere ($\sigma = 0$), the absolute refractive index of the atmosphere becomes

$$n = \sqrt{\frac{\varepsilon_2\mu_2}{\varepsilon_0\mu_0}} \simeq \sqrt{\varepsilon_2/\varepsilon_0}, \tag{2.42}$$

where ε_2 and μ_2 are the permittivity and the permeability of the medium 2, respectively, and $\mu_2 \simeq \mu_0$ for the atmosphere. If the continuity equation of the wave amplitude is applied to (2.36), the reflection coefficient E_r/E_i and the transmission coefficient E_t/E_i are derivable.

2.1.4 *Radiation*

In the following, the radiation field far from a wave source is derived. It is sometimes more convenient to introduce scalar and vector potentials than to express directly the electromagnetic field. This case is true for the radiation, and the aforementioned Maxwell's equation is rewritten by using potentials below.

Vector and Scalar Potentials

Magnetic flux density $\boldsymbol{B}$ is a rotational field and satisfies the continuity given by (2.4). The vector $\boldsymbol{A}$ defined as

$$\boldsymbol{B} = \mu \boldsymbol{H} \equiv \nabla \times \boldsymbol{A} \tag{2.43}$$

is called the vector potential. Substituting (2.43) into (2.1),

$$\nabla \times \boldsymbol{E} = -\frac{\partial}{\partial t}(\nabla \times \boldsymbol{A}) = \nabla \times \left(-\frac{\partial \boldsymbol{A}}{\partial t}\right) \tag{2.44}$$

is obtained. It leads to

$$\nabla \times \left(\boldsymbol{E} + \frac{\partial \boldsymbol{A}}{\partial t}\right) = 0. \tag{2.45}$$

From the vector formula, any function with zero rotation is given by gradient of scalar function. Applying this relation with scalar potential Ψ to (2.45),

$$\boldsymbol{E} + \frac{\partial \boldsymbol{A}}{\partial t} + \nabla \Psi = 0 \tag{2.46}$$

is derived. In the stationary state, $\boldsymbol{E} = -\nabla \Psi$, and hence Ψ is called a generalized electric potential. In the same way as the electrostatic field is only known within a constant Φ, either Ψ or $\boldsymbol{A}$ cannot be uniquely determined. For one solution of Ψ_0 and $\boldsymbol{A}_0$, a new solution which includes arbitrary functions χ and Φ,

$$\boldsymbol{A} = \boldsymbol{A}_0 - \nabla \chi, \tag{2.47}$$

$$\Psi = \Psi_0 + \frac{\partial \chi}{\partial t} + \Phi \tag{2.48}$$

is derivable for the same $\boldsymbol{E}$ and $\boldsymbol{B}$. Such a conversion between potentials is called gauge transformation.

The radiation fields $\boldsymbol{E}$ and $\boldsymbol{H}$ of a current source $\boldsymbol{J}$ are related as

$$\nabla\times\boldsymbol{H}=\varepsilon\frac{\partial}{\partial t}\boldsymbol{E}+\boldsymbol{J}, \tag{2.49}$$

which is derived from (2.2) and (2.5). Substituting (2.43) and (2.46) into (2.49),

$$\nabla\times\left(\frac{1}{\mu}\nabla\times\boldsymbol{A}\right)-\varepsilon\frac{\partial}{\partial t}\left(-\nabla\Psi-\frac{\partial\boldsymbol{A}}{\partial t}\right)=\boldsymbol{J} \tag{2.50}$$

is obtained. The above equation can be rewritten as

$$\nabla\times\nabla\times\boldsymbol{A}+\varepsilon\mu\frac{\partial^2\boldsymbol{A}}{\partial t^2}=\mu\boldsymbol{J}-\varepsilon\mu\nabla\frac{\partial\Psi}{\partial t}. \tag{2.51}$$

Applying the vector formula (2.10) to (2.51),

$$\nabla^2\boldsymbol{A}-\varepsilon\mu\frac{\partial^2\boldsymbol{A}}{\partial t^2}+\mu\boldsymbol{J}=\nabla\left(\nabla\cdot\boldsymbol{A}+\varepsilon\mu\frac{\partial\Psi}{\partial t}\right) \tag{2.52}$$

is obtained. Here, applying Lorentz gauge,

$$\nabla\cdot\boldsymbol{A}+\varepsilon\mu\frac{\partial\Psi}{\partial t}=0, \tag{2.53}$$

the following relation

$$\nabla^2\boldsymbol{A}-\varepsilon\mu\frac{\partial^2\boldsymbol{A}}{\partial t^2}+\mu\boldsymbol{J}=0 \tag{2.54}$$

is derived. Similarly, Ψ also satisfies the differential equation in the form

$$\nabla^2\Psi-\varepsilon\mu\frac{\partial^2\Psi}{\partial t^2}+\frac{\rho}{\varepsilon}=0. \tag{2.55}$$

Equations (2.54) and (2.55) are Maxwell's equations using vector and scalar potentials, which are accordingly called potential equations. Both equations are differential equations with respect to position $\boldsymbol{r}$ and time t, and are known to have the following solutions,

$$\boldsymbol{A}(\boldsymbol{r},\,t)=\frac{\mu}{4\pi}\int_V\frac{\boldsymbol{J}(\boldsymbol{r}',\,t-r/v)}{r}\mathrm{d}V, \tag{2.56}$$

$$\Psi(\boldsymbol{r},\,t)=\frac{\mu}{4\pi}\int_V\frac{\rho(\boldsymbol{r}',\,t-r/v)}{r}\mathrm{d}V, \tag{2.57}$$

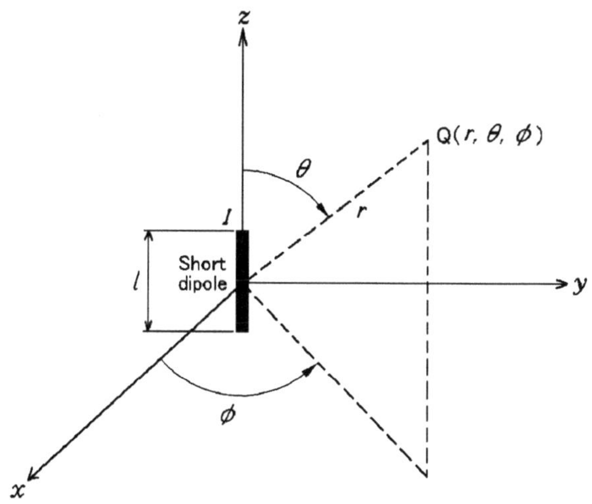

Fig. 2.5 The coordinates and a short dipole

where $r = |\boldsymbol{r}' - \boldsymbol{r}|$ and v is the phase velocity of the electromagnetic wave. Hence, the above two equations mean that the potentials observed on the point $\boldsymbol{r}$ at time t are contributed by the electric current and the electric charge on the point $\boldsymbol{r}'$ at time $t - r/v$. Such potentials which are radially radiated from a wave source with phase velocity v are called retarded potentials (e.g., Stratton 2007, p. 428).

Radiation Field

A short dipole (differential antenna) of length l ($l \ll \lambda$) is considered at the origin in the rectangular coordinates $(x,\ y,\ z)$ and polar coordinates $(r,\ \theta,\ \phi)$ as shown in Fig. 2.5. The current on the dipole is assumed to be directed along z-axis and to vary sinusoidally as

$$I = I_0 e^{j\omega t}, \tag{2.58}$$

where I_0 is the peak value. As the current vector is directed along z-axis, the vector potential $\boldsymbol{A}$ has only z-axis component and is given from (2.56) as (Ramo et al. 1965)

$$A_z = \frac{\mu I l}{4\pi r} \mathrm{e}^{-jkr}. \tag{2.59}$$

As the polar coordinates are more convenient than the rectangular coordinates to express actual electromagnetic field radiated from antenna, the polar coordinates are used to express three components of the vector potential $\boldsymbol{A}$ as follows;

$$A_r = A_z \cos\theta = \frac{\mu I l}{4\pi r} \mathrm{e}^{-jkr} \cos\theta, \tag{2.60}$$

$$A_\theta = -A_z \sin\theta = -\frac{\mu I l}{4\pi r} \mathrm{e}^{-jkr} \sin\theta, \tag{2.61}$$

$$A_\phi = 0. \tag{2.62}$$

Three components of the magnetic field $\boldsymbol{H}$ are derived from the vector potential, using (2.43) and (2.60) to (2.62), as

$$H_r = \frac{1}{\mu r \sin\theta}\left[\frac{\partial}{\partial\theta}(A_\phi \sin\theta) - \frac{\partial A_\theta}{\partial\phi}\right] = 0, \tag{2.63}$$

$$H_\theta = \frac{1}{\mu r}\left[\frac{1}{\sin\theta}\frac{\partial A_r}{\partial\phi} - \frac{\partial}{\partial r}(rA_\phi)\right] = 0, \tag{2.64}$$

$$H_\phi = \frac{1}{\mu r}\left[\frac{\partial}{\partial r}(rA_\theta) - \frac{\partial A_r}{\partial\theta}\right] = \frac{Il}{4\pi}\sin\theta\left(\frac{jk}{r} + \frac{1}{r^2}\right)\mathrm{e}^{-jkr}. \tag{2.65}$$

Similarly, the electric field $\boldsymbol{E}$ is expressed with vector potential using (2.46) and (2.53) as

$$\boldsymbol{E} = -\frac{j\omega}{k^2}\nabla(\nabla\cdot\boldsymbol{A}) - j\omega\boldsymbol{A}. \tag{2.66}$$

From the above equation, three components of $\boldsymbol{E}$ are given by (e.g., Stratton 2007, p. 436; Ulaby et al. 1981, p. 109)

$$E_r = \frac{Il\eta_\mathrm{i}}{2\pi}\cos\theta\left(\frac{1}{r^2} + \frac{1}{jkr^3}\right)\mathrm{e}^{-jkr}, \tag{2.67}$$

$$E_\theta = \frac{Il\eta_\mathrm{i}}{4\pi}\sin\theta\left(\frac{jk}{r} + \frac{1}{r^2} + \frac{1}{jkr^3}\right)\mathrm{e}^{-jkr}, \tag{2.68}$$

$$E_\phi = 0, \tag{2.69}$$

where η_i is the intrinsic impedance. The magnitude of the θ component of the electromagnetic field radiated by a short dipole is given by the sum of three components which are inversely proportional to distance, its square, and its third power. The components which strongly depend on the distance are predominant only in the vicinity of the current source (short dipole). The component that is predominant in the far distance r which satisfies the following relation

$$kr = \frac{2\pi r}{\lambda} \gg 1 \tag{2.70}$$

is called the radiation field, where λ is the wavelength. The radiation electric field is expressed from (2.58) and (2.68) as

$$E_\theta = j\frac{kI_0 l\eta_\mathrm{i}}{4\pi r}\sin\theta\,\mathrm{e}^{j(\omega t - kr)}. \tag{2.71}$$

Similarly, the radiation magnetic field is expressed from (2.58) and (2.65) as

$$H_\phi = j\frac{kI_0 l}{4\pi r}\sin\theta\,\mathrm{e}^{j(\omega t - kr)} = \frac{E_\theta}{\eta}. \tag{2.72}$$

Poynting Vector

The energy which passes through a unit area in a unit time, $\boldsymbol{S}$ [kg s^{-3}, i.e., J m^{-2} s^{-1} = W m^{-2}], where J and W denote "joule" and "watt", respectively, is called complex Poynting vector, and expressed as

$$\boldsymbol{S} = \frac{1}{2}\boldsymbol{E} \times \boldsymbol{H}^*, \tag{2.73}$$

where the superscript * denotes complex conjugate. The coefficient 1/2 is multiplied to the product because $\boldsymbol{E}$ and $\boldsymbol{H}$ are typically expressed in terms of their amplitude (not their rms value), whereas $\boldsymbol{S}$ is the average power.

In the case of the electromagnetic field radiated by a short dipole, the Poynting vector has only the r-component, which is radiated to the $+r$ direction as shown by

$$\boldsymbol{S} = \frac{1}{2}E_\theta H_\phi^* \mathrm{i}_r = \frac{k^2 I_0^2 l^2 \eta_\mathrm{i} \sin\theta^2}{32\pi^2 r^2}\mathrm{i}_r, \tag{2.74}$$

where i_r is the unit vector to r direction. The total radiation power P_s is derived from (2.74) as

$$\begin{aligned} P_\mathrm{s} &= \int_0^{2\pi}\left(\int_0^{\pi} \mathrm{Re}\left[\frac{1}{2}E_\theta H_\phi^*\right] r^2 \sin\theta \mathrm{d}\theta\right)\mathrm{d}\phi \\ &= \frac{k^2 I_0^2 l^2 \eta_\mathrm{i}}{32\pi^2}\int_0^{2\pi}\left(\int_0^{\pi} \sin^3\theta \mathrm{d}\theta\right)\mathrm{d}\phi = \frac{k^2 I_0^2 l^2 \eta_\mathrm{i}}{32\pi^2}\frac{8\pi}{3} \\ &= \frac{\pi\eta_\mathrm{i}}{3}\left(\frac{I_0 l}{\lambda}\right)^2. \end{aligned} \tag{2.75}$$

The radiation resistance R_R is given by

$$R_\mathrm{R} = \frac{2P_\mathrm{s}}{I_0^2} = \frac{2\pi\eta_\mathrm{i}}{3}\left(\frac{l}{\lambda}\right)^2. \tag{2.76}$$

Equation (2.76) means the easiness of radiation. In other word, it is shown that the longer the length of the antenna is, and/or the shorter the wavelength is, the easier to radiate.

Antenna Parameter

Degree of concentration of radiation power density into a specific direction of antenna is called directivity,[12] and is defined by the ratio of the radiation power

[12] There are two different definitions of antenna gain, directive gain and power gain. The former is generally called directive gain or directivity (the IEEE accepted notation), g_D, while the latter is called antenna power gain or simply gain, g. The directivity is defined as the maximum radiation intensity relative to the average intensity, whereas the antenna power gain involves antenna losses.

density in a specific direction to that power density of a non-directional antenna which isotropically radiates power in all directions. Since the radiation power density per unit area varies by distance from the source, it is common to take the power per unit solid angle instead of that per unit area as the radiation power density. When the radiation power density is normalized with the mean radiation power per unit solid angle, the directivity $g_D(\theta, \phi)$ is given by

$$g_D(\theta, \phi) = \frac{|E(\theta, \phi)|^2}{\frac{1}{4\pi}\int |E(\theta, \phi)|^2 d\Omega} = \frac{|E(\theta, \phi)|^2}{|E_0|^2}, \tag{2.77}$$

where Ω is the solid angle and $d\Omega = \sin\theta d\theta d\phi$. If a short dipole is assumed as the antenna, $E(\theta) \propto \sin\theta$ from (2.71), and therefore, (2.77) becomes

$$g_D(\theta) = \frac{\sin^2\theta}{\frac{1}{4\pi}\int \sin^2\theta \, d\Omega} = \frac{3}{2}\sin^2\theta. \tag{2.78}$$

Next, effective antenna aperture is derived from the view point of a matching circuit. If electric current I is supplied to the short dipole of the radiation resistance of R_R, the maximum output power is obtained under the condition that the internal resistance of the circuit is equivalent to R_R. Hence, if there is no ohmic power loss, the maximum power radiated from antenna P_R becomes

$$P_R = \frac{1}{2}|I|^2 R_R = \frac{(El)^2}{8R_R} = \frac{\eta_i l^2}{4R_R} S_r \equiv A_e S_r, \tag{2.79}$$

where E is the electric field strength (rms value), l the length of short dipole, S_r the electric power density $E^2/2\eta_i$, and A_e the effective antenna aperture. From (2.76) and (2.79), A_e of the short dipole becomes

$$A_e = \frac{\eta_i l^2}{4R_R} = \frac{3\lambda^2}{8\pi}. \tag{2.80}$$

In general, the following relation between antenna 1 of effective aperture A_{e1} and directivity g_{D1} and antenna 2 of A_{e2} and g_{D2} holds from the reciprocity theorem;

$$\frac{A_{e1}}{g_{D1}} = \frac{A_{e2}}{g_{D2}}. \tag{2.81}$$

Hence, if the efficiency of the antenna is η_l (≤ 1), $g = \eta_l g_D$. In this chapter and next chapter we treat $\eta_l = 1$ and $g = g_D$. In Chap. 8, the case $\eta_l < 1$ will be discussed.

The lower case g is used to denote the ratio directly (in linear unit), whereas the upper case G is used to express the ratio in decibel.

If antenna 2 is the short dipole antenna, (2.81) becomes by substituting (2.78) and (2.80) into (2.81),

$$\frac{A_{e1}}{g_{D1}} = \frac{3\lambda^2/(8\pi)}{3/2} = \frac{\lambda^2}{4\pi}, \tag{2.82}$$

i.e., the directivity g_D and A_e are related as follows;

$$g_D = \frac{4\pi A_e}{\lambda^2}. \tag{2.83}$$

The above discussion is based upon the short dipole which is the simplest antenna. In the case of general antenna (e.g., a parabolic reflector) of size much larger than the wavelength, the above discussion is extended by regarding that the antenna consists of a large number of short dipoles. The radiation field of the entire antenna is obtained by integrating the fields from all these short dipoles.

2.2 Electromagnetic Wave Propagation in the Atmosphere

2.2.1 Physical Property of the Atmosphere

The propagation speed v of the electromagnetic wave in the atmosphere is not equal to the speed of light c in a vacuum, but in fact is slightly slower. In space radiation travels in straight lines (i.e., along radial lines), but because the atmosphere's refractive index changes with height, the direction of propagation can curve from a straight line. This change in direction of propagation has significant consequences for the detection of targets near the horizon. The atmosphere's refractive index is significantly influenced by pressure, water vapor, and temperature, which considerably vary with altitude.

Height Profile of the Earth's Atmosphere

The Earth's atmosphere (the 1962 U.S. Standard Atmosphere, hereafter simply "the Standard Atmosphere") is subdivided into a number of spherical layers as shown in Fig. 2.6 (Ulaby et al. 1981, p. 257). The lowermost layer is known as the troposphere, in which the atmospheric temperature falls with the lapse late of about 6.5 K km^{-1}, and becomes well below 220 K at its upper boundary. The height of the upper boundary varies according to latitude, season, and weather. In general, it is between 8 and 10 km in arctic regions in winter and between 16 and 18 km in the tropical and equatorial regions (Cole et al. 1965). Most of the conspicuous

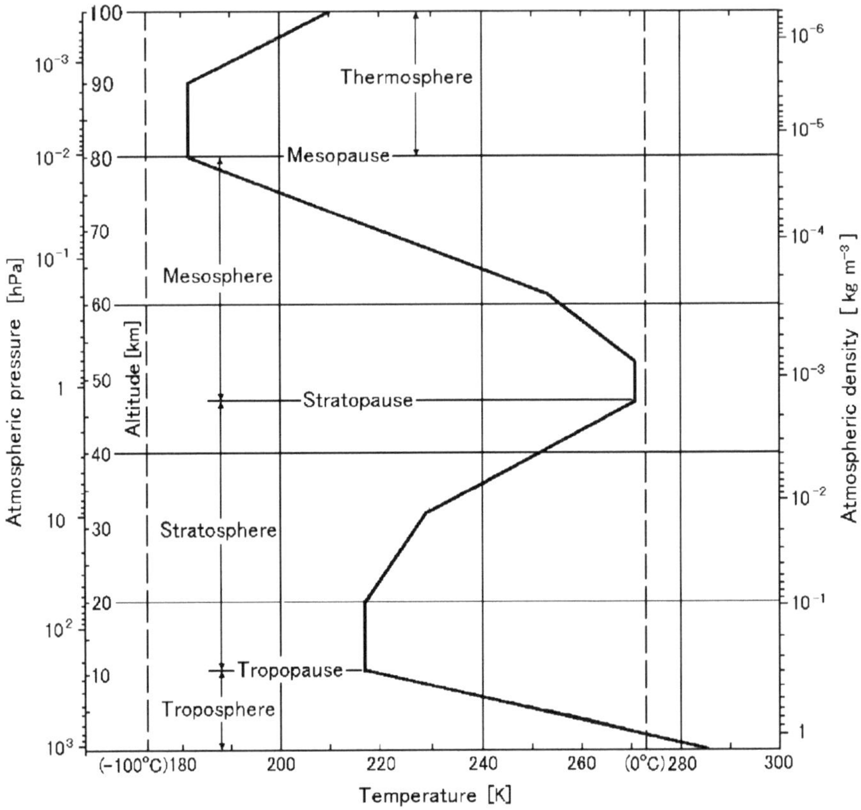

Fig. 2.6 Height profile of the temperature of the 1962 U.S. Standard Atmosphere [from Ulaby et al. 1981, p. 257]

meteorological phenomena such as extratropical cyclones, fronts, and typhoons accompanied by clouds and precipitations are generated in the troposphere. The lowest atmosphere in the troposphere is, in particular, directly influenced by the Earth's surface through the heat and friction. The layer to about 1,000 m altitude, where this influence reaches directly, is called the boundary layer.

The upper boundary of the troposphere is called the tropopause. The layer from the tropopause to approximately 50 km in altitude is called the stratosphere. In the lowest part of the stratosphere, the temperature hardly changes even if the altitude increases up to 20 km, Meanwhile, atmospheric ozone absorbs ultraviolet rays from the sun, and the temperature rises with height between 20 and 32 km above sea level with a gradient of 1 K km^{-1}, followed by a gradient of 2.8 K km^{-1} between 32 km and the upper boundary of the stratosphere, which is called the stratopause. The atmospheric temperature becomes approximately 270 K at the stratopause.

The layer above the stratopause to approximately 80 km altitude is called the mesosphere. In this layer, the atmospheric temperature decreases with altitude and it reaches the lowest of the typical value of about 180 K at the upper boundary of the layer, which is called the mesopause. The photochemical process is common in the stratosphere and mesosphere, which is determined, primarily, by heating due to absorption of ultraviolet radiation by ozone and cooling due to infrared radiation by carbon dioxide. Either cyclone or anticyclone is formed above tropopause. The circulations in the global scale are significantly influenced and determined by various atmospheric waves, which are dominant in these layers. With the above reason, the stratosphere and the mesosphere are collectively called the middle atmosphere and often treated together. The atmospheric temperature is constant at the mesopause to approximately 90 km, and above that altitude it increases again. The layer above the mesopause is called the thermosphere. The layer above the middle atmosphere is also known as the upper atmosphere.

The Neutral Atmosphere and the Ionized Atmosphere

The density of the Earth's atmosphere decreases exponentially as the altitude increases. Therefore, "the end" of the atmosphere cannot be precisely defined but the height of 1,000 km caps what is generally called the Earth's atmosphere. About three quarters of the gross weight of the atmosphere exists within the height of 10 km, whereas the gross weight of the atmosphere which exists above the height of 100 km is about 1/1,000,000 of the whole.

The characteristics of the atmosphere change remarkably around 100 km in altitude. First, the atmospheric motion changes from hydrodynamic motion to particle motion. Below approximately 100 km altitude, the mean free path (flight distance of the particle between successive collisions) among the gaseous particles which compose the atmosphere is as short as that of a fluid. Hence, the atmosphere is able to be considered as a fluid. On the other hand above 100 km, however, the atmosphere' molecules are no longer treated as an ensemble, but instead it becomes necessary to treat individual particles separately. In other words, the cause of vertical diffusion of the atmosphere in the lower part is due to hydrodynamic motion generated by turbulence and so on, whereas it is mostly due to molecular diffusion by the thermal motion of particles in the upper part. Therefore, the vertical distribution of the atmosphere in the thermosphere is determined by the balance between the downward Earth's gravitational force and the upward diffusion that varies according to the particle mass. Secondly, the atmosphere changes from neutral atmosphere to ionized one around this border. The atmosphere is neutral in the lower part. Whereas in the upper part, the atoms and molecules which compose the atmosphere are ionized to ions and free electrons by polar ultraviolet rays and X ray from the sun, and so on. Therefore, the thermosphere up to a few 100 km is partially ionized.

In the discussion hereafter, we treat the radar remote sensing for weather and the atmosphere in the neutral atmosphere. The primary function of a meteorological radar is to observe precipitation and clouds. On the other hand, atmospheric radars generally measure atmospheric properties in the troposphere, stratosphere, and mesosphere.

2.2.2 Propagation of Electromagnetic Wave

The refractive index n of the Earth's atmosphere is a function of three components, the dry air component, the water vapor component, and the component due to free electrons. The following formula (Balsley and Gage 1980) gives the refractive index in terms of these components,

$$n = 1 + 7.76 \times 10^{-5}\frac{p}{T} + 3.73 \times 10^{-1}\frac{e}{T^2} - 40.3\frac{N_e}{f^2}, \tag{2.84}$$

where p is atmosphere pressure [hPa], T atmospheric temperature [K], e is partial pressure of water vapor [hPa], N_e is the density of free electron $[\mathrm{m}^{-3}]$, f is the radar frequency [Hz]. The second term of the right-hand side is the contribution from dipole moment due to polarization field of neutral molecules. It contributes most to the refractive index between middle troposphere and stratopause. The third term is due to the permanent dipole moment of the water molecule, which is significant only in lower troposphere. Meanwhile, the forth term is the contribution from free electrons. It is significant only in the mesosphere and above, however it can be ignored in the stratopause or lower layer. Figure 2.7a shows the typical example of the height profile of these three components (Sato 1988). In general, n is almost 1 and the value is around 1.0003 or smaller, and its variance due to turbulence is extremely small as around 10^{-5} to 10^{-6}. Because of this reason, the scale of abscissa that shows the change of n is given by $\log_{10}(n-1)$.

There is an equivalent way of treatment which assumes that the surface of the Earth is a plane so as to facilitate the analysis of electromagnetic wave propagation in the troposphere, where the modified refractive index m or the refractive modulus M is used as follows (e.g., Battan 1973, p. 23);

$$m \equiv n + \frac{h}{a}, \tag{2.85}$$

$$M \equiv (m-1) \times 10^6, \tag{2.86}$$

where h is the height from sea level and a is the mean radius of the Earth (6,370 km). The curve which shows the vertical profile of M is called the M-curve. Because atmospheric temperature and partial pressure of water vapor change with height in the actual atmosphere, refractive index of the atmosphere shows complicated

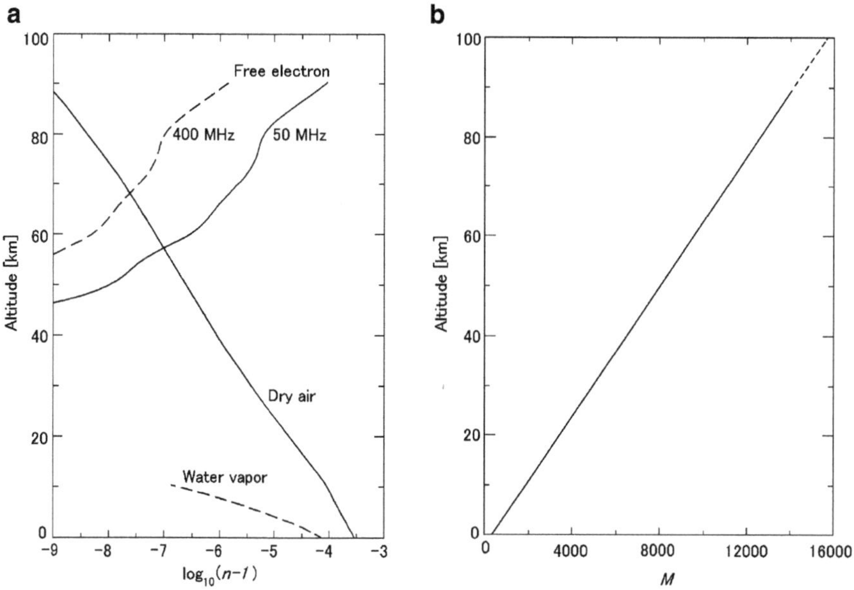

Fig. 2.7 Height profiles of (**a**) main components in the refractive index from Sato (1988), and (**b**) M-values for the dry air component in (**a**), where $\log_{10}(n-1)$ values are approximated by the linear curve between the altitude of 0 and 88 km

distribution. However, M gradually increases monotonously with height. Figure 2.7 (b) shows the height profile of M-values for the dry air component in (a), where $\log_{10}(n-1)$ values are approximated by the linear curve between the altitude of 0 and 88 km.

2.2.3 *Wave Path in the Spherically Stratified Atmosphere*

In the spherically stratified atmosphere, the permittivity ε decreases, in general, with increasing height from the ground, and approaches the permittivity in a vacuum ε_0. Thus, the propagation speed which is a little smaller than that in free space gradually approaches the speed of light c as the altitude increases. When a incident plane wave reaches the smooth boundary where the medium of refractive index n adjoins that of $n-\Delta n$, where $\Delta n \ll n$, Snell's law is satisfied along the wave path, and thus the following relation is obtained as (2.41);

$$\frac{\sin\theta_{\rm i}}{\sin\theta_{\rm t}} = \frac{n-\Delta n}{n}, \tag{2.87}$$

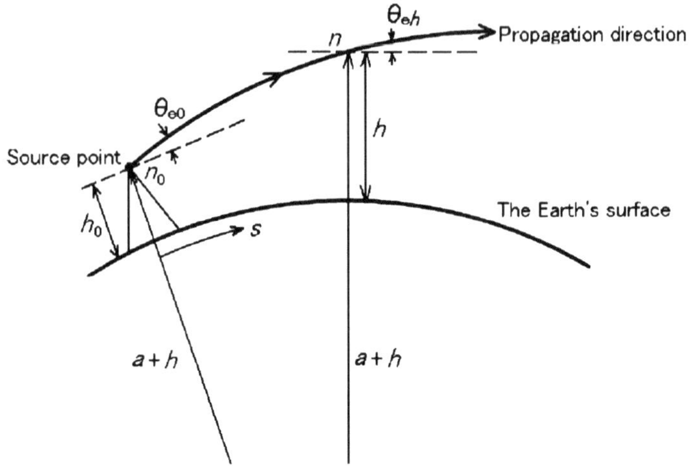

Fig. 2.8 Ray path through multiple atmospheric layers in which n decreases with height

where θ_i and θ_t are the incident angle and the refractive angle in Fig. 2.4, respectively. Also, n and $n-\Delta n$ are the refractive index of medium 1 and 2, respectively. From (2.87), $\theta_t > \theta_i$. It is found that if a electromagnetic wave is incident from a medium of larger refractive index (denser medium) to that of smaller one (rarefied medium), the radio path leans toward the denser medium. In other word, if multiple atmospheric layers are piled up and their radio refractive indexes change according to the change of permittivity of the atmosphere, the electromagnetic wave which is radiated upward from the ground does not propagate in a straight line, but is refracted on each boundary and leans downward as shown in Fig. 2.8. If the height gradient $\mathrm{d}n/\mathrm{d}h$ of the refractive index is small enough for the optical theory to be applicable, the propagation path of the electromagnetic wave through the atmosphere (ray path) can be expressed by the following differential equation (Hartree et al. 1946);

$$\frac{\mathrm{d}^2h}{\mathrm{d}s^2}-\left(\frac{2}{a+h}+\frac{1}{n}\frac{\mathrm{d}n}{\mathrm{d}h}\right)\left(\frac{\mathrm{d}h}{\mathrm{d}s}\right)^2-\left(\frac{a+h}{a}\right)^2\left(\frac{1}{a+h}+\frac{1}{n}\frac{\mathrm{d}n}{\mathrm{d}h}\right)=0, \tag{2.88}$$

where s is the horizontal distance, h the altitude, and a the Earth radius. If the elevation angle of the ray path θ_e is smaller than several degrees, $(\mathrm{d}h/\mathrm{d}s)^2 \ll 1$. In addition, n is approximately equal to 1, and $h \ll a$. Considering these conditions, (2.88) is simplified as

$$\frac{\mathrm{d}^2h}{\mathrm{d}s^2}=\frac{1}{a}+\frac{\mathrm{d}n}{\mathrm{d}h}. \tag{2.89}$$

Integrating the above equation with respect to h (Battan 1973, p. 23),

$$\left(\frac{\mathrm{d}h}{\mathrm{d}s}\right)^2=2\int\left(\frac{1}{a}+\frac{\mathrm{d}n}{\mathrm{d}h}\right)\mathrm{d}h+\mathrm{C}, \tag{2.90}$$

where C is a constant. Here, θ_e is small, and thus $dh/ds \simeq \theta_e$. Hence (2.90) is converted to

$$\begin{aligned}\frac{\theta_{eh}^2}{2}-\frac{\theta_{e0}^2}{2}&=\frac{h-h_0}{a}+n-n_0\\&=\left(\frac{h}{a}+n\right)-\left(\frac{h_0}{a}+n_0\right)\\&=(M-M_0)\times 10^{-6},\end{aligned}\tag{2.91}$$

where θ_{eh} is the elevation angle of the ray path at the height of h, and θ_{e0}, h_0, and n_0 are the elevation angle of the ray path, height, and refractive index at the source point, respectively. Moreover, M and M_0 are the refractive modulus at the height of h and h_0, respectively, and given by (2.86). Given the vertical profiles of M from (2.84) to (2.86), the elevation angle of the ray path at the height of h is obtained by (2.91).

If the atmosphere which surrounds the Earth does not exist or the value n is constant with height, the ray path does not curve and travels straight. In this case, the curvature of electromagnetic wave to the horizontal direction is $1/a$. Actually, the atmosphere exists and the ray path curves, and from (2.89) the curvature becomes

$$\frac{d\theta_e}{ds}=\frac{1}{a}+\frac{dn}{dh}.\tag{2.92}$$

Consider an effective Earth such that the curvature of the ray path is given so that its change in the Standard Atmosphere due to decrease of the atmospheric refractivity, dn/dh (<0), is added to that of the Earth's surface, the ray path can be treated as a straight line. Thus, if the curvature of the effective Earth is defined as (Bean and Dutton 1966, p. 57)

$$\frac{1}{a_e}=\frac{1}{a}+\frac{dn}{dh}\quad \text{i.e.,}\quad a_e=\frac{a}{1+\dfrac{dn}{dh}a},\tag{2.93}$$

the ray path becomes a straight line, where the effective Earth radius $a_e = k_e a$, and k_e is called the ratio of equivalent-to-real radius of the Earth. As dn/dh of the Standard Atmosphere is approximately constant and takes the value of approximately $-1/(4a)$ then it naturally follows that k_e equals 4/3. If the electromagnetic wave is radiated from sea level and the elevation angle is zero, the ray path height h_e from sea level on the effective Earth is given by

$$h_e=\frac{1}{2}\frac{s^2}{a_e}=\frac{3}{8}\frac{s^2}{a}\tag{2.94}$$

for the Standard Atmosphere, where s is the horizontal distance, i.e., the surface distance. The parameters h and s are not measurable with radar. Doviak and Zrnić (2006, pp. 20–21) derives them using radar-measurable parameters, the range r and θ_e, as shown in the following three equations. The height h of a ray leaving the radar at an elevation angle θ_e is expressed as

$$h = k_{\mathrm{e}}a\left[\frac{\cos\theta_{\mathrm{e}}}{\cos(\theta_{\mathrm{e}} + s/k_{\mathrm{e}}a)} - 1\right], \tag{2.95}$$

where the height h_0 of radar is not considered. The parameters h and s are related to r and θ_{e} as

$$h = [r^2 + (k_{\mathrm{e}}a)^2 + 2rk_{\mathrm{e}}a\sin\theta_{\mathrm{e}}]^{1/2} - k_{\mathrm{e}}a, \tag{2.96}$$

$$s = k_{\mathrm{e}}a\sin^{-1}\left(\frac{r\cos\theta_{\mathrm{e}}}{k_{\mathrm{e}}a + h}\right). \tag{2.97}$$

Without an atmosphere, ray paths are straight lines. With the Standard Atmosphere, the radio waves propagate according to Snell's law as shown by (2.41), thus ray paths are curved and the curvature of the ray paths is negative (i.e., the ray paths are bent toward the Earth) even though M-values are increasing with height. But if M-values increase more rapidly than that shown in Fig. 2.7b, the bending is less. If M-values, on the other hand, decrease with height, the ray paths increase negative curvature and bend more strongly toward the Earth. An inversion layer is formed at the Earth's surface if the atmospheric temperature increases with height. This causes M-values to decrease with height; M-values can also decrease if the partial pressure of water vapor strongly decreases with height. In this case, ray paths of electromagnetic radiation are bent toward the Earth, and if the temperature increase and/or pressure decrease is sufficiently strong, and the elevation angle sufficiently small, the ray paths can be so strongly bent so that they curve back to the Earth's surface. There the radiation is reflected and continues to repeat the curved propagation path to long distances. Doviak and Zrnić (2006, Fig. 2.11a) show an example of ducting of electromagnetic waves which propagate within a ground based inversion layer to strikingly long distances.

2.2.4 *The Profile of the Standard Atmosphere*

According to the Standard Atmosphere, the profile is defined as follows (Ulaby et al. 1981, p. 262).

Atmospheric Temperature Profile

The atmospheric temperature T[K] at the height of z[km] above sea level is given by

$$T(z) = \begin{cases} T(0) - a_{\mathrm{T}}z & 0 \le z \le 11\ \mathrm{km}, \\ T(11) & 11\ \mathrm{km} \le z \le 20\ \mathrm{km}, \\ T(11) + (z - 20) & 20\ \mathrm{km} \le 32\ \mathrm{km}, \end{cases} \tag{2.98}$$

where $T(0)$ and $T(11)$ are the atmospheric temperature at the height of 0 km and 11 km, respectively, and a_T is the temperature lapse rate. Here, $T(0) = 288.15$ K and $a_T = 6.5$ K km^{-1}.

Atmospheric Density Profile

The atmospheric density profile ρ_a [kg m^{-3}] in the height range from z[km] above sea level to 30 km is given by

$$\rho_a(z) = 1.225e^{-z/H_1}[1 + 0.3\sin(z/H_1)], \tag{2.99}$$

where $H_1 = 7.3$ km is called the scale height.

Atmosphere Pressure Profile

The atmospheric pressure p[hPa] in the height range from z[km] above sea level to 30 km is given by

$$p(z) = 2.87\rho_a(z)T(z), \tag{2.100}$$

where $\rho_a(z)$ and $T(z)$ are the values given in (2.99) and (2.98), respectively.

Water Vapor Density Profile

The water vapor density ρ_v [g m^{-3}] depends on various meteorological factors, but in particular depends on the atmospheric temperature. The ρ_v value at the height of z[km] above sea level is expressed as

$$\rho_v(z) = \rho_0\, e^{-z/H_2}, \tag{2.101}$$

where $\rho_0 = 7.72$ g m^{-3}, and H_2 is 2 to 2.5 km. The total water vapor content M_v [kg m^{-2}] is given by

$$M_v = \int_0^\infty \rho_v(z)\mathrm{d}z = \rho_0 H_2. \tag{2.102}$$

For example, if $\rho_0 = 7.72$ g m^{-3}, and $H_2 = 2$ km, $M_v = 15.4$ kg m^{-2}, or 1.54 g cm^{-2}.

It should be noted that it is not the Standard Atmosphere but the actual profiles of the atmosphere, e.g., atmospheric wind, temperature, and water vapor density, measured by atmospheric radar that will be discussed in this book.

Chapter 3
Radar Measurements and Scatterer Parameters

Parameters of scatterers which are directly measured by a radar are, in general, the received signal power, the mean Doppler shift (or mean Doppler frequency), and the Doppler frequency spectrum width. The echo intensity and radar reflectivity factor are deduced from the received signal power and the relevant radar system parameters in consideration of power attenuation of the radar wave on the way to and back from scatterers. The mean Doppler shift (or frequency) is caused by the radial velocity of scatterers, and the Doppler frequency spectrum width is caused by the velocity fluctuation of scatterers. The Doppler velocity and the Doppler velocity spectrum width are derived, respectively, from the mean Doppler shift (or frequency) and the Doppler frequency spectrum width in reference to the transmitted wave frequency. In this chapter, basic knowledge on radar measurements and methods of estimating the scatterer parameters are described. The main scatterers concerned are water and ice particles (they are precipitation if they fall and reach the ground, while cloud if they do not), and Bragg scatterers created by perturbations of the atmosphere's state variables (i.e., temperature, humidity, and pressure). Various physical quantities related to weather and the atmospheric phenomena will be described in Chaps. 6 and 7, respectively. Basic techniques to estimate the Doppler velocity will be discussed in Chap. 4.

3.1 Basics of Radar

3.1.1 System Parameters of Pulse Radars

Both meteorological and atmospheric radars commonly generate synchronizing signals, amplifying them as high power radio frequency (rf) pulses, and transmit them at the carrier frequency f_0. This significantly facilitates the setting of a variety of radar system parameters, e.g., maximum observable range and range resolution.

S. Fukao and K. Hamazu, *Radar for Meteorological and Atmospheric Observations*,
DOI 10.1007/978-4-431-54334-3_3,

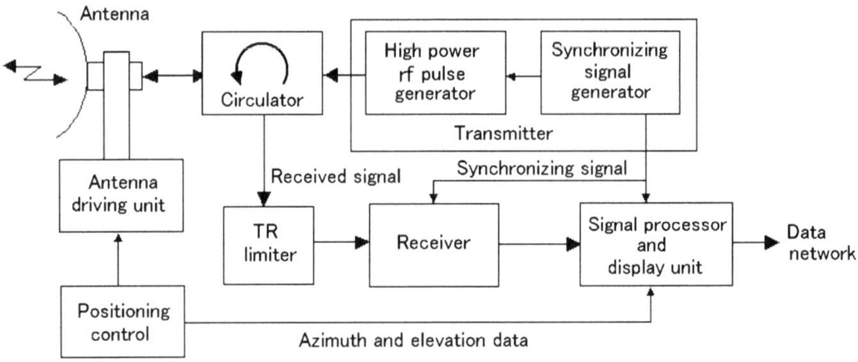

Fig. 3.1 Basic configuration of a pulse radar

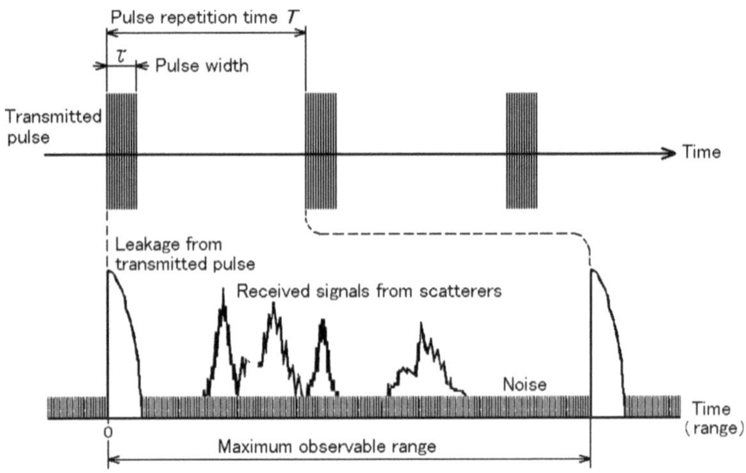

Fig. 3.2 Time sequence of transmitted pulses and received signals from scatterers

The basic configuration of a pulse radar is shown in Fig. 3.1. Meanwhile, Fig. 3.2 shows a conceptual diagram of the signal transmission and reception for a pulse radar.

Ranging

For a pulse radar, range or distance r between the radar and scatterers is determined from the time t between the transmission of a pulse and the receptions of a signal (or echo) backscattered by the scatterers. That is

$$r = \frac{ct}{2}, \tag{3.1}$$

where c is the speed of light. The speed of light in the atmosphere is slightly slower than that in a vacuum, but the difference is approximately 0.03 % or smaller.[1] Thus, hereafter we assume the speed of light in the atmosphere is equal to that in a vacuum and express it as c unless otherwise specified. Range resolution Δr is the minimum distance in which two adjoining scatterers in the same direction are resolvable, and is determined as

$$\Delta r = \frac{c\tau}{2}, \tag{3.2}$$

where τ is the transmitted pulse width. On the other hand, the range determined from the pulse repetition time (PRT) T is the maximum observable range r_a where the range to scatterers can be unambiguously determined as:

$$r_\mathrm{a} = \frac{cT}{2}. \tag{3.3}$$

The scattered signals (or echoes) are sometimes received from the distance beyond the unambiguous range. The echo which corresponds to the pulse that just precedes the present pulse is called an out-of-trip echo. The echoes corresponding to the preceding multiple pulses are generally called multiple echoes. They appear to come from scatterers within the unambiguous range even if they are actually outside r_a. Consequently, they would yield erroneous information on scatterer ranges when they would be intermingled with the primary echo. In recent years, new techniques which change transmission phases for successive pulses are widely used for identification and suppression of multiple echoes (e.g., Zrnić and Mahapatra 1985; Sachidananda and Zrnić 1999).

Direction Finding

The direction to a scatterer is determined from the angle where the received power becomes maximized as the narrow beam of the radar is scanned. The resolution with respect to azimuth and elevation angles is the minimum angle by which two scatterers at the same range from the radar can be identified. Therefore, the angular resolution is approximately equal to the beam width. The radar beam width is approximately determined by (8.11) in Sect. 8.2.1.

Radar Reflectivity and Radar Reflectivity Factor

Backscattered power (or echo) from a scatterer is received by the radar receiver. Radar reflectivity and radar reflectivity factor are calculated from the received power based upon the radar equation as will be discussed in detail in Sects. 3.3 and 3.4.

---

[1] It is derived from (2.40) and Fig. 2.7.

3.1.2 Characteristics of Scatter and Scatterers

Electromagnetic waves which are radiated from the antenna are scattered by particles such as raindrops and cloud droplets, and by perturbations of refractive index (hereafter refractive index perturbations) caused by turbulence. A part of the scattered power is received with the same antenna in the same way as a reflected power from aircrafts or ships. The main purpose of search radars is to detect aircrafts and ships. On the other hand, the main purpose of meteorological and atmospheric radars are to detect echoes from the hard scatterers, i.e., precipitation and cloud particles, and echoes from soft (i.e., Bragg) scatterers,[2] respectively. These radars have the following characteristics.

- Generally, search radars deal with the scatterers which are bigger than the radar wavelength. On the other hand, meteorological radars are principally applied to observe weather; echoes from objects other than precipitation and cloud are simply clutter. Also, in the case of the atmospheric radars, the radar wavelength decides the scale of scatterers. The main scatterer for the atmospheric radar is stochastic Bragg scatterer.
- Both refractive index perturbations on the scale of half the radar wavelength, and rainfall which is an aggregate of precipitation particles such as raindrops, cloud droplets, and snowflakes are not isolated scatterers but spread over a considerably wide volume of space. Furthermore, the spatial extent of the scatterers is generally much wider than the radar beam and there are many of these scatterers within the resolution volume (see Fig. 3.6). Therefore, the backscattered power is considered as an aggregate of powers principally scattered from individual scatterers in the whole radar resolution volume (or scattering/reflecting volume) which is determined by the radar beam width and transmitted pulse length. Scatterers outside the radar resolution volume also contribute power, but typically those powers are relatively small compared to echo power from scatterers within the resolution volume.
- The total received power from precipitation particles or refractive index perturbations is given as a vector sum of the electric field scattered from individual scatterers within a radar resolution volume. The relative positions of individual scatterers in the radar resolution volume always change even if the density of scatterers remains constant. Because the positions of the scatterers determine their phases, the vector sum of echo voltages contributed by the scatterers will vary from time to time as the scatterers' relative positions change. Therefore, the total received power by these scatterers changes its magnitude from transmitted pulse to pulse.

[2]Among the refractive index perturbations, only those which have a wavelength equal to one half the radar wavelength contribute substantially to the backscatter. These perturbations are defined as the Bragg scatterers as will be discussed in Sect. 3.4.

Table 3.1 Radar scatterers and systems utilized for monitoring

Category of scatterers		Typical scatterers	Radar systems
Hard scatterer	Isolated	Aircraft, ship	Search radar
	Distributed	Precipitation particle	Meteorological radar
			Cloud or fog radar
Soft scatterer	Distributed	Structure of the atmosphere	Atmospheric radar

Consequently, for getting an accurate signal power, it is common to take an average of signal powers received from multiple transmitted pulses. The characteristic radar scatterers are classified into two types as follows:

Hard Scatterer

The scatterers with an extent significantly larger than the radar wavelength such as aircrafts, and minute particles like raindrops and cloud droplets are called hard scatterers. The former type is named as isolated scatterers while the later type distributed scatterers. The main objectives of the meteorological radars are to monitor the strength of echoes from distributed hard scatterers.

Soft Scatterer

The scatterers associated with changes in the atmosphere's state variables such as refractive index perturbations of the atmosphere which contain no hard scatterers inside are called soft scatterers. There are a variety of mechanisms to generate soft scatterer such as Bragg scatterers and partial reflections as will be discussed in Sect. 3.4. The main objectives of the atmospheric radars are to monitor these soft scatterers.

Various radar scatterers and systems utilized for monitoring are summarized in Table 3.1. For isolated types of hard scatterer, the received power depends on the scatterer's size which scatters electromagnetic waves back, i.e., the cross section illuminated by the radar beam. On the other hand, the intensity of the signal power received by meteorological and the atmospheric radars is determined not only by the cross section of individual scatterers but also by the size of the radar resolution volume. This effect is taken into consideration for both meteorological radars which primarily observe the discretely distributed hard scatterers, and the atmospheric radars which mainly observe distributed soft scatterers.

3.2 Radar Observation of Isolated Scatterers

3.2.1 Radar Equation for an Isolated Scatterer

When a transmitted power P_t[W] is isotropically radiated from a point source to all directions, the power density $[\mathrm{W\,m^{-2}}]$ of the radiation at range r[m] is $P_t/4\pi r^2$. Therefore, the power P_σ[W] intercepted by a scatterer with the cross section of $\sigma[\mathrm{m^2}]$ at the same range is

$$P_\sigma = \frac{P_t \sigma}{4\pi r^2}. \tag{3.4}$$

The radar antenna has a gain so as to concentrate the power into a specific direction. Let the antenna gain[3] for the scatterer direction be $g_t(\theta,\phi)$, where θ and ϕ are the polar angles, and the received power $P_{\sigma g}$ is given by

$$P_{\sigma g} = g_t(\theta,\phi)P_\sigma = g_t(\theta,\phi)\frac{P_t\sigma}{4\pi r^2}. \tag{3.5}$$

If the scatterer isotropically reradiates the power with no attenuation along the propagation path, and the reradiated power is received by the antenna of the effective area of $A_e[\mathrm{m^2}]$, the received power P_r[W] is given by

$$P_r = \frac{P_{\sigma g}A_e}{4\pi r^2} = g_t(\theta,\phi)\frac{P_t A_e \sigma}{(4\pi r^2)^2}. \tag{3.6}$$

When the radar wavelength is λ[m], A_e and the gain g_r of the receiver antenna are related by previously cited (2.83). If A_e in (3.6) is replaced by (2.83), we can get the following equation:

$$P_r = \frac{P_t g_t(\theta,\phi) g_r(\theta,\phi)\lambda^2}{(4\pi)^3 r^4}\sigma. \tag{3.7}$$

Ordinary radars uses the same antenna for both transmission of pulses and reception of signals, and $g_t(\theta,\phi) = g_r(\theta,\phi) \equiv g(\theta,\phi)$. Therefore, the received power is expressed by

$$P_r = \frac{P_t A_e^2}{4\pi\lambda^2 r^4}\sigma = \frac{P_t g(\theta,\phi)^2\lambda^2}{(4\pi)^3 r^4}\sigma. \tag{3.8}$$

[3] Here we assume that the efficiency of the antenna $\eta_l = 1$ and power gain g is equal to directivity g_D. Refer to Sects. 2.1.4 and 8.2.1. The lower case g denotes the value in a pure number (in linear unit), whereas the upper case G denotes it in decibel.

Equation (3.8) is the radar equation for isolated scatterers. This equation shows that the received power P_r from an isolated scatterer decreases proportionally to r^{-4} for range r and proportional to A_e^2/λ^2.

Radars usually receive backscattered signals which return in the direction of transmitter.[4] The value of σ is called the radar cross section (RCS), which is a function of shape and properties of the scatterer, the radar wavelength and so on.

Radar Cross Section

The radar cross section is defined as the area of a supposed isotropic scatterer which radiates the same power as that reradiated by the particle concerned to the receiving antenna. In the case that the scattering from the particle occurs isotropically in all directions, and the scattered power density at range r from the particle is S_s, the total scattered power becomes $4\pi r^2 S_s$. On the other hand, the input power is $S_i\sigma$, when the incident power density S_i is intercepted by the particle with the cross section σ. The radar cross section is defined as that the two powers are identical at a sufficient long distance in the far field as

$$\sigma = \lim_{r\to\infty} 4\pi r^2 \frac{S_s}{S_i}, \tag{3.9}$$

where the far field region is commonly taken to exist at distances greater than $2D_a^2/\lambda$, D_a is the antenna diameter, and λ is the radar wavelength. (Refer to Sect. 8.2.1.) Theoretically, the radar cross section can be analytically obtained by solving Maxwell's equations under appropriate boundary conditions (e.g., Skolnik 1990, Chap. 11). However, it is limited only to the exceptional cases that the scatterers are in extremely simple shape, and the radar cross section is usually determined experimentally by measuring backscattered electric field intensity. Ordinary scatterers have complicated surface shape, and radar waves scattered from them often interfere with each other. Therefore, the radar cross section substantially varies depending upon the mutual angular position between the radar antenna and the scatterer. The radar cross sections for several typical scatterers at various radar wavelengths in the microwave band are shown in Table 3.2 (Nathanson 1991, p. 176).

Radar cross sections of precipitation particles and structure changes of the atmosphere (i.e., refractive index perturbations defined as the Bragg scatterers) will be discussed in Sect. 3.3 and 3.4, respectively. The radar cross sections of individual water drop are extremely small. For example, those of one water drop of 1 mm diameter in 10 °C for radar wavelength of 10 and 3 cm are approximately $2.8 \times 10^{-12}\,\mathrm{m}^2$ and $3.5 \times 10^{-10}\,\mathrm{m}^2$, respectively.[5] In general, the cross sections of these distributed scatterers are treated by the total amount per unit volume.

---

[4]Bistatic radars which will be discussed later in Sect. 4.2.3 utilizes the side scattered signals.

[5]These values are calculated using (3.58) and Table 3.4.

Table 3.2 Typical radar cross sections [m^2] for various scatterers [modified from Nathanson 1991, p. 176]

	Radar wavelength				
Category	30 – 15 cm	10 cm	6 cm	3 cm	1.8 cm
Aircraft, small propeller	2	3	2.5		
Jet skyliner, 747	70	70	70	70	
Standing man (180 lb)	0.3	0.5	0.6	0.7	0.7
Sea birds	0.002	0.001–0.004	0.004		
Sparrow, starling, etc.	0.001	0.001	0.001	0.001	0.001
Precipitation particle[a]		2.8×10^{-12}	2.2×10^{-11}	3.5×10^{-10}	2.7×10^{-9}

[a] 1 mm Diameter, 10°C

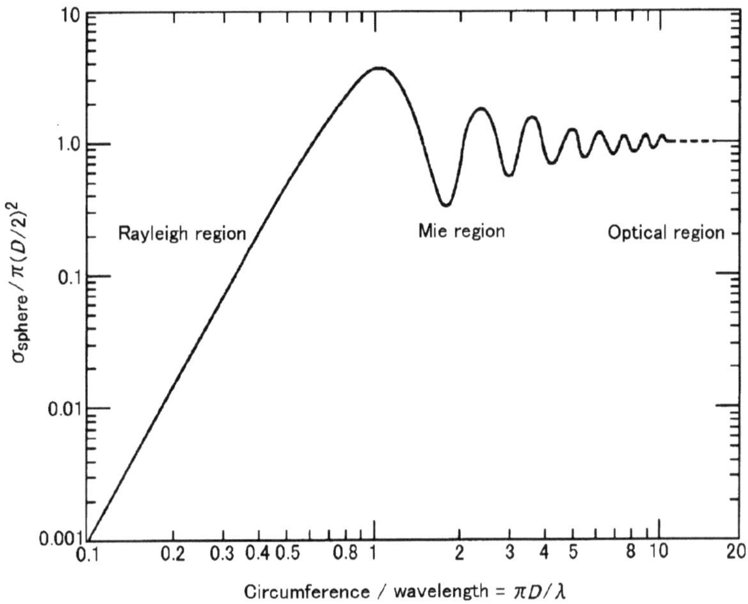

Fig. 3.3 Normalized radar cross section of the perfect conducting sphere [from Skolnik 1990, pp. 11–6]

3.2.2 Characteristics of Scattering

The radar cross sections of conducting and dielectric spheres can be calculated by the Mie scattering formula which will be discussed in Sect. 3.3.3. The radar cross section for the perfect conducting sphere derived from the formula is shown against diameter in Fig. 3.3 (Skolnik 1990, pp. 11–6). In the figure, the abscissa shows the circumference of the sphere of diameter D normalized by the radar wavelength λ,

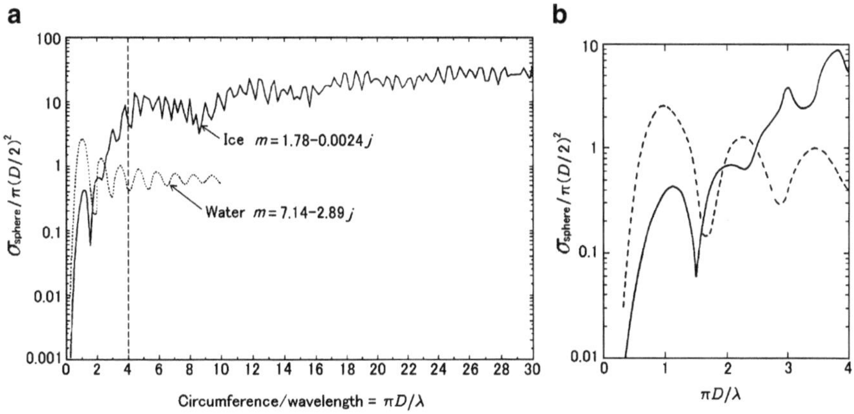

Fig. 3.4 (**a**) Normalized radar cross sections of water and ice spheres for $\lambda = 3.21\,\text{cm}$, where m represents the complex refractive index of the spheres [modified from Herman and Battan 1961]. (**b**) Expansion of the region $\pi D/\lambda < 4$ of (**a**) [from Sauvageot 1992, p. 94]

and the ordinate the radar cross section normalized by physical cross section of the sphere, with both axes on logarithmic scales. Figure 3.4a shows radar cross sections of the water and ice spheres for $\lambda = 3.21\,\text{cm}$ (Modified from Herman and Battan 1961), where m represents the complex refractive index of the spheres that will be given by (3.48). Panel (b) is an expansion of (a) (Sauvageot 1992, p. 94). The diameter in the abscissa and the cross section in the ordinate are normalized same as Fig. 3.3, and the ordinate is expressed on a logarithmic scale.

The region where the diameter of the sphere is large compared with radar wavelength, i.e., $\pi D/\lambda > 10$, is called the optical region (Skolnik 1990, pp. 11–5). In the optical region, the radar cross section of a conducting sphere approaches the physical cross section $\pi D^2/4$. On the other hand, the region where the diameter is small compared with radar wavelength ($\pi D/\lambda \ll 1$) is called the Rayleigh region.[6] The radar cross sections of scatterers in the Rayleigh region vary as λ^{-4}. The radar cross sections of raindrops and other precipitation particles fall within this range at the microwave and other ordinary meteorological radar frequencies. Therefore, the Rayleigh region is of great interest for radar meteorology. The region between the optical and the Rayleigh region is called the Mie region, or the resonance region. The radar cross section is oscillatory with frequency within this region. Sizes of typical scatterers relative to radar wavelengths are illustrated in Fig. 3.5. The horizontal extent of each vertical band illustrates the approximate size of scatterers ordinarily observed by radar. The slanted lines approximately delineate the three regions (Nathanson 1991, p. 152).

[6]Doviak and Zrnić (2006, p. 35) define the Rayleigh region as $\pi D/\lambda < \pi/16 \simeq 0.2$.

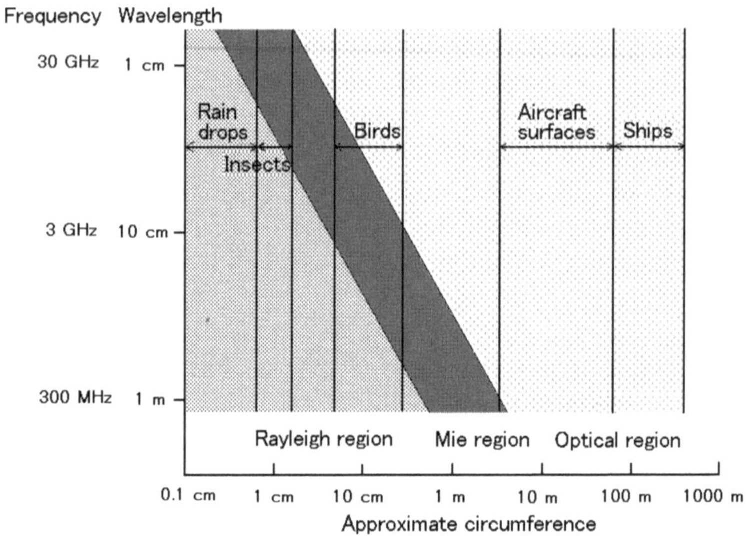

Fig. 3.5 Sizes of typical radar scatterers relative to radar wavelengths [from Nathanson 1991, p. 152]

3.3 Radar Theory for Hard Scatterers

Radio waves with wavelengths from several cm to mm have been widely utilized for radar observations of weather. They generate dielectric polarization in raindrops, since raindrops are a well-known dielectric. Thus the radio wave scattering by raindrops is caused by alternating electric fields due to the dipole moment of the dielectric polarization. First we discuss scattering by precipitation particles from the point of view of scattered field due to dielectric moment in the following subsection. The general solution of the scattering of electromagnetic waves due to particles, Mie scattering, and its approximation, the Rayleigh approximation, will be discussed in Sects. 3.3.3 and 3.3.4, respectively.

3.3.1 Scattering by Dielectric Spheres

If an electric field $\boldsymbol{E}$ is applied to a scatterer with the permittivity ε, the dielectric polarization $\boldsymbol{P}$ for the unit volume is

$$\boldsymbol{P} = (\varepsilon - \varepsilon_0)\boldsymbol{E}, \tag{3.10}$$

where ε_0 is the permittivity of propagation medium, i.e., the atmosphere. Then, the dipole moment $\boldsymbol{P}_\mathrm{V}$ of the dielectric in the volume V_D, becomes

$$\boldsymbol{P}_{\mathrm{V}} = V_{\mathrm{D}}\boldsymbol{P} = (\varepsilon - \varepsilon_0)V_{\mathrm{D}}\boldsymbol{E}, \tag{3.11}$$

and V_{D} for a raindrop of the diameter D is expressed as

$$V_{\mathrm{D}} = \frac{4}{3}\pi\left(\frac{D}{2}\right)^3. \tag{3.12}$$

Meanwhile, the amplitude of the far field vector at a place sufficiently distant from the scatterer is given by the following equation (Bringi and Chandrasekar 2001, Chap. 1);

$$\boldsymbol{f} = \frac{k_{\mathrm{s}}^2}{4\pi\varepsilon_0}\boldsymbol{P}_{\mathrm{V}}\sin\chi, \tag{3.13}$$

where k_{s}, the magnitude of scattering vector wave number $\boldsymbol{k}_{\mathrm{s}}$ in the direction of itself,[7] is $2\pi/\lambda$, and χ is the angle between the direction of polarization of the incident electric field and direction of $\boldsymbol{k}_{\mathrm{s}}$. In the case of backscattering, $\chi = \pi/2$. The electromagnetic field or scattered electric field $\boldsymbol{E}_{\mathrm{s}}$, generated by $\boldsymbol{f}$ at a range $\boldsymbol{r}_{\mathrm{s}}$ is given by (Bringi and Chandrasekar 2001, Chap. 1)

$$\boldsymbol{E}_{\mathrm{s}} = \boldsymbol{f}\frac{\mathrm{e}^{-j\boldsymbol{k}_{\mathrm{s}}\cdot\boldsymbol{r}_{\mathrm{s}}}}{r_{\mathrm{s}}}. \tag{3.14}$$

Substituting (3.11) and (3.13) into (3.14), $\boldsymbol{E}_{\mathrm{s}}$ becomes

$$\boldsymbol{E}_{\mathrm{s}} = \frac{k_{\mathrm{s}}^2 V_{\mathrm{D}}(\varepsilon/\varepsilon_0 - 1)}{4\pi r_{\mathrm{s}}}\boldsymbol{E}\sin\chi\,\mathrm{e}^{-j\boldsymbol{k}_{\mathrm{s}}\cdot\boldsymbol{r}_{\mathrm{s}}}. \tag{3.15}$$

Scattering by a Single Dielectric Sphere

The following discussion in this subsection is based upon Gossard and Strauch (1983, Chap. 2). The dielectric sphere with permittivity ε which is placed in a uniform external electric field is polarized in the direction of the external electric field. The electric field induced inside the dielectric sphere is expressed as (e.g., Gossard and Strauch 1983, p. 56)

$$\boldsymbol{E} = \frac{3}{\varepsilon/\varepsilon_0 + 2}\boldsymbol{E}_0\mathrm{e}^{j(\omega t - \boldsymbol{k}_0\cdot\boldsymbol{r}_0)}, \tag{3.16}$$

where $\boldsymbol{E}_0$ is the incident electric field and $\boldsymbol{r}_0$ is the position vector of the dielectric with respect to the transmitter point, and $\boldsymbol{k}_0$ is the wave number vector of incident electric field (k_0 is equal to $2\pi/\lambda$). Substituting (3.16) into (3.15), the scattered electric field at the distance $\boldsymbol{r}_s$ from the dielectric sphere is

[7] The distinction between the incident vector wave number $\boldsymbol{k}_0$ and scattering vector wave number $\boldsymbol{k}_{\mathrm{s}}$ is the direction of two vectors.

$$\boldsymbol{E}_{\mathrm{s}} = \frac{3k_{\mathrm{s}}^2 K V_{\mathrm{D}}}{4\pi r_{\mathrm{s}}} \boldsymbol{E}_0 \sin\chi \ \mathrm{e}^{j(\omega t - \boldsymbol{k}_{\mathrm{s}}\cdot \boldsymbol{r}_{\mathrm{s}} - \boldsymbol{k}_0 \cdot \boldsymbol{r}_0)}, \tag{3.17}$$

where K is written as

$$K = \frac{\varepsilon/\varepsilon_0 - 1}{\varepsilon/\varepsilon_0 + 2} = \frac{\varepsilon - \varepsilon_0}{\varepsilon + 2\varepsilon_0}. \tag{3.18}$$

Scattering by Multiple Dielectric Spheres

If dielectric spheres of the same size are distributed with density distribution $N(\boldsymbol{r}_0)$ on a propagation path, the scattered electric field is given by the following equation;

$$\boldsymbol{E}_{\mathrm{s}} = \frac{3k_{\mathrm{s}}^2 K V_{\mathrm{D}}}{4\pi r_{\mathrm{s}}} \boldsymbol{E}_0 \sin\chi \int_{\boldsymbol{r}_0} N(\boldsymbol{r}_0) \mathrm{e}^{j(\omega t - \boldsymbol{k}_{\mathrm{s}}\cdot \boldsymbol{r}_{\mathrm{s}} - \boldsymbol{k}_0 \cdot \boldsymbol{r}_0)} \mathrm{d}\boldsymbol{r}_0, \tag{3.19}$$

where the integration is performed within a resolution volume over which the incident electric field is uniform, assuming that all scatterers in the volume are illuminated by the same incident electric field strength. In the case of backscattering, $\boldsymbol{k}_0 = -\boldsymbol{k}_{\mathrm{s}}$, $k = |\boldsymbol{k}_0| = |\boldsymbol{k}_{\mathrm{s}}|$, $\boldsymbol{r}_0 = -\boldsymbol{r}_{\mathrm{s}}$, and $r = |\boldsymbol{r}_0| = |\boldsymbol{r}_{\mathrm{s}}|$.

The average power is proportional to $\overline{\boldsymbol{E}\cdot\boldsymbol{E}^*}$, where the over bar denotes time average and superscript * denotes complex conjugate. Individual signal is fluctuating, and the scattered power P_s is fluctuating in time. Consequently, the envelope of the expected scattered power $E[P_s]$ is given by

$$E[P_s] = \overline{\boldsymbol{E}_{\mathrm{s}} \cdot \boldsymbol{E}_{\mathrm{s}}^*}/\eta_{\mathrm{i}}, \tag{3.20}$$

where η_{i} is the intrinsic (or wave) impedance which is given by (2.24). Substituting (3.19) into (3.20), P_s is expressed as

$$E[P_{\mathrm{s}}] = \frac{9P_0 k_{\mathrm{s}}^4 |K|^2 V_{\mathrm{D}}^2 \sin^2\chi}{(4\pi r_{\mathrm{s}})^2} I, \tag{3.21}$$

$$I = \overline{\left| \int_{\boldsymbol{r}_0} N(\boldsymbol{r}_0) \mathrm{e}^{-j\boldsymbol{\kappa}\cdot\boldsymbol{r}_0} \mathrm{d}\boldsymbol{r}_0 \right|^2}, \tag{3.22}$$

where P_0 is the envelope of the incident power, and $P_0 = \overline{\boldsymbol{E}_0 \cdot \boldsymbol{E}_0^*}/\eta_{\mathrm{i}}$. The vector $\boldsymbol{\kappa} = \boldsymbol{k}_0 - \boldsymbol{k}_{\mathrm{s}}$, $\boldsymbol{k}_0 = -\boldsymbol{k}_{\mathrm{s}}$, and $\kappa = 2|\boldsymbol{k}_0|$ for backscatterer.

Scattering Intensity

The density distribution, i.e., the number density distribution function of dielectric, $N(\boldsymbol{r}_0)$ is composed of the stationary term and the random variation term. The stationary term is generated by large scale variation of permittivity or refractive index, and it will not be considered below. The random variation term causes the scattering of incident wave which will be the primary theme for discussions in this subsection.

Integrating (3.22) over time yields the time average of the integration in (3.22), which becomes the product of random variation terms. Thus (Gossard and Strauch 1983, p. 57),

$$I = \iint \overline{\delta N(\boldsymbol{r}_1)\delta N(\boldsymbol{r}_2)} e^{-j\boldsymbol{\kappa}\cdot(\boldsymbol{r}_1-\boldsymbol{r}_2)} d\boldsymbol{r}_1 d\boldsymbol{r}_2, \tag{3.23}$$

where I is the number density which has the dimensions of $(\text{length})^{-3}$. It is assumed that the scatterer concentration is random in the sense that the particle concentration in neighboring parcels is completely uncorrelated where the meaning of "parcel" is an atmospheric volume of size approximately $(\lambda/2)^3$. If the number of scatterers $\delta N(\boldsymbol{r}_1)d\boldsymbol{r}_1$ is independent of that of $\delta N(\boldsymbol{r}_2)d\boldsymbol{r}_2$, the time average of the product in (3.23) is zero except for $r_1 = r_2$; thus (Gossard and Strauch 1983, p. 60),

$$I = \int_V \overline{\delta N(\boldsymbol{r})^2} d\boldsymbol{r} = \int_V \overline{N(\boldsymbol{r})} d\boldsymbol{r} = N_T, \tag{3.24}$$

or the total number of raindrops in the volume. This context means that the scattering is not Bragg type, and particles are assumed to be in the Rayleigh region.[8] Substituting (3.24) and (3.12) into (3.21), the expected scattered power $E[P_s]$ becomes

$$E[P_s] = \frac{P_0 k_s^4 |K|^2}{r_s^2} \left(\frac{D}{2}\right)^6 \sin^2 \chi N_T, \tag{3.25}$$

where N_T is the total number of raindrops. Using the relation of (3.4), $E[P_s]$ is given by

$$E[P_s] = \frac{P_0}{4\pi r_s^2}\sigma. \tag{3.26}$$

Replacing σ in (3.26) by the total amount of minute scattering cross sections, $\Sigma\sigma$, and equating (3.25) and (3.26),

$$\Sigma\sigma = \frac{\pi}{16} k_s^4 |K|^2 D^6 \sin^2 \chi N_T. \tag{3.27}$$

Noting that $k_s = 2\pi/\lambda$ and $\chi = \pi/2$ for backscattering, $\Sigma\sigma$ is expressed as

---

[8] Thus, the following discussion leads to the same result as the Rayleigh approximation that will be discussed in Sect. 3.3.4.

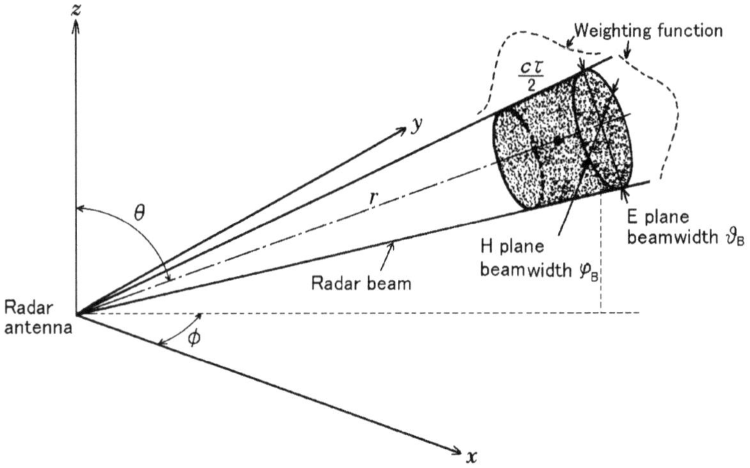

Fig. 3.6 The coordinate system and the scattering resolution volume of a pulse radar. The *shaded area* is assumed to be the radar resolution volume. However, the actual radar resolution volume is weighted as shown in the *broken lines* around the *shaded area*

$$\Sigma\sigma = \frac{\pi^5}{\lambda^4}|K|^2 D^6 N_{\mathrm{T}}. \tag{3.28}$$

In (3.28), it is assumed that the whole particles (raindrops) have the same diameter. However, actual rains have the own particular distributions with various raindrops different in size $N(D)$, where the number of raindrops in the size between D and $D + \mathrm{d}D$ is $N(D)$. Then, the total number of raindrops in the volume is given by

$$N_{\mathrm{T}} = \int_0^\infty N(D)\mathrm{d}D. \tag{3.29}$$

Substituting (3.29) into (3.28), $\Sigma\sigma$ is expressed as

$$\Sigma\sigma = \frac{\pi^5}{\lambda^4}|K|^2 \int_0^\infty D^6 N(D)\mathrm{d}D. \tag{3.30}$$

3.3.2 Radar Equation for Distributed Hard Scatterers

The primary scatterers for meteorological radars are, in general, distributed discrete hard scatterers such as raindrops and fog droplets. In the previous subsection, we discussed on the scattering intensity of multiple dielectric spheres. Its characteristic is the same as that of distributed discrete hard scatterers. The total radar cross section for these scatterers, $\Sigma\sigma$ of (3.30), corresponds to the total amount of minute scattering cross sections within the volume illustrated in Fig. 3.6. Thus, the radar equation (3.8) is replaced to

$$E[P_r] = \frac{P_t g^2 \lambda^2}{(4\pi)^3 r^4} \Sigma\sigma, \tag{3.31}$$

where $E[P_r]$ is the expected received power, P_t is the peak transmitted power, g is the antenna power gain (power gain, and $g = \eta_1 g_D$, where g_D is the directivity and η_1 is the efficiency of the antenna) at the direction of the maximum radiation pattern (main lobe), i.e., the maximum gain, λ[m] is the wavelength, and r[m] is the range between radar and the center of the resolution volume V which is expediently assumed to be the shaded truncated cone of the radius of radar beam width and length of $c\tau/2$, where τ is the pulse with, in Fig. 3.6. However, the actual radar resolution volume is weighted as shown in the broken lines around the shaded area of the same figure. The detail of the radar resolution volume will be discussed subsequently. Let η be the total amount of minute scattering cross sections per unit volume, $\Sigma\sigma$ is given by the product of V and η as

$$\Sigma\sigma = V\eta, \tag{3.32}$$

where $\eta\,[\mathrm{m^2\,m^{-3}} = \mathrm{m^{-1}}]$ is called radar reflectivity. Substituting (3.32) into (3.31), the radar equation for distributed hard scatterers is written as

$$E[P_r] = \frac{P_t g^2 \lambda^2}{(4\pi)^3 r^4} V\eta. \tag{3.33}$$

The derivative of V in the coordinates of Fig. 3.6 is given by

$$\mathrm{d}V = r^2 \mathrm{d}r \sin\theta \mathrm{d}\theta \mathrm{d}\phi. \tag{3.34}$$

Here we introduce V_6, which is circumscribed by the 6-dB contour of radar antenna beam width, the transmitted pulse width τ weighted with window function, and the receiver bandwidth, and replace V to V_6. Applying (3.34), V is given by

$$V \equiv V_6 = r^2 \int |W(r)|^2 \mathrm{d}r \int_0^{2\pi} \int_0^{\pi} f^4(\theta,\ \phi) \sin\theta \mathrm{d}\theta \mathrm{d}\phi, \tag{3.35}$$

where $|W(r)|$ is the range weighting function which is given by the function of receiver-filter bandwidth B_6 (6-dB width) and pulse width τ, and the integral of $|W(r)|^2$ approaches $c\tau/2$ if $B_6\tau$ gets large[9] (Doviak and Zrnić 2006, pp. 74–81).

The factor $f^4(\theta,\ \phi)$ is the two way antenna function, and the 6-dB width of $f^4(\theta,\ \phi)$ is often taken to be the angular width of V_6. Thus, $f^2(\theta,\ \phi)$ is the normalized one-way radiation pattern of the antenna. In general, the radiation pattern of the antenna can be represented by the Gaussian function. For convenience, putting z axis parallel to the direction of main lobe, and expressing zenith angles of the E- and H-plane of the radiation pattern by ϑ and φ, respectively, the radiation pattern of the antenna $f^2(\vartheta,\ \varphi)$ is written as follows (Nathanson and Reilly 1968);

[9] If $B_6\tau$ is 1.0, the integral of $|W(r)|^2$ normalized by $c\tau/2$ becomes 1.17.

$$f^2(\vartheta,\ \varphi) = \exp\left[-\left(\frac{\vartheta^2}{2\sigma_\vartheta^2} + \frac{\varphi^2}{2\sigma_\varphi^2}\right)\right], \tag{3.36}$$

where σ_ϑ and σ_φ are related to the full width at half maximum (FWHM) of the peak values of the Gaussian function.[10] Putting the FWHM values of ϑ and φ as ϑ_B and φ_B, respectively, where these values correspond to the half-power beam widths [rad] for the E plane and the H plane, respectively, $\sigma_\vartheta = \vartheta_B/\sqrt{8\ln 2}$ and $\sigma_\varphi = \varphi_B/\sqrt{8\ln 2}$. Considering (3.35) in the same coordinate system, $\sin\theta$ in the equation can be safely neglected. Therefore, using the relation

$$\int_{-\infty}^{\infty} \mathrm{e}^{-ax^2} \mathrm{d}x = \sqrt{\frac{\pi}{a}}, \tag{3.37}$$

the integration of $f^4(\vartheta,\ \varphi)$ with respect to ϑ and φ is approximated as (Probert-Jones 1962)

$$\int_{-\infty}^{\infty}\int_{-\infty}^{\infty} f^4(\vartheta,\ \varphi)\mathrm{d}\vartheta\mathrm{d}\varphi = \frac{\pi\vartheta_B\varphi_B}{8\ln 2}. \tag{3.38}$$

The values ϑ_B and φ_B are identical for circularly symmetric parabolic antennas, and $\vartheta_B = \varphi_B = \theta_1$, where θ_1 is the 3-dB width (in radians) of the one-way power pattern. Consequently, assuming $|W^2(r)|$ to be $c\tau/2$, the radar resolution volume can be written as

$$V_6 = r^2\frac{c\tau}{2}\frac{\pi\theta_1^2}{8\ln 2}. \tag{3.39}$$

Substituting (3.39) into (3.33), we finally obtain

$$E[P_r] = \frac{P_t g^2\lambda^2\theta_1^2 c\tau}{2^{10}(\ln 2)\pi^2 r^2}\eta. \tag{3.40}$$

This result exactly agrees with (4.14) of Doviak and Zrnić (2006) which is derived considering the particle size distribution. The above equation is called the radar equation for distributed hard scatterers when any loss is not taken into consideration. It is noted that the received power for a single isolated scatterer is proportional to r^{-4} as shown in (3.8). On the other hand, it is proportional to r^{-2} for distributed hard scatterers as shown in (3.40).

The signal power $E[P_r]$ for the typical 2.8-GHz radar, WSR-88D whose specifications will be listed later in Table 9.2, from the scatterers having the reflectivity

[10] The values σ_ϑ and σ_φ correspond to the standard deviations of the Gaussian distribution.

$\eta = 10^{-11}\ [\mathrm{m}^{-1}]$[11] at the range of 100 km can be calculated from (3.38). Assuming that $P_\mathrm{t} = 475$ kW $= 4.75 \times 10^5$ W, the gain $G = 44.5$ dB, i.e., $g \simeq 28{,}180$, $\lambda = 10.7\ \mathrm{cm} = 10.7 \times 10^{-2}\ \mathrm{m}$, $\theta_1 = 1.0° \simeq 1.75 \times 10^{-2}$ rad, $\tau = 1.57\ \mu\mathrm{s} = 1.57 \times 10^{-6}$ s, and $c = 3.00 \times 10^8\ \mathrm{m\,s}^{-1}$, we obtain $P_\mathrm{r} = 8.85 \times 10^{-14}$ W $= -100.5$ dBm.[12] Actually, the received power is reduced due to both power loss around 3 to 4 dB along the waveguide and propagation loss which will be discussed later in Sect. 6.3. The minimum received power capable of detecting signals is determined by the external noise which will be discussed in Sect. 5.1, receiver characteristics, and so on. The minimum detectable power of recent meteorological radars is smaller than 10^{-14} W or -110 dBm.

3.3.3 *Mie Scattering*

Precipitation particles (scatterers) absorb and change a part of the energy of the incident coherent wave into heat (i.e., incoherent infrared radiation), and scatter the rest in all directions as coherent radiation. Three quantities with the dimensions of area are derived for a scatterer in the path of an incident wave having a power density S_i.

The cross section σ_hd of a scatterer (e.g., hydrometeor) is an apparent area that intercepts a power $\sigma_\mathrm{hd} S_\mathrm{i}$, which, if radiated (i.e., scattered) isotropically, produces at the receiver a power density

$$S_\mathrm{r} = S_\mathrm{i}\sigma_\mathrm{hd}(\theta', \phi')/4\pi r^2, \tag{3.41}$$

equal to that scattered by the actual particle. The scatterer's cross section $\sigma_\mathrm{sd}(\theta', \phi')$ depends on the relative location of the transmitter and receiver, where the polar angles θ', ϕ' to the receiver are referenced to a polar axis drawn from the antenna to the scatterer (Doviak and Zrnić 2006, p. 35). When the direction of scattering is opposite to that of the incident wave, $\sigma_\mathrm{sd}(\theta', \phi')$ is called the backscattering cross section σ_b. As shown by (3.41) and (3.9), σ_b is identical to radar cross section σ.

The ratio of scattered power P_s to the incident power density S_i is called the scattering cross section σ_s, which gives, when multiplied by the incident power, the total power scattered by the particle. Likewise, the ratio of the absorbed power P_a to the incident power density is known as the absorption cross section σ_a, which gives, when multiplied by the incident power, the power dissipated as internal heat in the particle. The ratio of the total scattered and absorbed power P_tot to the incident power density is called the extinction (or attenuation) cross section σ_t, which gives,

---

[11]This value (i.e., cross section per unit volume) is equal to a radar the reflectivity factor of 6.6 dBZ as will be obtained by (3.67).

[12]The unit of dBm is the logarithmic expression of the power in regard to 1 mW. 1 mW = 0 dBm, hence 8.85×10^{-14} W $= 8.85 \times 10^{-11}$ mW $= -100.5$ dBm.

when multiplied by the incident power, gives the total power taken from the incident wave (Battan 1973, p. 67). These three cross sections are

$$\sigma_s \equiv P_s/S_i, \tag{3.42}$$

$$\sigma_a \equiv P_a/S_i, \tag{3.43}$$

$$\sigma_t \equiv P_{tot}/S_i, \tag{3.44}$$

respectively (Ulaby et al. 1981, p. 289). The total power removed from the incident wave is $P_s + P_a$, which is equal to P_{tot}, thus $\sigma_t = \sigma_s + \sigma_a$. The values of σ_b, σ_s, and thus σ_t are presented by the following Mie formulas, respectively (e.g., Stratton 2007, p. 569; Gunn and East 1954);

$$\sigma_b = \frac{\lambda^2}{4\pi}\left|\sum_{n=1}^{\infty}(-1)^n(2n+1)(a_n - b_n)\right|^2, \tag{3.45}$$

$$\sigma_s = \frac{\lambda^2}{2\pi}\sum_{n=1}^{\infty}(2n+1)(|a_n|^2 + |b_n|^2), \tag{3.46}$$

$$\sigma_t = \frac{\lambda^2}{2\pi}\sum_{n=1}^{\infty}(2n+1)\mathrm{Re}[a_n + b_n], \tag{3.47}$$

where λ is the wavelength in the medium, a_n and b_n are the nth Mie coefficients for positive integer n as will be discussed in the following paragraph, and Re[$*$] denotes the real part of the complex function [$*$]. The value of σ_a is derived from $\sigma_a = \sigma_t - \sigma_s$.

Mie Coefficients

The Mie coefficients a_n and b_n are expressed as the functions of x, given by $x = \pi D/\lambda$ with the diameter of precipitation particles D, and the complex refractive index of the drop (or particle) m which is given by

$$m = n_r - jk_a. \tag{3.48}$$

The value of n_r is the real part of the refractive index, the ordinary refractive index defined by (2.42), whereas k_a is the imaginary part of the refractive index which causes attenuation of radar waves.[13] The variation of the refractive index of water with temperature and wavelength is shown in Table 3.3. These values are taken from

[13]The attenuation is a complicated function of m, but in the special case that the radar wavelength is long compared to diameter of scatterers, i.e., in the Rayleigh region, it can be expressed simply as will be discussed in Sects. 3.3.4 and 6.3.3.

Table 3.3 Variation of the refractive index of water with respect to the temperature and the radar wavelength [from Gunn and East 1954]

		Radar wavelength			
	Temperature (°C)	10 cm	3.21 cm	1.24 cm	0.62 cm
	20	8.88	8.14	6.15	4.44
n_r	10	9.02	7.80	5.45	3.94
	0	8.99	7.14	4.75	3.45
	−8	–	6.48	4.15	3.10
	20	0.63	2.00	2.86	2.59
k_a	10	0.90	2.44	2.90	2.37
	0	1.47	2.89	2.77	2.04
	−8	–	–	2.55	1.77

Keer (1951, p. 675) for 10 cm wavelength, and from Lane and Saxton (1952) for the other wavelengths (Gunn and East 1954).

Using x and m, the Mie coefficients a_n and b_n can be expressed as (Van de Hulst 1957, p. 123)

$$a_n = \frac{A_n(y)\psi_n(x) - m\psi_n'(x)}{A_n(y)\zeta_n(x) - m\zeta_n'(x)}, \tag{3.49}$$

$$b_n = \frac{mA_n(y)\psi_n(x) - \psi_n'(x)}{mA_n(y)\zeta_n(x) - m\zeta_n'(x)}, \tag{3.50}$$

where $\psi_n(x)$ and $\zeta_n(x)$ are expressed in terms of Bessel functions of the first kind of order $(n+1/2)$, $J_{n+1/2}$, as

$$\psi_n(x) = \sqrt{\frac{\pi x}{2}} J_{n+1/2}(x), \tag{3.51}$$

$$\zeta_n(x) = \sqrt{\frac{\pi x}{2}} [J_{n+1/2}(x) + (-1)^n jJ_{-n-1/2}(x)], \tag{3.52}$$

and

$$A_n(y) \equiv \frac{\psi_n'(y)}{\psi_n(y)}, \tag{3.53}$$

$$y \equiv mx = n_r x - jk_a x. \tag{3.54}$$

Here the primed symbol denotes differentiation. As discussed above, m always takes complex values for such absorbing media as precipitation particles.

The Mie coefficients a_n and b_n are also derivable from the recursion formula (Deirmendjian 1969, Chap. 2) as the details will be discussed in Appendix A.

3.3.4 The Rayleigh Approximation

The Rayleigh approximation is given as an approximate solution to the Mie formulas. It can be applied for precipitation particles whose diameters are considerably smaller than the radar wavelengths. This approximation usually holds when raindrops are observed by ordinary meteorological radars. However, the Rayleigh approximation cannot be applied if precipitation particles are observed by millimeter wave radars whose wavelengths are comparable to particle diameters. Radars of 3-cm wavelength are often used for weather observations, but the hailstone often do not satisfy the Rayleigh approximation in this wavelength range. Furthermore, for extinction cross sections, the Rayleigh condition might not be satisfied even for wavelengths as long as 10 cm (Doviak and Zrnić 2006, pp. 40–41). In these cases, it is necessary to find out an exact solution for scattering from dielectric spheres based upon the Mie formulas.

In the Rayleigh region, $x = \pi D/\lambda \ll 1$, thus neglecting terms of higher than sixth power of x, only the Mie coefficients a_1, b_1, and b_2 are significant, and they are (Gunn and East 1954)

$$a_1 = -\frac{j}{45}(m^2-1)x^5, \tag{3.55}$$

$$b_1 = -\frac{2j}{3}\left(\frac{m^2-1}{m^2+2}\right)x^3\left[1+\frac{3}{5}\left(\frac{m^2-2}{m^2+2}\right)x^2-\frac{2j}{3}\left(\frac{m^2-1}{m^2+2}\right)x^3\right], \tag{3.56}$$

$$b_2 = \frac{j}{15}\left(\frac{m^2-1}{2m^2+3}\right)x^5. \tag{3.57}$$

Furthermore, as $x \ll 1$, only the leading term of b_1 need be considered and all higher order terms in the a's and b's can be neglected. Introducing a forth cross section, the cross sections σ_b, σ_s, and σ_a which is defined by $\sigma_t - \sigma_s$, are given by the Rayleigh approximation as follows (e.g., Gunn and East 1954; Ulaby et al. 1981, p. 295):

$$\sigma_b = \frac{\lambda^2}{\pi}x^6|K|^2 = \frac{\pi^5}{\lambda^4}|K|^2D^6, \tag{3.58}$$

$$\sigma_s = \frac{2\lambda^2}{3\pi}x^6|K|^2 = \frac{2}{3}\frac{\pi^5}{\lambda^4}|K|^2D^6, \tag{3.59}$$

$$\sigma_a = \frac{\lambda^2}{\pi}x^3\mathrm{Im}[-K] = \frac{\pi^2}{\lambda}\mathrm{Im}[-K]D^3, \tag{3.60}$$

$$\sigma_t = \sigma_s + \sigma_a. \tag{3.61}$$

where $|K|^2$ is given by

$$|K|^2 = \left|\frac{m^2-1}{m^2+2}\right|^2. \tag{3.62}$$

Table 3.4 Variations of $|K|^2$ and $\text{Im}[-K]$ of water with respect to the temperature and the wavelength [from Gunn and East 1954]

		Radar wavelength			
	Temperature (°C)	10 cm	3.21 cm	1.24 cm	0.62 cm
	20	0.9280	0.9275	0.9193	0.8926
$\|K\|^2$	10	0.9313	0.9282	0.9152	0.8726
	0	0.9340	0.9300	0.9055	0.8312
	−8	–	–	0.8902	0.7921
	20	0.00474	0.01883	0.0471	0.0915
$\text{Im}[-K]$	10	0.00688	0.0247	0.0615	0.1142
	0	0.01102	0.0335	0.0807	0.1441
	−8	–	–	0.1036	0.1713

From the definition $m = \sqrt{\varepsilon/\varepsilon_0}$, where ε is the complex permittivity of the medium, K is the same as that of (3.18). As evident from (3.58) and (3.59), the scattering cross section is proportional to $|K|^2$. On the other hand, σ_a accounts for the power taken from incident wave which is dissipated as heat and not scattered, and is proportional to $\text{Im}[-K]$. Although the values $|K|^2$ and $\text{Im}[-K]$ changes according to the surrounding atmospheric temperature and the radar wavelength, $|K|^2$ for water is practically constant. On the other hand, the value of $\text{Im}[-K]$ significantly varies at the temperatures ordinarily seen in the Earth's atmosphere and wavelengths in the centimeter range. These values are listed in Table 3.4 (Gunn and East 1954).

If the dielectric is loss-less, $\sigma_a = 0$ and consequently $\sigma_t = \sigma_s$. The value of σ_s is proportional to x^6 and accordingly to D^6, whereas that of σ_a is proportional to x^3 and to D^3. Hence, for minute particles such as cloud droplets, $\sigma_s \ll \sigma_a$, and in general $\sigma_t \simeq \sigma_a$. However, if the particles are raindrops which are large in size, σ_s and σ_a becomes comparable. In this case, both σ_s and σ_a should be taken into account in calculating σ_t. The relation between σ_t and attenuation of electric power of the radar wave propagating through precipitations will be described in Sect. 6.3.3.

3.3.5 *Radar Reflectivity Factor*

The scattering by distributed hard scatterers such as minute precipitation particles (raindrops, etc.) belongs to the Rayleigh region as illustrated in Fig. 3.5. Hence, the Rayleigh approximation can be applied for calculating the scattering properties. The radar cross section σ for a raindrop with diameter D can be expressed by (3.58), where $\sigma = \sigma_b$. Thus

$$\sigma = \frac{\pi^5}{\lambda^4}|K|^2 D^6. \tag{3.63}$$

Generally, it is not evident from radar observations of single polarized wave alone whether the precipitation particles are raindrops or ice crystals.[15] In practice, Z_e is changed to the value such as Z_i that looks most suitable according to the precipitation condition observed by other means.

As shown in Table 3.4, $|K_w|^2$ varies according to radar wavelength and atmospheric temperature. However, in practice, it is approximately constant for raindrops in the atmosphere where temperatures are above 0°C. In the following, it is assumed that $|K_w|^2 = 0.93$ unless otherwise mentioned. Also, the magnitude of $|K_i|^2$ is almost constant in radar wavelength, and $|K_i|^2 = 0.176$ (Battan 1973, p. 40; Smith 1984). Thus, Z_e for ice crystals is given from (3.70) as

$$Z_e = 0.189 Z_i. \tag{3.71}$$

A conversion of $|K|^2$ value is made in the following way in the case of snowflakes whose $|K_i|^2$ value is identical with that of ice spheres but the shape is more complicated. First, the snowflakes are melted, and diameter of equivalent water drop is estimated from the melted water. Then, the diameter of supposed ice sphere is compared with that of the water drops with same mass. As the density of ice is $0.92 \times 10^3\,\mathrm{kg\,m^{-3}}$, diameter of the ice spheres is $0.92^{-1/3}$ ($\simeq 1.028$) times that of water drops. Next, the radar reflectivity factor for water drops is proportional to D^6 as shown by (3.66). This effect is included to the value of $|K|^2$ to yield $|K_i'|^2$. Therefore, $|K_i'|^2$ becomes $\left(0.92^{-1/3}\right)^6 \simeq 1.18$ times $|K_i|^2$, or $|K_i'|^2 = 1.18|K_i|^2 = 0.208$. Finally the following relation is derived for snowflakes (Smith 1984);

$$Z_e = 0.224 Z_i. \tag{3.72}$$

The equivalent radar reflectivity factor for any non-spherical precipitation particles can be replaced with that of water drops in spherical shape. With the equivalent radar reflectivity factor Z_e and $|K_w|^2$, the radar equation for distributed hard scatterers with no loss included is expressed in more general form as

$$E[P_r] = \frac{\pi^3 P_t g^2 \theta_1^2 c\tau |K_w|^2 Z_e}{2^{10}(\ln 2)\lambda^2 r^2}. \tag{3.73}$$

[15] As will be discussed in Sect. 6.4, hydrometeor classifications (e.g., distinguishing ice crystals from rain drop) applying polarimetric parameters have become possible.

3.4 Radar Theory for Soft Scatterers

3.4.1 Backscattering Mechanisms

Refractive index perturbations arising from atmospheric turbulence and waves are the main sources of echoes for radars operating at wavelengths longer than a few centimeters (e.g., Woodman and Guillén 1974). At very high frequency (VHF) and lower frequencies, radars are insensitive to clouds and even to precipitations at the lowest frequencies. As discussed in Sect. 2.2.2, the refractive index of the atmosphere is determined by pressure, temperature, partial pressure of humidity, and electron density. Therefore, atmospheric turbulence and waves, which affect the distribution of these parameters, give rise to refractive index perturbations. According to the spatial distribution of these perturbations, the radar echoes will result either from scattering or from partial reflection of the transmitted electromagnetic wave, as shown schematically in Fig. 3.7.

Scattering from Refractive Index Perturbations in Turbulent Flow

Concerning a misleading phrase that suggests turbulence scatters radio waves, the authors did a little survey, and the following conclusion has been obtained (R.J. Doviak, 2011, personal communication).

The phrase “scattering of waves by atmospheric turbulence” commonly appears in the literature. This sentence is misleading because turbulence does not scatter radio waves, at least not for non-relativistic velocities. The origin of the misleading phrase might be traced to the 1971 translation of V. I. Tatarskii’s book “The Effects of the Turbulent Atmosphere on Wave Propagation”. For example the lead sentence of Chap. 2 of Tatarskii’s book reads: “Scattering of waves by atmospheric turbulence

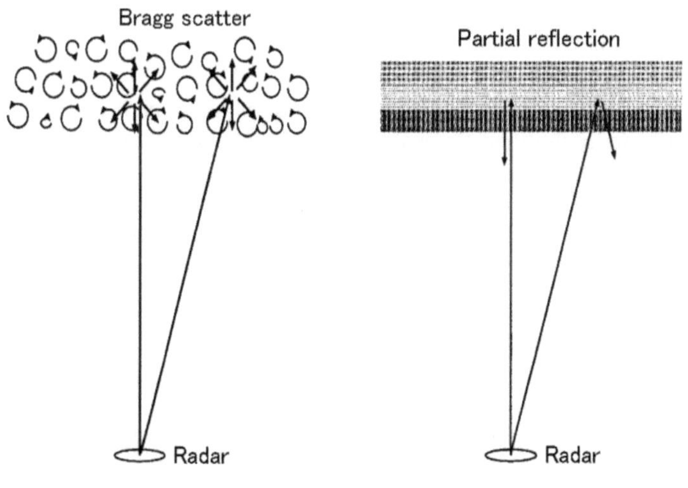

Fig. 3.7 Major scattering mechanism of soft scatterers

attracted considerable attention following the experimental discovery of the long-range tropospheric propagation of ultra-short waves." Perhaps this sentence might have been altered in translation because, as best as we can recall from reading Tatarskii's book many years ago, Tatarskii does not use this phrase elsewhere in his book when discussing scattering of ultra-short radio waves. But because the chapter heading is Scattering of Electromagnetic and Sound Waves in a Turbulent Atmosphere, and because sound waves can be scattered by turbulence, it is possible that Tatarskii had in mind the scattering of sound waves. His book also covers the topic of sound wave scatter. Perhaps the statement was meant to apply only to the scattering of sound waves by atmospheric turbulence, but was inadvertently applied scattering of radio waves by turbulence.

Bragg Scatter

Whereas scattering from precipitation is due to distributed hard scatterers (i.e., water drops, ice particles, hailstones, etc.), scattering from clear atmosphere is produced by randomly distributed refractive index perturbations, i.e., the refractive index $n \equiv n(\boldsymbol{r})$, where $\boldsymbol{r}$ is the distance between radar and a point in the region of refractive index perturbations, resulting from turbulent mixing of the (background) gradients of potential refractive index (e.g., Tatarskii 1971, Chap. 2; Doviak and Zrnić 2006, Chap. 11). For a monostatic radar,[16] scattering arises from refractive index perturbations caused by atmospheric turbulence having mainly the scale of half the radar wavelength along the direction of propagation. It is called "Bragg scatter". In the case of isotropic perturbations, the Bragg scatter is nearly isotropic. If the perturbations caused by turbulence uniformly fills the radar resolution volume, it is possible to define a distribution of Bragg scatterers so that the radar equation can be basically derived in the same way as the radar equation for distributed hard scatterers.

Partial Reflection

The mechanism of partial (or Fresnel) reflection is based on the concept of a (locally) horizontally stratified atmosphere; i.e., the refractive index of the atmosphere depends only on the vertical coordinate, $n \equiv n(z)$, where z is the height of scatter. As depicted in Fig. 3.7, the incident electromagnetic wave is partially reflected in the specular direction when encountering a sharp horizontally stratified refractive index gradient layer, as a mirror surface would do. It occurs when a single or

[16]Ordinary radars that share a single antenna for transmission and reception are called monostatic radars. On the other hand, the radars equipped with a pair of antennas that are separated, typically by large distances, where one antenna is used for transmitting and the other is used for receiving, are called bistatic radars as will be discussed in Sect. 4.2.3.

a few sharp gradient layers are embedded within the radar resolution volume. If multiple gradient layers are randomly distributed along the vertical direction then the backscattering process is referred to as Fresnel scattering (e.g. Gage 1990). If the surface of the gradient layer is corrugated or distorted (by surrounding turbulence or by gravity waves), specular reflection fades and the scattering pattern is broadened. It is referred to as "diffuse reflection" (e.g., Röttger 1980a). If the gradient layer has a limited extent along the horizon, diffraction at the edges can also occur (e.g. Luce et al. 1995).

By essence, partial reflection from gradient layers is based on a deterministic one-dimensional model whereas Bragg scatter from turbulence is based on three-dimensional statistical models. They thus rule out each other. However, Doviak and Zrnić (2006, Chap. 11) showed that it is possible to unify these two models. Mathematical expressions describing Bragg scatter caused by anisotropic turbulence tend to the expression derived for partial reflection mechanism if the horizontal coherence of the perturbations caused by turbulence tend to the limit case of the Fresnel zone diameter. In the realistic atmosphere, it is usually difficult, or even impossible, to distinguish diffuse reflection from corrugated gradient layers from Bragg scatter caused by anisotropic turbulence because they can produce similar scattering patterns.

Mathematical formulae for the Bragg scatter and the partial reflection will be given below in Sects. 3.4.2 and 3.4.3, respectively.

3.4.2 Bragg Scatter due to Refractive Index Perturbations

The refractive index field is assumed to be weakly stationary and locally homogeneous within the scattering volume (Ottersten 1969a), and the field is assumed to be isotropic. The relationship between refractive index perturbations and radar reflectivity in the isotropic field will be described in this subsection. First, the three-dimensional spectrum of refractive index perturbations $\Phi_n(\kappa)$, its observed one-dimensional (wave number energy) spectrum $S(\kappa)$, and the power spectral density (of the single quantity κ) $E(\kappa)$ will be discussed. Next, radar scattering intensity of soft (Bragg) scatterers due to refractive index perturbations will be shown using these spectra, and the relationship between the reflectivity and those spectra will be presented. Finally, the radar reflectivity of refractive index perturbations and its various properties will be shown.

From the point of view of radar observations from soft (Bragg) scatterers, only the fluctuating component with the spatial scale of a half transmitted wavelength in the direction of radar wave propagation, $\lambda/2$, contributes to the scattering. In other word, a radar ideally samples only the spectral component of refractive index perturbations with a half transmitted wavelength. Thus, the structure wave number κ for Bragg scatter becomes

$$\kappa = 2\pi/(\lambda/2) = 4\pi/\lambda = 2k, \tag{3.74}$$

where $\kappa = |\boldsymbol{\kappa}|$, and $\boldsymbol{k}$ is the radar wave number vector of scattered wave. In this subsection, we discuss the scattering from perturbations of $\boldsymbol{\kappa}$ given by (3.74) is effective.

Spectrum of Refractive Index Perturbations

The three-dimensional spectrum[17] of perturbations of refractive index n is expressed by $\Phi_n(\boldsymbol{\kappa})$. $\Phi_n(\boldsymbol{\kappa})$ has dimensions of (length)^3, where $\boldsymbol{\kappa}$ is the vector wave number of perturbations. The quantity $\Phi_n(\boldsymbol{\kappa})$ is obtained by a Fourier transform of the three-dimensional refractive index covariance function, and its coefficient is defined so that its integration for all $\boldsymbol{\kappa}$ becomes the variance of perturbations of n, that is $\overline{\delta n^2}$; i.e., the normalization used is $\int_{-\infty}^{\infty} \Phi_n(\boldsymbol{\kappa})\mathrm{d}\boldsymbol{\kappa} = \overline{\delta n^2}$.

In actual observations, the refractive index field is measured by i.e., radar waves traveling through the atmosphere, and thus the field is not described by the three-dimensional spectrum but the one-dimensional one. The one-dimensional spectrum $F_n^{\mathrm{x}}(\kappa_{\mathrm{x}})$, which describes the refractive index field along the x-axis in a coordinate system (x, y, z) of the real space, is obtained by integration of the three-dimensional spectrum $\Phi_n(\kappa)$ along the k_y and k_z axes in the coordinate system (k_x, k_y, k_z) in the wave number space. That is

$$F_n^{\mathrm{x}}(\kappa_{\mathrm{x}}) = \iint_{-\infty}^{\infty} \Phi_n(\boldsymbol{\kappa})\mathrm{d}\kappa_{\mathrm{y}}\mathrm{d}\kappa_{\mathrm{z}}. \tag{3.75}$$

However, the actually observed one-dimensional spectrum has values only for positive κ_{x}. Thus, normalizing same as $\Phi_n(\kappa)$, from (3.75) we have (Ottersten 1969a,b)

$$\int_0^{\infty} S_n^{\mathrm{x}}(\kappa_{\mathrm{x}})\mathrm{d}\kappa_{\mathrm{x}} = \int_{-\infty}^{\infty} F_n^{\mathrm{x}}(\kappa_{\mathrm{x}})\mathrm{d}\kappa_{\mathrm{x}} = 2\int_0^{\infty} F_n^{\mathrm{x}}(\kappa_{\mathrm{x}})\mathrm{d}\kappa_{\mathrm{x}} = \overline{\delta n^2}, \tag{3.76}$$

$$S_n^{\mathrm{x}}(\kappa_{\mathrm{x}}) = 2F_n^{\mathrm{x}}(\kappa_{\mathrm{x}}). \tag{3.77}$$

For an isotropic field, $\Phi_n(\boldsymbol{\kappa}) = \Phi_n(\kappa)$, and all one-dimensional spectra $S_n^{\alpha}(\kappa_{\alpha})$ in various directions are identical (Ottersten 1969a). Thus, hereafter we replace $S_n^{\alpha}(\kappa_{\alpha})$ to $S_n(\kappa)$.

Given the one-dimensional spectrum $S_n(\kappa)$, the relationship between $S_n(\kappa)$ and $\Phi_n(\kappa)$ is discussed below. From previously mentioned definition, the spectrum of refractive index perturbations is obtained by a Fourier transform of its covariance function. Thus, if the covariance function of refractive index perturbations along a

[17] From Gossard and Strauch (1983, p. 259). The nomenclature of $\Phi_n(\kappa)$ is not necessarily unique. It is either called the space spectrum or spatial power-spectral density (Ottersten 1969a), the space spectrum (Battan 1973, p. 256), or the spectral density (Doviak and Zrnić 2006, p. 390).

line is given by $C_n(l)$, the relation between $S_n(\kappa)$ and $C_n(l)$ is expressed as (Gossard and Strauch 1983, Appendix E)

$$S_n(\kappa) = \frac{2}{\pi}\int_0^\infty C_n(l)\cos\kappa l \mathrm{d}l. \tag{3.78}$$

The corresponding three-dimensional spectrum in vector $\boldsymbol{\kappa}$ is

$$\Phi_n(\boldsymbol{\kappa}) = \frac{1}{(2\pi)^3}\int_0^\infty C_n(l)\mathrm{e}^{-i\boldsymbol{\kappa}\cdot\boldsymbol{l}}\mathrm{d}\boldsymbol{l}. \tag{3.79}$$

To carry out the integration, it is simple to define a spherical coordinate system relative to the transmission direction, i.e., from transmitter to the scattering element, $\mathrm{d}\boldsymbol{l} = \mathrm{d}V$. Thus θ is the angle to the scattering element off the direction from transmitter to the center of the scattering volume, $\boldsymbol{l}$ is the separation of the scatterer from the volume center, and ϕ is the azimuthal angle about the transmission direction. Therefore, equating $\beta = -\cos\theta$, the differential volume element $\mathrm{d}\boldsymbol{l}$ becomes

$$\mathrm{d}\boldsymbol{l} = \mathrm{d}V = l^2\mathrm{d}l\sin\theta\mathrm{d}\theta\mathrm{d}\phi = l^2\mathrm{d}l\mathrm{d}\beta\mathrm{d}\phi\ . \tag{3.80}$$

Applying (3.80) to (3.79), (3.79) becomes

$$\Phi_n(\boldsymbol{\kappa}) = \frac{1}{(2\pi)^3}\int_0^\infty C_n(l)l^2\left(\int_{-1}^1 \mathrm{e}^{j\kappa l\beta}\mathrm{d}\beta\right)\mathrm{d}l\int_0^{2\pi}\mathrm{d}\phi. \tag{3.81}$$

Using

$$\int_{-1}^1 \mathrm{e}^{j\kappa l\beta}\mathrm{d}\beta = \frac{2\sin\kappa l}{\kappa l}, \tag{3.82}$$

(3.81) reduces to

$$\Phi_n(\boldsymbol{\kappa}) = \frac{2}{(2\pi)^2}\int_0^\infty C_n(l)\frac{l\sin\kappa l}{\kappa}\mathrm{d}l. \tag{3.83}$$

Inserting $l\sin\kappa l = -\dfrac{\mathrm{d}}{\mathrm{d}\kappa}\cos\kappa l$ into (3.83), we have

$$\Phi_n(\boldsymbol{\kappa}) = \frac{-2}{(2\pi)^2\kappa}\frac{\mathrm{d}}{\mathrm{d}\kappa}\int_0^\infty C_n(l)\cos\kappa l\ \mathrm{d}l. \tag{3.84}$$

Using (3.78) and (3.84), the three-dimensional spectrum of refractive index perturbations $\Phi_n(\boldsymbol{\kappa})$ and the one-dimensional spectrum $S(\kappa)$ are related as (Gossard and Strauch 1983, Appendix E)

$$\Phi_n(\boldsymbol{\kappa}) = -\frac{1}{4\pi\kappa}\frac{\mathrm{d}S_n(\kappa)}{\mathrm{d}\kappa} \qquad [\mathrm{m}^3]. \tag{3.85}$$

In an isotropic space, the three-dimensional spectrum in $\Phi_n(\boldsymbol{\kappa})$ can be related to the power spectral density[18] $E_n(\kappa)$ of the single quantity κ by integrating over all directions in $\boldsymbol{\kappa}$ space as

$$E_n(\kappa) \equiv \int_0^{\infty} \Phi_n(\boldsymbol{\kappa}) d\boldsymbol{\kappa}. \tag{3.86}$$

Analogous to (3.80), let

$$d\boldsymbol{\kappa} = \kappa^2 d\kappa \sin\theta d\theta d\phi. \tag{3.87}$$

Applying (3.87) to (3.86), then

$$\begin{aligned} E_n(\kappa) &= \Phi_n(\boldsymbol{\kappa}) \kappa^2 d\kappa \int_{-1}^{1} d\beta \int_0^{2\pi} d\phi \\ &= 4\pi\kappa^2 \Phi_n(\boldsymbol{\kappa}). \end{aligned} \tag{3.88}$$

Substituting (3.85) into (3.88), the relationship between the power spectral density $E_n(\kappa)$ and the one-dimensional spectrum $S_n(\kappa)$ is derived as (Gossard and Strauch 1983, Appendix E)

$$E_n(\kappa) = -\kappa \frac{dS_n(\kappa)}{d\kappa} \qquad [\mathrm{m}]. \tag{3.89}$$

The power spectral density $E_n(\kappa)$ contributes to radar scattering characteristic, whereas $S_n(\kappa)$ is the spatial spectrum measured by sensor such as sonde moving through the atmosphere.

Radar Reflectivity of Refractive Index Perturbations

Analogous to the scattering by spherical dielectrics (hard scatterers) discussed in Sect. 3.3.1, intensity characteristics of scattering by soft (Bragg) scatterers and the relations between radar reflectivity and spectrum of refractive index perturbations are discussed in this paragraph. First, assuming hard scatterers, the correlation of number distribution is discussed. Next, the number distribution is replaced by perturbations of permittivity by assuming soft (Bragg) scatterers, and its correlation is discussed. Finally, perturbations of permittivity is converted to refractive index perturbations.

---

[18] From Gossard and Strauch (1983, p. 59). The nomenclature of $E_n(\kappa)$ is not necessarily unique. It is alternatively called the three-dimensional spectrum (Ottersten 1969b) or the single spectral density (Doviak and Zrnić 2006, p. 389).

In the discussion of hard scatterers, we assumed that I in (3.23) is the function of number distribution, and the random variation terms $\delta N(\boldsymbol{r}_1)\mathrm{d}\boldsymbol{r}_1$ and $\delta N(\boldsymbol{r}_2)\mathrm{d}\boldsymbol{r}_2$ are mutually independent. Meanwhile, if the scatterers in a volume are correlated, it is convenient to define a spatial correlation function. In (3.23), replacing $\boldsymbol{r}_1$ and $\boldsymbol{r}_2$ to $\boldsymbol{r}_0$ and $\boldsymbol{r}_0+\boldsymbol{l}$, respectively, where $\boldsymbol{l}$ is the separation of a certain scatterer from reference position $\boldsymbol{r}_0$ in the scattering volume, I is expressed as

$$I = \iint \overline{\delta N(\boldsymbol{r}_0)\delta N(\boldsymbol{r}_0+\boldsymbol{l})}\mathrm{d}\boldsymbol{r}_0 \mathrm{e}^{-j\boldsymbol{k}\cdot\boldsymbol{l}}\mathrm{d}\boldsymbol{l}, \tag{3.90}$$

and a spatial correlation function

$$C(\boldsymbol{l}) = \frac{1}{\overline{\delta N^2}V}\int_V \overline{\delta N(\boldsymbol{r}_0)\delta N(\boldsymbol{r}_0+\boldsymbol{l})}\mathrm{d}\boldsymbol{r}_0 \tag{3.91}$$

is defined for the scattering resolution volume V, where δN is the variance of the number of scatterers N. Using (3.91), I is expressed as

$$I = \overline{\delta N^2}V\int_0^\infty C(l)\mathrm{e}^{-i\boldsymbol{k}\cdot\boldsymbol{l}}\mathrm{d}\boldsymbol{l}, \tag{3.92}$$

where $\boldsymbol{k}$ is the wave number vector of backscattered wave (which is the same as $\boldsymbol{k}_s$ in the case of backscattering discussed in Sect. 3.3.1).

Analogous to (3.78) and (3.79), the one-dimensional line spectrum of number distribution and the corresponding three-dimensional spectrum are expressed as

$$S(k) = \frac{2}{\pi}\overline{\delta N^2}\int_0^\infty C(l)\cos kl\,\mathrm{d}l, \tag{3.93}$$

$$\Phi(\boldsymbol{k}) = \frac{\overline{\delta N^2}}{(2\pi)^3}\int_0^\infty C(l)\mathrm{e}^{-i\boldsymbol{k}\cdot\boldsymbol{l}}\mathrm{d}\boldsymbol{l}, \tag{3.94}$$

respectively. Substituting (3.94) into (3.92), and applying (3.88), I is expressed as

$$I = 8\pi^3 V\Phi(\boldsymbol{k}) = 2\pi^2 V\frac{E(k)}{k^2}. \tag{3.95}$$

In the above, we have discussed hard scatterers and their number distribution which are assumed to be correlated. Comparing such a medium with spatial fluctuation in ε instead of droplets of constant $\delta\varepsilon$ but variable concentration N, it is ready seen that $\delta\varepsilon/\varepsilon$ is analogous to $3KNV_\mathrm{D}$ in (3.21), thus

$$E[P_\mathrm{s}] = \frac{P_0 k^4\sin^2\chi}{(4\pi r)^2}I, \tag{3.96}$$

$$I = \overline{\left|\int_{r_0}\frac{\delta\varepsilon(\boldsymbol{r}_0)}{\varepsilon_0}\mathrm{e}^{-jk\boldsymbol{r}_0}\mathrm{d}\boldsymbol{r}_0\right|^2}, \tag{3.97}$$

where χ is the angle between the direction of polarization of the incident electric field and direction of k_s, $\chi = \pi/2$ for the case of backscattering, and I has dimensions of $[m^6]$. Analogous to (3.95), I is replaced with

$$I = 8\pi^3 V \Phi_\varepsilon(\boldsymbol{k}) = 2\pi^2 V \frac{E_\varepsilon(k)}{k^2}, \tag{3.98}$$

where $\Phi_\varepsilon(\boldsymbol{k})$ and $E_\varepsilon(k)$ are the three-dimensional spectrum of perturbations of permittivity and its power spectral density, respectively.

By replacing ε_2 in (2.42) with ε, the refractive index n is expressed as $n \simeq \sqrt{\varepsilon/\varepsilon_0}$. In the atmosphere $n^2 = \varepsilon/\varepsilon_0 \simeq 1$, and consequently, a small variation of permittivity $\delta(\varepsilon/\varepsilon_0)$ becomes

$$\delta(\varepsilon/\varepsilon_0) \simeq 2n\delta n \simeq 2\delta n. \tag{3.99}$$

If $(\varepsilon/\varepsilon_0)$ and n fluctuate,

$$\overline{\delta\varepsilon/\varepsilon_0(r)\ \delta\varepsilon/\varepsilon_0(r-l)} \simeq 4\overline{\delta n(r)\ \delta n(r-l)}, \tag{3.100}$$

i.e., the amplitude ratio of these correlation functions becomes 4:1, thus

$$\Phi_\varepsilon(\boldsymbol{k}) = 4\Phi_n(\boldsymbol{k}), \tag{3.101}$$

$$E_\varepsilon(k) = 4E_n(k). \tag{3.102}$$

Substituting (3.101) and (3.102) into (3.98),

$$I = 32\pi^3 V \Phi_n(\boldsymbol{k}) = 8\pi^2 V \frac{E_n(k)}{k^2}. \tag{3.103}$$

Again substituting (3.103) to (3.96), and the estimated backscattered electric power (for $\sin^2\chi = 1$) is expressed as

$$E[P_s] = \frac{2P_0 k^4 \pi V \Phi_n(\boldsymbol{k})}{r^2}. \tag{3.104}$$

From three equations (3.104), (3.26), and (3.32), the radar reflectivity η (= $\Sigma\sigma/V$) is given by

$$\eta = 8\pi^2 k^4 \Phi_n(\boldsymbol{k}). \tag{3.105}$$

Using the relation $\kappa = 2k$ for backscattering, the radar reflectivity is related with the three-dimensional spectrum of refractive index perturbations (e.g., Ottersten 1969a,b),

$$\eta = \frac{\pi^2}{2} \kappa^4 \Phi_n(\kappa). \tag{3.106}$$

Next, substituting (3.88) into (3.106), the radar reflectivity η and power spectral density $E(\kappa)$ for refractive index perturbations in an isotropic medium are related as

$$\eta = \frac{\pi}{8}\kappa^2 E_n(\kappa). \tag{3.107}$$

Furthermore, substituting (3.89) into (3.107), the relation of η and the one-dimensional spectrum $S(\kappa)$ of refractive index perturbations in an isotropic medium is expressed by

$$\eta = -\frac{\pi}{8}\kappa^3 \frac{\mathrm{d}S_n(\kappa)}{\mathrm{d}\kappa}. \tag{3.108}$$

Radar Reflectivity in the Inertial Subrange

The primary source of refractive index perturbations in both the middle atmosphere and the troposphere is free turbulence. The three-dimensional spectrum $\Phi_n(\kappa)$ is determined by the spectrum of perturbations which depend on the wave number κ of Bragg scatter. When such refractive index perturbations are caused by locally homogeneous and isotropic turbulence in the inertial subrange,[19] a half the radar wavelength $\lambda/2$ should exist within the inertial subrange. The one-dimensional spectrum $S_n(\kappa)$ in this region is, in general, approximated to $S_n(\kappa) \propto \kappa^{-p}$, where $p > 0$. Consequently, turbulence spectrum satisfies the following equation (Ottersten 1969b);

$$\frac{\mathrm{d}S_n(\kappa)}{\mathrm{d}\kappa} = -\frac{p}{\kappa}S_n(\kappa). \tag{3.109}$$

Substituting (3.109) into (3.89) and (3.108), we obtain

$$E_n(\kappa) = pS_n(\kappa) \tag{3.110}$$

$$\eta = \frac{\pi}{8}p\kappa^2 S_n(\kappa). \tag{3.111}$$

In the case of inertial subrange turbulence, $p = 5/3$, thus $S_n(\kappa) \propto \kappa^{-5/3}$. Using (3.88) and (3.89), $\Phi_n(\kappa) \propto \kappa^{-11/3}$ and $E_n(\kappa) \propto \kappa^{-5/3}$ are derived, respectively (Heisenberg 1948; Ottersten 1969a,b). Moreover, from (3.111) η becomes

$$\eta = \frac{5}{3}\left(\frac{\pi}{8}\right)\kappa^2 S_n(\kappa). \tag{3.112}$$

[19] The spatial scale and characteristics of the inertial subrange will be discussed in Sect. 7.3.2

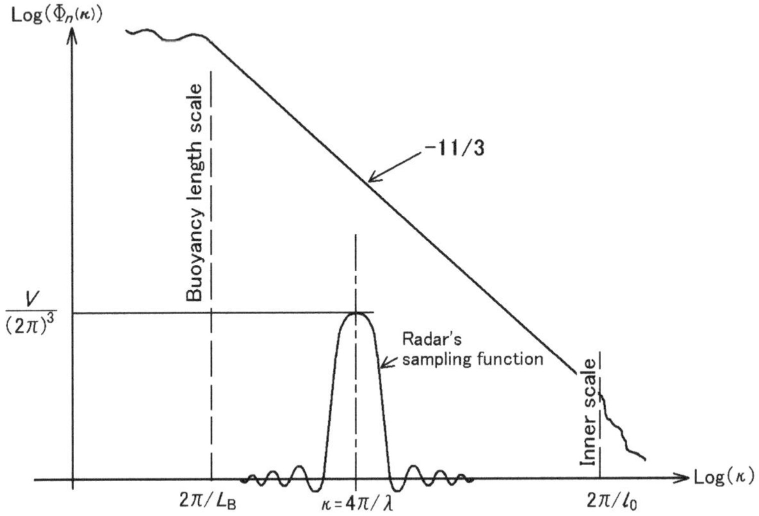

Fig. 3.8 A logarithm of an idealized spectrum of refractive index perturbations as a function of the logarithm of structure wave number for Bragg scatter κ, where l_0 and L_B are the internal scale of turbulence and the buoyancy length scale, respectively. Shown is the radar's sampling function which has the maximum value for $\kappa = 4\pi/\lambda$, where λ is the radar wavelength. The quantity V is the radars resolution volume [rewritten from Doviak and Zrnić 2006, p. 448]

Figure 3.8 shows the relation between κ and $\Phi_n(\kappa)$ of Bragg scatter within the inertial subrange. In the figure, a logarithm of an idealized spectrum of refractive index perturbations as a function of the logarithm of structure wave number for Bragg scatter $\kappa = 2\pi/\Lambda_s = 4\pi/\lambda$, where Λ_s is the structure wavelength of perturbations within inertial subrange, λ is the radar wavelength, l_0 and L_B are the internal scale of turbulence and the buoyancy length scale, respectively, whose height profile will be shown in Fig. 7.10. Shown is the radar's sampling function which has the maximum value for κ. The quantity V in the same figure is the radars resolution volume [rewritten from Doviak and Zrnić 2006, p. 448].

Structure Parameter for Refractive Index Perturbations

The structure constant for the perturbations (refractive index structure constant) C_n^2 $[\mathrm{m}^{-2/3}]$ is defined as

$$\overline{[n(r+\delta r)-n(r)]^2} \equiv C_n^2 |\delta r|^{2/3}, \tag{3.113}$$

where δr is a small distance between two space locations. The quantity C_n^2 is a measure of the mean-square refractive index perturbations over the distance δr within the inertial subrange. It is evident from (3.113) that C_n^2 is independent of the radar wavelength λ. As discussed in the preceding paragraph, $(-5/3)$th power law is applied to the one-dimensional spectrum $S_n(\kappa)$, and it is shown that $S_n(\kappa) = C_f \kappa^{-5/3}$. Here, the coefficient C_f is proportional to C_n^2, and both are related such that $C_f/C_n^2 \simeq 1/2.4$ (Silverman 1956) or $\simeq 1/4$ (Ottersten 1969a). According to the latter case, one dimensional spectrum $S_n(\kappa)$ can be expressed as follows using C_n^2;

$$S_n(\kappa) \simeq \frac{1}{4} C_n^2 \kappa^{-5/3}. \tag{3.114}$$

Substituting (3.114) into (3.85), $\Phi_n(\kappa)$ is expressed as

$$\Phi_n(\kappa) = 0.033 C_n^2 \kappa^{-11/3}. \tag{3.115}$$

Equation (3.115) is applicable for isotropic turbulence in the inertial subrange. The radar reflectivity of refractive index perturbations is presented by substituting (3.114) into (3.112) and utilizing the relation of the wave number κ and the radio wavelength λ in (3.74), as (e.g., Ottersten 1969a,b)

$$\eta \simeq 0.38 C_n^2 \lambda^{-1/3}. \tag{3.116}$$

Comparison of Radar Reflectivities for Refractive Index Perturbations

Backscattering by the refractive index perturbations also depends on the scattering resolution volume, as in the case of precipitation particles shown in Fig. 3.6 which is determined by beam cross section and pulse length. Thus, (3.40) applies to backscatter from refractive index perturbations. η is a general term for backscattering cross section per unit volume and can apply to many different scattering mechanisms. The only difference is how η is related to the distribution particle sizes (see (3.65)–(3.67)) and how η is related to the refractive index perturbations (see (3.116)).

The value of C_n^2 significantly varies in altitudes, seasons, and others, and approximately becomes 10^{-12} to 10^{-18} [$\mathrm{m}^{-2/3}$]. On the other hand, the radar reflectivity factor Z of distributed scatterers defined by (3.66) usually takes values around 10^0 to 10^4[$\mathrm{mm^6 m^{-3}}$] for rain and 10^{-3} to 10^{-1}[$\mathrm{mm^6 m^{-3}}$] for fog and cloud. Using these typical values for C_n^2 in (3.116) and Z values in (3.67), respectively, η becomes as shown in Fig. 3.9. As clearly demonstrated in the figure, the radar reflectivity due refractive index perturbations of generally becomes dominant in larger radar wavelengths, whereas the radar reflectivity by hard scatterers such as rain drops becomes larger at radar wavelengths of several-tens centimeter or shorter. As will be discussed in Sect. 7.3.3, refractive index perturbations of the length shorter than the high wave number limit of the inertial subrange are strongly damped by viscosity in the real atmosphere. Doviak and Zrnić (2006, p. 476) shows the

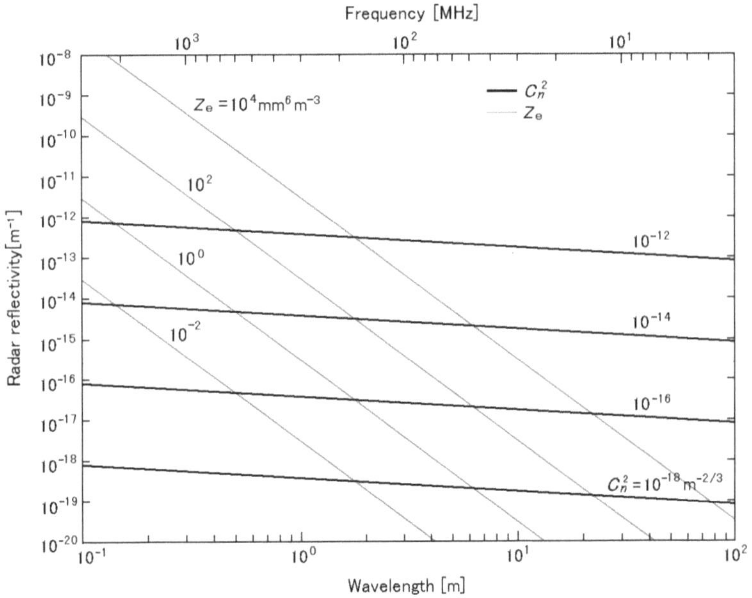

Fig. 3.9 Comparison of radar reflectivities between perturbations of refractive index caused by turbulence and precipitation particles. Note that actual extension of structure function curves (C_n^2) to high and low frequencies are limited to the inertial subrange (see Fig. 7.10)

relation between the maximum height to which echoes are returned from refractive index perturbations and radar wavelength for various energy dissipation rates.[20]

3.4.3 *Partial Reflection from a Stratified Atmosphere*

Radar Equation for Reflection

The Earth's atmosphere is well stratified, and accordingly radar waves at the comparatively low frequencies are partially reflected from the stratified layers. The radar equation for partial (or Fresnel) reflection is not the same as that for refractive index perturbations due to turbulence. The partial reflection is analogous to a specular reflection. The layer can be treated as a planar mirror with a small reflection coefficient ρ for incident electric field (Friend 1949), where ρ has a complex value of $|\rho|^2 \leq 1$.

The radar equation for this case becomes analogous to a one-way transmission from an antenna to its mirror image located at range $2r$ with a reflection coefficient

[20]See Sect. 7.3.3 for energy dissipation rate.

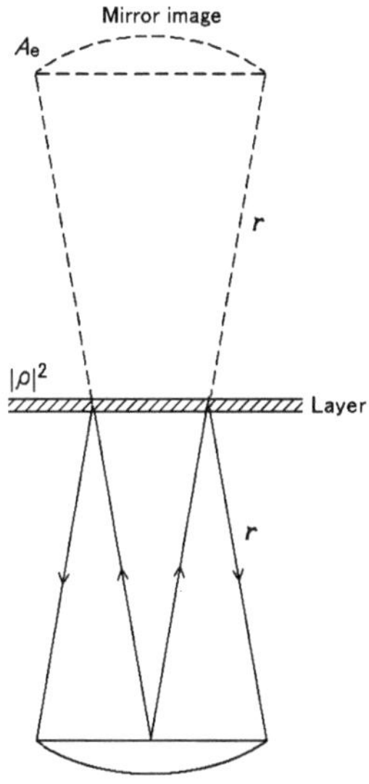

Fig. 3.10 Equivalent radiation and scattering for the partial reflection from the layer with a reflectivity ρ. The *broken line* indicates the mirror image of the radar antenna due to the layer

$|\rho|^2$ as shown in Fig. 3.10. Therefore, the expected received power $E[P_r]$ is given by replacing r and σ of (3.5) with $2r$ and $A_e|\rho|^2$, respectively, where A_e is the effective antenna aperture:

$$E[P_r] = \frac{P_t g_t}{4\pi(2r)^2} A_e|\rho|^2 = \frac{P_t A_e^2}{4\lambda^2 r^2}|\rho|^2. \tag{3.117}$$

It is clear from the above equation that the received power for partial reflection decreases by r^2 as in the case of distributed scatterers shown by (3.40), and is proportional to A_e^2/λ^2 as in the case of isolated scatterers shown in (3.8).

Reflectivity for Partial Reflection

Gage et al. (1981) have used the following formula for the partial reflection coefficient:

$$|\rho|^2 = \frac{1}{4}\left|\int_{-l/2}^{+l/2} \frac{1}{n}\frac{dn}{dz} e^{-j\kappa k} dz\right|^2, \tag{3.118}$$

where l is the thickness of the stratified layer, z the altitude, n the refractive index, and κ the wave number which is given as $\kappa = 4\pi/\lambda$. As $n \simeq 1$, (3.118) is equivalent to

$$|\rho|^2 = \frac{1}{4}\left|\int_{-l/2}^{+l/2} \frac{dn}{dz} e^{-j\kappa z} dz\right|^2, \tag{3.119}$$

which was given by e.g., Wait (1962) and Yeh and Liu (1972). Equation (3.119) shows that the intensity of the reflected power from such a layer depends on the Fourier component which has a half radar wavelength in the spatial spectrum of the refractive index perturbations. Therefore, if the vertical gradient of refractive index does not move vertically and does not change the shape of stratification either, both phase and amplitude of radar echo become constant in time, i.e., it becomes as a quasi-coherent process. As clearly shown by (3.118), the partial reflection becomes stronger if the inclination of refractive index becomes steeper. Also, the longer the radar wavelength is, the stronger the partial reflection for the same refractive index inclination (Röttger et al. 1978).

The partial reflection coefficient depends strongly on the height profile of refractive index. When the net change of refractive index in a layer with thickness l is δn, and the normalized thickness and height of the layer $l^* = l/\lambda$ and $z^* = z/\lambda$, respectively, (3.119) can be written as (Wait 1962, pp. 93–95; Yeh and Liu 1972, pp. 270–276)

$$|\rho|^2 = \frac{|\delta n|^2}{4} \left| \int_{-l^*/2}^{+l^*/2} \frac{\mathrm{d}(n/\delta n)}{\mathrm{d}z^*} e^{-4\pi j z^*} \mathrm{d}z^* \right|^2. \tag{3.120}$$

Some models for the shape of the layer are considered (Gage and Balsley 1980). For example, if the layer is expressed by a step function with δn, (3.120) becomes

$$|\rho|^2 = \frac{|\delta n|^2}{4}. \tag{3.121}$$

If the rate of change of the refractive index with height is constant, $|\rho|^2$ is expressed by (Friend 1949)

$$|\rho|^2 = \frac{|\delta n|^2}{16\pi^2} \left(\frac{\sin 2\pi l^*}{l^*} \right)^2 \tag{3.122}$$

for range resolution of many radar wavelengths. The oscillation which appears in the right hand side of (3.122) is due to the interference of reflections which occur by the discontinuities of $\mathrm{d}n/\mathrm{d}z$ at the upper and lower edges of the layer.

Aspect Sensitivity

In the Earth's atmosphere, it is known that the received power from near the zenith shows a conspicuous aspect sensitivity, especially, in the lower stratosphere (Gage and Green 1978; Röttger and Liu 1978; Fukao et al. 1979; Tsuda et al. 1986). Equation (3.120) gives the partial reflection coefficient when the radar beam is incident perpendicularly to the reflecting layer and the received power becomes maximum. The received power decreases rapidly when the beam leaves the zenith.

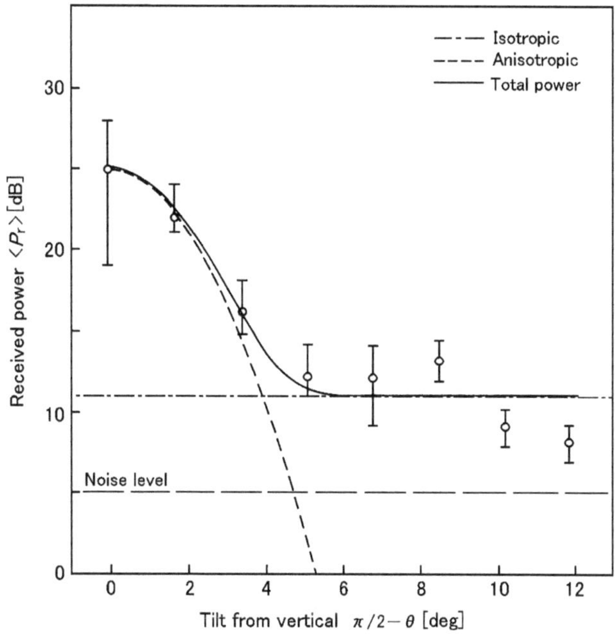

Fig. 3.11 Aspect sensitivity of the received power for a 47-MHz band atmospheric radar. *Open circles* show the observed mean backscatter power from anisotropic irregularities at the radar beam axis is tilted from the vertical (Röttger 1981). Fitted to the data is a model that consists of anisotropic irregularities of refractive index with a two-dimensional (horizontal) correlation function in an isotropic background. Height = 16.9–18.1 km [rewritten from Doviak and Zrnić 2006, p. 464]

The longer the horizontal correlation distance[21] becomes, the steeper the lapse rate. Figure 3.11 shows the zenith angle dependence of the received power which is observed with a 47-MHz band atmospheric radar fitted to the data of a model that consists of anisotropic irregularities of refractive index with a two-dimensional (horizontal) correlation function in an isotropic background (Doviak and Zrnić 2006, p. 464). In general, aspect sensitivity is predominant within a few degree from the zenith. When the zenith angle becomes more than 10°, isotropic scattering dominates. The reason why the received power becomes strong near the zenith angle can be explained that anisotropic scattering is more effective in the lower stratosphere (Doviak and Zrnić 2006, p. 463). There is also a different view that it is caused by undulation of the reflecting layer due to atmospheric gravity waves (e.g., Tsuda et al. 1997). Meanwhile the aspect sensitivity is not observed at ultra high frequency (UHF) and microwave bands.

---

[21] The horizontal distance over which the scattered waves have a strong correlation is called the horizontal correlation distance.

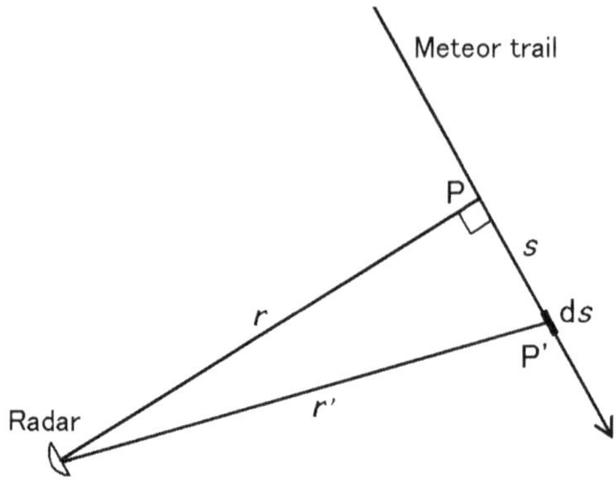

Fig. 3.12 Positional relation between meteor trail and radar

Magnitude of $|\rho|^2$ for the layer structure in the troposphere and the stratosphere was estimated by Gage and Green (1978). The heights of the tropopause and fronts accompanied with remarkable discontinuity in the reflection coefficient have been objectively determined by using partial reflection (Gage and Green 1979, 1982; Yamamoto et al. 2003a). Also, radar observations of the layer structure responsible for the partial reflection were concurrently conducted with in-situ measurements by high precision temperature sensors on board the GPS balloons (Luce et al. 2002).

3.4.4 Scattering by Linear Scatterers

Linear scatterers are classified into the isolated and volumetrically distributed scatterers. Meteor trails are one example of linear scatterers in which air molecules are linearly ionized by frictional heating due to meteor at the height of 80–110 km. The heating is caused when minute meteors with the diameter less than 10^{-3} m and the weight of 10^{-6} kg or so penetrate the atmosphere at high speed (e.g., 10–70 km s^{-1}). The regions of scatterers formed around meteor trails are sufficiently wider than half a radar wavelength and coherent. That is to say the Fresnel scatterers are formed.

As shown in Fig. 3.12, a meteor trail is considered to be rectilinear and sufficiently far from the radar. Let r be the distance to the point P where the radar beam orthogonally crosses the trail, and r' the distance to a line segment ds at P$'$ along the trail. The waves scattered from electrons that lie along the trail are dispersed as shown by the following integral, and the integrated value gives the effective length of the trail l_e.

$$l_e = \left| \int_{-\infty}^{\infty} e^{-j2k(r'-r)} ds \right|, \tag{3.123}$$

where $k = 2\pi/\lambda$. When the distance between P to P′ along the trail is s, $r' - r$ is expressed by $r' - r \simeq s^2/(2r)$. Thus, using the relation of (3.37) and applying the relation of

$$\int_{-\infty}^{\infty} \mathrm{e}^{-jat^2}\,\mathrm{d}t = \mathrm{e}^{-j\pi/4}\int_{-\infty}^{\infty} \mathrm{e}^{-ax^2}\,\mathrm{d}x, \tag{3.124}$$

the effective length l_e becomes

$$l_\mathrm{e} = \sqrt{\frac{r\lambda}{2}}, \tag{3.125}$$

which is identical to the radius of the first Fresnel zone. The meteor trails within the effective length cause a kind of Fresnel reflection where scattering waves return to the radar with the phase difference less than $\pi/2$, and efficiently contribute to the meteor echo (McKinley 1961). If this type of reflection occurs at a surface sufficiently wide compared with the radar wavelength, it is called partial reflection, while Fresnel reflection otherwise.

If the linear density of meteor trail is q_e $[\mathrm{m}^{-1}]$, the number of electrons on l_e is $l_\mathrm{e}q_\mathrm{e}$, and the received power becomes $(l_\mathrm{e}q_\mathrm{e})^2$ times that of single electron. Thus, the received power P_r is given by

$$P_\mathrm{r} = (l_\mathrm{e}q_\mathrm{e})^2\left(\frac{P_\mathrm{t}A_\mathrm{e}^2}{4\pi\lambda^2 r^4}\sigma_\mathrm{e}\right) = \frac{P_\mathrm{t}A_\mathrm{e}^2}{8\pi\lambda r^3}\sigma_\mathrm{e}q_\mathrm{e}^2, \tag{3.126}$$

where σ_e is the radar cross section of one electron $(= 9.98 \times 10^{-29}\,\mathrm{m}^2)$. From (3.126), it is clear that the received power of linear scatterer is proportional to r^{-3}. This is just in between the isolated and distributed scatterers; the former received power is proportional to r^{-4}, whereas the later proportional to r^{-2} .

The echo from meteor trails with linear density below $10^{14}\,\mathrm{m}^{-1}$ is generated by comparatively small meteors, and is called under-dense echo. It moves along the background wind, and therefore, the horizontal winds at the meteor heights can be estimated from moving velocity of the meteor trails. The meteor trail rapidly attenuates by diffusion within approximately 1 second after generation. This attenuation can be assumed to be caused by ambipolar diffusion, and diffusion coefficient depends on the background atmospheric temperature. Therefore, the atmospheric temperature can be estimated from the temporal change of echo intensity (Tsutsumi et al. 1994; Hocking et al. 1997). Received power from Fresnel reflection by meteor trails is comparatively strong, and the meteor trails are observable with small-scale radar system at the 30–60 MHz band, which is called a meteor radar.

Chapter 4
Principle of Doppler Velocity Measurement

Doppler radars apply the Doppler effect, the principle of which was first proposed by C.J. Doppler in 1842. In the present chapter, the principle of Doppler velocity measurement by single Doppler radar is discussed first, and then measurements by bistatic Doppler radar is described. They are followed by description on the similar measurements with multiple (primarily two) Doppler radars.

4.1 Doppler Velocity Measurements

4.1.1 Principles of Doppler Radar

Radar waves incident on a scatterer forces electromagnetic vibrations to the scatterer. If the scatterer moves toward the radar, the internal vibrations will be faster than that for a stationary scatterer because the wave's apparent propagation speed relative to the scatter is faster and thus the approaching scatterer experiences more rapid fluctuations of incident waves. Thus the backscattered radiation (i.e., echoes) received by radar will have higher frequency.

If the phase differences between transmitted pulses and received signals (echoes) from a scatterer are measured, the frequency shift (Doppler shift) of received signals from the transmitted pulses can be calculated. This functional capability is limited only to coherent radars, which has the stable transmitted pulse phase and handles phase information of every transmitted pulse and received echo signals.

If the radar transmitted wave length is λ, and the range to the scatterer is r, the two-way path from radar to scatterer normalized by λ, i.e., total number of wavelengths is $2r/\lambda$. It corresponds to the phase change of $(2r/\lambda) \times 2\pi = 4\pi r/\lambda$ radians. Thus, when the initial phase of transmitted signal is φ_0, the phase of received echo signal becomes

S. Fukao and K. Hamazu, *Radar for Meteorological and Atmospheric Observations*,
DOI 10.1007/978-4-431-54334-3_4, © Springer Japan 2014

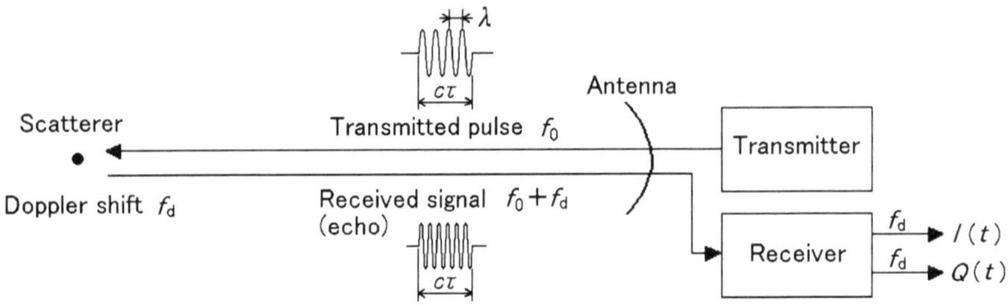

Fig. 4.1 Conceptual diagram of a Doppler radar. Refer to Figs. 5.6 and 8.19 for more details of the schematic configuration

$$\varphi = -\frac{4\pi r}{\lambda} + \varphi_0, \tag{4.1}$$

where the sign of minus means a phase delay. The phase change, which is a derivative with respect to time, is

$$\frac{\mathrm{d}\varphi}{\mathrm{d}t} = -\frac{4\pi}{\lambda}\frac{\mathrm{d}r}{\mathrm{d}t}. \tag{4.2}$$

If the movement of the scatterer is toward (away from) the radar beam, the radial (line-of-sight) velocity or Doppler velocity given by $v_\mathrm{d} = \mathrm{d}r/\mathrm{d}t$ becomes negative (positive). The value $\mathrm{d}\varphi/\mathrm{d}t$ is called Doppler angular frequency ω_d, which is equal to $2\pi f_\mathrm{d}$, where f_d is called Doppler shift or Doppler frequency. Substituting these values into (4.2), the basic relations of Doppler radar is given by

$$f_\mathrm{d} = \frac{\omega_\mathrm{d}}{2\pi} = -\frac{2v_\mathrm{d}}{\lambda}. \tag{4.3}$$

Configuration of Doppler Radar

The conceptual diagram of a Doppler radar which is capable of measuring f_d in (4.3) is illustrated in Fig. 4.1. If the radio frequency (rf) pulse of frequency f_0 (carrier frequency; $f_0 = c/\lambda$, where c is the speed of light and λ is the radar wave length) and pulse width τ is transmitted and the received signal from a moving scatterer is Doppler shifted by an amount f_d, the frequency of the received signal becomes $f_0 + f_\mathrm{d}$. It is difficult to make direct measurements of the rf signal, so most receivers shift the carrier frequency f_0 to zero (baseband) or an intermediate frequency (IF) so that the signal can be digitized and analyzed easily to obtain the Doppler frequency f_d. The detail of the frequency conversion from rf to IF and phase detection will be shown in Fig. 5.6. Refer to Fig. 8.19 for more details of the configuration of Doppler radar. In recent years with the advances of digital technology, modern receivers that digitize the signals at rf or IF, and translate them into baseband ($f_0 = 0$) are becoming quite common. These so-called digital receivers or software-defining receivers will be discussed in Sect. 8.3.5.

The complex received signal (i.e., echo) voltage $V(r,t)$ at the receiver's input is given by Doviak and Zrnić (2006, p. 50)

$$\begin{aligned} V(r,t) &= A\exp\left\{j\left[2\pi f_0\left(t-\frac{2r}{c}\right)+\varphi_0\right]\right\}U\left(t-\frac{2r}{c}\right) \\ &= A\left[\cos\left(2\pi f_0 t-\frac{4\pi r}{\lambda}+\varphi_0\right)+j\sin\left(2\pi f_0 t-\frac{4\pi r}{\lambda}+\varphi_0\right)\right]U\left(t-\frac{2r}{c}\right), \end{aligned} \tag{4.4}$$

where A is the echo amplitude, the cosine and sine are the real and imaginary part of $V(r,t)$ oscillating at the carrier frequency f_0, t is time when the echo is received after radiation of the transmitted pulse, φ_0 is the initial phase of transmitted signal, and $2\pi f_0(t-2r/c)+\varphi_0$ is the echo phase. The pulse function $U(t-2r/c)$ is unity when its argument $t-2r/c$ is between 0 and τ, and is zero otherwise.

The complex received signal after the signal at the carrier frequency f_0 is down converted to baseband (i.e., $f_0 = 0$) can be expressed as

$$V(r,t) = I(r,t) + jQ(r,t), \tag{4.5}$$

and the two components, the in-phase $I(r,t)$ and quadrature phase $Q(r,t)$ components of this complex signal are related by

$$I(r,t) = A\cos\varphi\, U\left(t-\frac{2r}{c}\right), \tag{4.6}$$

$$Q(r,t) = A\sin\varphi\, U\left(t-\frac{2r}{c}\right), \tag{4.7}$$

where the echo phase is defined as $\varphi = -4\pi r/\lambda + \varphi_0$, which becomes identical to (4.1). As shown in (4.5)–(4.7), the echo voltage is a function of both r and t.

The measurement of the Doppler shift in a single short received signal (i.e., an echo) is principally possible, but not a practical approach for short pulse Doppler radars. Instead, the change of the echo's phase angle within one transmitted pulse to the next is commonly measured. As stated in (4.1), the phase angle is proportional to the scatterer's range and equals to twice the number of wavelengths between the radar and the scatterer. Thus, the distance to the scatterer can be measured quite precisely with the phase angle. However, φ change is the more accurate measure of changes in scatterer location. For example, a change of r by $\lambda/4$ causes φ to change by 180°, a large angular change, whereas the change Δt along t is $\Delta t = \lambda/2c$ (e.g., 1.67×10^{-10} s) a extremely tiny value of t. Thus, scatterer motion is measured by change in φ (R.J. Doviak, 2011, personal communication).

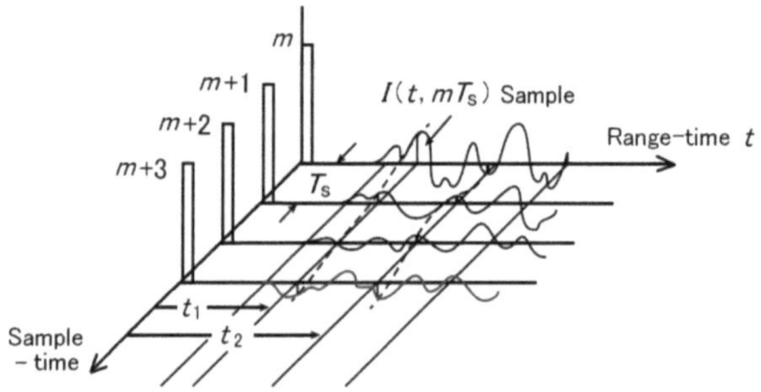

Fig. 4.2 Conceptual traces for $I(t,mT_\mathrm{s})$ of echoes from a distribution of scatterers. The *solid line traces* represent $I(t,mT_\mathrm{s})$ for the mth T_s interval. *Dashed lines* connect samples at each t [revised from Doviak and Zrnić 2006, p. 65]

There is ambiguity because the phase can only be determined within a wave length. Nevertheless, the change of the echo phase, from pulse to pulse, is used to measure the Doppler velocity v_d, the time rate-of-change of range to the scatter. Pulses of the time interval T_s are transmitted, thus echoes from a stationary scatterer will periodically appear at $t = 2r/c + mT_\mathrm{s}$, where $m = 1,2,3,\cdots$ defines each pulse. Conceptual traces for $I(t,mT_\mathrm{s})$ of echoes from a distribution of scatterers are shown in Fig. 4.2 (Doviak and Zrnić 2006, p. 65). The solid line traces represent $I(t,mT_\mathrm{s})$ for the mth T_s interval. Dashed lines connect samples (i.e., the vertical line segment) at each t, where t determines the range $r = ct/2$. In this sampling process, the sampled echo voltage is written as

$$V(t,mT_\mathrm{s}) = I(t,mT_\mathrm{s}) + jQ(t,mT_\mathrm{s}). \tag{4.8}$$

The sampled echo voltage of (4.8) can be represented as a vector which has the amplitude $|V(t)|$ and echo phase φ, where φ is positive when measured counterclockwise (ccw) from the $I(t,mT_\mathrm{s})$ axis. Figure 4.3 shows complex received signals for $m = 1$–5, (a) in-phase and quadrature phase signals where five range-time signals of successive intervals T_s are superimposed to show the signals' relative change for stationary and moving scatterers, and (b) vector presentation of five samples in (a) at the five intervals mT_s for the moving scatterers (Doviak and Palmer 2014).

If the scatterer moves, not only echo will change its location along t, but φ also change according to (4.1), its position along t and φ can be in principle be used to measure the change in scatterer location, and thus indirectly its radial velocity. From (4.2) and (4.3),

$$\frac{\Delta\varphi}{\Delta t} = \frac{\varphi_m - \varphi_{m-1}}{T_\mathrm{s}} = \omega_\mathrm{d} = 2\pi f_\mathrm{d}, \tag{4.9}$$

$$v_\mathrm{d} = -\frac{\lambda f_\mathrm{d}}{2}. \tag{4.10}$$

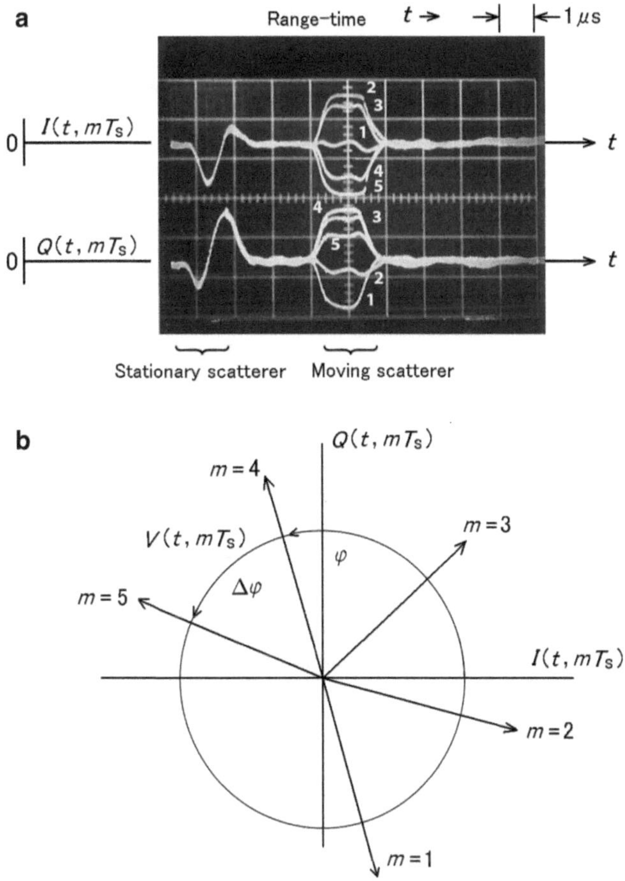

Fig. 4.3 Complex received signals. (**a**) In-phase and quadrature phase signals as a function of five successive intervals T_s superimposed to show the signals' relative change for stationary and moving scatterers. (**b**) Vector presentation of five samples in (**a**) at the five intervals mT_s for the moving scatterer [revised from Doviak and Palmer 2014]

where T_s is the sample time interval equal to the pulse repetition time (PRT). Negative (positive) f_d, where a scatterer moves away from (approaches toward) radar, defines velocity as positive (negative) by following the convention.

4.1.2 Measurable Limit of Doppler Velocity

Nyquist Limit

According to the Shannon and Someya's sampling theorem, the maximum measurable frequency is a half of the frequency of signal sampling, i.e., half of the

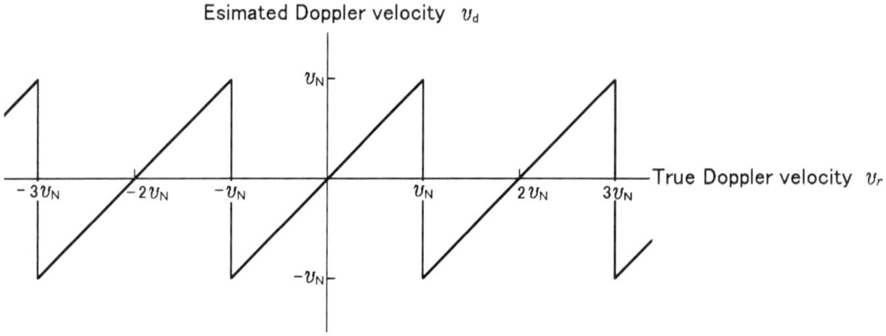

Fig. 4.4 Doppler velocity aliasing

pulse repetition frequency which is a reciprocal of the data interval. Therefore, the maximum measurable Doppler frequency $f_{\rm dmax}$ is given by

$$f_{\rm dmax} = \frac{1}{2T_{\rm s}}. \tag{4.11}$$

If the Doppler frequency of a scatterer $f_{\rm d}$ exceeds $f_{\rm dmax}$ or $-f_{\rm dmax}$, a frequency aliasing occurs and the measured Doppler frequency becomes ambiguous. The maximum velocity below which $f_{\rm d}$ is not affected by aliasing is the Nyquist limit $v_{\rm N}$ which is expressed as

$$v_{\rm N} = \frac{\lambda}{4T_{\rm s}}. \tag{4.12}$$

When the radial velocity v_r of a scatterer exceeds the Nyquist limit, the Doppler velocity is folded, and aliased as illustrated in Fig. 4.4, and the Doppler velocity $v_{\rm d}$ measured with a radar takes the value between $-v_{\rm N}$ and $+v_{\rm N}$. The values of v_r and $v_{\rm d}$ are related by the following equation;

$$v_r = v_{\rm d} + 2v_{\rm N}, \tag{4.13}$$

where the integers $N(= 0, \pm 1, \pm 2, \cdots)$ are Nyquist numbers, and $2v_{\rm N}$ is the Nyquist width.

The Maximum Observable Range

As discussed in Sect. 3.1.1, the maximum observable range $r_{\rm a}$, which is unambiguously determined without any range ambiguity, is given by (3.3) as

$$r_{\rm a} = \frac{cT_{\rm s}}{2}, \tag{4.14}$$

where c is the speed of light. From (4.12) and (4.14), v_N is related to r_a by

$$v_N r_a = \frac{c\lambda}{8}. \tag{4.15}$$

Equation (4.15) shows that the product of v_N by r_a becomes a constant value, and both v_N and r_a cannot be enlarged at the same time. In general, r_a is limited in order to enlarge v_N for suppressing the occurrence of velocity aliasing.

4.1.3 Expansion of Doppler Velocity Measurement Range

The maximum Doppler velocity below which aliasing does not occur is the Nyquist limit given by (4.12). One PRF determines one Nyquist limit. Two Nyquist limits are determined if two PRFs are available. Difference between the two Doppler velocities can be used as a clue for determining number of velocity aliasing, from which the true Doppler velocity is unambiguously determined in a wider velocity range. For example, the ratio of two PRFs (PRF1 and PRF2) is n_1 and n_2, where n_1 and n_2 are integers and the least common multiple is $n_1 \times n_2$. Suppose that the Nyquist limits for PRF1 and PRF2 are v_{N1} and v_{N2}, respectively, the respective Doppler velocities are v_1 and v_2, and the Doppler velocity difference between the two PRFs is Δv. The value Δv takes the unique value up to the true velocity of $v_{N1} \times n_2 = v_{N2} \times n_1$. Meanwhile, if the true velocity exceeds the above limit, Δv takes the same value as one of those in the lower velocity and the number of velocity aliasing becomes uncertain. As a result, the Doppler velocity aliasing can be corrected up to the value of $n_1 v_{N2}$ and $n_2 v_{N1}$. Figure 4.5 illustrates the case of $n_1 : n_2 = 4 : 3$; (a) Doppler velocity aliasing and (b) Doppler velocity difference.

4.2 Methods of Applying Doppler Radar

Doppler speeds directly obtained with single Doppler radar are the radial component of velocity vector of scatterer. Meteorological Doppler radars scan the three dimensional space. It is required to synchronously operate three or more meteorological Doppler radars, which are not aligned along the same baseline, to estimate precisely the velocity vectors (east-west, north-south, and vertical components). On the other hand, a variety of wind information such as radial and/or azimuthal changes of radial wind speeds can be obtained even with single meteorological Doppler radar. In this section, techniques for estimating two- or three-dimensional wind velocities by using single Doppler radar are described.

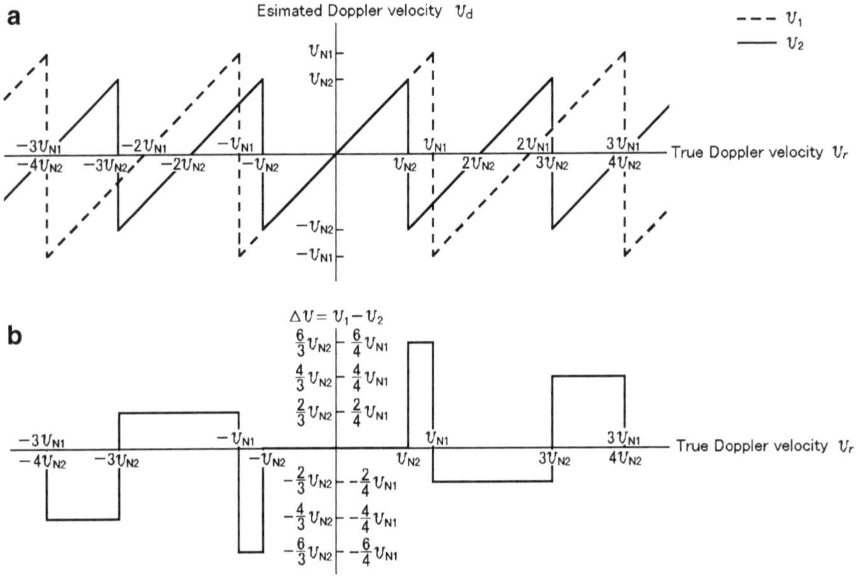

Fig. 4.5 (**a**) Doppler velocity aliasing and (**b**) Doppler velocity difference Δv between two PRFs with the ratio of 4:3. *Dotted lines* and *solid lines* in (**a**) are for v_1 and v_2, respectively

4.2.1 Volume Velocity Processing (VVP) Method

A three-dimensional volume in a suitable size where velocity changes linearly is assumed for estimation of velocities with single Doppler radar observations. The accuracy of estimation of wind and that of divergence and deformation of the wind field depend on the size and the shape of the volume, the measurement precision of radial wind speed, linearity of actual wind, etc. The volume velocity processing (VVP) method is a technique to estimate wind vectors directly in the volume with radial wind speeds inside obtained by single Doppler radar (Waldteufel and Corbin 1979; Koscielny et al. 1982). The volume is called the analysis volume. The typical dimensions of the volume are elevation angle width of 1–2°, azimuthal angle width of 30–40°, and range depth of 20–30 km (Fig. 4.6, Doviak and Zrnić 2006, p. 305). Consider that a radar located on the Earth's surface measures wind vector at a point $P(r, \theta_e, \phi)$ in Fig. 4.6. The radar measures the radial velocity of the scatterer. If the scatterer has significant terminal velocity, or is not passively carried by the wind, the wind measurements will be biased. In general, the observed Doppler velocity need to be corrected before the VVP method is applied. The radial component of wind vector after the above correction is derived as follows. In the spherical Earth coordinate with two orthogonal great circle paths, x and y are considered as arc distance from the radar, where the y axis is along a meridian and the north is positive. The vertical axis starts the center of the Earth, and goes through the point P. The height z is the distance between the Earth's surface and P. The effective Earth

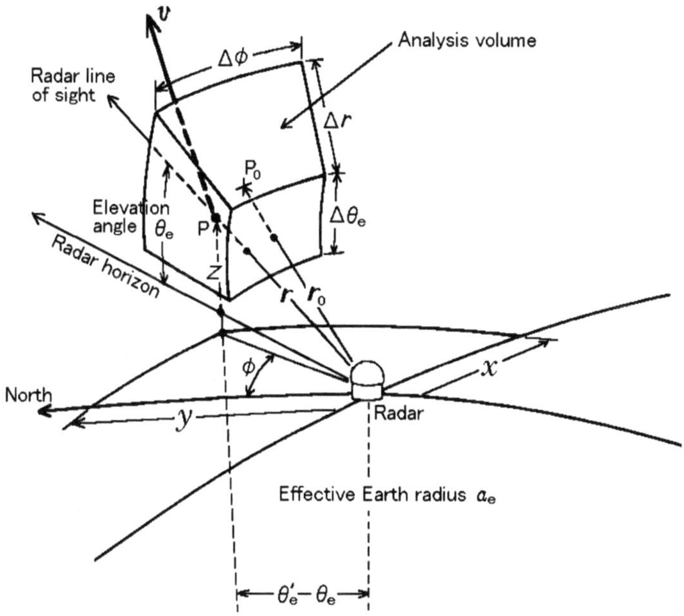

Fig. 4.6 The spherical coordinate system for the volume velocity processing (VVP) method. Typical analysis volume dimensions are $\Delta r \simeq 20$–$30\,\mathrm{km}$; $\Delta\theta_e \simeq 1$–$2°$; $\Delta\phi \simeq 30$–$40°$ [from Doviak and Zrnić 2006, p. 305]

radius a_e is determined by the mean vertical gradient of radio refractive index. The horizontal components u and v of the wind vector $\boldsymbol{v}$ are the tangents to the great circle arcs passing through P and parallel to x and y, respectively, where the east and north directions are positive. The vertical component w of $\boldsymbol{v}$ is along z-axis (upward positive).

The wind vector $\boldsymbol{v}$ at an arbitrary point (x, y, z) within the analysis volume is assumed to be expressed by the first-order Taylor series about a specific point $P_0(x_0, y_0, z_0)$ in the same analysis volume, where P_0 is usually chosen at the center of the analysis volume. Then,

$$\boldsymbol{v}(x, y, z) = \boldsymbol{v}(x_0, y_0, z_0) + \frac{\partial \boldsymbol{v}}{\partial x}(x - x_0) + \frac{\partial \boldsymbol{v}}{\partial y}(y - y_0) + \frac{\partial \boldsymbol{v}}{\partial z}(z - z_0). \tag{4.16}$$

In the ordinary radar observations, $r \ll a_e$, and the rectangular coordinates (x, y, z) can be converted to the spherical coordinates (r, θ_e, ϕ) as follows;

$$x \simeq r\cos\theta_e^{'}\sin\phi, \tag{4.17}$$

$$y \simeq r\cos\theta_e^{'}\cos\phi, \tag{4.18}$$

$$z = (a_e{}^2 + r^2 + 2ra_e\sin\theta_e)^{1/2} - a_e, \tag{4.19}$$

where θ_e' is the sum of elevation angle θ_e of the radar beam to point P and that subtended between the vertical lines at the point P and at the radar. The angle θ_e' is given by

$$\theta_e' = \theta_e + \tan^{-1}\left(\frac{r\cos\theta_e}{a_e + r\sin\theta_e}\right). \tag{4.20}$$

The radial velocity v_r is the projection of $\boldsymbol{v}$ onto $\boldsymbol{r}$, the vector from the radar to the point (r, θ_e, ϕ). Since $r \ll a_e$, v_r is expressed by

$$v_r = \boldsymbol{v}\cdot(\mathbf{i}_x\cos\theta_e'\sin\phi + \mathbf{i}_y\cos\theta_e'\cos\phi + \mathbf{i}_z\sin\theta_e'), \tag{4.21}$$

where $\mathbf{i}_x$, $\mathbf{i}_y$, and $\mathbf{i}_z$ are the unit vectors at $\boldsymbol{r}$ in the x-, y-, and z-directions, respectively. Substituting the first order expression of $\boldsymbol{v}$ from (4.16) with the relations of x, y, and z from (4.17) to (4.19) into (4.21) and rearranging individual terms, the following expression is obtained (Easterbrook 1974; Doviak and Zrnić 2006, pp. 307–309);

$$\begin{aligned}
v_r = {} & u_0'\cos\theta_e'\sin\phi \\
& + u_x\cos\theta_e'\sin\phi(r\cos\theta_e'\sin\phi - x_0) \\
& + u_z\cos\theta_e'\sin\phi(z - z_0) \\
& + v_0'\cos\theta_e'\cos\phi \\
& + v_y\cos\theta_e'\cos\phi(r\cos\theta_e'\cos\phi - y_0) \\
& + v_z\cos\theta_e'\cos\phi(z - z_0) \\
& + (u_y + v_x)\cos\theta_e'\left(r\cos\theta_e'\sin\phi\cos\phi - \frac{x_0\cos\phi + y_0\sin\phi}{2}\right) \\
& + w_0\sin\theta_e' \\
& + w_x\sin\theta_e'(r\cos\theta_e'\sin\phi - x_0) \\
& + w_y\sin\theta_e'(r\cos\theta_e'\cos\phi - y_0) \\
& + w_z\sin\theta_e'(z - z_0),
\end{aligned} \tag{4.22}$$

where the horizontal components of wind vector $\boldsymbol{v}$ at $(x_0,\ y_0,\ z_0)$ are u_0 and v_0. The values u_0' and v_0' are called the modified wind components, and given by $u_0' = \left[u_0 + \frac{1}{2}y_0(v_x - u_y)\right]$ and $v_0' = \left[v_0 - \frac{1}{2}x_0(v_x - u_y)\right]$, respectively. The values which are given by u_x, u_y, u_z $\cdots$ are the rate of linear increase of wind, and given by $\partial u/\partial x$, $\partial u/\partial y$, $\partial u/\partial z$ $\cdots$, respectively. A linear wind field is usually described by 12 parameters $(u_0,\ u_x,\ u_y,\ u_z,\ v_0,\ v_x,\ v_y,\ v_z,\ w_0,\ w_x,\ w_y,\ w_z)$. Equation (4.22) expresses this wind field with 11 unknowns, u_0', u_x, u_z, v_0', v_y, v_z, $u_y + v_x$, w_0, w_x, w_y, w_z, as a function of location, which can be solved by 11 independent equations.

To facilitate the manipulation of (4.22), the vector $\boldsymbol{K}_m$ is given by the 11×1 matrix, and its transpose $\boldsymbol{K}_m^{\mathrm{T}}$ by

$$\boldsymbol{K}_m^{\mathrm{T}} = (u_0', u_x, u_z, v_0', v_y, v_z, u_y + v_x, w_0, w_x, w_y, w_z), \tag{4.23}$$

where the subscript ^{T} denotes a transposed matrix. Next, $\boldsymbol{P}_m^{\mathrm{T}}$, the transpose of $\boldsymbol{P}_m$, that is the 1×11 matrix of parameters concerned with r, θ_e, and ϕ, is expressed as

$$\boldsymbol{P}_m^{\mathrm{T}} = \begin{bmatrix} P_1 \\ P_2 \\ P_3 \\ P_4 \\ P_5 \\ P_6 \\ P_7 \\ P_8 \\ P_9 \\ P_{10} \\ P_{11} \end{bmatrix} = \begin{bmatrix} \cos\theta_\mathrm{e}' \sin\phi \\ \cos\theta_\mathrm{e}' \sin\phi (r\cos\theta_\mathrm{e}' \sin\phi - x_0) \\ \cos\theta_\mathrm{e}' \sin\phi (z - z_0) \\ \cos\theta_\mathrm{e}' \cos\phi \\ \cos\theta_\mathrm{e}' \cos\phi (r\cos\theta_\mathrm{e}' \cos\phi - y_0) \\ \cos\theta_\mathrm{e}' \cos\phi (z - z_0) \\ \cos\theta_\mathrm{e}' \left(r\cos\theta_\mathrm{e}' \sin\phi \cos\phi - \dfrac{x_0 \cos\phi + y_0 \sin\phi}{2} \right) \\ \sin\theta_\mathrm{e}' \\ \sin\theta_\mathrm{e}' (r\cos\theta_\mathrm{e}' \sin\phi - x_0) \\ \sin\theta_\mathrm{e}' (r\cos\theta_\mathrm{e}' \cos\phi - y_0) \\ \sin\theta_\mathrm{e}' (z - z_0) \end{bmatrix}. \tag{4.24}$$

Thus, the radial velocity v_r is expressed by

$$v_r = \boldsymbol{P}_m \boldsymbol{K}_m. \tag{4.25}$$

The above equation indicates that the $m(= 11)$ unknowns related to v_r is uniquely determined by $m(= 11)$ equations (i.e., observations); however, the following equation obtained by the least squares method is usually applied to solve $\boldsymbol{K}_m$, utilizing redundancy of observed data. The number of data n obtained in one sequence of observation is generally more than m, and, at least, several hundreds of data are available, that is, the number of data is redundant. Radial velocity vector $\boldsymbol{v}_{rn}$, which is composed of n observed data of radial velocities, is given by the product of $n \times m$ matrix of $\boldsymbol{P}_{nm}$, which is composed of the corresponding n number of $P_j (j = 1, 2, \cdots, m)$, and $\boldsymbol{K}_m$, and therefore v_{rn} is expressed as

$$v_{rn} = \boldsymbol{P}_{nm} \boldsymbol{K}_m. \tag{4.26}$$

Equation (4.26) is based upon the observations, and the solution obtained by the least squares method for $\boldsymbol{K}_\mathrm{m}$ is given by

$$\boldsymbol{K}_m = (\boldsymbol{P}_{nm}^{\mathrm{T}} \boldsymbol{P}_{nm})^{-1} (\boldsymbol{P}_{nm}^{\mathrm{T}} \boldsymbol{v}_{rn}), \tag{4.27}$$

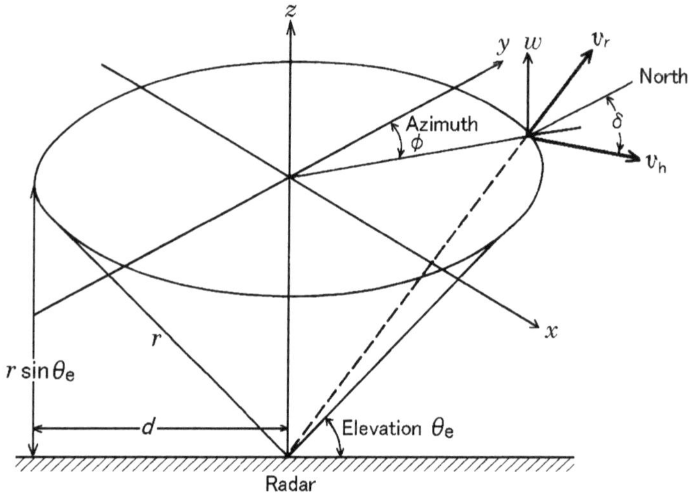

Fig. 4.7 The coordinate system for the VAD method

and every parameter of $\boldsymbol{K}_m$, i.e., $u_0^{'}$, u_x, u_z, $v_0^{'}$, v_y, v_z, u_y+v_x, w_0, w_x, w_y, w_z, is accordingly derived (Doviak and Zrnić 2006, pp. 307–309).

In the VVP method, it is assumed that the wind field linearly changes in the analysis volume. However, actual wind usually changes more complicatedly inside the analysis volume. In addition, it is necessary to be cautious about various errors inherently accompanied in radial velocity measurements.

4.2.2 Velocity Azimuth Display (VAD) Method

Figure 4.7 illustrates that the radar beam is conically steered in the wind field along a circle at a constant range r. If refraction of the radio ray path through a spherically stratified medium in the Earth's atmosphere (the curvature effect) is not taken into account. $x_0 = y_0 = 0$, and $z = z_0$. Since $\theta_e^{'} = \theta_e$ in the same figure, $z_0 = r \sin\theta_e$. Furthermore, if w is constant along the circle, or $w_x = w_y = 0$, v_r of (4.22) is simplified as follows;

$$\begin{aligned} v_r = {} & w_0 \sin\theta_e + \frac{1}{2}(u_x + v_y) r \cos^2\theta_e \\ & + v_0 \cos\theta_e \cos\phi + u_0 \cos\theta_e \sin\phi \\ & + \frac{1}{2}(v_y - u_x) r \cos^2\theta_e \cos 2\phi + \frac{1}{2}(v_x + u_y) r \cos^2\theta_e \sin 2\phi. \end{aligned} \tag{4.28}$$

As both r and θ_e are constant, (4.28) consists of the zeroth-, first- and second-order harmonics of the following Fourier series with respect to the azimuth angle ϕ in $0 \le \phi \le 2\pi$;

$$v_r = a_0 + \sum_{n=1}^{\infty} (a_n \cos n\phi + b_n \sin n\phi). \tag{4.29}$$

By comparing (4.28) and (4.29), Fourier coefficients a_0, a_1, b_1, a_2, and b_2 are given by

$$a_0 = w_0 \sin\theta_e + \frac{1}{2}(u_x + v_y) r \cos^2\theta_e, \tag{4.30}$$

$$a_1 = v_0 \cos\theta_e, \tag{4.31}$$

$$b_1 = u_0 \cos\theta_e, \tag{4.32}$$

$$a_2 = \frac{1}{2}(v_y - u_x) r \cos^2\theta_e, \tag{4.33}$$

$$b_2 = \frac{1}{2}(v_x + u_y) r \cos^2\theta_e, \tag{4.34}$$

respectively. The radial velocities v_r are generally observed at regular intervals along the steered circle. The average horizontal wind speed, wind direction, and divergence and deformation of the horizontal wind field are obtained by the following equations from the Fourier coefficients. This technique is called the velocity azimuth display (VAD) method (Browning and Wexler 1968).

Horizontal Wind Speed

$$v_h = \frac{\sqrt{a_1^2 + b_1^2}}{\cos\theta_e} = \sqrt{u_0^2 + v_0^2}. \tag{4.35}$$

Horizontal Wind Direction

$$\delta = \tan^{-1}\frac{b_1}{a_1} = \tan^{-1}\frac{u_0}{v_0}, \tag{4.36}$$

where the angle δ is the direction of v_h with respect to y-axis (North). The wind direction (from which the wind blows) D is related with δ by $D = \delta + \pi$.

Divergence of Horizontal Wind

$$\operatorname{div} v_h \equiv \left(\frac{\partial u}{\partial x} + \frac{\partial v}{\partial y}\right) = \frac{2}{r\cos^2\theta_e}(a_0 - w_0 \sin\theta_e). \tag{4.37}$$

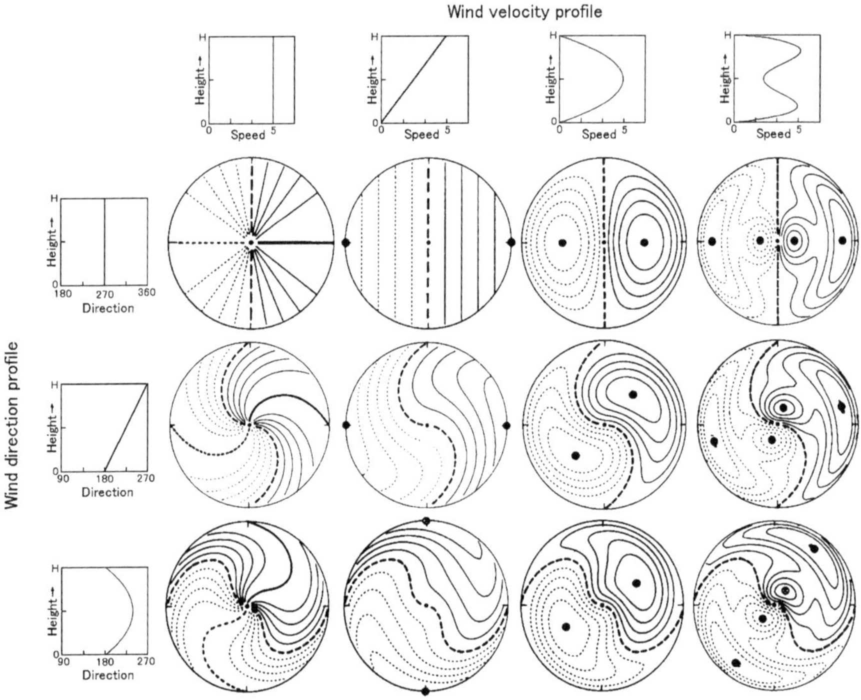

Fig. 4.8 The Doppler velocity patterns at a fixed elevation angle for various wind speed and direction profiles. *Solid lines*, *dotted lines*, and *thick dotted lines* are contour lines of positive, negative, and zero Doppler velocities, respectively. The interval of isolines is 0.2 times of the maximum velocity. The *black dots* indicate the point of maximum velocities. The radar is located at the center of the circles [from Wood and Brown 1983]

Deformations of Horizontal Wind

Stretching due to extension and shearing due to horizontal wind shear are expressed as

$$\frac{\partial u}{\partial x} - \frac{\partial v}{\partial y} = -\frac{2a_2}{r\cos^2\theta_e} \quad \text{Deformation due to extension,} \tag{4.38}$$

$$\frac{\partial u}{\partial y} + \frac{\partial v}{\partial x} = \frac{2b_2}{r\cos^2\theta_e} \quad \text{Deformation due to horizontal wind shear.} \tag{4.39}$$

If wind speed and direction are uniform within the circle ($u_x = u_y = 0$), v_r is given by

$$v_r = w\sin\theta_e + v_h\cos\theta_e\cos(\delta - \phi), \tag{4.40}$$

where w is the vertical component of v_r. The VAD method has been traditionally used by the meteorological radars, and the same method can be applied also to the atmospheric radars. This case will be discussed in Sect. 7.2.1. Figure 4.8 shows a variety of iso-contour lines of Doppler speed, which are observed by the VAD method at a constant elevation angle θ_e for various horizontal wind speeds and directions. Each outer circle represents radial Doppler speeds at the highest altitude H and the range $r = H/\sin\theta_e$ (Wood and Brown 1983). The pattern within the circle corresponds to wind speed at lower altitudes. The Doppler velocities within the circle are mainly determined by the wind speed, while the positions of the zero speed lines are uniquely decided by the wind direction profile. Thus, approximate height profile of the horizontal wind is inferred from this pattern.

4.2.3 Wind Observations with Bistatic Doppler Radar

Ordinary Doppler radars that share single antenna for transmission and reception are called monostatic Doppler radars. On the other hand, the radars equipped with a pair of antennas that are separately located, typically by a large distances, where one antenna is used exclusively for transmitting and the other is used for reception are called a bistatic Doppler radar. When a radar transmits pulses toward the scatterers, oblique (lateral) scatterings are generated to various directions, in addition to the backscattering toward the radar. In a bistatic Doppler radar system, the main radar (monostatic radar) antenna receives backscattered signals, while the separately located passive radar (receiver) antenna receives obliquely scattered signals (e.g., Doviak and Weil 1972; Doviak et al. 1972).

A bistatic multiple-Doppler radar which has a high angular resolution beam of the transmitting antenna that scans the common space, and has some receiving antennas which are relatively small in size and spaced far from the large transmitting antennas and far from each other, perform synchronized Doppler observations to measure two- or three-dimensional wind vectors (Doviak 1972; Wurman et al. 1993; de Elía and Zawadzki 2000, 2001; Satoh and Wurman 2003). A bistatic multiple-Doppler radar can be regarded as a configuration with the performance of several bistatic Doppler radar pairs, and a bistatic Doppler radar is composed of one transmitter and one receiver.

Geometry of Bistatic Radar

The performance of a bistatic Doppler radar depends on two basic geometrical characteristics, i.e., the transmitter–receiver antenna separation and transmitter–scatterer–receiver triangulation (e.g., Hanle 1986; Jackson 1986). For simplicity, in the following discussion we assume that elevation angle of the transmitting beam is zero. The geometry of bistatic Doppler radar is shown in Fig. 4.9a, where points T, R, and O denote the locations of the monostatic radar (transmitter), the bistatic radar (receiver), and the center of the bistatic scattering volume having a location determined by the bistatic receiver's range gate and the narrow beam of the transmitting antenna, respectively. The monostatic scattering volume location

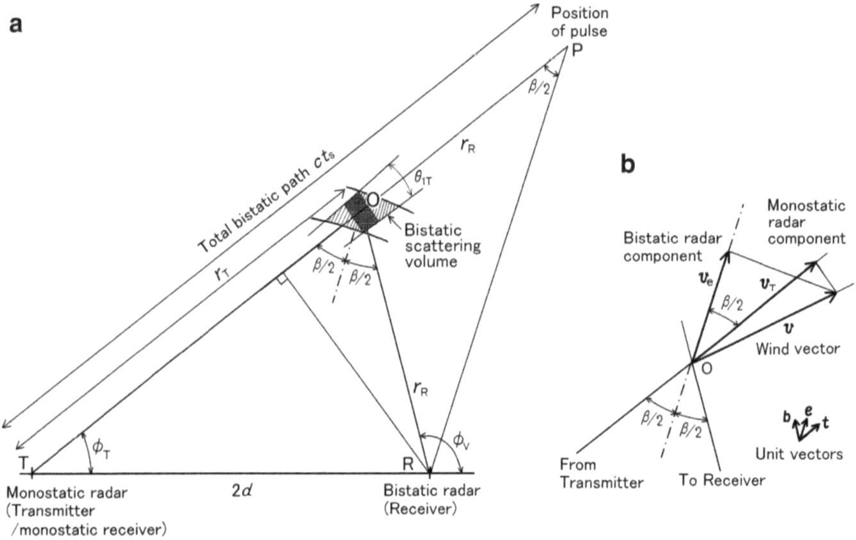

Fig. 4.9 (**a**) Geometry of a bistatic Doppler radar in an azimuthal plane, where elevation angle of the transmitting beam is zero. *Light* and *heavy shaded regions* indicate bistatic and monostatic scattering volumes, respectively. In (**b**), point O and surrounding geometry are the expansion of the same region of (**a**). β is the bistatic angle

is determined by the monostatic receiver's gate location and the receiver's field of view (i.e., the same as the transmitter's beamwidth θ_{1T}). The transmitting antenna is assumed to have a narrow beam (e.g., approximately 1°) directed at an elevation angle of 0°, and an azimuth angle of ϕ_T. The bistatic receiving antenna is assumed to have a large field of view that is fixed (i.e., the receiver's beam is fixed and very broad so there is no need to specify the receiver's elevation and azimuth angles) to observe echoes from anywhere within a large volume of scatterers (e.g., a thunderstorm) into which the transmitting beam is directed. Thus ϕ_V in Fig. 4.9 is not the receiving beam's azimuth angle, but the azimuth to the scattering volume (i.e., the shaded volume in Fig. 4.9, where light and heavy shaded regions indicate bistatic and monostatic scattering volume, respectively). The distance between the transmitter and the receiver is $2d$ (fixed value). The distances from the transmitter to the bistatic scattering volume and the distance from this volume to the receiver are r_T and r_R, respectively; thus $r_T + r_R$ is the total bistatic path. The angle β is the bistatic angle, i.e., the angle between r_T and r_R. Point P is the position of transmitted pulse when the bistatic echo is received by the bistatic radar. Although this point has no meaning for bistatic scattering, it will be used in the following paragraph to derive $\beta/2$ with simple geometry.

Here we assume that the receiver is phase locked to the transmitted rf signal, and the transmitted pulse and receiver's gates are synchronized. If an observer at R knows ϕ_T and the bistatic receiver's sampling gate times t_s (i.e., the delay time from the start of the transmitted pulse to the time t_s of any of the bistatic radar's sampling gates), the observer at R can determine the locations of the bistatic scattering volumes along the transmitter's beam. If the observer also knows the monostatic receiver's sampling gate locations, the locations of the backscattering volumes along the transmitting beam can also be calculated. The total bistatic path $(r_T + r_R)$ is equal to ct_s, where c is the speed of light. Thus, $\beta/2$ is expressed from the right triangle with vertices P and R in Fig. 4.9a as

$$\frac{\beta}{2} = \tan^{-1}\left(\frac{2d\sin\phi_T}{ct_s - 2d\cos\phi_T}\right). \tag{4.41}$$

Using (4.41) and measured t_s, the location of the scattering volume along the transmitting beam (in azimuth plane) can be determined. First, ϕ_V, i.e., the azimuth to the scattering volume at point R (receiver), is given by Jackson (1986)

$$\phi_V = \phi_T + \beta = \phi_T + 2\tan^{-1}\left(\frac{2d\sin\phi_T}{ct_s - 2d\cos\phi_T}\right). \tag{4.42}$$

Next, equating the total bistatic path ct_s to a given value, r_T takes the form (Jackson 1986; Doviak and Zrnić 2006, p. 431)

$$r_T = \frac{(ct_s)^2 - (2d)^2}{2(ct_s - 2d\cos\phi_T)} \qquad t_s \geq 2d/c. \tag{4.43}$$

Doppler Observations with Bistatic Radar

The bistatic radar receiver measures a Doppler velocity along the bisector of β, and the monostatic radar receiver measures a Doppler velocity along the transmitted beam. Thus the vector velocity in the plane can be computed from these two components. The ellipse of constant phase (surface of the bistatic scattering volume) in Fig. 4.9a has its foci at the locations of the transmitter and the receiver. If scatterers move along the ellipse, the receiver does not measure any phase change. Assuming the receiver is phase locked to the transmitted signal, all of the points on the ellipse (ellipse of constant phase) are seen by the receiver as equidistant from the transmitter, thus only motion perpendicular to the ellipse is measured (Protat and Zawadzki 1999). Doppler velocities measured by the radar in the azimuthal plane are shown in Fig. 4.9b. In (b), point O and surrounding geometry are the expansion of the same region of (a). Vector $\boldsymbol{v}$ is the wind velocity vector, $\boldsymbol{v}_T$ is the radial projection to the transmitter (Monostatic radar component), and $\boldsymbol{v}_e$ is the perpendicular component to the ellipse of constant phase (Bistatic radar component), i.e., the projection of the wind vector component onto the line that bisects the included angle between the lines drawn from the transmitter to "O" and from the receiver to "O". Vectors $\boldsymbol{t}$, $\boldsymbol{b}$, and $\boldsymbol{e}$ in (b) are unit vectors in the radial directions of the transmitter, the receiver, and the bistatic radar component, respectively.

If the time interval between successive pulses is Δt, the difference in the path length ΔS is expressed by

$$\Delta S = (\boldsymbol{v}\cdot\boldsymbol{t}+\boldsymbol{v}\cdot\boldsymbol{b})\Delta t, \tag{4.44}$$

where $|\boldsymbol{v}\cdot\boldsymbol{t}| = |\boldsymbol{v}_\mathrm{T}| = v_\mathrm{T}$ is the monostatic Doppler velocity. For a given true velocity along the beam or the true velocity $\boldsymbol{v}_\mathrm{e}$ along the bisector of β, the monostatic Doppler velocity is independent of the monostatic volume's location along the transmitter beam whereas the bistatic Doppler velocity is dependent of the location of the bistatic volume along the transmitter beam. The "apparent" velocity v_a measured by the receiver is defined by dividing the difference in path length by $2\Delta t$ as (Protat and Zawadzki 1999)

$$v_\mathrm{a} = \frac{\Delta S}{2\Delta t} = \frac{1}{2}\boldsymbol{v}\cdot(\boldsymbol{t}+\boldsymbol{b}). \tag{4.45}$$

Rewriting the vector sum in (4.45) as

$$(\boldsymbol{t}+\boldsymbol{b}) = 2\cos(\beta/2)\boldsymbol{e}, \tag{4.46}$$

$\boldsymbol{v}_\mathrm{e}$ can be expressed as (Protat and Zawadzki 1999; de Elía and Zawadzki 2001);

$$v_\mathrm{e} = \frac{v_\mathrm{a}}{\cos(\beta/2)}, \tag{4.47}$$

where $|\boldsymbol{v}\cdot\boldsymbol{e}| = |\boldsymbol{v}_\mathrm{e}| = v_\mathrm{e}$, and $\beta/2$ is given by (4.41). Equation (4.47) shows that the Doppler velocity of scatterers on a hyperbolic trajectory, which corresponds to radial direction in a monostatic Doppler radar, varies from the monostatic maximum value at large ranges to zero when it crosses the baseline (Hanle 1986). Figure 4.10 shows an example of curves of constant Doppler reduction ratio. The numeric values on the curves show that the Doppler velocity of the scatterer measured by the bistatic receiver is reduced from the monostatic maximum one by corresponding to the ratio. The value v_a has a unique Nyquist velocity (aliasing velocity) like a monostatic Doppler radar. Therefore, corrections of the aliasing velocity should be done before obtaining v_e by (4.47) (Satoh and Wurman 2003).

By using only a bistatic Doppler radar, the vector wind component can not be retrieved. To retrieve it in the azimuthal plane, another measurement is needed; this is obtained from the Doppler shift measured by the monostatic radar. As apparent from Fig. 4.9b, the wind component perpendicular to radial velocity can be shown by simple geometry using $\boldsymbol{v}_\mathrm{T}$, $\boldsymbol{v}_\mathrm{e}$ and $\cos(\beta/2)$ (de Elía and Zawadzki 2001). They can be used to calculate the total wind component in the azimuthal plane. To obtain the three-dimensional vector wind, another bistatic receiver is needed. The methods of synthesis of the monostatic Doppler velocity and the bistatic Doppler velocity to obtain three dimensional wind field (u,v,w) are discussed by e.g., Wurman et al. (1993) and Satoh and Wurman (2003).

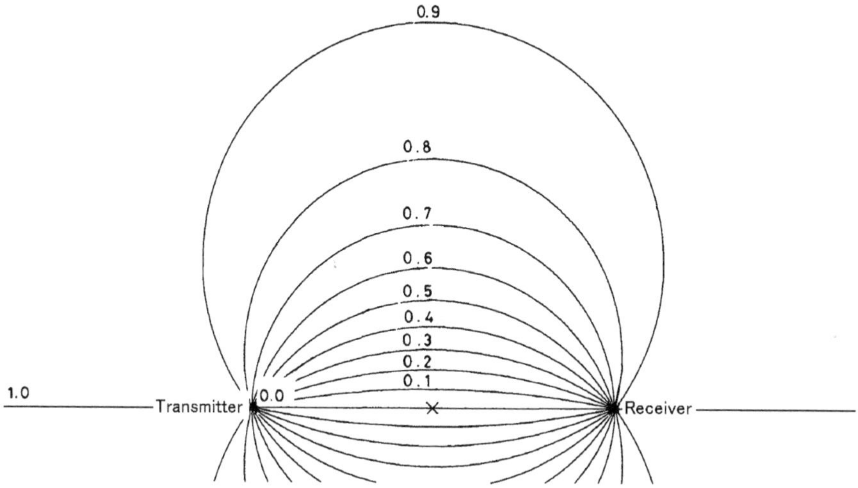

Fig. 4.10 Curves of constant Doppler reduction ratio. The numeric values on the curves show that the Doppler velocity of the scatterer measured by the bistatic receiver is reduced from the monostatic maximum one by corresponding to the ratio [revised from Hanle 1986]

Sample Volume and Radar Equation of Bistatic Radar

Applying (3.31), a bistatic Doppler radar equation is expressed by

$$E[P_r] = \frac{P_t g_T g_R \lambda^2}{(4\pi)^3 r_T^2 r_R^2} \Sigma\sigma_B \sin\chi, \tag{4.48}$$

where g_T and g_R are gains of the transmitting and receiving antenna, respectively, r_T and r_R are distances from the transmitter to the scatterer and from the scatterer to the receiver, respectively, σ_B is the bistatic scattering cross section, and χ is the angle between the incident and scattered electric field vectors. The bistatic scattering cross section of the scattering volume is given by

$$\Sigma\sigma_B = V_B \eta_B, \tag{4.49}$$

where V_B is the scattering volume of a bistatic Doppler radar and η_B is the bistatic reflectivity.

If the beam of the receiving radar is much larger than the region enclosed within the transmitting radar beam and two ellipsoid surfaces as shown in the shaded region in Fig. 4.9a, the bistatic resolution volume is determined by the shaded region. Using a cross sectional area of a transmitted beam defined by a Gaussian-shaped antenna pattern and the interval of two ellipsoid surface (separation) Δs, the bistatic scattering volume V_B is expressed in analogy with V_6 in (3.39) as

$$V_{\mathrm{B}} \simeq r_{\mathrm{T}}^2 \Delta s \frac{\pi \theta_{\mathrm{T}}^2}{8 \ln 2}, \tag{4.50}$$

where θ_{T} is the beamwidths of the transmitting radar in the E and H planes. The separation Δs is determined by the transmitted pulse width τ and the receiver's gating time t_{s} as[1] (Doviak and Zrnić 2006, p. 431)

$$\Delta s \simeq \frac{c\tau}{2\cos(\beta/2)} = \frac{c\tau}{2}\left[1 + \frac{4d^2 \sin^2 \phi_{\mathrm{V}}}{(ct_{\mathrm{s}} + 2d\cos\phi_{\mathrm{V}})^2}\right]^{1/2}. \tag{4.51}$$

Substituting (4.51) into (4.50), (4.50) into (4.49), and (4.49) into (4.48), the estimated received power $E[P_{\mathrm{r}}]$ of a bistatic Doppler radar is given by

$$E[P_{\mathrm{r}}] = \frac{P_{\mathrm{t}} g_{\mathrm{T}} g_{\mathrm{R}} \lambda^2 \theta_{\mathrm{T}}^2 c\tau}{2^{10} (\ln 2) \pi^2 r_{\mathrm{R}}^2}\left[1 + \frac{4d^2 \sin^2 \phi_{\mathrm{V}}}{(ct_{\mathrm{s}} + 2d\cos\phi_{\mathrm{V}})^2}\right]^{1/2} \eta_{\mathrm{B}} \sin\chi, \tag{4.52}$$

where all terms except for ϕ_{V} and r_{R} are parameters determined by the transmitter (monostatic radar) and the receiver (bistatic radar). The value ϕ_{V} is determined using measured receiver's gating time t_{s} and (4.42), and r_{R} is calculated by replacing $\cos\phi_{\mathrm{T}}$ to $-\cos\phi_{\mathrm{V}}$ in (4.43).

Applications of Bistatic Doppler Radar

Figure 4.11 shows the typical configuration of a bistatic multiple-Doppler radar network with one monostatic radar and two bistatic radars (receivers). The most important advantage of bistatic Doppler radars using wide receiving beams and combined with a narrow beam monostatic radar is Doppler velocities in more than one direction can be simultaneously obtained at each bistatic scattering volume $V_6(\mathrm{B})$ $(= V_{\mathrm{B}})$ for both bistatic radars, and at each monostatic scattering volume $V_6(\mathrm{M})$ along the transmitting beam (Fig. 4.11). That is, the vector wind is calculated from simultaneous observations of Doppler velocities at contiguous points along the transmitting beam without the need for temporal interpolation as with multiple monostatic measurements of vector winds (Sect. 4.3). Furthermore, this method is particularly effective from the viewpoint of frequency utilization, since it is possible to perform Doppler observations with only one operational frequency in the same way as that made by radar network operated with multiple operational frequencies. On the other hand, the bistatic Doppler radar is essentially a passive radar network, and the total scattered echo intensities are weaker[2] than those obtained with the same number of monostatic Doppler radar antennas. Actually, this considerably limits

---

[1] de Elía and Zawadzki (2000) and Satoh and Wurman (2003) adopts not $c\tau/[2\cos(\beta/2)]$ but $c\tau/[2\cos^2(\beta/2)]$ in (4.51).

[2] The scattered echoes are not weaker, the received echoes are weaker because of the low gain of the receiving antenna.

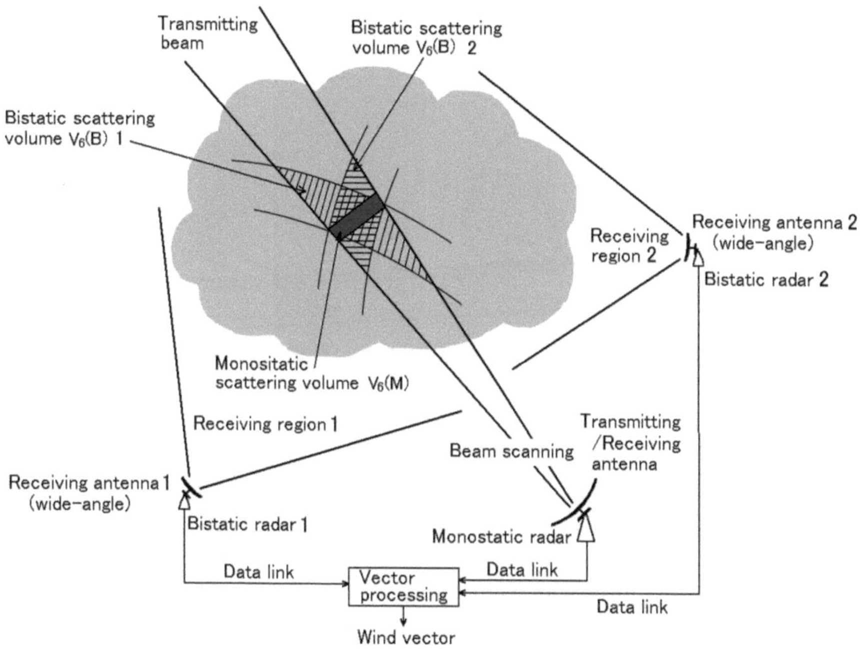

Fig. 4.11 Typical configuration of a bistatic multiple-Doppler radar network

the observation range for passive radar networks. This method is simple but quite effective for observations of heavy rainfalls.

4.3 Multiple Monostatic Doppler Radars

If multiple monostatic Doppler radars are located at a certain interval and applied for observation of the same scatterer at the same time, the number of wind vector components is expected to increase. In fact, as mentioned in the previous sections, single Doppler radar cannot measure the vector velocity of the scatterer. Nor two Doppler radars. At least, three non colinear radars are needed for precise measurement of the three components of the scatterer's velocity.

Presupposing that the wind field does not significantly change over the time it takes for both radars to scan the volume of interest (e.g., a storm), two Doppler radars are independently scanned to observe an identical three dimensional space as shown in Fig. 4.12a. Wind velocity data obtained with both radars in the polar coordinates are converted to vector $(u,\ v,\ w)$ in the Cartesian coordinate. This procedure is called either the independent scanning method (Brandes 1977; Ray et al. 1980) or the dual Doppler method. Because synchronization is not required and thus, the operations are comparatively simple, many observational results have

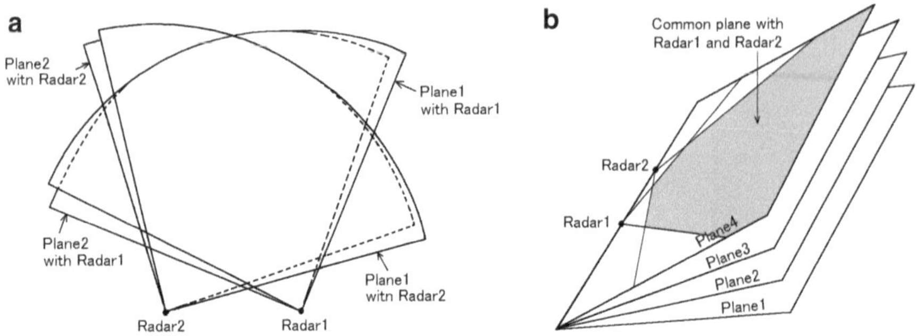

Fig. 4.12 Schematic of dual Doppler radar observations. (**a**) Independent scanning method and (**b**) COPLAN method

been reported (e.g., Ishihara et al. 1986; Fujiyoshi and Geng 1995). On the other hand, two radars scan the identical planes at the same time as shown in Fig. 4.12b. In this case, the elevation angles of the two radar antennas are varied according to the respective scanned planes where the data are derived from the two radars in the cylindrical coordinates with the polar axis connecting the two radars. The wind data are analyzed in the cylindrical coordinates. This method is called the COPLAN method (Lhermitte 1970; Lhermitte and Miller 1970).

4.3.1 Independent Scanning Method

In this method, wind vectors are synthesized by individual radial velocities observed independently by multiple Doppler radars. Because a single monostatic Doppler radar can obtain only the radial component of the wind vector at each point, the plural radars are needed to obtain radial wind components in plural directions. At least three non colinear components are needed to retrieve the wind vector. Furthermore, the wind field must not change significantly during the lapse time while each radar scans the common volume. In addition, when three components of wind vectors are estimated with two radars, the equation of continuity for incompressible medium should be used. The boundary conditions needed to solve this equation is the vertical velocity component w to start the calculation of the integral in the equation, e.g., $w = 0$ at the top of the echo or at the ground. The detail will be discussed later in this subsection.

The Doppler velocity $\boldsymbol{v}_i$ of precipitation particles at the point (x, y, z) observed with the ith radar, which is located at the point (x_i, y_i, z_i), is shown by the following equation, using the wind vector components $\boldsymbol{u}$, $\boldsymbol{v}$, $\boldsymbol{w}$ and the terminal velocity of precipitations (fall speed) $\boldsymbol{w}_{\mathrm{T}}$.

$$\boldsymbol{v}_i = \frac{1}{r_i}[(x-x_i)\boldsymbol{u} + (y-y_i)\boldsymbol{v} + (z-z_i)(\boldsymbol{w}-\boldsymbol{w}_{\mathrm{T}})], \tag{4.53}$$

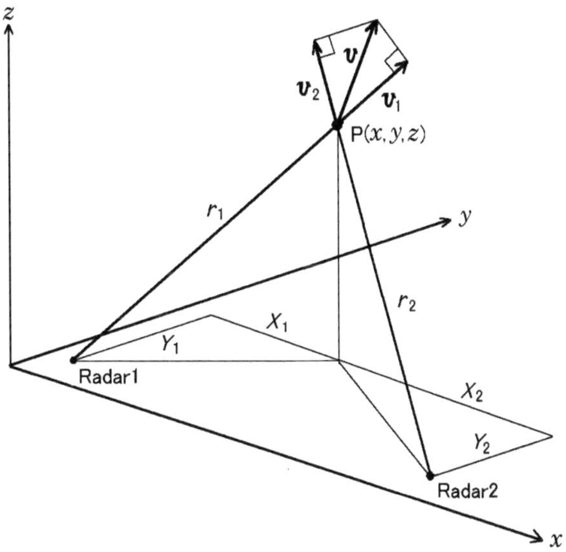

Fig. 4.13 The coordinate system for the independent scanning method with two Doppler radars

where r_i is the distance between the ith radar and the point (x, y, z), which is given by

$$r_i^2 = (x - x_i)^2 + (y - y_i)^2 + (z - z_i)^2. \tag{4.54}$$

Equation (4.53) is the Cartesian coordinate expression of (4.22). When two radars shown in Fig. 4.13 are employed, equations of (4.53) for $i = 1, 2$ are solved for $\boldsymbol{u}$ and $\boldsymbol{v}$ as follows:

$$\boldsymbol{u} = \frac{r_1 Y_2 \boldsymbol{v}_1 - r_2 Y_1 \boldsymbol{v}_2}{D} + \frac{Y_1 Z_2 - Y_2 Z_1}{D}(\boldsymbol{w} - \boldsymbol{w}_\mathrm{T}), \tag{4.55}$$

$$\boldsymbol{v} = \frac{r_2 X_1 \boldsymbol{v}_2 - r_1 X_2 \boldsymbol{v}_1}{D} + \frac{X_2 Z_1 - X_1 Z_2}{D}(\boldsymbol{w} - \boldsymbol{w}_\mathrm{T}), \tag{4.56}$$

where $X_i = x - x_i$, $Y_i = y - y_i$, $Z_i = z - z_i$, and $D = X_1 Y_2 - X_2 Y_1$ (Armijo 1969).

If the second terms of the right-hand side of both (4.55) and (4.56), which are related to the vertical velocity component, could be ignored to the first approximation, two vector components $(u,\ v)$ are uniquely determined. Meanwhile, $\boldsymbol{w}_\mathrm{T}$ and $\boldsymbol{w}$ must be separately known to obtain three components (u, v, w) of wind vector.

As for w_T(terminal velocity), various expressions have been presented as functions of the drop size distribution, radar reflectivity factor, and atmospheric density, or their combinations since Gunn and Kinzer (1949). Details will be shown later in Sect. 6.1.1. Here, we adopt, as an example, the following empirical formula, which is given as the function of radar reflectivity factor (Atlas et al. 1973). The standard error of this formula is known less than or equal to $1\,\mathrm{m\,s^{-1}}$ for rain of

$1 \leq Z \leq 10^5$ [$\mathrm{mm}^6\,\mathrm{m}^{-3}$] (Joss and Waldvogel 1970). It increases to several $\mathrm{m\,s}^{-1}$ for ice crystal, hail and so on.

$$w_{\mathrm{T}} = 2.65Z^{0.1114}\left(\frac{\rho_{\mathrm{a}_0}}{\rho_{\mathrm{a}}}\right)^{0.4} \quad [\mathrm{m\,s}^{-1}], \tag{4.57}$$

where Z is the radar reflectivity factor [$\mathrm{mm}^6\,\mathrm{m}^{-3}$]. The parameters ρ_{a_0} and ρ_{a} are atmospheric density on the ground and that at the height of the observed point, respectively.

Another equation should be required to obtain three components of wind vector $(u,\ v,\ w)$, in addition to (4.55) and (4.56). When the ordinary precipitations are considered, the equation of continuity for incompressible medium is applied as follows (Ogura and Phillips 1962);

$$\frac{\partial u}{\partial x} + \frac{\partial v}{\partial y} + \frac{\partial w}{\partial z} = -\frac{w}{\rho_{\mathrm{a}}}\frac{\partial \rho_{\mathrm{a}}}{\partial z}. \tag{4.58}$$

The above equation can be transformed to

$$\frac{\partial(\rho_{\mathrm{a}} w)}{\partial z} = -\rho_{\mathrm{a}}\left(\frac{\partial u}{\partial x} + \frac{\partial v}{\partial y}\right), \tag{4.59}$$

and therefore vertical wind w at the height z_n is given by

$$w = \frac{\rho_{\mathrm{a}1}}{\rho_{\mathrm{a}n}} w_1 - \frac{1}{\rho_{\mathrm{a}n}} \int_{z_1}^{z_n} \rho_{\mathrm{a}}\left(\frac{\partial u}{\partial x} + \frac{\partial v}{\partial y}\right) \mathrm{d}z, \tag{4.60}$$

where $\rho_{\mathrm{a}n}$ is the atmospheric density at the height z_n, and w_1 is the boundary condition for w at the height z_1 where the above integration starts.

Usually, the following relaxation (iterative) method (e.g., Saad 2003, Chap. 4) is applied to derive the unknown quantities u, v, and w from (4.55), (4.56), and (4.60). First, given the boundary condition of w_1 at the altitude z_1, u_1 and v_1 are calculated at the same altitude by (4.56) and (4.57). Next, a temporary w_2 at the subsequent altitude z_2 is estimated by (4.60) using the above u_1 and v_1. Then, u_2 and v_2 are calculated using the temporary w_2. Next, w_2 is computed again using u_2 and v_2, and the difference δw from the previously obtained temporary value of w is calculated. This procedure is repeated until $|\delta w|$ becomes smaller than the default value. If $|\delta w|$ becomes smaller than the default value, the relaxation is considered to be converged, and the computation at the subsequent altitude is performed. In this way three dimensional distributions of u, v and w are derived in the common volume which is common to the two volumes scanned by the two radars. If the boundary condition is given at the ground that $w = 0$, z is increased upward from $z_1 = 0$. Whereas z is decreased if the calculation is started at the echo top. The errors in the calculation of (4.60) are discussed by e.g., Testud (1982) and Yamada (1997). The difference of accumulated error in between increase and decrease of z is discussed by e.g., Ray et al. (1980) and Nelson and Brown (1987).

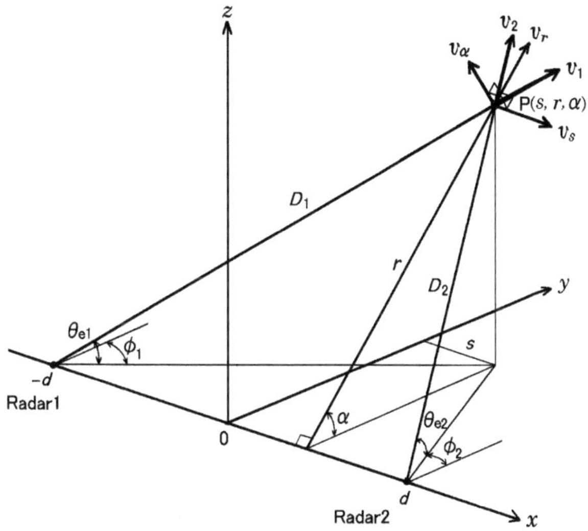

Fig. 4.14 The cylindrical coordinates for the COPLAN method

In this processing, a stationarity is assumed in the object space during the period of one sequence of three dimensional scanning. The total period required to scan the three dimensional space is decided by the antenna azimuthal rotation speed, number of elevation angles, and the dwell time (i.e., the time specified to make a measurement of radial velocity). The dwell time in turn depends on the specified accuracy of the radial velocity measurement. It is necessary to select wisely the time period of observations in the balance with the stationarity of scatterers.

4.3.2 *COPLAN*

In the above-mentioned independent scanning method, a wind vector (u, v, w) is derived from (4.55), (4.56), and (4.60) using two radial velocities obtained independently with the two radars. Here, the wind field is assumed to be quasi-stationary over the time it takes for both radars to scan the volume of interest (e.g., a storm). In the COPLAN method, on the other hand, two out of three components of wind vector in the cylindrical coordinates are directly obtained as follows;

In the cylindrical coordinates shown in Fig. 4.14, two radars 1 and 2 are located at the distances $-d$ and $+d$, respectively, from the coordinate origin along the base line (i.e, the cylinder axis; x-axis). Two radars are synchronously scanned in the plane which contains the x-axis and the observation point P. If the angle between the ground which contains the radar axis and the scanning plane is α, the elevation angle of P from the ith radar $(i = 1,2)$ is θ_{ei}, and the azimuthal angle from the direction which is perpendicular to the radar axis is ϕ_i, the antenna of each radar is scanned restrictedly to satisfy thc following relations (Lhermitte and Miller 1970);

$$\tan\alpha = \frac{\tan\theta_{e1}}{\cos\phi_1} = \frac{\tan\theta_{e2}}{\cos\phi_2}. \tag{4.61}$$

If x coordinate of the observation point P is s, and if the distance between P and the cylinder axis is r, the wind vector at $P(r,\ s,\ \alpha)$ are expressed by the following three components; v_r and v_s which are perpendicular and parallel to the base line, respectively, and v_α which is perpendicular to the scanning plane. They are related to the components $u,\ v,\ w$ in the Cartesian coordinate as (Ray et al. 1975; Testud 1982).

$$v_r = v\cos\alpha + w\sin\alpha, \tag{4.62}$$

$$v_s = u, \tag{4.63}$$

$$v_\alpha = -v\sin\alpha + w\cos\alpha. \tag{4.64}$$

The analysis of the COPLAN method requires to express v_r, v_s, and v_α with the radial velocities observed by the radars and their coordinates. First in (4.55), $Y_1 = Y_2$ and $Z_1 = Z_2$, and thus

$$u = \frac{r_1 v_1 - r_2 v_2}{X_1 - X_2}. \tag{4.65}$$

Next, $X_1 = s + d$, $X_2 = s - d$, and $X_1 - X_2 = 2d$. Thus, (4.63) and (4.64) give

$$v_s = \frac{r_1 v_1 - r_2 v_2}{2d}. \tag{4.66}$$

Similarly, (4.56) becomes

$$v = \frac{1}{2dY_1}[r_2 v_2 (s+d) - r_1 v_1 (s-d) - 2dZ_1 (w - w_T)]. \tag{4.67}$$

Substituting (4.67) into (4.62),

$$v_r = \frac{\cos\alpha}{2dY_1}[r_2 v_2 (s+d) - r_1 v_1 (s-d) - 2dZ_1 (w - w_T)] + w\sin\alpha \tag{4.68}$$

is obtained. Furthermore, using the relations $Y_1 = r\cos\alpha$, and $Z_1 = r\sin\alpha$, (4.68) yields

$$v_r = \frac{r_2 v_2 (s+d) - r_1 v_1 (s-d)}{2rd} + w_T \sin\alpha. \tag{4.69}$$

The component v_α, which is perpendicular to the scanning plane, is obtained by the equation of continuity in the same way as that for the independent scanning mode. The equation of continuity for an incompressible medium in the cylindrical coordinates is expressed by

$$\frac{\partial}{\partial s}(\rho_{\mathrm{a}} v_s) + \frac{1}{r}\frac{\partial}{\partial r}(r\rho_{\mathrm{a}} v_r) + \frac{1}{r}\frac{\partial}{\partial \alpha}(\rho_{\mathrm{a}} v_\alpha) = 0, \tag{4.70}$$

where ρ_{a} is the atmospheric density. If (4.70) is integrated with respect to α,

$$v_\alpha = -\frac{r}{\rho_{\mathrm{a}}}\int_0^\alpha \left[\frac{\partial}{\partial s}(\rho_{\mathrm{a}} v_s) + \frac{1}{r}\frac{\partial}{\partial r}(r\rho_{\mathrm{a}} v_r)\right] \mathrm{d}\alpha \tag{4.71}$$

is derived (Ray et al. 1975; Testud 1982).

Equations (4.66) and (4.69) indicate that v_s and v_r are directly obtained from Doppler velocity components in the scanning plane and the coordinate of P. Thus, it is apparent that scatterers should be quasi-stationary during observations in one scanning plane to calculate v_s and v_r. Therefore, the lapse time of scanning required for the COPLAN method is an order of magnitude smaller than that for the independent scanning method. Although the antenna scanning sequence is rather complicated in the COPLAN method, it is more suitable than the independent scanning method to the observations of the real atmosphere which changes rapidly.

4.3.3 Distance of Two Doppler Radars

When two monostatic Doppler radars 1 and 2 are used to determine the wind field inside storms, the distance between them determines the volume within which accurate wind fields can be calculated. The error variances of the wind components u and v are given by ${\varepsilon_u}^2$ and ${\varepsilon_v}^2$, respectively, and those of Doppler velocity which are inherent in the two radars are ${\varepsilon_1}^2$ and ${\varepsilon_2}^2$, respectively. The relation between these error variances and the between-beam angle β (see Fig. 4.15a) are derived as follows.

If the radial velocities observed with Doppler radars 1 and 2 are v_{r1} and v_{r2}, respectively, the azimuth angles between the direction to P (the point where two radar beams intersect in Fig. 4.15a) and the base line of two radars are θ_1 and θ_2, respectively (these are not shown in Fig. 4.15), and the elevation angle is negligibly small, v_{r1}, v_{r2} and u, v are related by the following equations;

$$v_{r1} = u\sin\theta_1 + v\cos\theta_1, \tag{4.72}$$

$$v_{r2} = u\sin\theta_2 + v\cos\theta_2, \tag{4.73}$$

$$u = \frac{\cos\theta_2}{A}v_{r1} - \frac{\cos\theta_1}{A}v_{r2}, \tag{4.74}$$

$$v = \frac{-\sin\theta_2}{A}v_{r1} + \frac{\sin\theta_1}{A}v_{r2}, \tag{4.75}$$

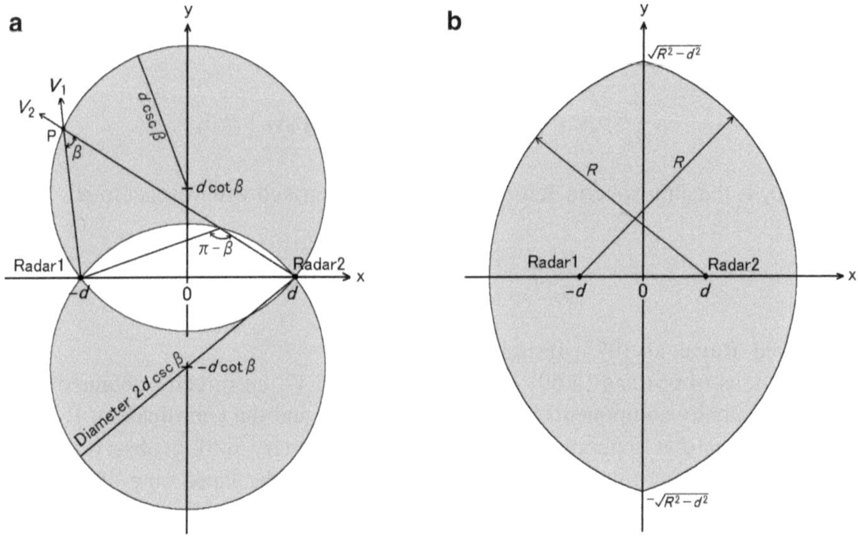

Fig. 4.15 Regarding dual Doppler radar observation by Radar 1 at $-d$ and Radar 2 at d, (**a**) the area where the sum of error variances ${\varepsilon_u}^2 + {\varepsilon_v}^2$ is smaller than a specified value, and (**b**) the area where the distance from two radars is equal to R and less [from Davies-Jones 1979]

where $A = \sin\theta_1 \cos\theta_2 - \cos\theta_1 \sin\theta_2 = \sin(\theta_1 - \theta_2) = \sin(-\beta) = -\sin\beta$. As v_{r1} and v_{r2} are independent, the error variances of u and v, ${\varepsilon_u}^2$ and ${\varepsilon_v}^2$, respectively, become

$$ {\varepsilon_u}^2 = \frac{\cos^2\theta_2}{A^2}{\varepsilon_1}^2 + \frac{\cos^2\theta_1}{A^2}{\varepsilon_2}^2, \tag{4.76} $$

$$ {\varepsilon_v}^2 = \frac{\sin^2\theta_2}{A^2}{\varepsilon_1}^2 + \frac{\sin^2\theta_1}{A^2}{\varepsilon_2}^2, \tag{4.77} $$

$$ {\varepsilon_u}^2 + {\varepsilon_v}^2 = \frac{{\varepsilon_1}^2 + {\varepsilon_2}^2}{\sin^2\beta}, \tag{4.78} $$

and finally the following relation is derived (Lhermitte and Miller 1970).

$$ \frac{{\varepsilon_u}^2 + {\varepsilon_v}^2}{{\varepsilon_1}^2 + {\varepsilon_2}^2} = \csc^2\beta. \tag{4.79} $$

In Fig. 4.15a, the contour where the sum of error variances ${\varepsilon_u}^2 + {\varepsilon_v}^2$ becomes a constant value is the two circular arcs which have two radars on their circumferences. The area where this sum becomes smaller than the constant value is the shaded region surrounded by the two circular arcs, where the crossing angles of two radar beams are β and $\pi - \beta$, respectively. The sum ${\varepsilon_u}^2 + {\varepsilon_v}^2$ becomes minimum on the circumference of the circles whose diameter is determined by the base line of the two radars. The sum of the error variance increases both inside and outside the circumference. On the other hand, the resolution which depends on the range is $R\theta$. where the distance from radar and the beamwidth are R and θ [rad], respectively. The shaded region in Fig. 4.15b indicates the area where the distance from the two radars is less than or equal to R, i.e., the area where the resolution becomes geometrically less than or equal to $R\theta$. Note that Doppler velocity data collected with equal spacing of radar range and beam position result in dual Doppler radar fields having no data location coincidence. Thus, interpolation is required to locate data on a common grid (Doviak and Ray 1976). The interpolation volume is usually much larger than the above value.

Because the resolution deteriorates with increasing distance, it generally worsens the error variance of Doppler velocities by vertical shear of horizontal wind to expand the distance between two radars. Actually, the common area of Fig. 4.15a, b becomes the effective observation area with two Doppler radars from the points of view of error variance and resolution to be allowed. The more quantitative evaluations are discussed in detail by Davies-Jones (1979) and others. In the above discussion, the beamwidths of two radars are assumed to be equal. However, the similar discussion can be also extended when their beamwidths are not identical.

The lapse time required to conduct a sequence of observations in three dimensional space[3] should be decided by taking the characteristic time of the wind field into consideration. The narrower the interval of elevation angles between scanning planes is, the smaller the error variance of vertical wind. Hence, it is desirable to set elevation scanning planes as many as possible, whereas the time to scan the volume increases. Thus, there is a tradeoff between exceeding the characteristic time of the wind field and the accuracy of the vertical wind calculation. For example, Ushiyama et al. (2005) discusses an idealized dual Doppler radar observation based on the numerical simulation for tropical cloud systems. Usually, radars are operated according to a set of predefined coverage patterns. On the contrary, Wang et al. (2008) discusses dynamic scan strategy for dual Doppler radar observation so that the coverage patterns can be optimized for acquisition of data in a networked radar system.

4.3.4 Wind Velocity Observations with Three or More Radars

For wind observations by two monostatic Doppler radars, three components of wind vector (u, v, w) are obtained from (4.55) and (4.56), when supplemented by the equation of continuity (4.60), and using the radial wind velocities which are measured by each radar. If three radars are implemented, wind vector can be uniquely determined from the radial wind velocities, which are measured by these radars, and the equation of continuity is not necessary. In this case, we can

[3] The scan strategies or volume coverage patterns (VCPs) of typical monostatic radars will be discussed in Sect. 8.1.4.

obtain three equations which correspond to (4.53) simultaneously at one observation point, and (u, v, w) is derived by computing the inverse matrix for the matrix with coefficients of the above three equations. If the continuity equation is used and there are three radars, the terminal velocity can then be retrieved. When more than three radars are used, there are more number of equations than that of the unknowns, and thus they are determined by the least squares method. Multi-beam observations by one single atmospheric radar to be described later in Sect. 7.2.1 are regarded equivalent to the above-mentioned single beam observations by multiple radars.

Chapter 5
Reception and Processing of Signals

Signals returned from scatterers are received by antenna and sent to receiver through feeders such as waveguide and coaxial cable. At the receiver, signals are discriminated from noise, and various signal processing is performed to obtain the desired signals. In recent years, conversion of analog to digital signals tends to be done in the earlier stage than before, where various filter operations for discrimination of received signals are widely conducted by digital processing. In this chapter, the characteristics of radar signals, the receiver system, and processing techniques of discrete signals will be discussed.

5.1 Receiver Sensitivity

The received power from cloud and precipitation particles which are the major scatterers for meteorological radars, and that from Bragg scatterers which are major ones for atmospheric radars is often as small as the noise power or much smaller. Thus, it is indispensable to know the characteristics of internal and external noise which affect receiver sensitivity.

5.1.1 Noise Power

Here, we assume that undesired signals like ground clutter have been properly removed or suppressed,[1] and discuss the characteristics of noise power that is originated by internal and external sources. Noise of radar includes receiver thermal noise, which is of internal origin, and noises due to thermal emissions from the

[1]Removal or suppression of the undesired signals will be discussed later in Sect. 8.4.2.

S. Fukao and K. Hamazu, *Radar for Meteorological and Atmospheric Observations*,
DOI 10.1007/978-4-431-54334-3_5,

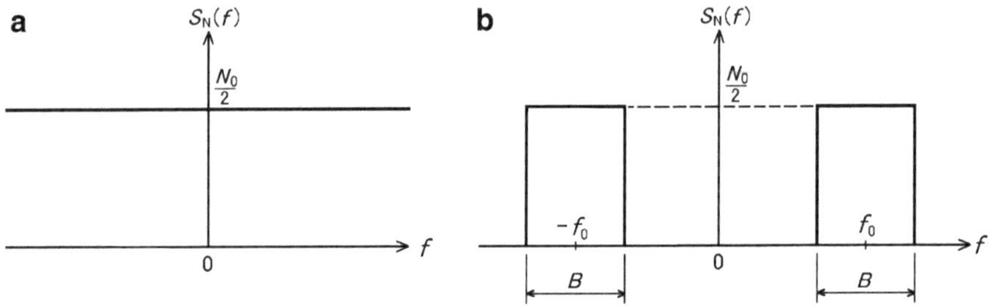

Fig. 5.1 Power spectral density of white noise (**a**) spreading over the whole frequency band and (**b**) existing within the bandwidth B

sun, galactic radio emission, atmospheric emission, clouds,[2] and ground, which are of external origin. Thermal emission is called white noise because power spectral density is almost uniform over the whole frequency band as shown in Fig. 5.1a. The radar receiver noise can be considered to be the white noise. Such a noise would have the infinite power if the receiver bandwidth were infinite but all practical receivers have finite bandwidth. For finite bandwidth receivers, the noise power is typically white over the bandwidth B (Fig. 5.1b) of the receiver. If the noise temperature of the noise source is T [K], the power spectral density $S_N(f)$ [W Hz^{-1}] is expressed by

$$S_N(f) = \frac{N_0}{2} = \frac{kT}{2} \qquad -\infty < f < \infty, \tag{5.1}$$

where the Boltzmann's constant, k, is 1.38×10^{-23} [J K^{-1}]. Practically, the power spectral density of noise is defined in the positive frequency domain whereas spectrum analysis deals with it in both positive and negative frequencies, and consequently the coefficient $1/2$ is needed here. When defining it only in the positive frequency, the power spectral density of noise is expressed as $S_N(f) = N_0 = kT$ $(0 < f < \infty)$. If the noise is received by receiver with bandwidth B as shown in Fig. 5.1b, the input power to the receiver, or the noise power P_N becomes

$$P_N = \frac{N_0}{2} \times 2B = kTB. \tag{5.2}$$

To estimate the noise power, the base level is generally set in the stage before the received signal is converted from radio frequency (rf) to intermediate frequency (IF). The bandwidth at the base level is shown by B. The optimal value of B is equal to the bandwidth of the matched filter as shown later. However, it is generally set, from a

[2]Two sources of radiation from cloud and ground are often observed with weather radars (e.g., Fang et al. 2004).

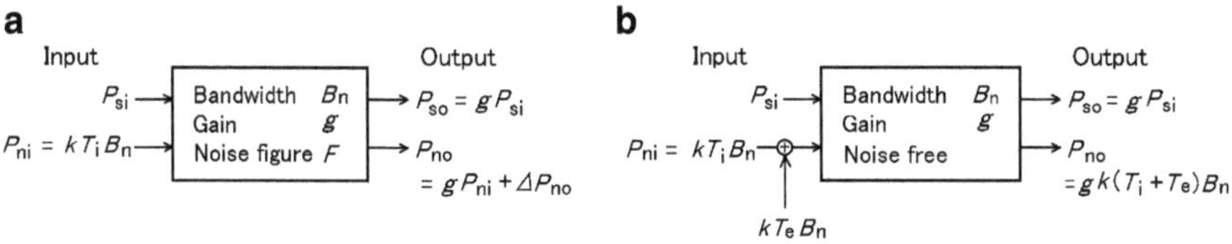

Fig. 5.2 (**a**) Input power and output power of a noisy circuit. (**b**) Equivalent noise-free circuit to (**a**)

practical standpoint, to the value which is near the reciprocal of the transmitted pulse width (for example, if the pulse width is 1 μs, the bandwidth is approximately 1 MHz).[3]

Noise Figure

Ulaby et al. (1981, p. 347) discussed in detail characteristics of noise. The following discussion in this subsection is based upon their description. The noise figure (NF) F of a system (or circuit) is defined by

$$F \equiv \frac{P_{si}/P_{ni}}{P_{so}/P_{no}}, \tag{5.3}$$

where P_{si} and P_{ni} are the input signal power and the input noise power, respectively, while P_{so} and P_{no} are the output signal power and the output noise power, respectively. Furthermore, the actual input noise power available at the input terminal of the circuit, P_{ni}, is given as (5.2) by

$$P_{ni} = kT_iB_n, \tag{5.4}$$

where T_i is the input noise temperature, and B_n is the noise bandwidth.

If P_{si} is inputted to a noisy circuit of gain g, where the lower case symbol g denotes gain values in a pure number, and NF is F as shown in Fig. 5.2a, P_{so} and P_{no} become

$$P_{so} = gP_{si}, \tag{5.5}$$

$$P_{no} = gP_{ni} + \Delta P_{no}, \tag{5.6}$$

respectively, where ΔP_{no} is the noise power generated by the noisy circuit. Substituting (5.4) to (5.6) into (5.3), F becomes

$$F = 1 + \frac{\Delta P_{no}}{gkT_iB_n}. \tag{5.7}$$

[3]The actual receiver bandwidth will be discussed in Sect. 8.3.5.

From (5.7), ΔP_{no} is given by

$$\Delta P_{\mathrm{no}} = (F-1) g k T_{\mathrm{i}} B_{\mathrm{n}}, \tag{5.8}$$

and the output noise power P_{no} is expressed as

$$P_{\mathrm{no}} = g k T_{\mathrm{i}} B_{\mathrm{n}} + (F-1) g k T_{\mathrm{i}} B_{\mathrm{n}} = F g k T_{\mathrm{i}} B_{\mathrm{n}}. \tag{5.9}$$

The above equation shows that the noise performance of the noisy circuit in Fig. 5.2a is equivalent to that of an ideal noise-free circuit with input noise temperature equal to $F T_{\mathrm{i}}$. An equivalent noise-free circuit to Fig. 5.2a is shown in b. Defining the equivalent input noise temperature T_{e} in b which is referred to the input terminal of the circuit, ΔP_{no} and P_{no} are expressed as

$$\Delta P_{\mathrm{no}} = 7 g k T_{\mathrm{e}} B_{\mathrm{n}}. \tag{5.10}$$

$$P_{\mathrm{no}} = g k (T_{\mathrm{i}} + T_{\mathrm{e}}) B_{\mathrm{n}}, \tag{5.11}$$

respectively. Furthermore, from (5.8) and (5.10),

$$T_{\mathrm{e}} = (F-1) T_{\mathrm{i}}, \tag{5.12}$$

$$F = 1 + \frac{T_{\mathrm{e}}}{T_{\mathrm{i}}}, \tag{5.13}$$

are derived. Both quantities, F and T_{e}, describe the same property of the circuit. To measure F in an actual circuit, it is necessary to specify the value of T_{i}, thus the definition of noise figure has been standardized by choosing T_{i} to be $T_0 = 290\,\mathrm{K}$ (room temperature) (Ulaby et al. 1981, p. 353).

Noise of Cascaded Circuits

If noisy two circuits such as amplifiers are connected in cascade, the output noise powers of circuit 1 and 2 are given using (5.11) as

$$P_{\mathrm{no1}} = g_1 k (T_{\mathrm{i}} + T_{\mathrm{e1}}) B_{\mathrm{n}}, \tag{5.14}$$

$$\begin{aligned} P_{\mathrm{no2}} &= g_2 k [g_1 (T_{\mathrm{i}} + T_{\mathrm{e1}}) + T_{\mathrm{e2}}] B_{\mathrm{n}} \\ &= g_1 g_2 k \left(T_{\mathrm{i}} + T_{\mathrm{e1}} + \frac{T_{\mathrm{e2}}}{g_1} \right) B_{\mathrm{n}}, \end{aligned} \tag{5.15}$$

respectively, where T_{e1} and T_{e2} are the equivalent input noise temperatures of circuits 1 and 2, respectively. Using (5.13) and putting $T_{\mathrm{i}} = T_0$, (5.15) is rewritten as

$$P_{\mathrm{no2}} = g_1 g_2 k T_0 B_{\mathrm{n}} \left(1 + F_1 - 1 + \frac{F_2 - 1}{g_1} \right), \tag{5.16}$$

where F_1 and F_2 are the noise figures of circuits 1 and 2, respectively. By generalizing this 2-stage circuit to a system consisting of n stages, the equivalent input noise temperature of the overall system, T_e, becomes

$$T_e = T_{e1} + \frac{T_{e2}}{g_1} + \frac{T_{e3}}{g_1 g_2} + \cdots + \frac{T_{en}}{g_1 g_2 \cdots g_{n-1}}. \tag{5.17}$$

In the same way, the noise figure of the overall system is given by

$$F = F_1 + \frac{F_2 - 1}{g_1} + \frac{F_3 - 1}{g_1 g_2} + \cdots + \frac{F_{n-1}}{g_1 g_2 \cdots g_{n-1}}. \tag{5.18}$$

Equations (5.17) and (5.18) shows that if the gain of the first stage, i.e., low noise amplifier (LNA) in radar receiver, is sufficiently high to amplify the equivalent noise power at its input terminal to levels far above the additional noise due to subsequent amplifiers, then the noise contributed by them can be ignored (Doviak and Zrnić 2006, p. 55).

Noise of Lossy Circuit

Noise power internally generated by a lossy circuit with the loss factor l will be discussed, where l is defined[4] as $l \equiv P_{in}/P_{out}$ where P_{in} and P_{out} are the input power and output power of the circuit, respectively. If the physical temperature of the lossy circuit is T_p, the noise power P_{no} that appears at the output port of the circuit is given by (Dicke 1946; Ulaby et al. 1981, p. 353)

$$P_{no} = kT_p B_n, \tag{5.19}$$

where B_n is the noise bandwidth of the circuit. Figure 5.3a shows a single-stage circuit with loss factor l. In this circuit, P_{no} consists of the noise power flowing into the input terminal toward the output terminal, attenuated by l as it passes through the circuit, plus noise generated internally by the circuit and denoted ΔP_{no} as (Ulaby et al. 1981, p. 353)

$$P_{no} = \frac{1}{l} kT_p B_n + \Delta P_{no}. \tag{5.20}$$

Substituting (5.19) into (5.20), ΔP_{no} is expressed as

$$\Delta P_{no} = kT_p \left(1 - \frac{1}{l}\right) B_n. \tag{5.21}$$

---

[4] The lower case symbol l denotes loss value in a pure number, whereas the upper case symbol L denotes it in dB. For instance, $l \simeq 1.58$ for circuit with L of 2 dB loss.

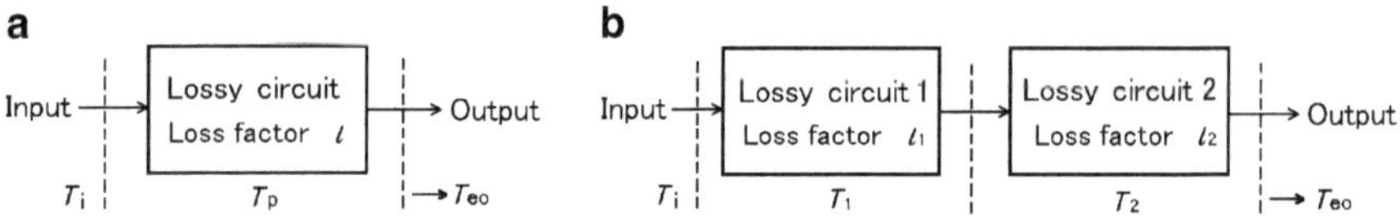

Fig. 5.3 Noise temperature of lossy circuits. (**a**) One- and (**b**) two-stage circuits in cascade

This internally generated noise power at the lossy circuit is equal to the noise power that would appear at the output of an equivalent noise-free loss circuit with an input noise power P_e given by

$$P_e = l\Delta P_{no} = kT_p(l-1)B_n, \tag{5.22}$$

and the equivalent input noise temperature T_e is expressed as

$$T_e = T_p(l-1). \tag{5.23}$$

Thus, the effective output noise temperature T_{eo} becomes (Ulaby et al. 1981, p. 353)

$$T_{eo} = \frac{T_e}{l} = T_p\left(1 - \frac{1}{l}\right). \tag{5.24}$$

From (5.12) and (5.24), the noise figure of a lossy circuit is expressed as

$$F = 1 + (l-1)\frac{T_p}{T_0}, \tag{5.25}$$

and $F = l$ for $T_p = T_0$.

If the noise temperature of input signal is T_i, the noise temperature of this circuit is expressed by Ulaby et al. (1981, p. 355)

$$T_e = T_i\frac{1}{l} + T_p\left(1 - \frac{1}{l}\right). \tag{5.26}$$

In the same way, when a two-stage lossy circuit consists of cascade connection as shown in Fig. 5.3b, the equivalent noise temperature of the circuit is given by

$$T_e = T_i\frac{1}{l_1 l_2} + T_1\frac{1}{l_2}\left(1 - \frac{1}{l_1}\right) + T_2\left(1 - \frac{1}{l_2}\right). \tag{5.27}$$

5.1.2 Receiver Noise

Total noise of a receiver is the sum of external noise which enters through the antenna from the surrounding environment, antenna noise which is generated in the antenna, feeder noise which is caused in the path from the antenna to the receiver, and receiver noise which is generated inside the receiver. The noise generated inside receiver includes the thermal agitation due to the electric current which flows through the electronic circuit. Noise up to the receiver input is caused by lossy circuits (i.e., the feeder lines, antenna, etc.). On the other hand, the amplifier is also a noise source with loss, and therefore the receiver, a cascaded connection of multiple amplifiers and lossy circuits, is a noise source.

Sky Noise

The major external noises are cosmic noise, atmospheric noise, and atmospheric absorption. Cosmic noise is radiated from the Galaxy, extragalactic nebulae, discrete radio sources and so on. Therefore, the noise intensity depends on the direction of the antenna beam; it becomes the maximum in the direction of the Galactic center while the minimum in the direction of its pole. The sky noise temperature shown in Fig. 5.4 is expressed by temperature due to cosmic and atmospheric radiation[5] (Skolnik 1990, p. 2.29; Doviak and Zrnić 2006, p. 57). Cosmic noise predominates from very high frequency (VHF) to several MHz band, and increases at the gradient of the approximately 2.5th power toward the lower frequency side. The average noise temperature becomes 10,000–15,000 K in the frequency band between 40 and 50 MHz (Alvarez et al. 1997; Maeda et al. 1999). Cosmic noise decreases rapidly in the higher frequency band. On the other hand, the atmospheric absorption becomes dominant at frequencies higher than 10 GHz. Because the noise power due to the atmospheric absorption depends on propagation losses, which the radar beam passes through, the noise temperature becomes the maximum in the horizontal direction, and the minimum in the direction toward zenith. The peaks that are seen around 22 and 60 GHz are caused by the resonance absorption of water vapor and oxygen, respectively.

Atmospheric noise due to lightning predominates in the frequency band below high frequency (HF; 3–30 MHz). Solar noise is also a strong noise source (noise temperature is about 6,000 K at the quiet period), and sometimes increases by several figures with solar flares and so on. However, solar noise does not become a serious problem except if the radar beam points in the direction or vicinity of the sun. As clearly shown in Fig. 5.4, cosmic noise is not negligible for atmospheric radars which generally use operational frequency less than 1,000 MHz. On the other

[5]The warm atmosphere emits thermal radiation and leave out the absorption as a noise.

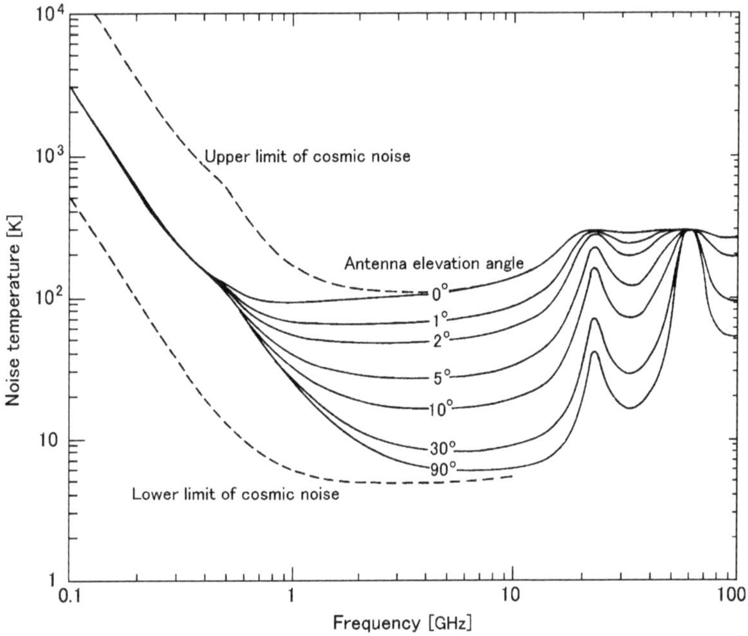

Fig. 5.4 Sky noise temperature, due to cosmic and atmospheric radiation, of an idealized antenna (lossless, no earth-direction side lobes) located on the earth's surface [from Skolnik 1990, p. 2.29; Doviak and Zrnić 2006, p. 57]

hand, cosmic noise temperature becomes around 100 K or less for meteorological radars which use operational frequency of 3–9 GHz.

Figure 5.5 shows the noise generation elements through the path of received signals in actual observations.[6] As shown in the figure, a ground noise temperature contribution should be accounted because part of the total antenna pattern is directed toward the ground. Meanwhile, the apparent sky noise temperature, T_s', given by Fig. 5.4, must be reduced somewhat because part of the total antenna pattern is not then at the sky. The reduction factor is $(1 - T_{ag}/T_g)$, where T_{ag} is the ground noise temperature contribution to the antenna temperature and T_g is the effective noise temperature of the ground. As a result, effective sky noise temperature T_s is expressed using T_s', the reduction factor, and T_g as (Skolnik 1990, p. 2.30)

$$T_s = T_s' \left(1 - \frac{T_{ag}}{T_g}\right) + T_{ag}. \tag{5.28}$$

[6]The temperature of the clouds and precipitations that produce losses along the propagation path is not discussed here. These losses will be discussed in Sect. 6.3 as attenuations of radio waves in the atmosphere.

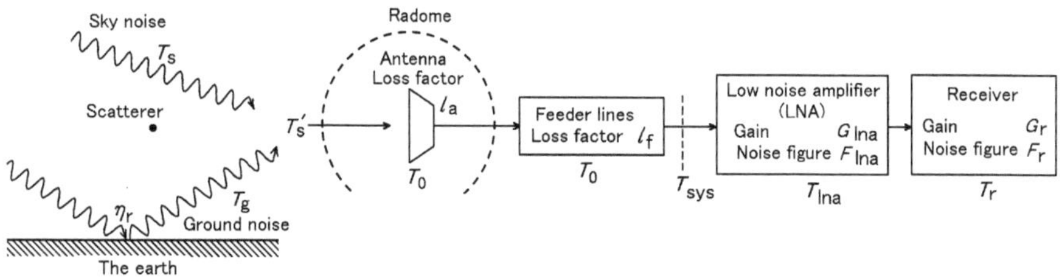

Fig. 5.5 Noise generation elements through the path of received signals in actual observations

As suggested conventional values,[7] $T_{ag} = 36\,\text{K}$, and $T_g = T_0 = 290\,\text{K}$ (equal to room temperature), respectively (Skolnik 1990, p. 2.30). For example, if T_s' is 100 K from Fig. 5.4, the effective sky noise temperature T_s becomes 124 K.

Antenna Noise

At the antenna, thermal noise which is caused by the lossy circuit in the antenna, and in the radome if equipped, is added to the effective sky noise. Thus, noise temperature T_a at the received signal output terminal of the antenna becomes the sum of effective sky noise temperature T_s with antenna loss factor l_a and newly added noise temperature through antenna with loss factor l_a at ambient temperature T_0. From (5.26), T_a is given by

$$T_a = T_s \frac{1}{l_a} + T_0 \left(1 - \frac{1}{l_a}\right). \tag{5.29}$$

Noise Generated in the Feeder Lines and Receiver Circuits

Thermal noise is additionally generated because of the power attenuation through the feeder lines such as waveguides and coaxial cables from antenna to the feeder lines output terminal, i.e., the low noise amplifier (LNA) input terminal which is assumed to be the reference point. Effective noise temperature T_f of the feeder lines output at the point is expressed as the same manner to (5.29) by

$$T_f = T_a \frac{1}{l_f} + T_0 \left(1 - \frac{1}{l_f}\right), \tag{5.30}$$

[7]The value $T_{ag} = 36\,\text{K}$ would result if a $T_g = T_0 = 290\,\text{K}$ earth were viewed over a π-steradian solid angle by sidelobes and backlobes averaging 0.5 gain ($-3\,\text{dB}$). These sidelobes are typical of a "good" radar antenna but not a ultralow-noise variety (Skolnik 1990, p. 2.30).

where l_f is the loss factor of feeder lines. The antenna and the feeder lines are not heated by themselves, and thus the ambient temperature is equal to their physical temperature T_0. Meanwhile, at the reference point, the equivalent input noise temperature T_r of the LNA and the receiver is given from (5.12), assuming that they are noise free circuits, as

$$T_r = T_0(F-1), \tag{5.31}$$

where F is NF at the reference point, and expressed from (5.18) as

$$F = F_{lna} + \frac{F_r - 1}{g_{lna}}, \tag{5.32}$$

where F_{lna} and F_r are the NF of the LNA and that of the receiver, respectively.

System Noise Temperature

The system noise temperature, which is a significant factor to determine the sensitivity of a radar, is generally defined at the reference point, i.e., the input terminal of the LNA. The system noise temperature T_{sys} at the reference point is given by

$$T_{sys} = T_f + T_r. \tag{5.33}$$

Substituting (5.31) and (5.32) into (5.33) and using (5.29) and (5.30), T_{sys} is expressed as

$$\begin{aligned} T_{sys} &= T_a\frac{1}{l_f} + T_0\left(1 - \frac{1}{l_f}\right) + T_0\left(F_{lna} - 1 + \frac{F_r - 1}{g_{lna}}\right) \\ &= (T_s - T_0)\frac{1}{l_a l_f} + T_0\left(F_{lna} + \frac{F_r - 1}{g_{lna}}\right). \end{aligned} \tag{5.34}$$

Furthermore, as discussed in Sect. 5.1.1, if the gain of the LNA, g_{lna}, is sufficiently high, the noise contribution by the following circuits can be ignored, and the system noise temperature becomes

$$T_{sys} = (T_s - T_0)\frac{1}{l_a l_f} + T_0 F_{lna}. \tag{5.35}$$

From the above, it is considered that the noise power is added to the reference point by the noise source of T_{sys}, where the attenuation of the received power in the feeder lines from antenna to receiver is included in T_{sys}. Thus, the noise power of a receiver becomes

$$P_N = kT_{sys}B, \tag{5.36}$$

where B is the receiver bandwidth. The ratio of received signal power from scatterers to noise power is called the signal-to-noise ratio (SNR). Generally, the SNR is the parameter which has a significant influence on probability for signal detection and accuracy for estimation of physical parameters.

Example of Noise Power

The MU radar, a representative example of the MST (mesospheric–stratospheric–tropospheric) radar of which detail will be shown in Sect. 10.2.3, operates at the frequency of 46.5 MHz, and the apparent sky noise is about 10,000 K. Thus, the effective sky noise at the antenna input port is $10000(1-36/290)+36 \simeq 8795$ K. Assuming NF of 5 dB ($\simeq 3.16$), antenna loss of 2 dB ($\simeq 1.58$), and feeder lines loss of 8 dB ($\simeq 6.31$), the system noise temperature T_{sys} at the receiver becomes $(8795-290) \times 1/(1.59 \times 6.31)+290 \times 3.16 \simeq 1768$ K.

5.2 Receiver System

5.2.1 *Matched Filter*

Noise power increases and the SNR deteriorates if the receiver bandwidth B is wider than that of the bandwidth of the received signals. On the other hand, if B is narrower than the required bandwidth to receive an relatively undistorted echo from a point scatterer, noise power decreases and range resolution worsens. However, the signal power decreases too, causing distortion and lack of information of signals. If it is constrained that the resolution to be a constant, the SNR is an optimum when the filter response is matched to the transmitted pulse (Doviak and Zrnić 2006, p. 83). Therefore, it is crucial to determine B by keeping the balance between the bandwidth which the desired received signals occupy and that which makes SNR best. The optimum solution for this subject is the filter bandwidth which maximizes SNR to the received signals for pulse radar, and the filter is called matched filter.

Response of Matched Filter

The filter characteristic (transfer function) which maximizes SNR are discussed in this paragraph (Nathanson 1991, pp. 355–360). If the time dependence of the signal at the receiver input is $s(t)$, the total energy E which is included in $s(t)$ is expressed as

$$E=\int_{-\infty}^{\infty} s^2(t)\mathrm{d}t=\int_{-\infty}^{\infty} |S(f)|^2\mathrm{d}f, \tag{5.37}$$

where E should be finite. The complex amplitude spectrum $S(f)$ is the Fourier transform[8] of $s(t)$ and given by

$$S(f) = \int_{-\infty}^{\infty} s(t)\mathrm{e}^{-j2\pi ft}\mathrm{d}t. \tag{5.38}$$

The filter output signal $s_{\mathrm{O}}(t)$ is expressed by $S(f)$ and the filter transfer function $H(f)$ as

$$s_{\mathrm{O}}(t) = \int_{-\infty}^{\infty} S(f)H(f)\mathrm{e}^{j2\pi ft}\mathrm{d}f. \tag{5.39}$$

Meanwhile, the power spectral density (or simply power spectrum)[9] of output signal for linear filter becomes a product between the power spectrum of filter input signal and the squared absolute value of filter transfer function. Thus, using noise power spectral density $N_0/2$ from (5.1), the noise power of the filter output P_{N} is given by

$$P_{\mathrm{N}} = \frac{N_0}{2}\int_{-\infty}^{\infty} |H(f)|^2\mathrm{d}f. \tag{5.40}$$

The optimum condition for receiver filter, i.e., the transfer function which maximizes SNR is one which maximizes the ratio of the maximum value of the signal to noise power at the output port. From (5.39) and (5.40), the value of SNR is given by Skolnik (2001, p. 279)

$$\frac{|s_{\mathrm{O}}(t_{\mathrm{M}})|^2}{P_{\mathrm{N}}} = \frac{\left|\int_{-\infty}^{\infty} S(f)H(f)\,\mathrm{e}^{j2\pi ft_{\mathrm{M}}}\,\mathrm{d}f\right|^2}{\frac{N_0}{2}\int_{-\infty}^{\infty}|H(f)|^2\mathrm{d}f}, \tag{5.41}$$

where t_{M} denotes the time t at which the output $|s_{\mathrm{O}}(t)|^2$ is a maximum. The function $|H(f)|$ which maximizes (5.41) is that of the matched filter, and is derived by applying Schwarz's inequality to two complex functions $a(x)$ and $b(x)$ as follows.

$$\left|\int_{-\infty}^{\infty} a(x)b(x)\mathrm{d}x\right|^2 \leq \int_{-\infty}^{\infty}|a(x)|^2\mathrm{d}x\int_{-\infty}^{\infty}|b(x)|^2\mathrm{d}x. \tag{5.42}$$

Substituting $a(x)$ and $b(x)$ with $H(f)$ and $S_{\mathrm{v}}(f)\mathrm{e}^{j2\pi ft_{\mathrm{M}}}$, respectively, and using (5.37), (5.41) and (5.42),

$$\frac{|s_{\mathrm{O}}(t_{\mathrm{M}})|^2}{P_{\mathrm{N}}} \leq \frac{2E}{N_0} \tag{5.43}$$

[8]The Fourier transform will be described in Sect. 5.4.1.

[9]Power spectrum will be defined in Sect. 5.4.2.

is obtained. The equality condition applies only when the ratio of $a(x)$ to $b^*(x)$ is constant, or $H(f)$ and $S(f)$ are in the following relation (Skolnik 2001, p. 280).

$$H(f) = KS^*(f)\,\mathrm{e}^{-j2\pi f t_\mathrm{M}}, \tag{5.44}$$

where K is an arbitrary gain constant, and the superscript * denotes complex conjugate. The filter whose transfer function has the characteristics of (5.44) is the matched filter, and (5.43) becomes

$$\frac{|s_\mathrm{O}(t_\mathrm{M})|^2}{P_\mathrm{N}} = \frac{2E}{N_0}, \tag{5.45}$$

where $|s_\mathrm{O}(t_\mathrm{M})|^2$ is the peak power of signal, and P_N is noise power. As shown in (5.45), the ratio of the peak signal power to noise power for the matched filter output depends only on the spectral densities of signal power which is the infinite integral of power spectra of receiver input and noise. It should be noted that the ratio does not depend on the signal waveform. The time domain expression for $H(f)$ is the inverse Fourier transform[10] of (5.44), and expressed as

$$h(t) = K\int_{-\infty}^{\infty} S^*(f)\,\mathrm{e}^{-j2\pi f(t_\mathrm{M}-t)}\mathrm{d}f = Ks^*(t_\mathrm{M}-t). \tag{5.46}$$

Equation (5.46) indicates that the impulse response of a matched filter is the same as that for delayed input signal on the reversed time axis. Here, it should be noted that a matched filter maximizes SNR of output signal; however, it does not reproduce the waveform of input signal.

Matched Filter and Correlation Function

The correlation function and the impulse response of a linear system are utilized to explain (5.46). Here, we simplify $K = 1$. The cross-correlation function of two different signals $x_1(t)$ and $x_2(t)$ for finite delay time τ is defined as follows for signals whose energy is finite as shown by

$$R_{x_1x_2}(\tau) = \int_{-\infty}^{\infty} x_1(t)x_2{}^*(t-\tau)\mathrm{d}t. \tag{5.47}$$

In signal analysis, the integration in the form of (5.47) is often called correlation function. However, it is applicable only to the signal that the total energy of (5.37) is finite. In general, correlation function indicates the one to be described in Sect. 5.4.2, and thus it is necessary not to confuse both.

[10]The inverse Fourier transform will be discussed later in Sect. 5.4.1 as the Fourier transform.

When the input signal $y_{\mathrm{i}}(t)$ is applied to the filter whose impulse response is $h(t)$, the output signal $y_{\mathrm{o}}(t)$ is obtained as the response of the linear system

$$y_{\mathrm{o}}(t) = \int_{-\infty}^{\infty} y_{\mathrm{i}}(\tau)h(t-\tau)\mathrm{d}\tau. \tag{5.48}$$

If the filter is matched filter, $h(t) = s^*(t_{\mathrm{M}} - t)$ from (5.46), and consequently (5.48) is expressed as

$$y_{\mathrm{o}}(t) = \int_{-\infty}^{\infty} y_{\mathrm{i}}(\tau)s^*[\tau - (t - t_{\mathrm{M}})]\mathrm{d}\tau. \tag{5.49}$$

Using the relation of (5.47), (5.49) is written as

$$y_{\mathrm{o}}(t) = R_{ys}(t - t_{\mathrm{M}}). \tag{5.50}$$

The above equation means that the matched filter output $y_{\mathrm{O}}(t)$ is given by the cross-correlation function of input signal $y_{\mathrm{i}}(t)$ and the receiver input pulse waveform $s(t)$. In the case that the receiver noise can be ignored, $y_{\mathrm{O}}(t)$ is equivalent to autocorrelation function of receiver input pulse waveform.

5.2.2 *Frequency Conversion and Phase Measurement*

Generally, the received signal is detected as continuous signal. There are two methods for detection; one is the envelope detection that treats only amplitude of signals, and the other is synchronous detection which treats both amplitude and phase of signals. In synchronous detection, it is necessary to keep the phase information of the transmitted signal until the scattered signal is received. Consequently, the system configuration for synchronous detection becomes more complicated than that for amplitude detection. The synchronous detection, however, enables to obtain the power spectrum which contains the wind velocity information in addition to the echo intensity information. Therefore, recent meteorological and atmospheric radars almost universally adopt synchronous detection. The choice of the carrier frequency for practical operational meteorological and atmospheric radars depends principally on the type of scatterers to be detected (e.g., Bragg, insects, cloud particles, precipitation particles, etc.), the needed angular resolution, and cost. The carrier frequencies range from several MHz to 100 GHz (Table 1.1). Because signal processing cannot, for most meteorological and atmospheric radars, be performed on signals at the carrier frequency (especially if the carrier frequency f_0 is above 1 GHz), the received signal (at the carrier frequency f_0) is down converted to a lower IF from several MHz to approximately 100 MHz. The phase information of the received signal whose carrier frequency is down converted to an IF frequency is preserved, and thus the phase of the transmitted pulse can be compared to the phase

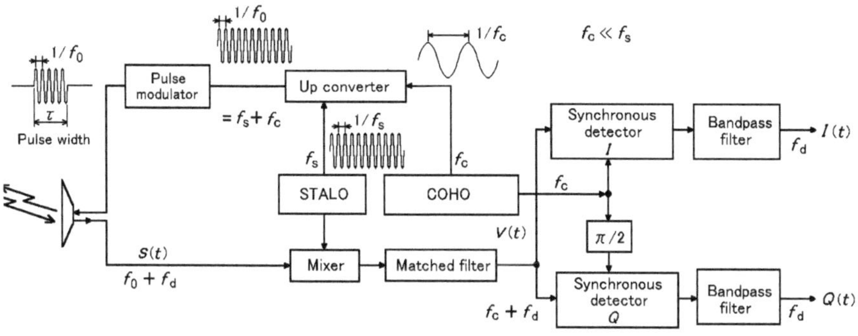

Fig. 5.6 The process for frequency conversion of radar received signals and phase measurements. The transmitted carrier frequency is denoted by f_0, while the local (LO) and the IF by f_s and f_c, respectively

of the received IF signal. The typical scheme to compare the phase of the transmitted pulse and that of the received signal is to generate both the carrier frequency of transmitted pulse and receiver frequency necessary for phase measurement[11] from a common standard signal, i.e., the coherent oscillator (COHO) in Fig. 5.6. In general, the desired carrier frequency of transmitted pulse is generated by up-converting the COHO frequency to f_0, whereas, the COHO signal is used as is for phase measurement.

A basic process for the frequency conversion of received signals and phase measurements is illustrated in Fig. 5.6. The local frequency (LO) signal with frequency f_s is generated by a highly stabilized local oscillator (STALO). The LO signal is mixed with the IF signal with frequency f_c which is generated by the COHO, and becomes $f_0 = f_s + f_c$ where f_0 is the transmitted frequency. The up converted continuous wave (CW) at f_0 is pulse modulated and amplified by amplifying devices such as transistor and klystron, and sent to antenna. Here the CW is not the "carrier" before pulse modulation because it is not carrying any information. It becomes "carrier" only after pulse modulation.

The frequency of the received signal from a moving scatterer is shifted by an amount f_d due to Doppler effect and becomes $f_0 + f_d$. When this signal is mixed with LO frequency f_s, an IF signal with $f_c + f_d$ that possesses the phase information of the scatterer is generated. Subsequently, the IF signal is mixed with the COHO frequency to detect phase, and Doppler frequency f_d is obtained by measuring the change of the echo's phase angle from one transmitted pulse to the next as was discussed in Sect. 4.1.1. From (4.11), it is derived that the maximum measurable Doppler frequency f_{dmax} is the half of the pulse repetition frequency (PRF), thus the values of f_d is between 0 and around several hundred Hz to kHz depending on the PRF of the radar.

[11]The phase measurement function in synchronous detection.

A received signal $x(t)$ at time t is expressed by operational (or carrier) frequency f_0, amplitude $a(t)$, and phase $\varphi(t)$ as

$$x(t) = a(t)\cos[2\pi f_0 t + \varphi(t)]. \tag{5.51}$$

Phase $\omega_d t = 2\pi f_d t$ in (4.9) is generalized to $\varphi(t)$. The complex signal $s(t)$, which corresponds to real signal $x(t)$ in (5.51), is given by

$$s(t) = a(t)e^{j[2\pi f_0 t + \varphi(t)]}. \tag{5.52}$$

As described in Sect. 4.1.1, the complex received signal is expressed as follows by performing frequency conversion for $s(t)$ and removing carrier frequency term,

$$v(t) = a(t)e^{j\varphi(t)}, \tag{5.53}$$

where $v(t)$ is the complex envelope signal of $s(t)$. If in-phase and quadrature-phase signals are expressed as

$$I(t) = a(t)\cos\varphi(t), \quad Q(t) = a(t)\sin\varphi(t), \tag{5.54}$$

respectively, (5.53) becomes

$$v(t) = I(t) + jQ(t). \tag{5.55}$$

As the basic function of a Doppler radar which is shown in Fig. 4.1, the procedure to convert received signal into IF signal and subsequently to obtain I and Q signals by synchronous detection is shown in Fig. 5.6. Two kinds of signals, I and Q, with the phase difference 90°, are given as the output of bipolar receiver. The $I(r,t)$ and $Q(r,t)$ terms previously shown in (4.6) and (4.7), respectively, are equivalent to the I and Q signals which mean the real and imaginary part of the received envelope signal, respectively.

5.3 Characteristics of Received Signal

5.3.1 Signals Received from Precipitation Particles and the Atmosphere

As for the scatterers such as precipitation particles that are randomly distributed in the scattering resolution volume, which is delineated with the beamwidth of radar antenna and pulse duration, the relative positions of individual scatterers in the scattering resolution volume always vary even if the density of scatterers remains constant. Therefore, scattered radar signal becomes random and is treated as a

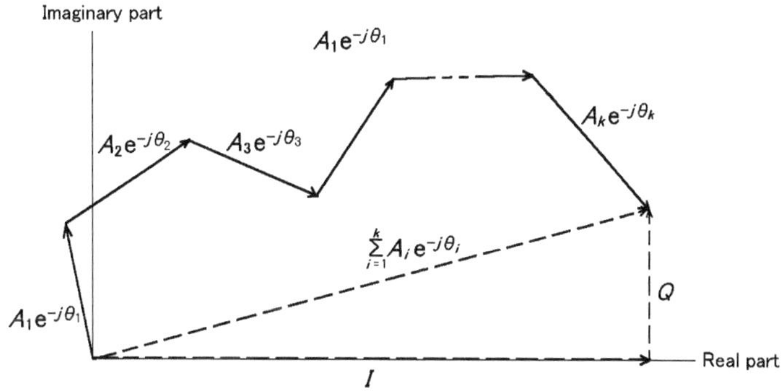

Fig. 5.7 A schematic for complex signal obtained by adding multiple instantaneous phase-detected signals

stochastic signal, i.e., phase of a signal returned from one scatterer is uncorrelated with those from other scatterers. Thus, the phase of scattered signal is considered as being uniformly distributed over $[-\pi, \pi]$. Also, amplitude of scattered signal changes randomly. Hence, the statistical approach becomes indispensable to extract any meaningful information from radar signals. An example of complex signal which is obtained by adding multiple instantaneous phase-detected signals given by (5.55) is schematically shown in Fig. 5.7.

According to the central limit theorem, summation of multiple independent randomly scattered signals leads both real and imaginary parts to Gaussian distribution with the zero mean value, if there is no bias in both parts[12] (e.g., Papoulis 1991, p. 214). Generally, the properties of the received radar signals given by (5.55) resemble those of narrowband Gaussian noise encountered in statistical theory. The real part (in-phase component) I and the imaginary part (quadrature phase component) Q have the following characteristics (e.g., Bringi and Chandrasekar 2001, pp. 262–263).

1. The real and imaginary parts of the received voltage at the same time are uncorrelated,

$$\mathrm{cov}[I(t)Q(t)] = 0, \tag{5.56}$$

---

[12]If the average of signal $x[n]$ (or bias) is $\bar{x}$, and the standard deviation is σ, the probability density function $p(x)$ for $x[n]$ is given by

$$p(x) = \frac{1}{\sqrt{2\pi}\sigma}\exp\left[-\frac{(x-\bar{x})^2}{2\sigma^2}\right].$$

If the process is ergodic, the time average value is equal to the ensemble mean value. In this book, we use the expression $\overline{*}$ for the mean value obtained from temporal or spatial averages on the premise that the radar received signal is ergodic.

The above distribution is called Gaussian distribution or the normal distribution. The variance is given by $\sigma^2 = \overline{(x-\bar{x})^2}$. The probability that x comes within the limit of $\pm 1.0\sigma$, $\pm 2.0\sigma$, $\pm 3.0\sigma$, and $\pm 3.333\sigma$, is 68.2 %, 95.4 %, 99.7 %, 99.9 %, respectively.

where cov[ ] denotes covariance[13] value.

2. The mean values of real and imaginary parts are zero,

$$E[I(t)] = E[Q(t)] = 0, \tag{5.57}$$

where $E[\ ]$ denotes ensemble mean value, which is also called expectation or expected value.

3. The variance of real and imaginary parts is identical to the value σ_{IQ}^2, which becomes

$$E[I^2(t)] = E[Q^2(t)] = \sigma_{IQ}^2. \tag{5.58}$$

4. The real and imaginary parts have the same autocorrelation functions $\rho(t)$, thus,

$$E[I(t_1)I(t_2)] = E[Q(t_1)Q(t_2)] = \sigma_{IQ}^2\,\rho(t); \quad t = t_2 - t_1. \tag{5.59}$$

5. Real and imaginary parts are statistically uncorrelated at an identical time. On the other hand, both components are correlated at different time, and satisfy the following relation,

$$E[I(t_1)Q(t_2)] = -E[I(t_2)Q(t_1)]. \tag{5.60}$$

5.3.2 Probability Density Functions

Real part I and imaginary part Q of received radar signals are mutually independent random variables. Hence, the probability density function $p(I,Q)$, where I and Q are combined, is given by the product of individual probability density functions $p(I)$ and $p(Q)$. Since both $p(I)$ and $p(Q)$ become Gaussian distribution as described in the previous paragraph, $p(I,Q)$ is given by

$$\begin{aligned} p(I,Q) &= \frac{1}{\sqrt{2\pi}\sigma_{IQ}}\exp\left(-\frac{I^2}{2\sigma_{IQ}^2}\right)\frac{1}{\sqrt{2\pi}\sigma_{IQ}}\exp\left(-\frac{Q^2}{2\sigma_{IQ}^2}\right) \\ &= \frac{1}{2\pi\sigma_{IQ}^2}\exp\left(-\frac{I^2+Q^2}{2\sigma_{IQ}^2}\right), \end{aligned} \tag{5.61}$$

where σ_{IQ}^2 is the variance of I and Q.

[13]The product of two stochastic variables that are statistically averaged is called covariance. Stochastic signals contain time as a parameter, the value is called covariance function or correlation function. Correlation function averaged over time will be defined in Sect. 5.4.2.

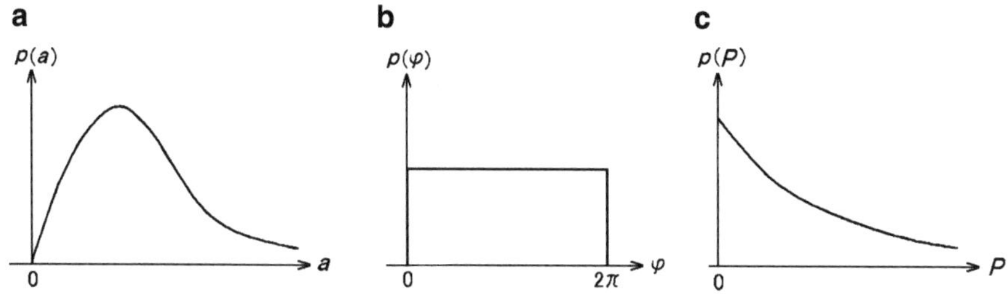

Fig. 5.8 Probability density functions of the received radar signal; (**a**) amplitude, (**b**) phase, and (**c**) power

Probability Density Functions of Amplitude and Phase

If time t is removed, for simplicity, from (5.53) and (5.55), the received radar signal is expressed as

$$v = ae^{j\varphi} = I + jQ. \tag{5.62}$$

Transforming variables of (5.61) by using the relations $a^2 = I^2 + Q^2$ and $\mathrm{d}I\mathrm{d}Q = a\mathrm{d}a\mathrm{d}\varphi$, $p(a,\varphi)$ becomes

$$p(a,\varphi) = \frac{1}{2\pi}\frac{a}{\sigma_{IQ}^2}\exp\left(-\frac{a^2}{2\sigma_{IQ}^2}\right) \tag{5.63}$$

$$= p(a)p(\varphi), \tag{5.64}$$

where $p(a)$ and $p(\varphi)$ are probability density functions of amplitude and phase, respectively, and expressed as

$$p(a) = \frac{a}{\sigma_{IQ}^2}\exp\left(-\frac{a^2}{2\sigma_{IQ}^2}\right) \qquad a > 0, \tag{5.65}$$

$$p(\varphi) = \frac{1}{2\pi} \qquad 0 < \varphi \leq 2\pi. \tag{5.66}$$

The probability density function of signal amplitude follows the Rayleigh distribution as shown in Fig. 5.8a. Meanwhile that of signal phase follows a uniform distribution as shown in Fig. 5.8b. Here, the amplitude and phase of the received signal are independent.

Probability Density Function of Signal Power

The signal power P is proportional to I^2+Q^2, where noise power is assumed to be zero, and thus the probability density function $p(P)$ of the signal power is expressed as

$$p(P)=\frac{1}{2\sigma_{IQ}^2}\exp\left(-\frac{P}{2\sigma_{IQ}^2}\right), \tag{5.67}$$

where the signal mean power is given by $\overline{P}=E[P]=E[v^2]=2\sigma_{IQ}^2$. Using $\overline{P}$, (5.67) can also be rewritten as

$$p(P)=\frac{1}{\overline{P}}\exp\left(-\frac{P}{\overline{P}}\right) \qquad P>0. \tag{5.68}$$

The probability density function of signal power follows an exponential distribution as shown in Fig. 5.8c.

5.4 Fundamentals of Radar Signal Processing

Received radar signals are random variables; it is possible, however, to estimate a wide range of physical properties of scatterers by processing a number of received radar signals statistically. One of the fundamental processing techniques is to analyze frequency components which are contained in received radar signals, i.e., to perform the spectral analysis. Also, the correlation function analysis is normally applied for the radar signal processing, e.g., they are the effective means to analyze the relation among more than one signal on the time axes.

The Fourier transform is a well known technique for the spectral analysis, which is the basic technique to change a time-series signal into the signal in the frequency domain. On the other hand, the autocorrelation function analysis is applied for processing time-series signals. It shows to what extent one signal is similar to the original signal when the one signal is shifted to a certain time with respect to the original signal. The crosscorrelation function analysis is also an efficient mean which is used to deduce the time delay between two similar signals. The mathematical definition of the autocorrelation function and the crosscorrelation function will be discussed in Sect. 5.4.2.

Here, the basic characteristics of the Fourier transform will be shown first, and then, the characteristics of the Fourier transform and the crosscorrelation function for discrete signals will be described. Finally, physical quantities of scatterers will be estimated by the spectral analysis.

5.4.1 *Fourier Transform and Its Characteristics*

Fourier Transform

Time series signal $v(t)$ and the corresponding frequency domain signal $V(f)$ are related by the following equations with the Fourier transform and the inverse Fourier transform,

$$V(f) = \int_{-\infty}^{\infty} v(t)e^{-j2\pi ft}\mathrm{d}t, \tag{5.69}$$

$$v(t) = \int_{-\infty}^{\infty} V(f)e^{j2\pi ft}\mathrm{d}f. \tag{5.70}$$

Using angular frequency $\omega = 2\pi f$, where f is frequency, $V_\omega(\omega) = V_f(f)/2\pi$. In the following, Fourier transform will be expressed with f unless otherwise mentioned. $V(f)$ is a complex function, and $V(-f) = V^*(f)$, where superscript * denotes complex conjugate. From Parseval's theorem,

$$\int_{-\infty}^{\infty} |v(t)|^2\mathrm{d}t = \int_{-\infty}^{\infty} |V(f)|^2\mathrm{d}f \tag{5.71}$$

is given. The above equation shows the energy in the time domain is equal to that in the frequency domain.

If the Fourier transform pair $v(t)$ and $V(f)$ is expressed by $v(t) \Longleftrightarrow V(f)$, the Fourier transform pair has the following characteristics;

Superposition

$v_1(t) \Longleftrightarrow V_1(f)$, $v_2(t) \Longleftrightarrow V_2(f)$. When c_1 and c_2 are constant,

$$c_1v_1(t) + c_2v_2(t) \Longleftrightarrow c_1V_1(f) + c_2V_2(f). \tag{5.72}$$

The received radar signal is the sum of desired signal and noise, and consequently the Fourier transform of the received signal is the sum of the Fourier transform of desired signal and that of noise, which are independently derived.

Time Delay

The signal after time delay t' is shown as $v(t-t')$, and the Fourier transform of $v(t-t')$ becomes the product of $V(f)$ and $\exp(-j2\pi ft')$. The relationship is expressed as

$$v(t-t') \Longleftrightarrow V(f)\mathrm{e}^{-j2\pi ft'}. \tag{5.73}$$

Frequency Translation

When a frequency translation f' occurs to the function $V(f)$ in the frequency domain, the corresponding time-series signal becomes the product of the original function and $\exp(-j2\pi f't)$

$$v(t)\mathrm{e}^{-j2\pi f't} \Longleftrightarrow V(f-f'). \tag{5.74}$$

The above equation shows that the modulation in the time domain becomes spectral translation in the frequency domain.

Response of Linear System

If the response of a linear system to an impulse, or impulse response, is $h(t)$, the system response $y(t)$ to an arbitrary input function $x(t)$ is expressed as

$$y(t) = \int_{-\infty}^{\infty} x(\tau)h(t-\tau)\mathrm{d}\tau. \tag{5.75}$$

The above integration is called convolution integral, which is commutative, and thus $y(t)$ is also given by

$$y(t) = \int_{-\infty}^{\infty} h(\tau)x(t-\tau)\mathrm{d}\tau. \tag{5.76}$$

Using operator $*$, which denotes convolution, (5.75) and (5.76) are expressed as

$$y(t) = x(t) * h(t) = h(t) * x(t). \tag{5.77}$$

Applying the Fourier transform to both sides of (5.77),

$$Y(f) = X(f)H(f) \tag{5.78}$$

can be given, where $y(t) \Longleftrightarrow Y(f)$, $x(t) \Longleftrightarrow X(f)$, and $h(t) \Longleftrightarrow H(f)$. The function $H(f)$ is called transfer function or frequency response of linear system. $H(f)$ corresponds to the output for the input signal $x(t) = \exp(-j2\pi ft)$ at a single frequency f.

The product of the Fourier integrals of two independent output signals of linear system is equivalent to the Fourier integral for the convolution of two signals, and consequently, if $v_1(t) \Longleftrightarrow V_1(f)$ and $v_2(t) \Longleftrightarrow V_2(f)$,

$$v_1(t) * v_2(t) \Longleftrightarrow V_1(f)V_2(f). \tag{5.79}$$

Equation (5.79) shows that the convolution in the time domain is equivalent to the product in the frequency domain. The reverse relationship is also valid, and the convolution in the frequency domain is equivalent to the product in the time domain, and

$$v_1(t)v_2(t) \Longleftrightarrow V_1(f) * V_2(f). \tag{5.80}$$

5.4.2 *Signals in a Linear System*

Correlation Function of Continuous and Stationary Signal

The total energy given by Parseval's theorem in (5.71) is divergent for infinitely continuing stationary signal or noise. Therefore, the following integration similar to (5.47),

$$\rho(\tau) = \int_{-\infty}^{\infty} v(t)v^*(t+\tau)\mathrm{d}t, \tag{5.81}$$

is divergent, where τ is a time lag.

If the integration can be obtained as time averaging as follows;

$$R(\tau) = \lim_{T\to\infty} \frac{1}{T} \int_0^T v(t)v^*(t+\tau)\mathrm{d}t. \tag{5.82}$$

$R(\tau)$ is called the autocorrelation function[14] of $v(t)$. In particular, when $\tau = 0$,

$$R(0) = \lim_{T\to\infty} \frac{1}{T} \int_0^T |v(t)|^2\mathrm{d}t \tag{5.83}$$

indicate the (time) mean power. If $v(t)$ is the stationary process with the mean value equal to 0, the time average by integration of the right-hand side of (5.82) can be replaced with the ensemble mean.[15] Thus,

$$R(\tau) = E[v(t)v^*(t+\tau)] = [v(0)v^*(\tau)]. \tag{5.84}$$

The above function is covariance between $v(t)$ and $v^*(t+\tau)$. Since the covariance is stationary, it does not depend on t but on τ. Especially, $R(0) = E[|v(0)|^2]$ is equal to the variance.

[14] This is the correlation function in general harmonic analysis. Occasionally, the integral interval is taken between $-T/2$ and $T/2$. Note that the definitions (5.81) and (5.82) are essentially different in nature.

[15] They are called ergodic when signals have such a property.

Power Spectral Density

If the correlation function of a stationary complex signal $v(t)$, $R(\tau)$, approaches zero sufficiently fast when $|\tau| \to \infty$, and if

$$\int_{-\infty}^{\infty} |R(\tau)| \mathrm{d}\tau < \infty \tag{5.85}$$

is satisfied,

$$S(f) = \int_{-\infty}^{\infty} R(\tau) \mathrm{e}^{-j2\pi f\tau} \mathrm{d}\tau; \qquad -\infty < f < \infty \tag{5.86}$$

is the power spectral density or power spectrum. It is the function of frequency f $(-\infty < f < \infty)$. Using $S(f)$, the correlation function is expressed as

$$R(\tau) = \int_{-\infty}^{\infty} S(f) \mathrm{e}^{j2\pi f\tau} \mathrm{d}f. \tag{5.87}$$

This indicates that the correlation function and the power spectra are mutually in the relationship of the Fourier transform, which is called the Wiener–Khinchine's theorem. Especially, the following relation can be formed.

$$R(0) = \int_{-\infty}^{\infty} S(f) \mathrm{d}f. \tag{5.88}$$

It means that the total sum of power spectral density is equal to average power.

Power Spectral Density and Periodogram

Designating the function $v_T(t)$, which is taken out from $v(t)$ between 0 and T (the period can also be set between $-T/2$ and $T/2$), as

$$v_T(t) = \begin{cases} v(t) & 0 < t < T \\ 0 & \text{otherwise,} \end{cases} \tag{5.89}$$

then,

$$V_T(f) \equiv \int_{0}^{T} v(t) \mathrm{e}^{-j2\pi ft} \mathrm{d}t; \qquad -\infty < f < \infty \tag{5.90}$$

is the Fourier transform of $v_T(t)$ whose period is T, i.e., complex frequency component. Defining the function $S_T(f)$ as

$$S_T(f) \equiv \frac{1}{T} |V_T(f)|^2, \tag{5.91}$$

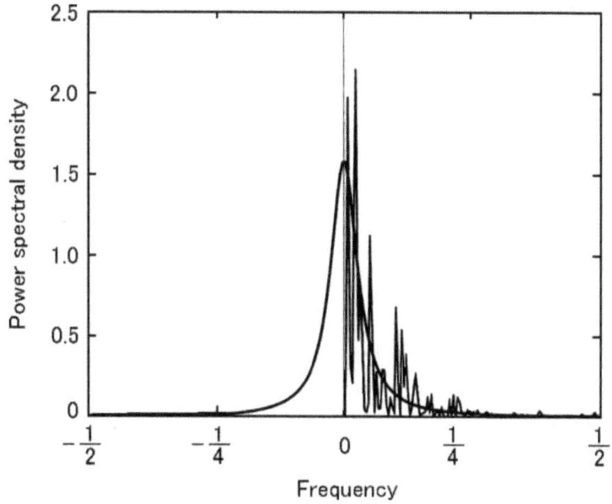

Fig. 5.9 Example of the periodogram (*zigzag line*; $f > 0$) and expected power spectral density (*smoothed curve*). The frequency unit of the abscissa is $2\pi fT$ [from Ogura 1998, p. 74]

which is called the periodogram with period T. Equation (5.91) means the power density of a signal whose frequency is f and length is T. Figure 5.9 shows an examples of the periodogram $S_T(f)$ with sequential line and power spectral density $S(f)$ with smoothed curve (Ogura 1998, p. 74). The resolution of $v_T(t)$ increases proportionally to the period T, and the number of fluctuation grows as T increases, when $S_T(f)$ is considered as function of frequency f. Therefore, it should be noted that periodogram $S_T(f)$ does not converge[16] to the power spectral density $S(f)$ as $T \to \infty$. Averaging of multiple periodograms or smoothing of it along the frequency axis is required to converge periodograms to the power spectral density.

The function $V_T(f)$ is a stochastic variable as is $v_T(t)$, which is a piece of a stationary signal, the right-hand side of (5.91) is also a stochastic variable which changes with data. The averaged periodogram (e.g., Sauvageot 1992, p. 189) (ensemble mean value or expectation) of $S_T(f)$ is

$$E[S_T(f)] = \frac{1}{T} E[|V_T(f)|^2]. \tag{5.92}$$

The averaging is called incoherent integration.[17] It can be easily proved that the average periodogram for a stationary signal is converged to the power spectral density $S(f)$ for $T \to \infty$. That is

$$\lim_{T \to \infty} E[S_T(f)] = S(f). \tag{5.93}$$

[16] It is sometime stated "converge" by mistake.

[17] The detail will be discussed in Sect. 5.6.3.

Accordingly, the power spectral density can be approximately calculated by ensemble averaging of the periodogram. Long-term data are usually divided into several segments, which are used for data ensembles.

As discussed above, the periodogram $S_T(f)$ dose not converge to the power spectral density $S(f)$. On the other hand, it is mathematically confirmed that the cumulative periodogram obtained by integrating the periodogram $S_T(f)$ as

$$F_T(f) = \int_0^f S_T(f')\mathrm{d}f' \tag{5.94}$$

converges to the corresponding cumulative power spectral density as $T \to \infty$, that is

$$\lim_{T\to\infty} F_T(f) = \int_0^f S(f')\mathrm{d}f'. \tag{5.95}$$

This mathematically means that the intense oscillation of the periodogram $S_T(f)$ at $T \to \infty$ is equalized by integration. Therefore, the periodogram $S_T(f)$ for a long period oscillates considerably, but the power spectral density can be obtained by applying integration and averaging of them within suitable frequency window f (Ogura 1998).

5.4.3 Power Spectral Moments and Basic Radar Parameters

A schematic illustration for the power spectrum of a received radar signal is shown in Fig. 5.10. Basic parameters which are obtained by radar observations are the radar echo intensity (radar reflectivity factor), the mean Doppler frequency, and the frequency spectrum width that are related to the power spectrum $S(f)$ and calculated with power spectral moments.

The value of kth moment of a power spectrum (power spectral moment) is given by

$$\mu_k = \int f^k S_\mathrm{n}(f)\mathrm{d}f, \tag{5.96}$$

where $S_\mathrm{n}(f)$ is a normalized power spectrum, and defined as

$$S_\mathrm{n}(f) = \frac{S(f)}{\displaystyle\int_{-\infty}^{\infty} S(f)\mathrm{d}f}. \tag{5.97}$$

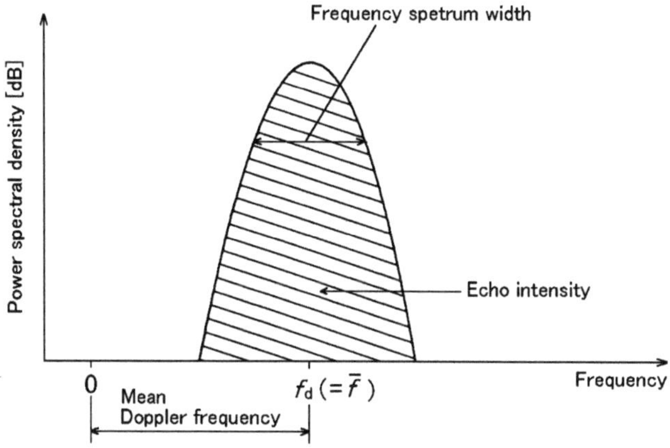

Fig. 5.10 Schematic for the power spectrum of a received radar signal

Echo Intensity

The echo intensity (received power) P_r is given by the zeroth moment (f^k for $k = 0$) of the power spectrum $S(f)$ at the respective scattering resolution volumes as

$$P_r = \mu_0 = \int_{-\infty}^{\infty} S(f)\mathrm{d}f, \tag{5.98}$$

where noise and undesired signals are assumed, for simplicity, to have been removed prior to this procedure.[18]

Mean Doppler Frequency

The mean frequency shift of power spectrum (mean Doppler frequency) is given by the 1st moment (f^k for $k = 1$) of the power spectrum or the power spectral density which is weighted with f, as

$$\mu_1 = \int f S_n(f)\mathrm{d}f = E[f] = \overline{f}, \tag{5.99}$$

where $E[f]$ is the ensemble mean value (expectation). Because the mean value obtained using temporal or spatial averaging are equal to the ensemble mean value for ergodic processes, we use $\overline{f}$ for $E[f]$, and the mean Doppler velocity or the mean radial velocity is calculated by converting the Doppler frequency to the radial speed as

$$\overline{v} = -\frac{\lambda \overline{f}}{2}. \tag{5.100}$$

[18]Type and removal method of undesired signals will be described later in Sect. 8.4.2.

Here $\overline{f}$ and $\overline{v}$ correspond to Doppler frequency f_d and Doppler velocity v_d, respectively, which were discussed in Sect. 4.1.1. If v_d exceeds the Nyquist limit v_N, sampling interval of power spectrum is deficient, and aliasing occurs in the true velocity. In this case, f_d contains a bias.

Spectrum Width

Variance from the mean Doppler frequency, σ_f^2, is given, from its definition, by the 2nd moment of power spectrum as

$$\sigma_f^2 = \int (f-\overline{f})^2 S_n(f)\mathrm{d}f. \tag{5.101}$$

From the definition it follows that

$$\sigma_f^2 = \overline{f^2} - (\overline{f})^2 = \mu_2 - \mu_1^2, \tag{5.102}$$

where μ_2 is given by (5.96) for $k = 2$. When the power spectrum is given by Gaussian distribution,[19] its standard deviation is called frequency spectrum width (spectral width is also often used for the same meaning in a variety of literatures) σ_f unless otherwise mentioned. Since velocities are often discussed in this book, it is convenient to convert the frequency spectrum width to that of radial velocity, i.e., the velocity spectrum width σ_v, in the same way as (5.100). Consequently,

$$\sigma_v = \frac{\lambda \sigma_f}{2} \tag{5.103}$$

is obtained.

Relationship Between Moments of Power Spectral Density and Autocorrelation Function

The nth differential coefficient of an autocorrelation function $R(\tau)$ is expressed by power spectral density $S(f)$ as (e.g., Sauvageot 1992, p. 184)

$$\frac{\mathrm{d}^n R(\tau)}{\mathrm{d}\tau^n} = (j2\pi)^n \int_{-\infty}^{\infty} f^n S(f) \mathrm{e}^{j2\pi f\tau}\mathrm{d}f, \tag{5.104}$$

where τ is time lag. From (5.96) and (5.104), it is found that derivative of the nth order of $R(\tau)$ at $\tau = 0$ and the nth moment of the power spectral density are related by the following equation;

[19] As will be discussed in Sect. 5.6.1, power spectrum for distributed scatterers, in general, follows Gaussian distribution.

$$\mu_n = \frac{1}{(j2\pi)^n R(0)} \frac{\mathrm{d}^n R(0)}{\mathrm{d}\tau^n}. \tag{5.105}$$

Using the above equation, the received radar power given by (5.98) and the mean Doppler frequency expressed by (5.99) are expressed, respectively, by

$$P_\mathrm{r} = R(0) \tag{5.106}$$

$$\mu_1 = \frac{1}{j2\pi R(0)} \frac{\mathrm{d}R(0)}{\mathrm{d}\tau}. \tag{5.107}$$

The relation of (5.106) is also obviously verified from (5.88) and (5.98). In the same way, the 2nd moment is given by

$$\mu_2 = \frac{1}{(j2\pi)^2 R(0)} \frac{\mathrm{d}^2 R(0)}{\mathrm{d}\tau^2}. \tag{5.108}$$

Thus, using (5.107) and (5.108), both the mean Doppler frequency obtained by (5.99) and the frequency spectrum width obtained by (5.102) are expressed by autocorrelation functions. The above discussion on estimation method for the mean Doppler frequency and the spectrum width based on the autocovariance analysis is applied to sampled data of received radar signals as shown in Appendix B.

5.5 Processing of Sampled Signals

A pulse radar repeatedly transmits pulse signals at a constant interval, and received signals from a constant distance becomes discrete time-series signals on the time axis. Based upon the discussions in the preceding section, handling of discrete signals, primarily, spectral analysis of discrete signals, which are used by actual radar signal processing, will be discussed in this section.

5.5.1 Waveform of Transmitted Pulse and Series of Signal Waves

The transmitted pulse waveform of a pulse radar is composed of a series of pulses at a constant time period (pulse repetition time) T, as illustrated in Fig. 5.11. Backscattered signals (echoes) which contribute to received radar signals are received after the time delay while transmitted signal hits scatterer and returns to the receiver. The time t_1 in the same figure indicates that the scatterer is located at the distance $ct_1/2$.

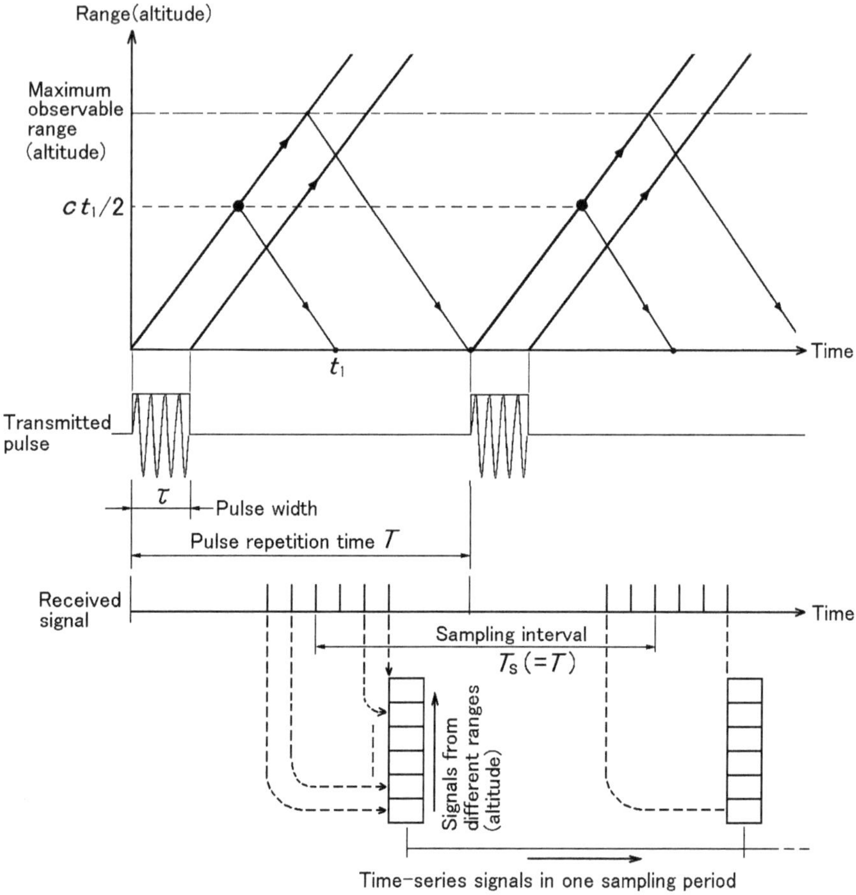

Fig. 5.11 Range-time sequence of transmitted pulses and received signals. A time-series signals collected by sampling in time are re-arranged to two-dimensional data of range (altitude) versus time

By sampling received signals for one transmitted pulse successively at a constant sampling interval, information from scatterers located at different distances (or altitudes) can be obtained. In the discrete signal processing of the received signal to be mentioned later, the time-series signals collected every sampling interval T_S (which is equal to T) as shown in the lower panel of Fig. 5.11 are used.

The conceptual diagram of the signal processing after conversion of carrier or microwave signal to IF signal in Fig. 5.6 is shown in Fig. 5.12. I and Q signals are sampled and applied spectral analysis and integrations. The processing of sampled signals are described in this section. The coherent integration and the incoherent integration will be discussed in Sects. 5.6.2 and 5.6.3, respectively.

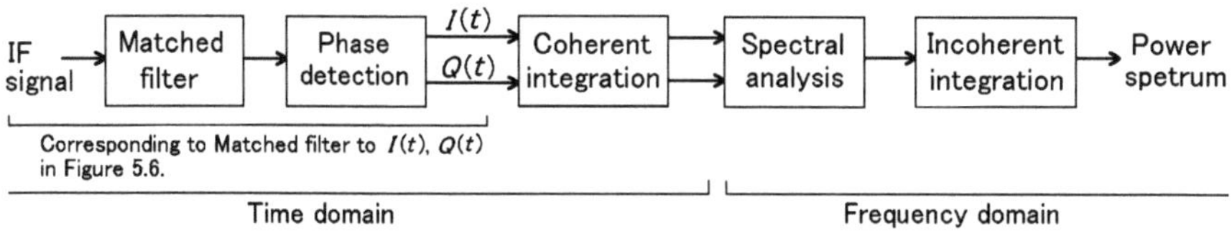

Fig. 5.12 Conceptual diagram of signal processing from IF signal to power spectrum. Coherent integration is performed in the time domain for a number of transmitted pulses within the correlation time. Incoherent integration is the averaging processing of periodograms in the frequency domain

If a rectangular pulse of amplitude A and duration (pulse width) τ is consecutively transmitted at the repetition time T as shown in Fig. 5.13a, the frequency spectrum of a pulse train becomes as shown in the same figure (b), where the dotted line is the envelope, and f_0 is the carrier frequency. Assuming that the envelope of the transmitted rectangular pulse is given by $g(t)$, and if its Fourier spectrum is $G(f)$, they are expressed as

$$g(t) = \begin{cases} A & -\dfrac{\tau}{2} \le t \le \dfrac{\tau}{2} \\ 0 & |t| > \dfrac{\tau}{2}, \end{cases} \tag{5.109}$$

and

$$G(f) = \int_{-\tau/2}^{\tau/2} g(t)\mathrm{e}^{-j2\pi f t}\mathrm{d}t = A\frac{\sin(\pi f \tau)}{\pi f}. \tag{5.110}$$

Here the frequency for $G(f) = 0$ is $f = m/\tau$ $(m = \pm 1,\ \pm 2 \cdots)$. Actually, the transmitted signal is a sinusoidal wave of carrier frequency f_0 which is amplitude modulated by a square wave of (5.109). Thus, using (5.110), the Fourier spectrum $G_\mathrm{t}(f)$ is expressed as

$$G_\mathrm{t}(f) = G(f - f_0) = A\frac{\sin[\pi(f - f_0)\tau]}{\pi(f - f_0)}, \tag{5.111}$$

and $|G_\mathrm{t}(f)|$ corresponds to the dotted lines in Fig. 5.13b.

As shown in Fig. 5.13b, the frequency spectral harmonics of a rectangular pulse extends both side of the main spectrum ($< -1/\tau$, $> 1/\tau$). These expansion of frequency spectrum becomes one of the cause of mutual interference with neighboring radars. Meanwhile in actual transmitted wave form, the rise time and fall time are not zero, thus perfectly rectangular pulse cannot be formed. This fact effects to reduce the expansion of the frequency spectrum of outside the main band. In other word, by deforming the transmitted pulse waveform, the spectral harmonics can be adjusted.

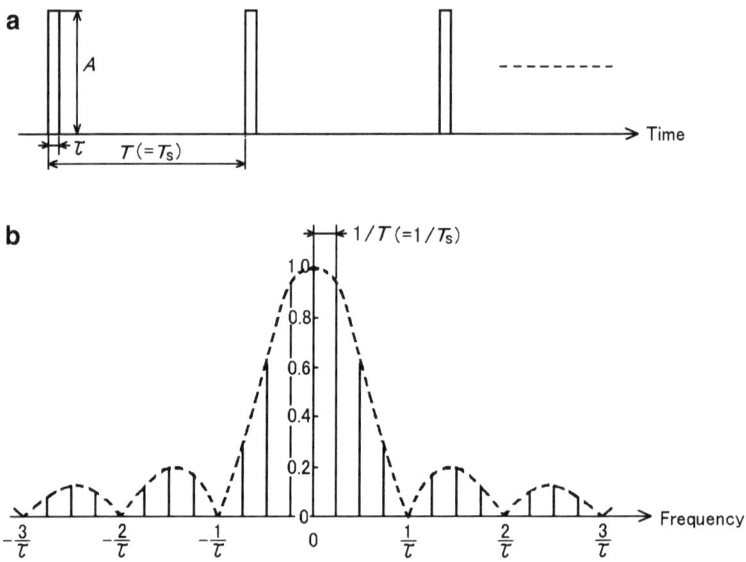

Fig. 5.13 (**a**) Rectangular pulse train in time region, where carrier in the pulse is abbreviated. (**b**) The frequency spectrum of the pulses. *Dotted line* in (**b**) is the envelope, and the frequency at the point 0 in the x-axis is f_0, the carrier frequency. The y-axis shows relative value in the linear scale

Figure 5.14 shows various examples of transmitted pulses in time region (upper panels) and their absolute frequency power spectrum envelopes in frequency region (bottom panels), (a) a cosine pulse, (b) cosine-squared pulse, and (c) Gaussian pulse, where σ_t and σ_f are the standard deviations of t and f, respectively, and $\sigma_f = (2\pi\sigma_t)^{-1}$.

In the upper panels, only the envelopes are shown, and the carriers in the pulses are abbreviated. In the lower panels, the frequency at the point 0 in the x-axis is f_0, the carrier frequency. The values of y-axis are normalized by the maximum values and are shown in the linear scale. As shown in (c), The shape of a Gaussian pulse in time region and its frequency spectrum envelope are the same shape, i.e., the frequency transform of the Gaussian distribution also becomes the Gaussian distribution. The time functions in the upper panels and the frequency functions in Fig. 5.14a–c are listed on Table 5.1.

The above discussion is concerned with the spectrum of transmitted pulses, but the same is applicable to the received signal, since the bandwidth for received signal never changes. Consequently, sampling the received signal at the sampling frequency more than $2/\tau$, the received signal can be reproduced by Shannon and Someya's sampling theorem. This frequency becomes the standard for the sampling interval in digitalizing received signals.

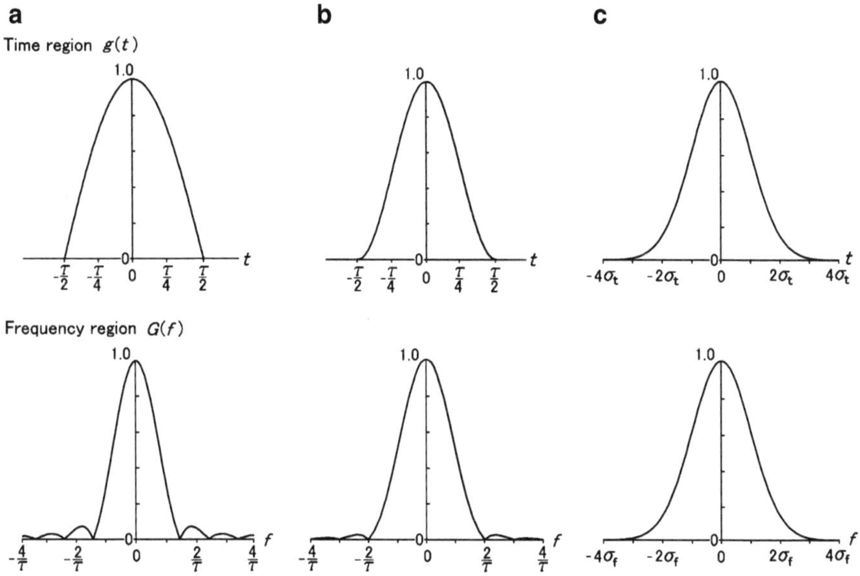

Fig. 5.14 Various examples of transmitted pulses in time region (*upper panel*) and their frequency power spectrum envelopes in frequency region (*bottom panels*). (**a**) Cosine pulse, (**b**) cosine-squared pulse, where τ is the duration that $g(t)$ becomes 0. (**c**) Gaussian pulse, where σ_t and σ_f are the standard deviations of t and f, respectively, and $\sigma_f = (2\pi\sigma_t)^{-1}$. In the *upper panels*, carriers in the pulses are abbreviated. In the *lower panels*, line spectrum is abbreviated and the frequency at the point 0 in the x-axis means f_0, the carrier frequency. The values of y-axis are normalized by the maximum values and are the shown in the linear scale

Table 5.1 Time functions $g(t)$ and frequency functions $G(f)$ which are shown in Fig. 5.14. Amplitudes of these functions are normalized by the maximum values at $t = 0$ and $f = 0$

	Time functions	Frequency functions
cosine pulse	$g(t) = \cos\pi\frac{t}{\tau},\ -\frac{\tau}{2} < t < \frac{\tau}{2}$ $= 0$, otherwise	$G(f) = \frac{\cos\pi\alpha}{1-4\alpha^2},\ \alpha = \tau f$
cosine-squared pulse	$g(t) = \cos^2\pi\frac{t}{\tau},\ -\frac{\tau}{2} < t < \frac{\tau}{2}$ $= 0$, otherwise	$G(f) = \frac{\sin\pi\alpha}{\pi\alpha(1-\alpha^2)},\ \alpha = \tau f$
Gaussian pulse	$g(t) = \exp\left[-\frac{1}{2}\left(\frac{t}{\sigma_t}\right)^2\right]$ σ_t: standard deviation	$G(f) = \exp\left[-\frac{1}{2}\left(\frac{f}{\sigma_f}\right)^2\right]$ $\sigma_f = 1/(2\pi\sigma_t)$

5.5.2 *Sampling of a Received Signal*

If the scattered signal $v(t)$ is periodically received at the same time period T (equal to the sampling interval T_s), a periodic waveform

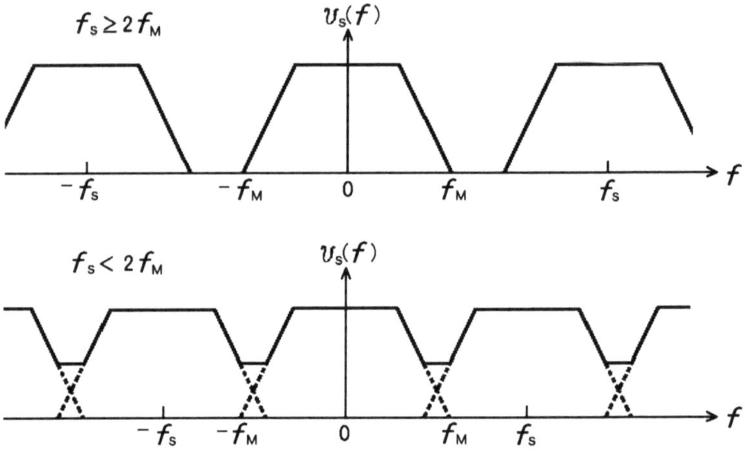

Fig. 5.15 Radar spectra for (*top*) $f_s \geq 2f_M$, and (*bottom*) $f_s < 2f_M$

$$v_s(t) = \sum_{m=-\infty}^{\infty} v(t - mT_s) \tag{5.112}$$

results with period T_s. This pulse train is expressed as a convolution product

$$v_s(t) = v(t) * \left[\sum_{m=-\infty}^{\infty} \delta(t - mT_s) \right], \tag{5.113}$$

where $\delta(t)$ is the unit impulse which is called Dirac's Delta function or simply Delta function, and is defined by

$$\begin{aligned} \delta(t) &= \frac{1}{\Delta t} \qquad t = 0 \\ \delta(t) &= 0 \qquad\quad t \neq 0. \end{aligned} \tag{5.114}$$

Applying (5.77) and (5.78), the Fourier transform of $v_s(t)$ is given by

$$V_s(f) = \frac{1}{T_s} V(f) \left[\sum_{m=-\infty}^{\infty} \delta(f - mf_s) \right] = \frac{1}{T_s} \sum_{m=-\infty}^{\infty} V(f - mf_s), \tag{5.115}$$

where $V(f)$ is the spectrum of $v(t)$. Equation (5.115) shows that the spectrum $V_s(f)$ for discrete signals obtained by sampling $v(t)$ becomes a periodic function which contains $V(f)$ at every interval of f_s on the frequency axis. Then, the maximum frequency which is contained in $v(t)$ is expressed as f_M. In the case that $f_s \geq 2f_M$ as shown in upper panel of Fig. 5.15, aliasing (overlapping) of spectra is avoidable, and the original analog signal can be correctly reproduced by taking out the spectral component $|f| < f_s$. On the other hand, in the case that $f_s < 2f_M$, the spectral aliasing occurs as shown in the lower panel of the same figure, and the spectra of the original analog signal is not preserved. The maximum frequency $f_N = f_s/2$ at which the spectrum of the original signal is correctly preserved is called the Nyquist frequency.

5.5.3 *Processing of Discrete Signal*

The major part of the signal processing for pulse radar is to perform spectral analysis of discrete signals. This technique includes the discrete Fourier transform (DFT), and, in particular, the fast Fourier transform (FFT) which efficiently executes DFT by applying the circular convolution theorem. In this subsection, the expression of discrete signals is described first. Then, the technique to express Doppler shift in the frequency domain using the DFT and that to express it in the time domain using the time-series signal will be described.

Discrete Fourier Transform

If the mth complex time series signal which is sampled at a constant sampling interval $T_s \leq 1/2f_N$ is denoted by $v(mT_s)$, a discrete complex time series signal $v_s(t)$ is expressed by

$$v_s(t) = \sum_{m=-\infty}^{\infty} v(mT_s)\delta(t - mT_s), \tag{5.116}$$

and the Fourier transform of $v_s(t)$ becomes

$$\begin{aligned} V_s(f) &= \int_{-\infty}^{\infty} \left[\sum_{m=-\infty}^{\infty} v(mT_s)\delta(t - mT_s) \right] e^{-j2\pi ft} dt \\ &= \sum_{m=-\infty}^{\infty} v(mT_s) \left[\int_{-\infty}^{\infty} \delta(t - mT_s) e^{-j2\pi ft} dt \right] \\ &= \sum_{m=-\infty}^{\infty} v(mT_s) e^{-j2\pi f mT_s}, \end{aligned} \tag{5.117}$$

where $V_s(f)$ is a periodic function with period $f_s = 1/T_s$. The discrete periodic spectrum in the frequency domain which is obtained by sampling $V_s(f)$ with the fundamental frequency $f_b = 1/MT_s$ corresponds to the discrete periodic time series signal in time domain that the discrete signal $v_s(t)$ recurs with period MT_s. This correspondence is the discrete Fourier transform.

Abbreviating T_s and denoting the time domain signal with a period of M as $\tilde{v}(m)$ hereafter, the Fourier series expansion of a periodic signal $\tilde{v}(m)$ can be obtained in analogy to a continuous time series signal. The basic function for the Fourier series expansion is composed of M complex sinusoidal signal series

$$\mathrm{e}_k(m) = \mathrm{e}^{j2\pi mk/M}, \qquad k = 0, 1, 2, \cdots, M-1. \tag{5.118}$$

Replacing the exponent with

$$W_M = \mathrm{e}^{-j2\pi/M}, \tag{5.119}$$

the periodic function $\tilde{v}(m)$ is expressed as

$$\tilde{v}(m) = \frac{1}{M} \sum_{k=0}^{M-1} \tilde{V}(k) W_M^{-km}, \tag{5.120}$$

where the Fourier coefficient $\tilde{V}(k)$ is given by

$$\tilde{V}(k) = \sum_{m=0}^{M-1} \tilde{v}(m) W_M^{km}, \tag{5.121}$$

where $\tilde{V}(k)$ is also a periodic signal series with period M.

The DFT which relates the discrete time series signal with discrete spectra of a fundamental period is expressed as follows. First, a discrete time series signal $v(m)$ $(0 \le m \le M-1)$ with a limited length M is defined using the periodic function $\tilde{v}(m)$ as

$$v(m) = \tilde{v}(m) P_M(m), \tag{5.122}$$

where $P_M(m)$ is a pulse signal in the time domain and given by

$$P_M(m) = \begin{cases} 1 & 0 \le m \le M-1 \\ 0 & \text{otherwise.} \end{cases} \tag{5.123}$$

Then, the discrete spectrum $V(k)$ with limited length M is defined using the Fourier coefficient $\tilde{V}(k)$ as

$$V(k) = \tilde{V}(k) P_M(k). \tag{5.124}$$

Accordingly, the DFT is expressed as follows:

$$V(k) = \sum_{m=0}^{M-1} v(m) W_M^{km} \qquad 0 \le k \le M-1. \tag{5.125}$$

On the other hand, the inverse discrete Fourier transform (IDFT) is given by

$$v(m) = \frac{1}{M}\sum_{k=0}^{M-1} V(k)W_M^{-km} \qquad 0 \le m \le M-1. \tag{5.126}$$

Using $f = kf_b = k/(MT_s)$, (5.125) becomes

$$V(f) = \sum_{m=0}^{M-1} v(m)\mathrm{e}^{-j2\pi f T_s m}. \tag{5.127}$$

Discrete Correlation Function

Since the discrete time series signal $v(m)$ is a steady signal, the discrete autocorrelation function $R_{vv}(l)$ is defined in the same manner as that for continuous time series (5.84) by the following equation:

$$R_{vv}(l) = E[v^*(m)v(m+l)] = E[v^*(m-l)v(m)], \tag{5.128}$$

where superscript * denotes complex conjugate, m is the number of discrete time series signal, and l is a time lag. As is obvious from the above equation,

$$R_{vv}(-l) = R_{vv}^*(l). \tag{5.129}$$

Using (5.129), the discrete autocorrelation function for stationary (i.e., time independent statistical) signal, which corresponds to the autocorrelation function for continuous signal (5.82), is given by Doviak and Zrnić (e.g. 2006, p. 95)

$$R_{vv}(l) = \lim_{M\to\infty} \frac{1}{M} \sum_{m=0}^{M-|l|-1} v^*(m)v(m+l), \tag{5.130}$$

where M is the number of samples. As M is a limited number, the estimate, ^ denotes the estimate, for (5.130) is given by

$$\hat{R}_{vv}(l) = \begin{cases} \dfrac{1}{M}\displaystyle\sum_{m=0}^{M-|l|-1} v^*(m)v(m+l) & |l| \le M-1 \\ 0 & \text{otherwise.} \end{cases} \tag{5.131}$$

Discrete Power Spectrum and Periodogram

As aforementioned by (5.86), the power spectrum is equivalent to the Fourier transform of the autocorrelation function. Likewise the discrete power spectrum $S(f)$ is given by the Fourier transform of the discrete autocorrelation function in the following form;

$$S(f) = \lim_{M\to\infty} T_s \sum_{l=-(M-1)}^{M-1} R_{vv}(l) e^{-j2\pi f T_s l}. \tag{5.132}$$

If (5.131) is used as the estimate of the discrete autocorrelation function for signals having a finite number M of samples, the estimate of discrete power spectrum for (5.132) is given by

$$\hat{S}_a(f) = T_s \sum_{l=-(M-1)}^{M-1} \hat{R}_{vv}(l) e^{-j2\pi f T_s l}. \tag{5.133}$$

Meanwhile, the estimate of the periodogram that corresponds to (5.91) is expressed using the discrete spectrum $V(f)$ as

$$\hat{S}_b(f) = \frac{T_s}{M} |V(f)|^2 \tag{5.134}$$

Using (5.127) for $V(f)$, $\hat{S}_b(f)$ is expressed as

$$\begin{aligned} \hat{S}_b(f) &= \frac{T_s}{M} \left[\sum_{m=0}^{M-1} v^*(m) e^{j2\pi f T_s m} \sum_{n=0}^{M-1} v(n) e^{-j2\pi f T_s n} \right] \\ &= \frac{T_s}{M} \left[\sum_{m=0}^{M-1} \sum_{n=0}^{M-1} v^*(m) v(n) e^{-j2\pi f T_s l} \right], \end{aligned} \tag{5.135}$$

where $l = n - m$. Here, the term in the parenthesis in the right side is transformed as

$$\sum_{m=0}^{M-1} \sum_{n=0}^{M-1} v^*(m) v(n) e^{-j2\pi f T_s l} = \sum_{l=-(M-1)}^{M-1} e^{-j2\pi f T_s l} \sum_{m=0}^{M-|l|-1} v^*(m) v(m+l). \tag{5.136}$$

Using (5.131),

$$\sum_{m=0}^{M-|l|-1} v^*(m) v(m+l) = M \hat{R}_{vv}(l). \tag{5.137}$$

Substituting (5.137) into (5.136), and substituting it into (5.135),

$$\hat{S}_b(f) = T_s \sum_{l=-(M-1)}^{M-1} \hat{R}_{vv}(l) e^{-j2\pi f T_s l} = \hat{S}_a(f) \tag{5.138}$$

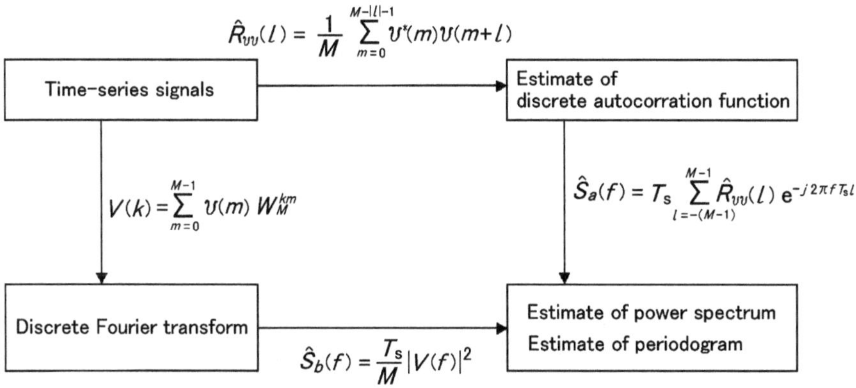

Fig. 5.16 Relationship between the Fourier transform and the autocorrelation function

is obtained (Doviak and Zrnić 2006, pp. 97–98). The above relationship is schematically shown in Fig. 5.16. From the figure it is evident for discrete time series signal too that the estimate of the power spectral density which is obtained by the Fourier transform of an autocorrelation function is equivalent to the estimate of the periodogram which is given by the Fourier transform of a discrete time series signal.

5.5.4 Estimation of Mean Doppler Frequency

Spectral Analysis in Frequency Domain

Using estimate of the discrete power spectral density, the estimate of mean Doppler frequency $\hat{f}_\mathrm{f}$ (which is $\hat{f}_\mathrm{d}$ obtained in the frequency domain) is given, in analogy to (5.99), by

$$\hat{f}_\mathrm{f} = \frac{1}{MT_\mathrm{s}} \frac{\sum_{k=-M/2}^{M/2} k\hat{S}(k)}{\sum_{k=-M/2}^{M/2} \hat{S}(k)}, \tag{5.139}$$

where $\hat{S}(k)$ is the estimate of the kth sample of the periodogram which is given by

$$\hat{S}(k) = \left| \frac{1}{M} \sum_{m=0}^{M-1} v(m) \mathrm{e}^{-j2\pi km/M} \right|^2. \tag{5.140}$$

The above relation becomes erroneous, producing a biased estimate, when a part of the original Doppler spectrum is displaced due to aliasing (Mahapatra 1999, p. 198). The following expression improves the estimate of the mean Doppler frequency and minimizes the bias (Zrnić 1979);

$$\hat{f}_{\mathrm{f}} = \frac{1}{M}\left[\frac{k_{\mathrm{m}}}{T_{\mathrm{s}}} + \frac{1}{\hat{P}T_{\mathrm{s}}}\sum_{k=k_{\mathrm{m}}-M/2}^{k_{\mathrm{m}}+M/2}(k-k_{\mathrm{m}})\hat{S}_{\mathrm{mod}_M}(k)\right], \tag{5.141}$$

where k_{m} $(-M/2 \le k_{\mathrm{m}} \le M/2)$ is argument of the sample, which yields the maximum Fourier coefficient. $\hat{P}$ is estimate of the sum of periodograms, i.e., total power, while $\hat{S}_{\mathrm{mod}_M}(k)$ is the estimate of $\mathrm{mod}_M(k)$th periodogram, where $\mathrm{mod}_M(k)$ is the remainder from dividing k by M. It is assumed in (5.141) that power spectral density is maximized in the vicinity of the mean Doppler frequency. Hence, the power spectrum is weighted according to the distance $k-k_{\mathrm{m}}$ from the maximum spectral point, and the correction is made for aliased components of the power spectrum around the Nyquist frequency.

The estimate of the mean Doppler velocity is given from (5.100) as $\hat{v}_{\mathrm{f}} = -\lambda\hat{f}_{\mathrm{f}}/2$.

Autocovariance Analysis in Time Domain

Using (5.107), the mean Doppler frequency f_{t} (which is f_{d} obtained in time domain) can be precisely derived from the autocorrelation data as

$$f_{\mathrm{t}} = \frac{1}{j2\pi R(0)}\frac{\mathrm{d}R(0)}{\mathrm{d}\tau}. \tag{5.142}$$

Using the estimate $\hat{R}(T_{\mathrm{s}})$ for the autocorrelation function $R(T_{\mathrm{s}})$, the estimate $\hat{f}_{\mathrm{t}}$ for f_{t} can be expressed, as details are discussed in Appendix B.1, by

$$\hat{f}_{\mathrm{t}} = \frac{1}{2\pi T_{\mathrm{s}}}\arg\hat{R}(T_{\mathrm{s}}) = \frac{1}{2\pi T_{\mathrm{s}}}\arctan\left[\frac{\mathrm{I_m}[\hat{R}(T_{\mathrm{s}})]}{\mathrm{R_e}[\hat{R}(T_{\mathrm{s}})]}\right], \tag{5.143}$$

where $\mathrm{R_e}[\hat{R}(T_{\mathrm{s}})]$ and $\mathrm{I_m}[\hat{R}(T_{\mathrm{s}})]$ are the real and imaginary parts of the autocorrelation function $\hat{R}(l)$ with time lag (sampling interval) T_{s}, respectively. If M pulses are transmitted at a constant period in actual radar operation, $\hat{R}(T_{\mathrm{s}})$ can be obtained by successive pair of received signal, v_m and v_{m+1}, as

$$\hat{R}(T_{\mathrm{s}}) = \frac{1}{M}\sum_{m=0}^{M-1} v_m^* v_{m+1}. \tag{5.144}$$

From (5.143) and (5.144), $\hat{f}_{\mathrm{t}}$ becomes

$$\hat{f}_{\rm t} = \frac{1}{2\pi T_{\rm s}} \arctan \left[\frac{\sum\limits_{m=0}^{M-1} {\rm Im}(v_m^* v_{m+1})}{\sum\limits_{m=0}^{M-1} {\rm Re}(v_m^* v_{m+1})} \right]$$

$$= \frac{1}{2\pi T_{\rm s}} \arctan \left[\frac{\sum\limits_{m=0}^{M-1} |v_m^* v_{m+1}| \sin(\phi_m - \phi_{m+1})}{\sum\limits_{m=0}^{M-1} |v_m^* v_{m+1}| \cos(\phi_m - \phi_{m+1})} \right], \qquad (5.145)$$

where ϕ_m is the vector angle of v_m in the complex plane. If v is given by $v = a + jb$, (5.145) is expressed by

$$\hat{f}_{\rm t} = \frac{1}{2\pi T_{\rm s}} \arctan \left[\frac{\sum\limits_{m=0}^{M-1} a_m b_{m+1} - \sum\limits_{m=0}^{M-1} a_{m+1} b_m}{\sum\limits_{m=0}^{M-1} a_m a_{m+1} + \sum\limits_{m=0}^{M-1} b_m b_{m+1}} \right]. \qquad (5.146)$$

In (5.146), M samples of $R(T_{\rm s})$ are added vectorially, and the mean Doppler frequency is estimated from the average of the phase shift weighted by the power of samples (Sauvageot 1992, p. 194).

Using $\hat{f}_{\rm t}$, mean Doppler velocity can be estimated as $\hat{v}_{\rm t} = -\lambda \hat{f}_{\rm t}/2$.

5.5.5 Estimation of Spectrum Width

Spectral Analysis in Frequency Domain

If the Doppler velocity spectrum width is small compared with the Nyquist velocity, estimate of the Doppler frequency spectrum width $\hat{\sigma}_f$, which is free from the bias due to aliasing, can be obtained from the power spectrum as (Doviak and Zrnić 2006, p. 140);

$$\hat{\sigma}_f^2 = \frac{1}{\hat{P} T_{\rm s}^2} \left[\sum_{k=k_{\rm m}-M/2}^{k_{\rm m}+M/2} \left(\frac{k}{M} - \hat{f}_{\rm f} T_{\rm s} \right)^2 \hat{S}_{{\rm mod}_M}(k) \right]. \qquad (5.147)$$

The estimate of the Doppler velocity spectrum width $\hat{\sigma}_v$ can be obtained from the estimate of the Doppler frequency spectrum width $\hat{\sigma}_f$ as $\hat{\sigma}_v = \lambda \hat{\sigma}_f/2$.

Autocovariance Analysis in Time Domain

Substituting (5.108) and (5.107) into (5.102), the squared Doppler frequency spectrum width $\overline{\sigma}_f^2$ is expressed using time series signal as

$$\overline{\sigma}_f^2 = \mu_2 - \mu_1^2 = \frac{1}{(j2\pi)^2 R(0)} \frac{\mathrm{d}^2 R(0)}{\mathrm{d}\tau^2} - \left[\frac{1}{j2\pi R(0)} \frac{\mathrm{d}R(0)}{\mathrm{d}\tau}\right]^2. \tag{5.148}$$

In (5.148), only the first several time lags are important. Using estimate $\hat{R}(T_\mathrm{s})$, the estimate $\hat{\sigma}_f^2$ for $\overline{\sigma}_f^2$ is obtained, as details are discussed in Appendix B.2, by

$$\hat{\sigma}_f^2 = \frac{1}{2\pi^2 T_\mathrm{s}^2}\left[1 - \frac{|\hat{R}(T_\mathrm{s})|}{\hat{R}(0)}\right]. \tag{5.149}$$

It is also shown as the same manner as (5.145) by

$$\hat{\sigma}_f^2 = \frac{1}{2\pi^2 T_\mathrm{s}^2}\left(1 - \frac{\left|\sum_{m=0}^{M-1} v_m^* v_{m+1}\right|}{\left|\sum_{m=0}^{M-1} v_m^* v_m\right|}\right). \tag{5.150}$$

If the received signal spectra follow a Gaussian shape as will be discussed in Sect. 5.5.6, $\hat{\sigma}_f^2$ can be expressed by using autocorrelation functions of time lags 0 and 1, $\hat{\sigma}_f^2 = (1/2\pi^2 T_\mathrm{s}^2) \ln |\hat{R}(0)/\hat{R}(T_\mathrm{s})|$ (Doviak and Zrnić 2006, p. 136), and using lags 1 and 2, $\hat{\sigma}_f^2 = (1/6\pi^2 T_\mathrm{s}^2) \ln |\hat{R}(T_\mathrm{s})/\hat{R}(2T_\mathrm{s})|$ (Srivastava et al. 1979).

The estimate of the Doppler velocity spectrum width $\hat{\sigma}_v$ is given with the estimate of the Doppler frequency spectrum width $\hat{\sigma}_f$ by the same equation as that in the frequency domain, $\hat{\sigma}_v = \lambda \hat{\sigma}_f / 2$.

5.5.6 Estimation of Spectral Parameter by Fitting

In the case that the signal is contaminated with high level noise or statistical fluctuations, parameter estimation by the direct moment method mentioned in Sects. 5.5.4 and 5.5.5 becomes inaccurate. Hence, the following method is generally adopted, in particular, by atmospheric radars. That is, power spectrum is first supposed to be expressed by an analytic function determined by several spectral parameters, and then, the parameters are estimated by fitting the supposed analytic spectrum to observed one. Because the power spectrum is often well described by a Gaussian function (e.g., Janssen and van der Spek 1985; Doviak and Zrnić 2006, p. 112), the following model function of the Gaussian distribution is fitted[20]

$$S_t(v) = \frac{S_0}{\sqrt{2\pi}\sigma_v} \exp\left[-\frac{(v-\overline{v})^2}{2\sigma_v^2}\right], \tag{5.151}$$

where S_0 is the scattered power, $S_0/(\sqrt{2\pi}\sigma_v)$ is the peak value of power spectral density, v is the radial velocity, $\overline{v}$ is the mean radial velocity (i.e., expected value $E[v]$), and σ_v is the Doppler velocity spectrum width . The spectrum width (full width, or half width times 2) is defined as the width of spectrum where spectral power becomes half of the peak value, and given by $2\sqrt{2\ln 2}\,\sigma_v$.

The above function is nonlinear, and so the fitting is performed by nonlinear least-square method. The principle of nonlinear least-square fitting is to decide spectral parameters that minimize residual sum of squares $\varepsilon(\boldsymbol{P})$ between model function and observation. The value of $\varepsilon(\boldsymbol{P})$ is given by

$$\varepsilon(\boldsymbol{P}) = \sum_{k=1}^{N} [S_t(k;\boldsymbol{P}) - y(k)]^2, \tag{5.152}$$

where $S_t(k;\boldsymbol{P})$ is the model function given by (5.151), and $y(k)$ is the observed value. Note that frequency f is replaced with the frequency point k, and N is the total number of frequency points. The function $\boldsymbol{P}$ is the parameter vector, and its elements are S_0, $\overline{v}$, and σ_v. The necessary condition to minimize (5.152) yields

$$\sum_{k=1}^{N} [S_t(k;\boldsymbol{P}) - y(k)] \frac{\partial S_t(k;\boldsymbol{P})}{\partial P_i} = 0 \qquad (i = 1,\ 2,\ 3). \tag{5.153}$$

$S_t(k;\boldsymbol{P})$ is linearized by applying the Taylor series expansion to $S_t(k;\boldsymbol{P})$ around a proper initial value $\boldsymbol{P}_0$, and taking only the zeroth and first order terms as follows;

$$S_t(k;\boldsymbol{P}_0 + \delta\boldsymbol{P}) \simeq S_t(\boldsymbol{P}_0) + \sum_{j=1}^{3} \delta P_j \frac{\partial S_t(k;\boldsymbol{P})}{\partial P_j}. \tag{5.154}$$

Substituting (5.154) into (5.153),

$$\boldsymbol{A} \cdot \delta\boldsymbol{P} = \boldsymbol{b} \tag{5.155}$$

is obtained, where the elements of $\boldsymbol{A}$ and $\boldsymbol{b}$ are

[20]Refer to Sect. 5.6.1 for the details on Gaussian functions.

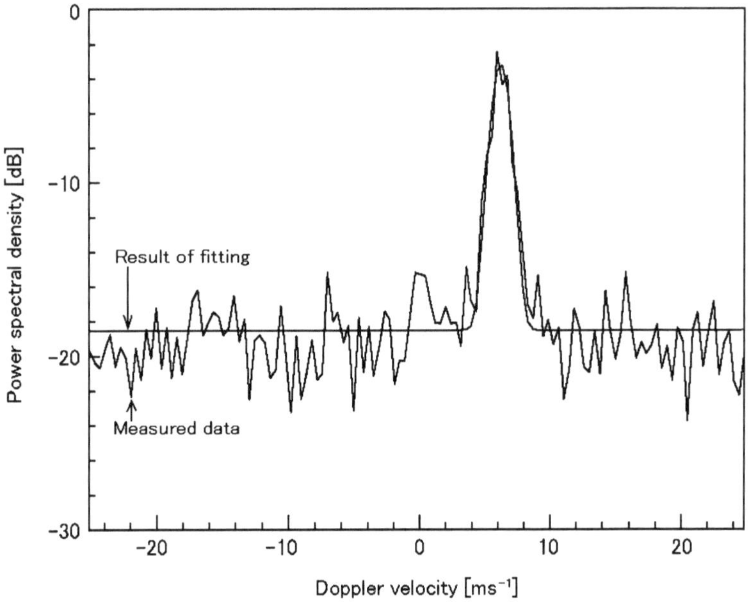

Fig. 5.17 A result of Gaussian function fitting for measured power spectrum by the MU radar at the direction of (azimuth, zenith) = (90°, 10°) and 6.2 km height between 0001:03 and 0002:03 LT on 8 March 2011. *Thick lines* show measured data, and *thin line* the result of fitting. Courtesy of Dr. Hashiguchi, Kyoto University

$$a_{ij} = \sum_{k=1}^{N} \frac{\partial S_t(k;\boldsymbol{P})}{\partial P_i} \frac{\partial S_t(k;\boldsymbol{P})}{\partial P_j}, \tag{5.156}$$

$$b_j = \sum_{k=1}^{N} [y(k) - S_t(k;\boldsymbol{P}_0)] \frac{\partial S_t(k;\boldsymbol{P})}{\partial P_i}, \tag{5.157}$$

respectively. The correction value $\delta\boldsymbol{P}$ is obtained from (5.155) as

$$\delta\boldsymbol{P} = \boldsymbol{A}^{-1} \cdot \boldsymbol{b}. \tag{5.158}$$

Replacement of $\boldsymbol{P}_0$ with $\boldsymbol{P}_0 + \alpha \cdot \delta\boldsymbol{P}$ is recurred until $\boldsymbol{P}$ converges, where α $(0 \leq \alpha \leq 1)$ is called reduction factor, which is used to stabilize convergence. The value of $\boldsymbol{P}$ is supposed to be convergent if the change of $\varepsilon(\boldsymbol{P})$ and the correction become below the specified values.

Figure 5.17 shows a result of Gaussian function fitting using the above method for measured power spectrum by the MU radar (Sect. 10.2.3) at the direction of (azimuth, zenith) = (90°, 10°) and 6.2 km height between 0001:03 and 0002:03 LT on 8 March 2011.

5.5.7 Estimation Based on Prediction Theory

As will be discussed later in Sect. 8.4.2, received signals are practically contaminated with various undesired signals. As a result, the power spectrum is considerably distorted in shape and/or very low frequency noise intermingles with received signals, that often makes it more difficult to accurately estimate spectral parameters. In these cases, the estimation method for time series signal based on prediction theory such as the maximum entropy method (MEM) is adopted.

In the MEM, $\{v_n\}$ is assumed as a complex stationary time series signal, and the value of one step advanced v_n is estimated with the linear combination of N data in the past data set $\{v_m;\ m = n-1,\ n-2,\ \ldots\}(m < n)$ as follows;

$$v_n = \sum_{m=1}^{N} a_m v_{n-m}, \tag{5.159}$$

where N indicates the dimension of estimation.

When $\{a_m\}$ that minimizes estimated error $e_n{}^2 = \langle |v_n - \sum_{m=1}^{N} a_m v_{n-m}|^2 \rangle$ is determined, power spectral density of dimension N is given by

$$S(f) = \frac{e_n{}^2}{\left| 1 - \sum_{m=1}^{N} a_m e^{-j2\pi fm} \right|^2}. \tag{5.160}$$

To obtain $\{a_m\}$, a simultaneous equation of unknowns $\{e_N{}^2,\ a_1,\ \ldots,\ a_N\}$ is derived from (5.159) as;

$$\begin{bmatrix} R_0 & R_{-1} & R_{-2} \ldots & R_{-N} \\ R_1 & R_0 & \ldots\ldots & R_{-N+1} \\ \vdots & \vdots & \ddots & \vdots \\ R_N & R_{N-1} & \ldots\ldots & R_0 \end{bmatrix} \begin{bmatrix} 1 \\ a_1 \\ \vdots \\ a_N \end{bmatrix} = \begin{bmatrix} e_N{}^2 \\ 0 \\ \vdots \\ 0 \end{bmatrix}, \tag{5.161}$$

where the autocorrelation function $R(l)$ is indicated as R_l. From Crammer's formula, (5.161) is solved as follows;

$$e_N{}^2 = \frac{D_{N+1}}{D_N}, \tag{5.162}$$

$$a_n = \frac{A_n}{D_N} \qquad (n = 1,\ \cdots,\ N), \tag{5.163}$$

$$D_{N+1} \equiv \begin{vmatrix} R_0 & R_{-1} & R_{-2} \ldots & R_{-N} \\ R_1 & R_0 & \ldots\ldots & R_{-N+1} \\ \vdots & \vdots & \ddots & \vdots \\ R_N & R_{N-1} & \ldots\ldots & R_0 \end{vmatrix}. \tag{5.164}$$

D_N and A_n are the cofactor matrices of $N \times N$ regarding R_0 and R_{-n} in the Toeplitz matrix[21] D_{N+1} of $(N+1) \times (N+1)$, respectively.[22]

If R_ls are true correlation functions, the above relation holds strictly. However, $N+1$ estimated values are used for R_l in the MEM, and their fluctuations are not taken into consideration in actual calculations. As a result, the power spectral density estimated by (5.160) fluctuates more conspicuously with increasing N. Hence, the dimension N must be determined in a suitable way. If N becomes large, the power spectral density estimated by (5.160) approaches periodogram.

In the improved MEM method by Ogura and Yoshida (1981), it is possible to estimate power spectral density in the higher precision by calculating $\{e_n{}^2, a_1, \ldots, a_n\}$ with considering the fluctuation of estimated correlation functions.

5.6 Correlation and Accuracy of Sampled Signal

Parameters such as the echo intensity, the mean Doppler frequency (velocity), and the Doppler frequency spectrum width (velocity spectrum width) are estimated from either power spectral density or correlation function. Here, it should be noted that both the power spectral density and the correlation function are estimates of the true values which are obtained by actual observation. Therefore, it is important to know accuracy of the estimated parameters. First, assuming that these parameters are obtained by ideal radar which is free from any error of observation, we discuss correlation function, coherent integration, and incoherent integration. Next, we theoretically evaluate the errors which originate in the SNR and the Doppler frequency spectrum width.

5.6.1 Correlation Function and Correlation Time

As discussed in Sect. 5.5.6, the power spectra are often well described by Gaussian functions. Meanwhile, if the time correlation function $g(t)$ is Gaussian function whose variance about t is given by σ_t^2, it is expressed as

$$g(t) = \frac{1}{\sqrt{2\pi}\sigma_t} \exp\left(-\frac{t^2}{2\sigma_t^2}\right), \tag{5.165}$$

[21] The matrix whose elements in the diagonal line are identical is called the Toeplitz matrix, and its determinant is called the Toeplitz determinant.

[22] In actual numerical calculations, the simple Levinson algorithm which suites sequential computation is adopted.

and its Fourier transform $G(f)$ becomes

$$G(f) = \int_{-\infty}^{\infty} g(t)\mathrm{e}^{-j2\pi ft}\mathrm{d}t = \exp\left[-\frac{4\pi^2 f^2}{2(1/\sigma_t^2)}\right]. \tag{5.166}$$

If $\sigma_f = (2\pi\sigma_t)^{-1}$ is applied, (5.166) is expressed as

$$G(f) = \exp\left(-\frac{f^2}{2\sigma_f^2}\right). \tag{5.167}$$

From (5.165) and (5.167), it is apparent that the Fourier transform of a Gaussian function also becomes a Gaussian function. It means that the autocorrelation function of a time function whose power spectral density is expressed by a Gaussian function also becomes the Gaussian function in the similar form (see Fig. 5.14c).

Correlation Function

Applying the relations $\overline{v} = -\lambda\overline{f}/2$ and $\overline{\sigma}_v = -\lambda\overline{\sigma}_f/2$ to (5.151), a Doppler-shifted frequency spectrum $S(f)$ becomes

$$S(f) = \frac{S_0}{\sqrt{2\pi}\sigma_f}\exp\left[-\frac{(f-\overline{f})^2}{2\sigma_f^2}\right], \tag{5.168}$$

where f is Doppler frequency and $\overline{f}$ is the expectation (mean value of f, and is also shown by f_d). Accordingly, (5.165) and (5.167) are used in the relation of (5.74), and the autocorrelation function $R(T_\mathrm{s})$ is expressed as

$$R(T_\mathrm{s}) = S_0 \exp(-2\pi^2\sigma_f^2 T_\mathrm{s}^2)\exp(j2\pi\overline{f}T_\mathrm{s}) \tag{5.169}$$

$$= S_0 \exp\left(-\frac{8\pi^2\sigma_v^2 T_\mathrm{s}^2}{\lambda^2}\right)\exp\left(-j\frac{4\pi\overline{v}T_\mathrm{s}}{\lambda}\right) + N, \tag{5.170}$$

where T_s is the time lag, i.e., sampling interval, and N is defined as the mean white noise power. From (5.170), the normalized signal correlation, or the correlation coefficient $\rho_\mathrm{n}(T_\mathrm{s})$ is given by

$$\rho_\mathrm{n}(T_\mathrm{s}) = \exp\left[-8\left(\frac{\pi\sigma_v T_\mathrm{s}}{\lambda}\right)^2\right]. \tag{5.171}$$

Correlation Time

The time lag T_c that the correlation coefficient of (5.171) falls to e^{-1} is called a correlation time. In this case, $8(\pi\sigma_v T_c/\lambda)^2 = 1$, and

$$T_c = \frac{\lambda}{2\sqrt{2}\pi\sigma_v}. \tag{5.172}$$

The longer the correlation time is, the more moderate the change. If a number of samples are obtained within the correlation time, it is possible to assume that the individual samples have correlation and are sampled during an identical physical process. Thus, in signal processing to calculate power spectral density, samples obtained within the correlation time are used. Also, coherent integration which will be mentioned later is performed within the correlation time.

Independent Samples

The elapsed time that mutual correlation between received signal samples approximately dissolves changes depending on the condition of scatterers and observed altitude. The independent sample time in which each sample becomes independent is defined by the time lag T_i that the correlation coefficient of (5.171) falls to $e^{-\pi}$. At this time, $8(\pi\sigma_v T_i/\lambda)^2 = \pi$, and

$$T_i = \frac{\lambda}{2\sqrt{2\pi}\sigma_v}. \tag{5.173}$$

Mean receiver output value of M velocity samples $v_1, v_2, \cdots, v_M$, i.e., expected value $\overline{v}$, is given by

$$\overline{v} = \frac{1}{M}\sum_{i=1}^{M} v_i. \tag{5.174}$$

If M samples are mutually independent, and the variance of $v_1, v_2, \cdots, v_M$ are expressed as σ_v^2, then the receiver output velocity variance is reduced to $1/M$, and its expected value $E[\sigma_v^2]$ becomes

$$E[\sigma_v^2] = \frac{\sigma_v^2}{M}. \tag{5.175}$$

Meanwhile, if samples have correlation, total number of independent samples M_I becomes smaller than M, and is given by the function of M and the normalized

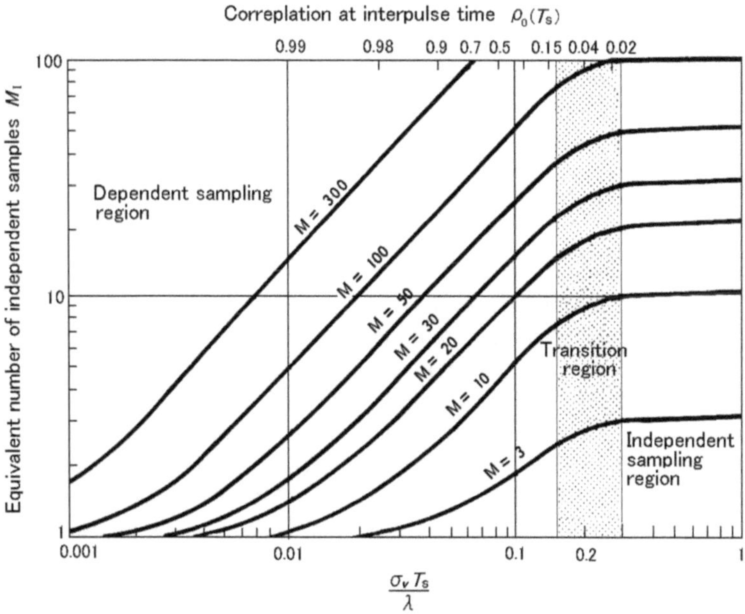

Fig. 5.18 Number of independent samplcs versus the correlation coefficient for a Gaussian distribution [from Nathanson 1991, p. 97]

autocorrelation function $\rho_0(T_s)$. If the receiver transfer function is square law,[23] M_I is given by Zrnić (1979)

$$\frac{E[\sigma_v^2]}{\sigma_v^2} = \frac{1}{M_I} = \sum_{m=-(M-1)}^{M-1} \frac{M-|m|}{M^2} \rho_0(mT_s), \tag{5.176}$$

where

$$\rho_0(mT_s) = \exp\left[-8\left(\frac{\pi\sigma_v mT_s}{\lambda}\right)^2\right], \tag{5.177}$$

and T_s is the sampling interval.

Figure 5.18 shows the relationship between the correlation coefficient and number of independent samples calculated from (5.176), assuming the correlation function has a Gaussian shape (Nathanson 1991, p. 97). As apparent from the figure, samples are transferred to the independent condition at the time around $\sigma_v T_s/\lambda = 0.2$. The value of T_s is approximately equivalent to T_i, which is given

[23] The output sample Q_k is proportional to the input power P_k, i.e., $Q_k \propto P_k \propto V|k|^2$ (Doviak and Zrnić 1984, p. 95).

by (5.173). The signals sampled at the interval longer than this value are independent ($M = M_\mathrm{I}$). It is noted that the averaging of independent samples contributes to suppressing the variance of signals.

5.6.2 Coherent Integration

Because there is a statistical fluctuation in signals from Bragg scatterers, averaging (or integration) of sampled signals which are independent with each other is indispensable to obtain precise amplitude and phase information. The correlation time of signals from Bragg scatterers at very long wavelengths is much longer than the pulse repetition time (PRT). If that is the case the coherent averaging can be instituted to reduce the signal processing load. In addition, vertically pointed beams typically require a small unambiguous range and thus pulses can be transmitted at a high pulse repetition frequency (PRF). Using high PRF spreads the noise power over a large Nyquist interval and thus received complex echoes can be time averaged (i.e., the so called coherent integration) before Doppler processing as shown in Fig. 5.12.

The processing is performed in the coherent integrator which unites the signal phase at the identical range point (or range bin). Because signals are coherent over a time interval short compared to the signal's correlation time, samples of I and Q for the signal do not change significantly over time periods less than the correlation time. But the noise samples are incoherent and change randomly from sample to sample. Thus the sum of I and Q signal samples progressively increases (i.e., a coherent sum) whereas the sum of the I and Q components of noise weakens on average. If a complex voltage of the coherent integrator $v(k)$ is comprised of a signal $s(k) = A\mathrm{e}^{j\varphi}$ (A: amplitude; φ: phase) and complex white Gaussian noise $n(k)$ of variance σ^2,

$$v(k) = s(k) + n(k) = A\mathrm{e}^{j\varphi} + n(k), \tag{5.178}$$

and the SNR of $v(k)$, denoted $(S/N)_1$ is

$$(S/N)_1 = \frac{A^2}{\sigma^2}. \tag{5.179}$$

By adding M samples, the coherent integrator output becomes

$$\sum_{k=0}^{M-1} v(k) = \sum_{k=0}^{M-1} s(k) + \sum_{k=0}^{M-1} n(k) = MA\mathrm{e}^{j\varphi} + \sum_{k=0}^{M-1} n(k) \tag{5.180}$$

which keeps the form of the sum of a signal component and a noise component. The power in the signal component is $(MA)^2$, whereas that of the noise component is expressed as

$$E\left[\left|\sum_{k=0}^{M-1} n(k)\right|^2\right] = E\left\{\left[\sum_{k=0}^{M-1} n(k)\right]\left[\sum_{l=0}^{M-1} n^*(l)\right]\right\} = \sum_{k=0}^{M-1}\sum_{l=0}^{M-1} E\left[n(k)n^*(l)\right] = M\sigma^2. \tag{5.181}$$

Thus the SNR of the coherent integrator for M data samples $(S/N)_M$ becomes

$$(S/N)_M = M\frac{A^2}{\sigma^2} = M(S/N)_1. \tag{5.182}$$

It means that coherent integration within the correlation time, from (5.172) about $0.113\lambda/\alpha_v$,[24] provides a factor of M gain in SNR.

As shown in Fig. 5.12, the actual coherent integrator is inserted following to the matched filter and phase detector in receiver. If it is implemented as a band-pass filter, then the Gain M corresponds to B_N/B, where B_N is the effective noise bandwidth and B is the receiver bandwidth (the band-pass filter width). If the coherent integrator is implemented with a FFT, then M is the number of taps in the FFT.

A conceptual diagram of the coherent integration of the FFT is shown in Fig. 5.19. Panel (a) shows the frequency spectrum of received signal similar to Fig. 5.13b. Panel (b) is the expansion of the line spectrum of (a). In these panels, the origin of the horizontal axis f is the center frequency of the carrier, and is zero after phase measurement, T_s is the sampling interval, f_d is the Doppler frequency, and σ_f is the width of Doppler frequency perturbation (standard deviation). The frequency spectrum of the received signal is the line spectrum of period $1/T_s$, and the envelope corresponds to the frequency spectrum of each single pulse. The coherent integration is equivalent to extracting the spectrum of the signal with the comb filter (pulse Doppler filter) shown as the solid lines of Panel (c), where the filter bandwidth is equivalent to $1/MT_s$. If the received signals are integrated within the correlation time, the spectrum are add in the identical comb filter, and the noise power increases or decreases in proportion to the filter bandwidth.

5.6.3 *Incoherent Integration*

The estimates of power spectrum obtained by the FFT or other means in the frequency domain fluctuate around the true spectrum, and have a statistical variance at each frequency. As previously described in Sect. 5.4.2, the periodogram does not converge to the true power spectral density even if the period $T \to \infty$, or the sampling number $m \to \infty$. Therefore, averaging (or integration) of periodograms

[24] Smith (1986) shows the time available for coherent integration as approximately $0.14\lambda/\alpha_v$ or neighborhood of $\lambda/(6\langle v\rangle)$, where $\langle v\rangle$ is the mean Doppler velocity.

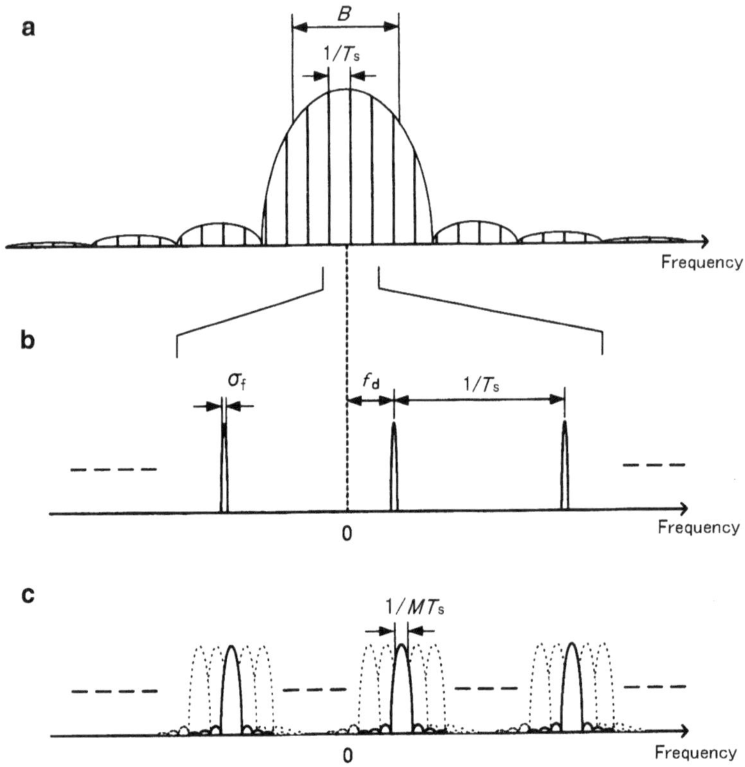

Fig. 5.19 Conceptual diagram of the coherent integration with a FFT. (**a**) Frequency spectrum of received signal, (**b**) expansion of line spectrum in (**a**), and (**c**) Comb filters of coherent integration. In (**c**), the coherent integration is equivalent to extracting the solid lines

is needed for estimating the true power spectral density. That is, the average approaches the true power spectral density if multiple independent periodograms are added at each frequency. Such averaging processing in the frequency domain is called incoherent integration. Since phase information is discarded in incoherent integration, separation of signal component and noise component such as shown in (5.180) cannot be expected. Figure 5.20 shows schematic power spectrum averaging of signal plus noise. (a) No averaging, $M_s = 1$, (b) averaging of $M_s = 10$, and (c) $M_s = 100$, where M_s is the number of spectra being averaged (adapted from Lyons 2004, Chap. 11).

To discuss averaging (or integration) of power, it might be useful to take a general view of the χ^2-distribution first. We may assume that the amplitude of a random variable x has a Gaussian distribution expressed as

$$a(x) = \frac{1}{\sqrt{2\pi}} \exp\left(-\frac{x^2}{2}\right), \tag{5.183}$$

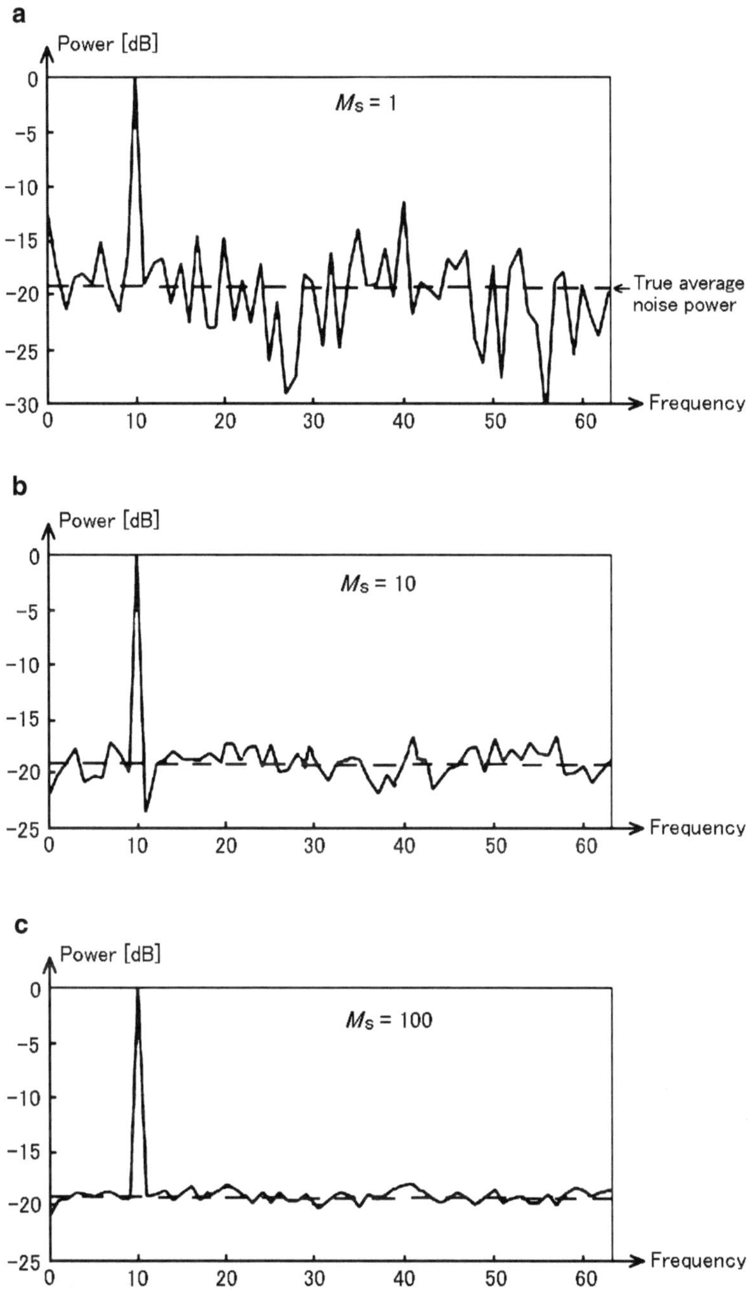

Fig. 5.20 Schematic power spectrum averaging of signal plus noise. (**a**) No averaging, $M_s = 1$, (**b**) averaging of $M_s = 10$, and (**c**) $M_s = 100$, where M_s is the number of spectra being averaged [adapted from Lyons 2004, Chap. 11]

where the mean value and standard deviation are assumed to be 0 and 1, respectively. When independent estimates of power x^2 are added m times as

$$y = \sum_{k=1}^{m} x_k^2, \tag{5.184}$$

it can be shown that y follows a χ^2-distribution given by

$$b(y) = \frac{(y/2)^{m/2-1} \exp(-y/2)}{2\Gamma(m/2)}, \tag{5.185}$$

where $\Gamma(m/2)$ is defined as

$$\Gamma(m/2) = \begin{cases} \left(\frac{m}{2} - 1\right)! & \text{when } m \text{ is even integer} \\ \left(\frac{m}{2} - 1\right)\left(\frac{m}{2} - 2\right) \cdots \frac{1}{2}\sqrt{\pi} & \text{when } m \text{ is odd integer and } m \geq 3 \\ \sqrt{\pi} & \text{when } m = 1. \end{cases} \tag{5.186}$$

The mean value μ and standard deviation σ of the χ^2-distribution are

$$\mu = m, \tag{5.187}$$

$$\sigma = \sqrt{2m}. \tag{5.188}$$

Therefore, when the square of random variable is added m times, the resultant mean value or the integrated power increases m times. The number of averaging (or integration) m is called the degree of freedom of the χ^2-distribution.

As discussed in Sect. 5.3.2, I and Q components are both assumed to follow the Gaussian distribution, and signal power which is proportional to $I^2 + Q^2$ follows an exponential function, and its degree of freedom is 2. When M_s power spectra are averaged (integrated; $m = 2M_s$), the degree of freedom becomes $2M_s$. From (5.187) and (5.188), the ratio σ/μ is given by

$$\frac{\sigma}{\mu} = \frac{1}{\sqrt{M_s}}. \tag{5.189}$$

It means that the fluctuation of signal and noise decreases to $1/\sqrt{M_s}$ by a incoherent integration of M_s independent samples, but such improvement of SNR as achieved by the coherent integration cannot be expected.

The incoherent integration gain depends upon the values specified for the false-alarm probability P_{fa} and the probability of detection P_d in a complicated way

(Smith 1986). It is estimated by considering the SNR required to achieve a specified performance when multiple data samples by incoherent integration are applied, and comparing that to the SNR to achieve the same performance when only single sample is used. Thus, the incoherent integration gain $I_{\mathrm{inc}}(M_{\mathrm{s}})$ is expressed as

$$I_{\mathrm{inc}}(M_{\mathrm{s}}) = \frac{(S/N)_1}{(S/N)'_{M\mathrm{s}}}, \tag{5.190}$$

where $(S/N)_1$ is the SNR for one pulse, and $(S/N)'_{M\mathrm{s}}$ is the SNR for one pulse of the same P_{d} and the same P_{fa} with the integration M_{s}. From (5.189), $I_{\mathrm{inc}}(M_{\mathrm{s}}) = \sqrt{M_{\mathrm{s}}}$ is expected.

5.6.4 Standard Deviation of Radar Reflectivity Factor

Doviak and Zrnić (2006, errata[25]) have discussed in detail the accuracy of the observational parameter estimates. Based upon their description, estimate accuracy for the received power, the mean Doppler velocity, and the Doppler velocity spectrum width are subsequently discussed in Sects. 5.6.4–5.6.6.

The estimated value for the radar reflectivity factor at the receiver output is given by

$$\hat{Z} = \alpha\hat{S} = \alpha(\hat{P} - N), \tag{5.191}$$

where symbol $(\hat{\ })$ denotes estimates, whereas expected value otherwise, α is a constant determined from radar equation, $\hat{S}$ is estimate for signal power, N noise power, and $\hat{P}$ the mean value of M samples of output power P_k, which is given by

$$\hat{P} = \bar{P} = \frac{1}{M}\sum_{k=0}^{M-1} P_k. \tag{5.192}$$

Radar reflectivity factor Z $[\mathrm{mm}^6\,\mathrm{m}^{-3}]$ of (5.191) is expressed in decibels by

$$\hat{Z}\ [\mathrm{dBZ}] = 10\log_{10}\hat{Z} = 10\log_{10}\alpha\hat{S}. \tag{5.193}$$

Expressing $\hat{S} = S + \delta S$, where S is the expected value, and δS is the displacement of $\hat{S}$ from S, (5.193) becomes (Doviak and Zrnić 2006, errata, p. 128, (6.13b))

[25] The errata and supplements for the third and fourth printings of Doviak and Zrnić (2006, Chap. 6). http://www.nssl.noaa.gov/papers/books.html.

$$\hat{Z}\,[\mathrm{dBZ}] = 10\log_{10}\alpha S + 10\log_{10}\left(1+\frac{\delta S}{S}\right) = Z + \delta Z. \tag{5.194}$$

If M is large enough, $\delta S/S$ becomes smaller than 1. Therefore, the second term of (5.194) is approximated by applying Taylor series expansion to it:

$$10\log_{10}\left(1+\frac{\delta S}{S}\right) \simeq \frac{10}{\ln 10}\frac{\delta S}{S}. \tag{5.195}$$

Substitution of (5.195) into (5.194) leads to

$$\hat{Z}\,[\mathrm{dBZ}] \simeq 10\log_{10}\alpha S - 4.34 + 4.34\frac{\hat{S}}{S}. \tag{5.196}$$

As the variable contained in (5.196) is the third term only, the estimate error, i.e., the standard deviation (SD) of $\hat{Z}[\mathrm{dBZ}]$ is simplified as

$$\mathrm{SD}[\hat{Z}\,[\mathrm{dBZ}]] = 4.34\ \mathrm{SD}[\hat{S}/S], \tag{5.197}$$

where $\hat{S} = \hat{P} - N$, and N is assumed to be a given fixed value. Thus,

$$\mathrm{SD}[\hat{S}] = \mathrm{SD}[\hat{P}] = \frac{P}{\sqrt{M_\mathrm{I}}} = \frac{S+N}{\sqrt{M_\mathrm{I}}}, \tag{5.198}$$

where M_I is the number of samples which are independent out of M samples that are used for the averaging. From (5.197) and (5.198),

$$\mathrm{SD}[\hat{Z}\,[\mathrm{dBZ}]] = \frac{4.34(S+N)}{S\sqrt{M_\mathrm{I}}} \tag{5.199}$$

is derived (Doviak and Zrnić 2006, errata, p. 128, (6.13c)). If sampling is made at a constant interval T_s in a steady state, M_I is determined in the similar way as (5.176)

$$\frac{1}{M_\mathrm{I}} = \sum_{m=-(M-1)}^{M-1} \frac{M-|m|}{M^2}\rho_{S+N}(mT_\mathrm{s}), \tag{5.200}$$

where $\rho_{S+N}(mT_\mathrm{s})$ is the autocorrelation coefficient for the power sample with noise included at time lag mT_s.

Here, it is assumed, for simplicity, that the receiver has a square-law characteristic.[26] In this case, the correlation coefficient of square-law receiver output signal (power sample) becomes square of the correlation coefficient of input signal sample.

[26] The square law receiver is typically the receiver of choice because it does not produce any bias and has the minimum variance (Doviak and Zrnić 1984, Fig. 6.2b).

If a signal spectrum of the Gaussian distribution is superposed with white noise spectrum, the correlation coefficient of the signal S with white noise N included, where S and N are normalized by $S+N$, respectively, is expressed as (Doviak and Zrnić 2006, errata, p. 128, (6.13d))

$$\rho_{S+N}(mT_s) = \left[\frac{S}{S+N}\mathrm{e}^{-2(\sigma_{vn}\pi m)^2} + \frac{N}{S+N}\delta_m\right]^2, \tag{5.201}$$

where $\delta_m = 1$ for $m = 0$ and $\delta_m = 0$ otherwise. The parameter σ_{vn} is the Doppler velocity spectrum width which is normalized with the Nyquist width $2v_N = \lambda/2T_s$, and given by

$$\sigma_{vn} = \frac{2\sigma_v T_s}{\lambda}, \tag{5.202}$$

where λ is radar wavelength.

According to Doviak and Zrnić (2006, errata, p. 128), alternative to a numerical solution of (5.200), there are conditions whereby we can obtain a useful analytical solution. It can be shown the condition $\sigma_{vn} \leq 0.25$ still applies to replace the sum of (5.200) by an integral even if $N \neq 0$; furthermore, $\sigma_{vn} \geq 2.4/M$ becomes a sufficient condition to ignore the term m/M in that equation. Because the dwell time MT_s is typically much longer than the correlation time of the weather signal samples, $\rho_{s+n}(MT_s) \ll 1$ and thus the limits on the integral can be extended to infinity. Evaluation of this integral under these condition yields (Doviak and Zrnić 2006, errata, p. 128, (6.13e))

$$M_I = \frac{\left(1+\frac{S}{N}\right)^2 M}{1+2\frac{S}{N}+\frac{(S/N)^2}{2\sigma_{vn}\sqrt{\pi}}}. \tag{5.203}$$

The formula for calculating the standard error in estimating Z[dBZ] as a function of S/N is obtained by substituting (5.203) into (5.199) yielding (Doviak and Zrnić 2006, errata, p. 128, (6.13f))

$$\mathrm{SD}[\hat{Z}\ [\mathrm{dBZ}]] = \frac{4.34}{M^{1/2}}\frac{N}{S}\left[1+2\frac{S}{N}+\frac{(S/N)^2}{2\sigma_{vn}\sqrt{\pi}}\right]^{1/2} \quad [\mathrm{dB}], \tag{5.204}$$

a result agreeing with that obtained from simulation (Melnikov et al. 2011). Although $\mathrm{SD}[\hat{Z}\ [\mathrm{dBZ}]]$ increases without bound as S/N decreases, it can be shown the SD of reflectivity factor estimates, expressed in linear units (i.e., $\mathrm{mm^6\,m^{-3}}$), decreases as S/N decreases and reaches a minimum value at $S = 0$ (i.e., $\mathrm{SD}[\hat{Z}\ [\mathrm{mm^6\,m^{-3}}]] = \alpha N/\sqrt{M}$ if $S = 0$; in this case $\hat{Z} = 0$). This apparent dichotomy is explained by the fact that the slope of the logarithmic function increases without

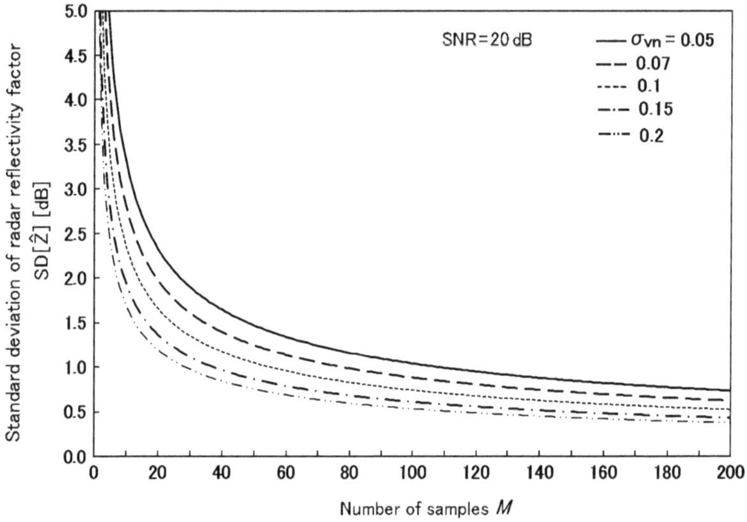

Fig. 5.21 Relationship between the standard deviation of the radar reflectivity factor and the number of samples M for various normalized Doppler velocity spectrum widths σ_{vn}. The plotted curves are calculated based on (5.204), which are normalized with the Nyquist width $2v_{\mathrm{N}}$. Curves apply if effective for $\sigma_{vn} \geq 2.4/M$

bound as its argument decreases. That is, a constant width of a distribution on the linear scale increases on the logarithmic scale as the mean decreases.

The relationship between the standard deviation of the radar reflectivity factor [dBZ], and the number of samples M for normalized Doppler velocity width, when SNR is large ($=20\,\mathrm{dB}$), is shown in Fig. 5.21.

5.6.5 Standard Deviation of Mean Doppler Velocity

Spectral Analysis in Frequency Domain

The variance of the mean Doppler frequency $\mathrm{var}[\hat{f}_{\mathrm{f}}]$ can be derived from the signal power spectral density. That is given by integrating the mean Doppler frequency which is weighted with power spectral density of signal within the Nyquist frequency width, $-1/2T_{\mathrm{s}}$ to $1/2T_{\mathrm{s}}$, as (Berger and Groginsky 1973)

$$\mathrm{var}[\hat{f}_{\mathrm{f}}] = \frac{1}{MT_{\mathrm{s}}S^2} \int_{-1/2T_{\mathrm{s}}}^{1/2T_{\mathrm{s}}} f^2 S^2(f + \hat{f}_{\mathrm{f}})\mathrm{d}f, \tag{5.205}$$

where $S(f)$ is the power spectral density of signal with noise included and given by $S(f) = S_{\mathrm{S}}(f) + S_{\mathrm{N}}(f)$. Note that $S_{\mathrm{S}}(f)$ and $S_{\mathrm{N}}(f)$ are the power spectral density of signal and noise, respectively, and $\hat{f}_\mathrm{f}$ is the mean Doppler frequency obtained by spectral analysis. If $S(f)$ is the signal spectrum of the Gaussian distribution expressed by (5.168) and its Doppler frequency spectrum width σ_f is narrow enough ($\sigma_f T_\mathrm{s} \ll 1$), the variance of mean Doppler velocity $\mathrm{var}[\hat{v}_\mathrm{f}]$ is given from $\mathrm{var}[\hat{v}_\mathrm{f}] = (\lambda/2)^2 \mathrm{var}[\hat{f}_\mathrm{f}]$, as (Doviak and Zrnić 2006, p. 135)

$$\mathrm{var}[\hat{v}_\mathrm{f}] = \frac{\lambda^2}{4MT_\mathrm{s}^2}\left[\frac{\sigma_{vn}}{4\sqrt{\pi}} + 2(\sigma_{vn})^2\frac{N}{S} + \frac{1}{12}\left(\frac{N}{S}\right)^2\right], \tag{5.206}$$

where σ_{vn} is the Doppler velocity spectrum width which is normalized by the Nyquist width $2v_\mathrm{N}$. The standard deviation of mean Doppler velocity is given by $\mathrm{SD}[\hat{v}_\mathrm{f}] = \sqrt{\mathrm{var}[\hat{v}_\mathrm{f}]}$.

Autocovariance Analysis in Time Domain

If number of samples M and the Doppler velocity spectrum width σ_{vn} normalized by the Nyquist width satisfies the following condition,

$$\frac{1}{2\pi M} \ll \sigma_{vn} \ll 1, \tag{5.207}$$

the variance of the mean Doppler velocity $\mathrm{var}[\hat{v}_\mathrm{t}]$ is expressed as (Doviak and Zrnić 2006, p. 134)

$$\mathrm{var}[\hat{v}_\mathrm{t}] \simeq \frac{\lambda^2}{32\pi^2 M\rho^2(T_\mathrm{s})T_\mathrm{s}^2}\left\{\frac{1-\rho(T_\mathrm{s})}{2\sigma_{vn}\sqrt{\pi}} + \left(\frac{N}{S}\right)^2 + 2\frac{N}{S}[1-\rho(2T_\mathrm{s})]\right\}, \tag{5.208}$$

where $\rho(T_\mathrm{s})$ is the correlation coefficient which was shown in (5.171) and given by

$$\rho(T_\mathrm{s}) = \exp[-8(\pi\sigma_v T_\mathrm{s}/\lambda)^2]. \tag{5.209}$$

If samples are independent, $\mathrm{var}([\hat{v}_\mathrm{t}])$ is expressed as (Doviak and Zrnić 2006, p. 134)

$$\mathrm{var}[\hat{v}_\mathrm{t}] \simeq \frac{\lambda^2}{32\pi^2 M\rho^2(T_\mathrm{s})T_\mathrm{s}^2}\left[(1+N/S)^2 - \rho^2(T_\mathrm{s})\right]. \tag{5.210}$$

The standard deviation of the mean Doppler velocity is given by $\mathrm{SD}[\hat{v}_\mathrm{t}] = \sqrt{\mathrm{var}[\hat{v}_\mathrm{t}]}$.

The relationship between the standard deviation of mean Doppler velocity and Doppler velocity spectrum width for various SNR calculated based on (5.206) and (5.208) is shown in Fig. 5.22.

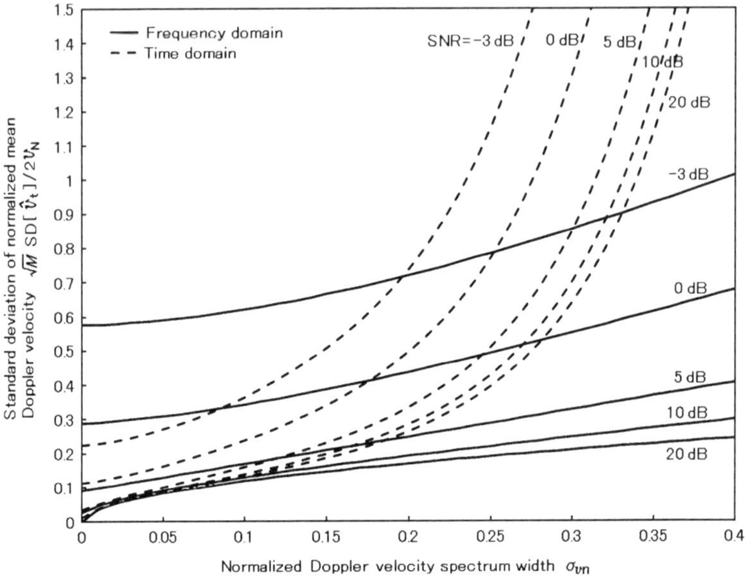

Fig. 5.22 Relationship between the standard deviation of the mean Doppler velocity and the Doppler velocity spectrum width for various SNR, which are normalized with the Nyquist width $2v_N$. Calculated based on (5.206) and (5.208). Estimation error increases in the region $\sigma_{vn} > 0.25$

5.6.6 Standard Deviation of Spectrum Width

Spectral Analysis in Frequency Domain

If the Doppler frequency spectrum is of a Gaussian distribution and the spectrum width is narrow enough, the variance of spectrum width $\mathrm{var}[\hat{\sigma}_f]$ for the Doppler frequency spectrum width which is given by (5.147) can be obtained in the similar way as (5.205) and (5.206). Thus, the variance of the Doppler velocity spectrum width $\mathrm{var}[\hat{\sigma}_v]$ is given by Doviak and Zrnić (2006, p. 140)

$$\mathrm{var}[\hat{\sigma}_v] = \frac{\lambda^2}{4MT_s^2}\left[\frac{3\sigma_{vn}}{32\sqrt{\pi}} + \sigma_{vn}^2\frac{N}{S} + \left(\frac{1}{320\sigma_{vn}^2} - \frac{1}{24} + \frac{\sigma_{vn}^2}{4}\right)\left(\frac{N}{S}\right)^2\right]. \quad (5.211)$$

The standard deviation of the Doppler velocity spectrum width is given by $\mathrm{SD}[\hat{\sigma}_v] = \sqrt{\mathrm{var}[\hat{\sigma}_v]}$.

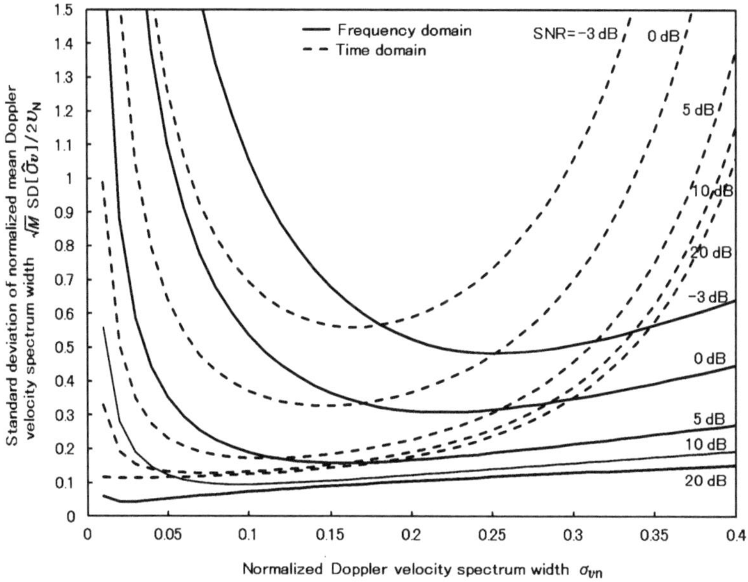

Fig. 5.23 Relationship between the standard deviation of the Doppler velocity spectrum width and the Doppler velocity spectrum width for various SNR, which are normalized with the Nyquist width $2v_{\mathrm{N}}$. Calculated based on (5.211) and (5.212). Estimation error increases in the region $\sigma_{vn} > 0.25$

Autocovariance Analysis in Time Domain

Under the condition of (5.207), the variance of the Doppler velocity spectrum width $\mathrm{var}[\hat{\sigma}_v]$ is given by Doviak and Zrnić (2006, p. 137)

$$\mathrm{var}[\hat{\sigma}_v] = \frac{\lambda^2}{128M\pi^4\sigma_{vn}^2\rho^2(T_s)T_s^2} \times \left\{[2-2\rho^2(T_s)+\rho^4(T_s)]\frac{N}{S}+[1+2\rho^2(T_s)]\left(\frac{N}{S}\right)^2+F\right\}, \quad (5.212)$$

where F is

$$F=\rho^2(T_s)\sum_{m=-(M-1)}^{M-1}\left\{3\rho^2(mT_s)+\frac{\rho^2(mT_s)}{\rho^2(T_s)}-4\frac{\rho[(m+1)T_s]\rho(mT_s)}{\rho(T_s)}\right\}\left(1-\frac{|m|}{M}\right), \quad (5.213)$$

σ_{vn} is given by (5.202), and $\rho(T_s)$ is given by (5.209). If samples are independent, $\mathrm{var}[\hat{\sigma}_v]$ is given by Doviak and Zrnić (2006, p. 137)

$$\mathrm{var}[\hat{\sigma}_v] = \frac{\lambda^2}{128M\pi^4\sigma_{vn}^2\rho^2(T_\mathrm{s})T_\mathrm{s}^2} \times \left\{ \left[1-\rho^2(T_\mathrm{s})\right]^2 + 2\left[1-\rho^2(T_\mathrm{s})\right]\frac{N}{S} + \left[1+\rho^2(T_\mathrm{s})\right]\frac{N^2}{S^2} \right\}. \quad (5.214)$$

The standard deviation of the Doppler velocity spectrum width is given by $\mathrm{SD}[\hat{\sigma}_v] = \sqrt{\mathrm{var}[\hat{\sigma}_v]}$. Relationship between the Doppler velocity spectrum width and its standard deviation for various SNR calculated based on (5.211) and (5.212) is shown in Fig. 5.23.

Chapter 6
Radar Observations of Precipitation

Among various physical quantities which are measured with meteorological and atmospheric radars, precipitation is widely considered the most important quantity from the viewpoint of the practical use. In this chapter, we first discuss the relation between the basic radar quantities which are obtained by a single radar, including radar reflectivity factor, mean Doppler frequency, and mean Doppler spectrum width, and the typical physical parameters which represent characteristics of precipitation, including water contents, the rainfall rate, and the drop size distribution (DSD). Then, we describe the specific techniques to derive these physical parameters from the basic radar quantities. As shown in Fig. 3.9, the radar reflectivity by precipitation particles becomes larger than that by refractive index perturbations in the frequency range higher than 2–3 GHz. Therefore, as for discussions in this chapter, meteorological radar and the related subjects become the main topics. However, the relation between basic radar quantities and physical parameters concerned with precipitation is common to both meteorological and atmospheric radars. Meanwhile, for high precision estimation of DSD which is discussed in Sect. 6.2, it is crucially important to separate particle's falling velocity from the vertical motion of the background atmosphere, and thus the estimation is usually performed as the part of atmospheric radar observations.

Next, the attenuation of radar wave in the atmosphere which sometimes significantly affects observations by the meteorological radar is discussed. The attenuation is not significant in the frequency range which is generally used for atmospheric radar, and thus, this discussion is limited primarily to meteorological radar.

Finally, techniques to estimate precipitation rate and the DSD in higher precision by using polarimetric radar (or dual polarization radar) are described.

S. Fukao and K. Hamazu, *Radar for Meteorological and Atmospheric Observations*,
DOI 10.1007/978-4-431-54334-3_6, © Springer Japan 2014

6.1 Parameters of Precipitation

6.1.1 *Parameters of Drop Size*

Drop Size Distributions of Precipitations

Precipitation particles include raindrops, graupels, hails, clouds, fogs, and so on. Thus, meteorological radars, in general, choose the wavelengths where the Rayleigh approximation becomes effective. Because radar cross section is proportional to the sixth power of particle diameter D as discussed in Sect. 3.3, the rainfall rate is related closely to the DSD when rainfall rate is given as the function of the radar cross section (or radar reflectivity factor). A variety of functions have been so far proposed for the DSD, and among them, the following gamma distribution is known as the most popular one which expresses a general DSD (e.g., Ulbrich 1983; Willis 1984).

$$N(D) = N_0 D^{\mu} \exp(-\Lambda D). \tag{6.1}$$

This distribution depends on three parameters, N_0, μ, and Λ. If small drop size is predominant and the distribution monotonically decreases as drop size becomes large, the values of $\mu \leq 0$ fit better. On the other hand, when a specific drop size which is not the minimum diameter is predominant, the values of $\mu > 0$ represent the distribution better (Ulbrich 1983). Theoretical models of the DSD generated by breakup and coalescence have been presented by e.g., Srivastava (1967, 1971).

The parameter N_0 is an intercept parameter of the distribution curve, and its unit becomes $[\mathrm{m}^{-1-\mu}\ \mathrm{m}^{-3}]$ if the unit of diameter is expressed in [m]. The parameter Λ $[\mathrm{m}^{-1}]$ is a slope parameter which shows the gradient of distribution curve. Figure 6.1 illustrates the drop size distributions for the gamma distribution for $\mu = -2,\ 0,\ +2$ in (6.1). When $\mu = 0$, $N(D)$ becomes an exponential distribution expressed as (Marshall and Palmer 1948)

$$N(D) = N_0 \exp(-\Lambda D), \tag{6.2}$$

The Marshall–Palmer (M–P) distribution belongs to the group of (6.2) with the following values for N_0 and Λ;

$$N_0 = 8.0 \times 10^6\ [\mathrm{m}^{-1}\mathrm{m}^{-3}], \tag{6.3}$$

$$\Lambda = 4100 R^{-0.21}\ [\mathrm{m}^{-1}], \tag{6.4}$$

where R is the rainfall rate $[\mathrm{mm\ h}^{-1}]$.

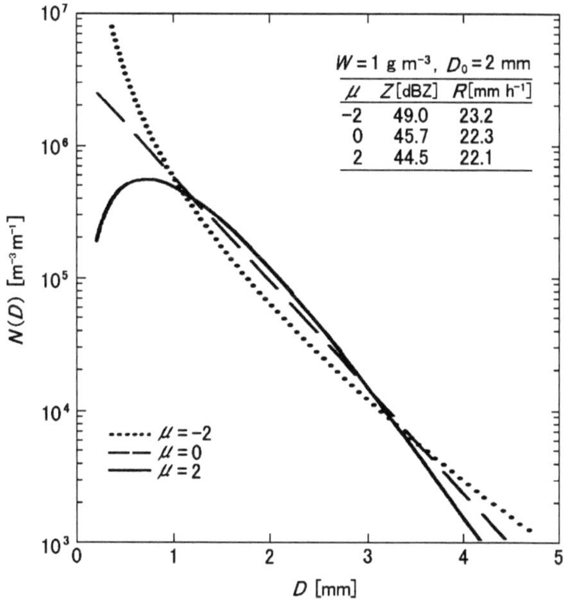

Fig. 6.1 Drop size distribution of the gamma distribution for the case where $W = 1$ [g m^{-3}] (see Sect. 6.1.3) and the median value of diameters $D_0 = 2$ mm are assumed. $N(D)$ is calculated for $\mu = -2,\ 0,\ +2$. The inset table shows the corresponding values of radar reflectivity factor [from Ulbrich 1983]

Higher Moments of DSD

It is convenient to use the higher moment of the DSD to express various physical quantities of precipitation. The nth moment of $N(D)$ is defined as

$$m_n \equiv \int_0^\infty D^n N(D) \mathrm{d}D. \tag{6.5}$$

Although the actual integration range is between $D_{\min}$ and $D_{\max}$, it is generally assumed, for simplicity, that $D_{\min} = 0$ and $D_{\max} = \infty$. The error caused by this approximation can be ignored to practical use. We adopt the following formula which is useful in expressing the DSD (Doviak and Zrnić 2006, p. 219);

$$\int_0^\infty x^{a-1} \mathrm{e}^{-bx} \mathrm{d}x = \frac{1}{b^a} \Gamma(a), \tag{6.6}$$

where $a > 1$ and $b > 1$. The function Γ is the complete gamma function, and $\Gamma(n+1) = n\Gamma(n)$. If n is an integer, $\Gamma(n+1) = n!$. Substituting (6.1) into (6.5) and utilizing (6.6), the nth moment of the DSD can be expressed as

$$m_n = N_0 \frac{\Gamma(n + \mu + 1)}{\Lambda^{n+\mu+1}}. \tag{6.7}$$

Terminal Velocity of Precipitation Particle

Precipitation particles are falling at a constant speed in the atmosphere, which is determined by the balance between the downward force of gravity and the upward resistance exerted by the atmosphere. The speed is called the terminal velocity. A variety of formulae have been proposed to express the terminal velocity w_T [m s^{-1}] of precipitation particles in the atmosphere. An empirical equation was given by (4.57) as the function of radar reflectivity factor and atmospheric density. In addition, it is possible to express it by the function of drop diameter D [m] in the range around 3×10^{-4} m $< D < 6 \times 10^{-3}$ m as follows (Atlas et al. 1973);

$$w_T(D) = 9.65 - 10.3 \exp(-600D). \tag{6.8}$$

The error from (6.8) has been confirmed less than 2 % by comparison with the actual measurement in the range of drop diameter 6×10^{-4} m $< D < 5.8 \times 10^{-3}$ m. The more simplified expression is given by power law fit to the actual measurement as follows (Atlas and Ulbrich 1977);

$$w_T(D) \simeq 386.6 D^{0.67}. \tag{6.9}$$

It is said that the above equation meets well with the actual measurement in the range 5×10^{-4} m $< D < 5 \times 10^{-3}$ m. However, the difference with the result of (6.8) becomes large gradually as shown in Fig. 6.2.

Theoretical terminal velocity of hailstones can be obtained through Newton's third law by equating the total drag force to the weight. The most uncertain parameter for determining the terminal velocity is the drag coefficient (Doviak and Zrnić 2006, p. 217). Matson and Huggins (1980) determined the best-fit relationship between the diameter and terminal velocity of hailstones with diameters less than 2.5×10^{-2} m as

$$w_T(D) \simeq 114.5 D^{0.5}, \tag{6.10}$$

where w_T [m s^{-1}] and D [m], respectively.

As for aggregate snowflakes, Gunn and Marshall (1958) used the formula for

$$w_T(D) \simeq 8.34 D^{0.31}, \tag{6.11}$$

where w_T [m s^{-1}] and D [m] is the diameter of a water sphere with the same mass as the snowflake less than 2.5×10^{-3} m.

Figure 6.2 shows the terminal velocities as a function of the particle diameters of raindrops, hailstone, and snowflake calculated by (6.8) to (6.11).

The falling speed of precipitation particles is affected by air resistance, and strictly speaking, the terminal velocity changes according to the atmospheric density. If atmospheric densities at sea level and at the level of observation are ρ_{a0} and ρ_a, respectively (Sect. 4.3.1), the terminal velocity of precipitation particles in diameter D becomes

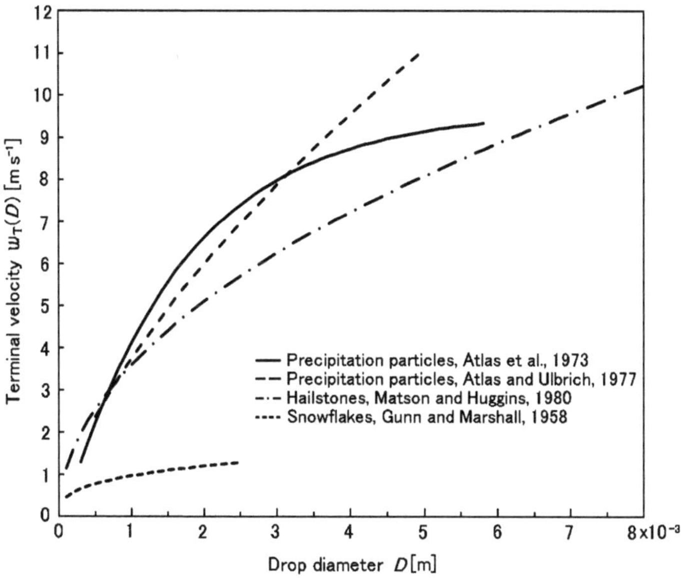

Fig. 6.2 Terminal velocities as a function of the particle diameters of raindrops, hailstone, and snowflake calculated by (6.8) to (6.11)

$$w_T(D,\rho_a) = w_T(D)\left(\frac{\rho_{a0}}{\rho_a}\right)^{0.4}. \tag{6.12}$$

Hereafter, we assume that the correction term due to atmospheric density in (6.12) is unity and so there is no change of terminal velocity due to atmospheric density unless otherwise mentioned.

6.1.2 Relations Between Basic Radar Parameters and DSD

Radar Reflectivity Factor and DSD

The radar reflectivity factor is the sixth moment of the DSD as given in the form of (3.66), and expressed by using the complete gamma function as

$$Z = \int_0^\infty D^6 N(D)\mathrm{d}D = m_6 = N_0 \frac{\Gamma(\mu+7)}{\Lambda^{\mu+7}}, \tag{6.13}$$

where the unit of Z is [m^{6}m^{-3}], N_0 [m$^{-1-\mu}$m^{-3}], and Λ[m^{-1}]. It should be noted that N_0 has a physical interpretation for the exponential distribution, i.e., it is the number density per unit drop diameter. However, for the gamma distribution, as shown in Sect. 6.1.1, it depends on μ and does not seem to have any physical meaning.

Doppler Velocity and DSD

The mean Doppler velocity $\overline{w}_{\mathrm{d}}$ in the vertical direction which is measured with ground based radar is expressed by the following equation as the sum of the terminal velocity w_{T} of precipitation particles and the vertical component w of the background wind velocity, where the velocity away from radar is assumed to be positive.[1]

$$\overline{w}_{\mathrm{d}} = -\frac{\displaystyle\int_{D_{\min}}^{D_{\max}} w_{\mathrm{T}}(D) D^6 N(D)\mathrm{d}D}{\displaystyle\int_{D_{\min}}^{D_{\max}} D^6 N(D)\mathrm{d}D} + w. \tag{6.14}$$

Using the higher moment of (6.7) and the terminal velocity of (6.8), the equation (6.14) can be expressed as

$$\begin{aligned}\overline{w}_{\mathrm{d}} &= -\frac{\displaystyle\int_0^{\infty} (9.65 - 10.3\mathrm{e}^{-600D}) D^6 N_0 D^{\mu} \mathrm{e}^{-\Lambda D}\mathrm{d}D}{\displaystyle\int_0^{\infty} D^6 N_0 D^{\mu}\mathrm{e}^{-\Lambda D}\mathrm{d}D} + w \\ &= -9.65 + 10.3\left(\frac{\Lambda}{\Lambda + 600}\right)^{\mu+7} + w. \end{aligned} \tag{6.15}$$

Doppler Velocity Spectrum Width and DSD

The Doppler velocity spectrum width for vertical direction σ_w can be expressed using the Doppler velocity w_{d} as

$$\sigma_w = (\overline{w_{\mathrm{d}}^2} - \overline{w}_{\mathrm{d}}^2)^{1/2}. \tag{6.16}$$

If the background wind velocity can be ignored, it is possible to express σ_w of (6.16) as follows by using (6.8) and (6.7);

$$\begin{aligned}\sigma_w &= \left[\left(\frac{\displaystyle\int_0^{\infty} w_{\mathrm{T}}^2 D^6 N(D)\mathrm{d}D}{\displaystyle\int_0^{\infty} D^6 N(D)\mathrm{d}D}\right) - \left(\frac{\displaystyle\int_0^{\infty} w_{\mathrm{T}} D^6 N(D)\mathrm{d}D}{\displaystyle\int_0^{\infty} D^6 N(D)\mathrm{d}D}\right)^2\right]^{1/2} \\ &= 10.3\left[\left(\frac{\Lambda}{\Lambda + 1200}\right)^{\mu+7} - \left(\frac{\Lambda}{\Lambda+600}\right)^{2\mu+14}\right]^{1/2}. \end{aligned} \tag{6.17}$$

[1] According to the convention, the terminal velocities are in (6.8) to (6.11), and therefore, it is necessary to reverse the sign when making the direction away from radar positive.

Equations (6.15) and (6.17) show that $\overline{w}_d$ and σ_w are independent of N_0. It suggests that these two parameters can determine μ and Λ. If the constrained gamma DSD (Vivekanandan et al. 2004) is based on valid physical grounds, then the DSD can be well determined simply from the Doppler velocity and the spectrum width for vertically directed beams. Meanwhile, N_0 is determined from (6.13). Above discussion shows that three basic radar parameters, Z, $\overline{w}_d$, and σ_w, can be related with relatively simple analytical equations to the three parameters of a gamma DSD. First, N_0, Λ and μ are determined, as will be shown in the subsequent sections, by the least squares method from a lot of samples of Z, $\overline{w}_d$, and σ_w, which are obtained by radar observations. Next, using these parameters of the DSD, it is possible to know important physical quantities which show the characteristics of precipitation, cloud water content, rainfall rate, and median volume diameter. However, this method is not necessarily suitable for real time processing. Therefore, rainfall rates, cloud water content, etc., are generally obtained from empirically obtained parameters.

The discussion in this section is based on the parameters observable with one single polarization radar. On the other hand, the precipitation parameter estimation by the above mentioned empirical transformation equation becomes more sophisticated by using polarimetric radars, which can transmit multiple polarization waves. The detail of these methods will be described in Sect. 6.4.

6.1.3 Physical Quantities Concerned with Precipitation

Cloud Water Content

The cloud water content W is the water amount contained in a unit volume, which is proportional to the third moment of the DSD. As the volume of one particle in diameter D [m] is $\frac{4}{3}\pi\left(\frac{D}{2}\right)^3$, W [g m^{-3}] is given by

$$W = \frac{\pi\rho_w}{6}\int_0^\infty D^3 N(D)\mathrm{d}D, \tag{6.18}$$

where the density of precipitation particle ρ_w is in [g m^{-3}]($= 10^6$ for water), and $N(D)\mathrm{d}D$ is in [m^{-3}]. Using (6.5) and (6.7),

$$W = \frac{\pi\rho_w}{6} m_3 = \frac{\pi\rho_w N_0 \Gamma(\mu+4)}{6\Lambda^{\mu+4}}. \tag{6.19}$$

If $\mu = 0$, i.e., $N(D)$ follows an exponential distribution, (6.19) becomes

$$W = \frac{\pi\rho_w N_0 \Gamma(4)}{6\Lambda^4} = \frac{\pi\rho_w N_0}{\Lambda^4}. \tag{6.20}$$

Total Number of Precipitation Particles

Total number of precipitation particles per unit volume $N_{\mathrm{T}}[\mathrm{m}^{-3}]$ is the 0th moment of the DSD, and given by

$$N_{\mathrm{T}} = \int_0^\infty N(D)\mathrm{d}D = m_0 = N_0 \frac{\Gamma(\mu+1)}{\Lambda^{\mu+1}}. \tag{6.21}$$

Mass Weighted Mean Drop Diameter

Mass weighted mean drop diameter $D_{\mathrm{m}}[\mathrm{m}]$ is given by

$$D_{\mathrm{m}} = \frac{\displaystyle\int_{D_{\min}}^{D_{\max}} D^4 N(D)\mathrm{d}D}{\displaystyle\int_{D_{\min}}^{D_{\max}} D^3 N(D)\mathrm{d}D} \simeq \frac{\displaystyle\int_0^\infty D^4 N_0 D^\mu \mathrm{e}^{-\Lambda D}\mathrm{d}D}{\displaystyle\int_0^\infty D^3 N_0 D^\mu \mathrm{e}^{-\Lambda D}\mathrm{d}D} = \frac{4+\mu}{\Lambda}. \tag{6.22}$$

Median Volume Diameter

For the cloud water content of W, the median volume diameter is given by the value of $D_0[\mathrm{m}]$, which satisfy the following equation;

$$\frac{\pi\rho_{\mathrm{w}}}{6}\int_0^{D_0} D^3 N(D)\mathrm{d}D = \frac{1}{2}\frac{\pi\rho_{\mathrm{w}}}{6}\int_0^\infty D^3 N(D)\mathrm{d}D = \frac{W}{2}. \tag{6.23}$$

If $N(D)$ follows an exponential distribution, substitution of (6.2) and (6.20) into (6.23) leads to

$$\int_0^{D_0} D^3 \exp(-\Lambda D)\mathrm{d}D = \frac{3}{\Lambda^4}. \tag{6.24}$$

Applying the relation

$$\int x^n \mathrm{e}^{-ax}\mathrm{d}x = \frac{n}{a}\int x^{n-1}\mathrm{e}^{-ax}\mathrm{d}x - \frac{x^n}{a}\mathrm{e}^{-ax} \tag{6.25}$$

to (6.24), and solving it for D_0 leads to[2]

[2]By integrating (6.24), the following equation is given;

$$(6+6\Lambda D_0 + 3\Lambda^2 D_0^2 + \Lambda^3 D_0^3)\exp(-\Lambda D_0) = 3.$$

Solving the above equation numerically for D_0, $D_0 \simeq 3.67/\Lambda$ is obtained.

$$D_0 \simeq \frac{3.67}{\Lambda}. \tag{6.26}$$

If $N(D)$ follows a gamma distribution,

$$\int_0^{D_0} D^{\mu+3} \exp(-\Lambda D) \mathrm{d}D = \frac{3}{\Lambda^4}. \tag{6.27}$$

Applying the relation (6.25) to (6.27) and solving it for D_0 in a similar way as (6.24) to (6.26), and using (6.22), D_0 is given by

$$D_0 \simeq \frac{\mu+3.67}{\Lambda} = \frac{\mu+3.67}{\mu+4} D_{\mathrm{m}}. \tag{6.28}$$

Drop Size Distribution

The parameter D_0 has an easier physical interpretation than Λ, and thus, we will use D_0 for expressing the DSD.[3] Substituting Λ which is obtained by (6.28) into (6.1), $N(D)$ is expressed as

$$N(D) = N_0 D^{\mu} \exp\left[-(\mu+3.67)\frac{D}{D_0}\right]. \tag{6.29}$$

Rainfall Rate

The rainfall rate R is defined as the amount of precipitation, or volume of precipitation, which passes through a horizontal unit area in a unit time. Therefore, R has the unit of height of precipitation volume per unit time [m s^{-1}], which is expressed by the following equation (e.g., Sauvageot 1992, p. 72);

$$R = \frac{\pi}{6} \int_{D_{\mathrm{min}}}^{D_{\mathrm{max}}} D^3 N(D)(w_{\mathrm{T}}(D) - w) \mathrm{d}D, \tag{6.30}$$

where $w_{\mathrm{T}}(D)$ is the terminal velocity of precipitation particle which was given by (6.8), while w is the vertical component of the background wind velocity (upward

---

[3] Ulbrich (1983) describes the advantages of using D_{m} rather than D_0, i.e., the use of D_{m} in analyses of precipitation data is more convenient computationally and avoids interpolation of particle size spectra to find D_0. D_{m} is a very good approximation to D_0 for $\mu \geq -2$.

positive). When the precipitation particles are observed in the stationary atmosphere which has no vertical motion, w is assumed to be zero, and (6.30) is expressed as

$$R = \frac{\pi}{6} \int_{D_{\min}}^{D_{\max}} D^3 N(D) w_{\mathrm{T}}(D) \mathrm{d}D. \tag{6.31}$$

Using the higher moment of the DSD in (6.7) and the terminal velocity in (6.8), (6.31) becomes

$$\begin{aligned} R &= \frac{\pi}{6} \int_0^{\infty} (9.65 - 10.3\mathrm{e}^{-600D}) N_0 D^{\mu+3} \mathrm{e}^{-\Lambda D} \mathrm{d}D \\ &= \frac{\pi N_0}{6} \left[9.65 \left(\frac{\Gamma(4+\mu)}{\Lambda^{4+\mu}} \right) - 10.3 \left(\frac{\Gamma(4+\mu)}{(\Lambda+600)^{4+\mu}} \right) \right]. \end{aligned} \tag{6.32}$$

If $\mu = 0$, the above equation is simplified to

$$R = \pi N_0 \left[\frac{9.65}{\Lambda^4} - \frac{10.3}{(\Lambda+600)^4} \right], \tag{6.33}$$

where the MKS units are assumed throughout. It should be noted that (6.8) is applicable in the range D about 3×10^{-4} m or larger, and R in (6.32) and (6.33) takes positive values in the above range. From a practical view point, however, the fall rate of precipitation is expressed in units of [mm h^{-1}]. In this case, coefficient $10^3 \times 3{,}600$ is multiplied to the right side of (6.30) to (6.33). This unit conversion is made to R in (6.42) to (6.45) which will be mentioned later.

6.1.4 *Radar Reflectivity Factor and Rainfall Rate*

General Expression

Using the general expression for the higher moment of the DSD, $N(D)$ is expressed by D and another physical quantity S, e.g., the cloud water content W, the rainfall rate R, or moments of the DSD as (Sempre-Torres et al. 1994; Bringi and Chandrasekar 2001, pp. 421–422)

$$N(D;S) = S^{\alpha} g\left(\frac{D}{S^{\beta}}\right) = S^{\alpha} g(x), \tag{6.34}$$

where α and β are the constants, and $g(x)$ is a general distribution function which is independent from S. The M–P DSD in (6.2) to (6.4) corresponds to $\alpha = 0$, $\beta = 0.21$, and $g(x) = 8 \times 10^6 \exp(-4100x)$.

The cloud water content W in (6.18) is given by

$$W = \frac{\pi\rho_w}{6}\int_0^\infty D^3 N(D;W)\mathrm{d}D = \frac{\pi\rho_w}{6}\int_0^\infty D^3 W^\alpha g\left(\frac{D}{W^\beta}\right)\mathrm{d}D$$
$$= \frac{\pi\rho_w}{6} W^{\alpha+4\beta}\int_0^\infty x^3 g(x)\mathrm{d}x. \tag{6.35}$$

As is clear from (6.35), α and β are not independent with each other, and $g(x)$ is, in addition, constrained so that they have to satisfy the following equations;

$$\alpha + 4\beta = 1, \tag{6.36}$$

$$\frac{\pi\rho_w}{6}\int_0^\infty x^3 g(x) = 1. \tag{6.37}$$

Applying (6.34) to the nth moment in (6.5), we obtain

$$m_n = \int_0^\infty D^n N(D;S)\mathrm{d}D = \int_0^\infty D^n S^\alpha g\left(\frac{D}{S^\beta}\right)\mathrm{d}D$$
$$= S^{\alpha+\beta(n+1)}\int_0^\infty x^n g(x)\mathrm{d}x$$
$$= a_n S^{b_n}, \tag{6.38}$$

where

$$a_n = \int_0^\infty x^n g(x)\mathrm{d}x, \tag{6.39}$$

$$b_n = \alpha + \beta(n+1). \tag{6.40}$$

Using the rainfall rate R as the physical quantity for S, the radar reflectivity factor Z given by the sixth moment of the DSD in (6.13) is related to R by applying (6.38) as

$$Z = a_6 R^{b_6}, \tag{6.41}$$

where a_6 and b_6 are precipitation constants.

Relation Between Rainfall Rate and Radar Reflectivity Factor

It is necessary for estimating R from (6.41) to know the DSD. However, actual drop size distributions varies in every precipitation event, and consequently, it is difficult to fix R only from Z. The efforts which empirically define the values of a_6 and b_6 based on a lot of precipitation events have been continued. For example, Battan (1973, pp. 90–92) studied 69 patterns of the Z–R relation for various types of precipitations. Figure 6.3 shows these relations plotted by Doviak and Zrnić (2006, p. 224). Through these efforts, a_6 and b_6 pairs have been prepared to

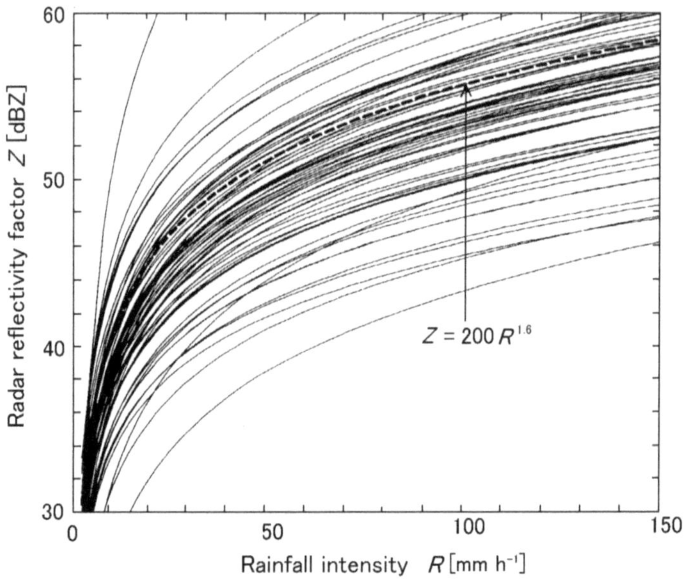

Fig. 6.3 *Z–R* relations for various types of precipitations from Battan (1973, pp. 90–92). The *thick broken line* represents the M–P distribution for stratiform rain. Plotted by Doviak and Zrnić (2006, p. 224)

represent various types of precipitations. The most popular Z–R relation is the M–P distribution for stratiform rain, which is given by the following equation (Marshall et al. 1955);

$$Z = 200R^{1.6}, \tag{6.42}$$

where the units Z and R are [mm^6 m^{-3}] and [mm h^{-1}], respectively. The Japan Meteorological Agency adopts it as the measure of echo intensity. The following equation is an example of the Z–R relation for precipitations concerned with typhoon and convective cloud such as thunderstorm (Sekhon and Srivastava 1971; Jorgensen and Willis 1982).

$$Z = 300R^{1.35}. \tag{6.43}$$

The Doppler radar WSR-88D(NEXRAD; refer to Sect. 9.2.2.), which is the most typical meteorological radar in the U.S.A., adopts the following relation;

$$Z = 300R^{1.4}. \tag{6.44}$$

Rewriting the above equation to express the rainfall rate R for a radar reflectivity factor Z, the following relation is obtained;

$$R = 0.017Z^{0.714}. \tag{6.45}$$

For snowfall, the following relation is represented, based upon a series of observations at different places (Sekhon and Srivastava 1970);

$$Z = 1780R^{2.21}. \tag{6.46}$$

Equations (6.42) to (6.45) use Z for water which is equal to the equivalent radar reflectivity factor Z_e. On the other hand, Z in (6.46) is Z_i for snow, and R refers to the rainfall rate for melted water. Thus, using (3.72), (6.46) becomes $Z_e = 399R^{2.21}$.

6.2 Estimation of DSD

For precise estimation of the DSD, it is necessary to remove vertical motion of the background atmosphere from the observed velocity of precipitation particle. Therefore, in general, atmospheric radar which can directly measure vertical wind component simultaneously with precipitation particle velocity with higher precision is utilized. In this section, we discuss the technique to estimate the DSD of precipitation particles, which are discretely distributed scatterers, with atmospheric radar (e.g., Wakasugi et al. 1986; Sato 1990). For simplicity, our discussion is limited to the terminal velocity of precipitation particles and the vertical component of turbulent velocity. More strictly, the horizontal wind component gives a Doppler shift for a vertically pointed beam of finite width. This issue is discussed in the paper by Fang et al. (2012). Furthermore, Fang et al. derive a generalized analytical expression for the Doppler spectra of scatter from precipitation particles observed with a vertically pointed beam which incudes the spectral broadening mechanisms of mean wind, turbulence, particle oscillation/wobble, and the dispersion of particle terminal velocity, all of which are not necessarily spatially uniform.

Doppler Spectrum of Precipitation Particles

The relation between the normalized mean Doppler spectrum (power spectral density or power spectrum) $S_p(w_T)$ of precipitation observed with a vertically pointed beam and the DSD $N(D)$ is derived by (Doviak and Zrnić 1984, p. 231)

$$S_p(w_T)\mathrm{d}w_T(D) = \sigma_b(D)N(D)\frac{\mathrm{d}D}{\eta}, \tag{6.47}$$

where w_T is the terminal velocity of a raindrop of diameter D which is expressed by e.g., (6.8), σ_b the backscattering cross section, and η the radar reflectivity, respectively. In the above equation, it is assumed that the resolution volume is much larger than the spacing between drops of all diameters. Using the relation $\sigma(D)/\eta = D^6/Z$ from (3.63) and (3.67), (6.47) is expressed as

$$S_p(w_T)\mathrm{d}w(D) = D^6 N(D)\frac{\mathrm{d}D}{Z}, \tag{6.48}$$

where Z is the radar reflectivity factor. Thus, the power spectrum of a distribution of drops is expressed as

$$S_\mathrm{p}(w_\mathrm{T}) = \frac{D^6 N(D)}{Z} \Big/ \left[\frac{\mathrm{d}w_\mathrm{T}(D)}{\mathrm{d}D}\right]. \tag{6.49}$$

Meanwhile, the probability distribution of turbulent velocity (i.e., the radial component of the wind velocity) is assumed to have a Gaussian form, thus the normalized turbulence spectrum of vertical wind $S_\mathrm{t}(w)$ can be written as

$$S_\mathrm{t}(w) = \frac{1}{\sqrt{2\pi}\sigma_w} \exp\left(-\frac{w^2}{2\sigma_w^2}\right), \tag{6.50}$$

where σ_w is the spectral broadening (Wakasugi et al. 1986).

If radars of larger wavelengths, e.g., very high frequency (VHF) radars, are used to measure the Doppler spectrum, the sum of the two spectra expressed by (6.49) and (6.50), corresponding precipitation and refractive index perturbations caused by atmospheric turbulence, respectively, are obtained. Furthermore, if raindrops of any size instantly respond to the turbulence, the Doppler spectrum of the precipitation component can be expressed by the convolution of S_p and S_t. Then, the observed Doppler spectrum S_o is written as (Wakasugi et al. 1986)

$$S_\mathrm{o}(w) = P_\mathrm{rp} S_\mathrm{p}(w_\mathrm{T}) * S_\mathrm{t}(w) + P_\mathrm{rt} S_\mathrm{t}(w), \tag{6.51}$$

where P_rp and P_rt are echo powers associated with precipitation and perturbation of refractive index caused by atmospheric turbulence, respectively, and $*$ denotes convolution.

In the presence of the mean (up- or downdraft) velocity $\overline{w}$, the velocity w in (6.47) to (6.50) must be changed to $w - \overline{w}$ corresponding to a shift of the spectra. Then, the Doppler spectrum S_o can be written as

$$S_\mathrm{o}(w - \overline{w}) = P_\mathrm{rp} S_\mathrm{p}(w_\mathrm{T} - \overline{w}) * S_\mathrm{t}(w) + P_\mathrm{rt} S_\mathrm{t}(w - \overline{w}), \tag{6.52}$$

where the first term of S_t should have the argument w because the precipitation spectrum has been shifted by the argument $w_\mathrm{T} - \overline{w}$ of S_p (Wakasugi et al. 1986).

The value of $S_\mathrm{o}(w)$ is determined by eight parameters; that of background atmosphere, i.e., the echo intensity, the vertical component of wind velocity, and the Doppler velocity spectrum width, and in addition, the parameters of the DSD which were discussed in Sect. 6.1.1, i.e., N_0, Λ, and μ, terminal velocity and noise. First, the parameter μ is fixed, and the least square fitting[4] is performed by changing other seven parameters to find the set of parameters that minimize the estimation error. Next, the value of μ is changed successively, and finally, the set of parameters that minimizes the least square error is derived as the optimum estimate (Wakasugi et al. 1986). The existence of spectral broadening σ_w leads to an underestimate of

[4] As for fitting, refer to Sect. 5.5.6.

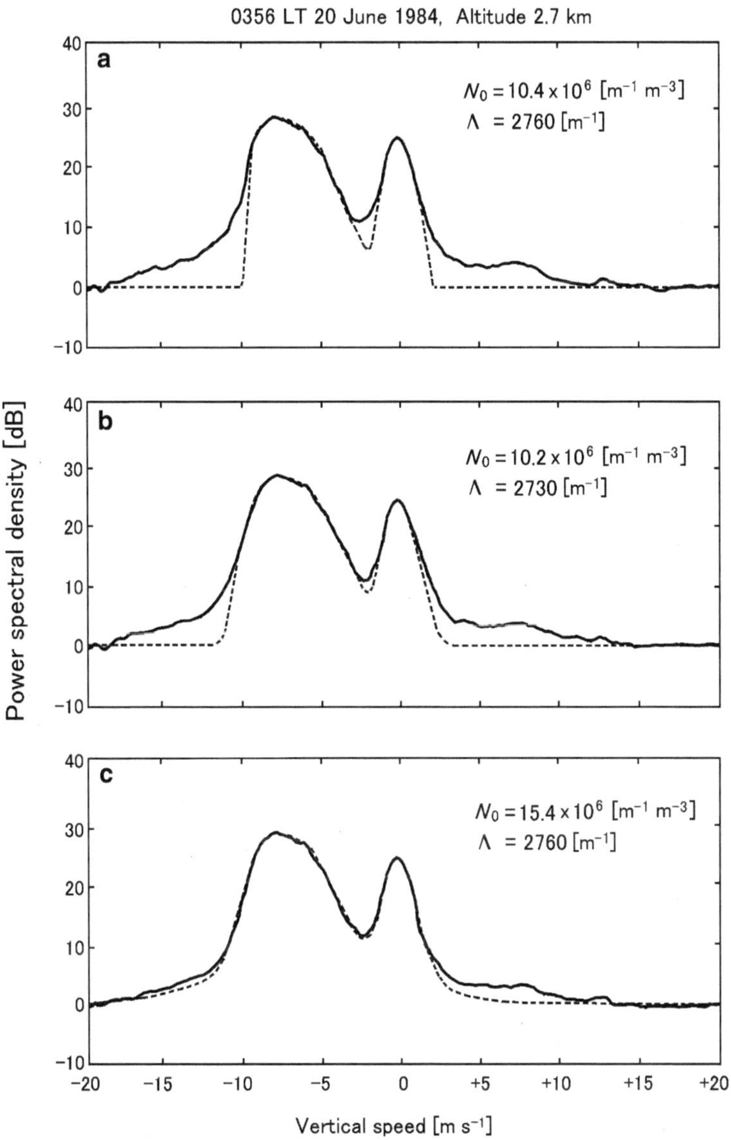

Fig. 6.4 Observed Doppler spectrum (*solid curve*) by the MU radar during precipitation at the height of 2.7 km and the fitted theoretical spectra (*dashed curve*). (**a**) A simple summation of precipitation and air components, (**b**) the effect of air turbulence on raindrops are accounted, and the function defined by (6.52) is used to calculate the spectrum, and (**c**) fitting is performed with (6.53), where the simple uniform weighting is used [adapted from Wakasugi et al. 1986]

the N_0 and Λ values, and an overestimate of $\overline{w}$, thus, the broadening σ_w should be less than half width of the precipitation spectrum (Hauser and Amayenc 1981; Wakasugi et al. 1986).

Figure 6.4 shows the Doppler spectra obtained by the MU radar, a 46.5 MHz MST radar which will be mentioned in Sect. 10.2.3, during precipitation (Wakasugi et al. 1986). There are two components in the Doppler spectra. One is a symmetrical form with a vertical speed of around $0\,\mathrm{m\,s^{-1}}$ which corresponds to the echo from air (scattering due to perturbations of radio refractive index), and the other a skewed form with a speed of around -10 to $-5\,\mathrm{m\,s^{-1}}$, where "–" denotes downward, which is originating from raindrops. In Panel (a), a simple summation of two spectral components $P_{\mathrm{rp}}S_{\mathrm{p}}(w-\overline{w})+P_{\mathrm{rt}}S_{\mathrm{t}}(w-\overline{w})$ is used as a fitting function, whereas in Panel (b), the function defined by (6.52) is used. The fitted spectra in Panels (a) and (b) shows considerable difference at the skirts of each component, which is because the sampled data are truncated when calculating the periodogram to estimate the Doppler spectrum (power spectral density; Balsley and Gage 1980). If the observed spectrum results in a convolution of S_{o} and a window function $W(w)$ as

$$S_{\mathrm{o}}'(w) = S_{\mathrm{o}}(w) * W(w), \tag{6.53}$$

difference due to the truncate of sampled data can be made smaller. Various types of window functions have been discussed by e.g., Harris (1978). Figure 6.4c shows the results when fitting is performed with (6.53), where the simple uniform weighting of sidelobe fall of $-6\,\mathrm{dB\,oct^{-1}}$ is used. In the case of (c), a spillover of the spectral component throughout the entire spectral range should be accounted.

6.3 Attenuation of Radio Waves in the Atmosphere

The intensity of radio wave that propagates through the atmosphere is attenuated by various factors in the atmosphere. The attenuation rate depends, to a great extent, on the atmospheric condition and radar frequency. As will be discussed in detail below, the attenuation rate of radio wave whose frequency is not larger than approximately 3 GHz and is generally applied to atmospheric radar is comparatively small. Furthermore, as atmospheric radars radiate radio wave aloft near the zenith, the distance for the radio wave to pass through the height range where the atmospheric density is comparatively dense and consequently the attenuation is more significant is limited. Thus, attenuation of radar wave can be safely ignored in most cases of atmospheric radar. On the other hand, meteorological radars use the radar waves from several GHz to 95 GHz whose attenuation rate is comparatively large. Also, meteorological radar is usually steered at low elevation angles to obtain horizontal distribution of precipitation particles. Therefore, the attenuation of radio wave in the atmosphere becomes significant for meteorological radar, often limiting the observation range. Specifically, it is noted that the influence of the attenuation becomes more significant at the frequencies of millimeter waves and higher.

6.3.1 Attenuation Rate

The attenuation rate a of the power density S is defined as

$$\frac{\mathrm{d}S}{\mathrm{d}r} \equiv -aS, \tag{6.54}$$

where r is range, and a [m^{-1}] is the attenuation rate when the radio wave propagates a unit distance. For simplicity, the effect that the received power density decreases in proportion to $1/r^2$ is not taken into consideration. Integrating (6.54), power density $S(r_2)$ at an arbitrary range r_2 is expressed as (Doviak and Zrnić 1984, p. 29)

$$S(r_2) = S(r_1)\exp\left(-\int_{r_1}^{r_2} a\mathrm{d}r\right), \tag{6.55}$$

where $S(r_1)$ is the power density at the reference point r_1.

Taking the logarithm of (6.55) leads to

$$\ln\frac{S(r_2)}{S(r_1)} = -\int_{r_1}^{r_2} a\mathrm{d}r. \tag{6.56}$$

Generally, the attenuation coefficient (or specific attenuation) A [dB km^{-1}] is given as the attenuation per unit kilometer, and thus the above equation becomes, by changing natural to common logarithms, to

$$10\log\frac{S(r_2)}{S(r_1)} = \frac{10}{\ln 10}\ln\frac{S(r_2)}{S(r_1)} = -4.34\times 10^3\int_{r_1}^{r_2} a\mathrm{d}r = -\int_{r_1}^{r_2} A\mathrm{d}r, \tag{6.57}$$

and

$$A = 4.34\times 10^3\, a \quad [\mathrm{dB\ km^{-1}}]. \tag{6.58}$$

The value obtained by (6.57) is the one-way attenuation. In general, radar wave makes a round-trip, and thus the actual attenuation is twice the above value. It is possible to describe the total attenuation per unit distance by adding the attenuation due to the atmosphere A_{atms} and that due to precipitation A_{prp}, i.e., $A = A_{\mathrm{atms}} + A_{\mathrm{prp}}$.

The major factor of attenuation of radar wave in the atmosphere which should be considered in practical use is the absorption by resonance with oxygen and water molecules at the specific wavelengths as mentioned in Sect. 5.1.2 (Van Vleck 1947a,b). On the other hand, the major factor of attenuation due to precipitation is the absorption and scattering by individual precipitation particles. The attenuation by these particles is determined by the ratio of the size of precipitation particles to the radar wavelength. As the detail will be discussed in Sect. 6.3.3, the extinction cross section σ_{t}, which was previously mentioned in Sects. 3.3.3 and 3.3.4, determines the attenuation.

6.3.2 Attenuation by the Atmosphere

Attenuation of radio waves at wavelength longer than approximately 3 cm (at frequency lower than 10 GHz) is primarily caused by oxygen, and the attenuation rate (or coefficient) is nearly constant, approximately 0.01 dB km^{-1}, irrespective of wavelength. At the higher frequency, an additional attenuation due to water vapor increases, and moreover, large attenuation is caused in specific frequency bands due both to oxygen and water vapor. Attenuation of radio waves within a designated range in the atmosphere is the sum of attenuations by oxygen and water vapor. The attenuation coefficient $A_{\rm atms}$ is given by the sum of the attenuation coefficient of oxygen $A_{\rm oxy}$ and that of water vapor $A_{\rm vpr}$, i.e., $A_{\rm atms} = A_{\rm oxy} + A_{\rm vpr}$.

Attenuation of radio waves in the atmosphere depends on atmospheric pressure, temperature, water vapor density and so on. Various models which include these values to calculate the attenuation rate are proposed. Among them, ITU-R computational model (ITU-R 2001) is based on spectroscopic study, while the millimeter-wavelength propagation model (MPM) takes atmospheric pressure, temperature, humidity, and tiny precipitation particles that intercept the propagation path into consideration for frequency less than 1,000 GHz (Liebe 1985). On the other hand, a simplified expression for the standard atmosphere pressure and temperature by the International Radio Consultative Committee (CCIR) is also available (CCIR 1991), although the applicability is limited. Both the simplified models by the CCIR and the MPM show the precision high enough for purpose of the actual radar measurement, and therefore, only the CCIR will be presented and used in the calculation of results below, while the MPM is used for comparison.

Attenuation by Oxygen

Attenuation by oxygen in the atmosphere is caused by magnetic interaction between magnetic moment of oxygen molecule and radio wave. There are a variety of absorption bands of radio wave around 60 GHz and one isolated band at 118 GHz. The simplified model of the CCIR gives the attenuation coefficient of oxygen $A_{\rm oxy}$ [dB km^{-1}] at frequency f [GHz], atmospheric temperature 15°C, and atmospheric pressure 1,013 hPa as follows; for $f \leq 57$ GHz

$$A_{\rm oxy} = \left[7.19 \times 10^{-3} + \frac{6.09}{f^2 + 0.227} + \frac{4.81}{(f-57)^2 + 1.50} \right] f^2 \times 10^{-3}, \tag{6.59}$$

and for 63 GHz $\leq f \leq$ 350 GHz

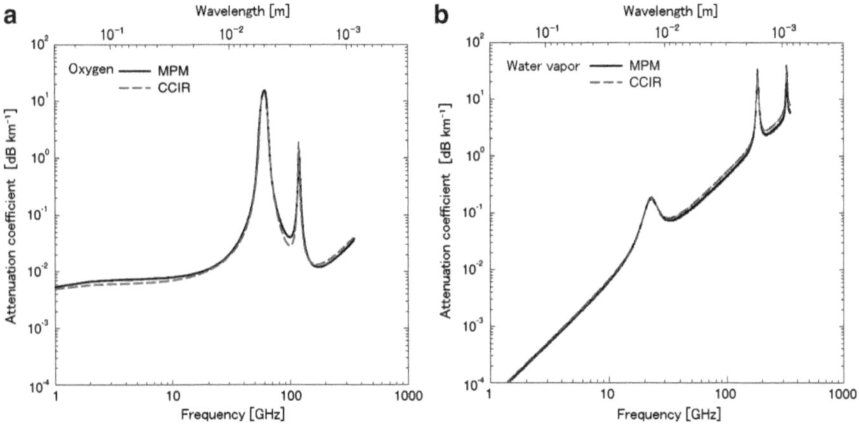

Fig. 6.5 Attenuation coefficients of radio waves due to (**a**) oxygen and (**b**) water vapor in the atmosphere calculated by the CCIR simplified model (*broken line*) and the MPM (*solid line*)

$$A_{\mathrm{oxy}} = \left[3.79\times 10^{-7} f + \frac{0.265}{(f-63)^2+1.59} + \frac{0.028}{(f-118)^2+1.47}\right] \times (f+198)^2 \times 10^{-3}. \tag{6.60}$$

The attenuation coefficients of oxygen calculated with this model and the MPM are shown in Fig. 6.5a.

Attenuation by Water Vapor

Attenuation by water vapor in the atmosphere is caused by the interaction between the electric moment of water vapor molecule and radio wave. There are two isolated absorption bands of radio wave at 22.5 and 183.3 GHz (Ulaby et al. 1981, p. 270). The simplified model of the CCIR gives the attenuation coefficients of water vapor A_{vpr} [dB km^{-1}] for the frequency range $1 \le f \le 350$ GHz, at atmospheric temperature 15°C, and atmospheric pressure 1,013 hPa as follows;

$$A_{\mathrm{vpr}} = \left[0.050 + 0.0021\rho_{\mathrm{v}} + \frac{3.6}{(f-22.2)^2+8.5} + \frac{10.6}{(f-183.3)^2+9.0} + \frac{8.9}{(f-325.4)^2+26.3}\right] f^2 \rho_{\mathrm{v}} \times 10^{-4}, \tag{6.61}$$

where ρ_{v} is the density of water vapor [g m^{-3}]. The attenuation coefficients of water vapor calculated with this model and the MPM are shown in Fig. 6.5b.

6.3.3 *Attenuation by Water Particle*

Attenuation of radio waves by cloud droplets and precipitation particles (rain, snow, hail, etc.) is related with the extinction cross section of the particles. The attenuation coefficient A_{prp} [dB km^{-1}] is expressed with $N(D)$ and the extinction cross section $\sigma_{\mathrm{t}}(D)$, where $\sigma_{\mathrm{t}}(D)$ is defined by (3.44) and given by (3.47) and (3.61), as (e.g., Doviak and Zrnić 2006, p. 40)

$$A_{\mathrm{prp}} = 4.34 \times 10^3 \int_0^\infty \sigma_{\mathrm{t}}(D)N(D)\mathrm{d}D, \tag{6.62}$$

where D is the diameter of water particles in [m].

Attenuation by Cloud

Cloud droplets are water particles or ice particles whose diameters are not larger than 100 μm. In general, the diameter of particle D and the radar wavelength λ satisfy the Rayleigh approximation $\pi D/\lambda \ll 1$ for ordinary radars. Therefore, the absorption cross section σ_{a} is much larger than the scattering cross section σ_{s} as mentioned in Sect. 3.3.4, and therefore, σ_{s} can be ignored. From (6.62) and (3.60), the attenuation coefficient of cloud, A_{cld} [dB km^{-1}] is expressed as

$$\begin{aligned} A_{\mathrm{cld}} &\simeq 4.34 \times 10^3 \int_0^\infty \sigma_{\mathrm{a}}(D)N(D)\mathrm{d}D \\ &\simeq 4.34 \times 10^3 \frac{\pi^2}{\lambda}\mathrm{Im}|-K| \int_0^\infty D^3 N(D)\mathrm{d}D. \end{aligned} \tag{6.63}$$

Substituting (6.18) into (6.63), A_{cld} becomes

$$A_{\mathrm{cld}} \simeq 4.34 \times 10^3 \left(\frac{6\pi}{\lambda}\right) \mathrm{Im}|-K| \,\frac{W}{\rho_{\mathrm{w}}}. \tag{6.64}$$

As shown in Table 3.4, $\mathrm{Im}|-K|$ depends on the temperature and the radar wavelength. Benoit (1968) formalized these parameters and derived empirical equations for calculating A_{cld}. According to Benoit, A_{cld} for the cloud water content W [g m^{-3}] is expressed at radar frequency f [GHz] as follows;

$$A_{\mathrm{cld}} = f^{a_1} \exp[a_2(1+a_3T)]\, W, \tag{6.65}$$

for water cloud and fog. But for ice cloud,

$$A_{\mathrm{cld}} = f^{b_1} \exp[b_2(1+b_3T+b_4T^2)]\, W, \tag{6.66}$$

where T is the atmospheric temperature [°C]. Constants $a_i(i=1-3)$ and $b_j(j=1-4)$ are

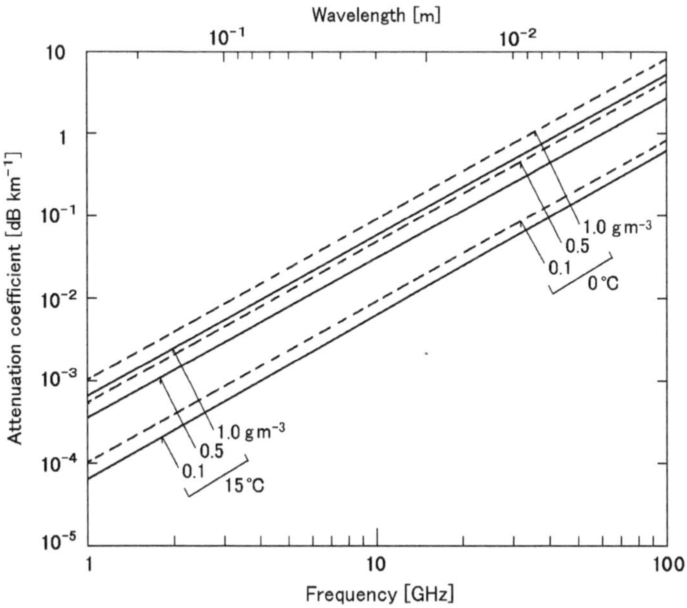

Fig. 6.6 Attenuation coefficients of radio waves calculated by (6.65) for different cloud water contents at the atmospheric temperatures 0°C (*dashed lines*) and 15°C (*solid lines*)

$$a_1 = 1.95,\ a_2 = -6.866,\ a_3 = 4.5\times 10^{-3},$$

$$b_1 = 1.006,\ b_2 = -8.261,\ b_3 = -1.767\times 10^{-2},\ b_4 = -4.374\times 10^{-4},$$

respectively. The attenuation coefficients of radio waves calculated by (6.65) for different cloud water contents at the atmospheric temperatures 0°C and 15°C are shown in Fig. 6.6.

Attenuation by Precipitation

As discussed in Sect. 3.3.4, both the scattering cross section σ_s and the absorption cross section σ_a should be taken into account in calculating σ_t for attenuation by precipitation. From (6.58) and (6.62), and applying (3.61), the attenuation coefficient of precipitation A_{prp} [dB km^{-1}] is given by

$$\begin{aligned} A_{prp} &= 4.34\times 10^3 \int_0^\infty \sigma_t(D)N(D)\mathrm{d}D \\ &= 4.34\times 10^3 \left(\int_0^\infty \sigma_s(D)N(D)\mathrm{d}D + \int_0^\infty \sigma_a(D)N(D)\mathrm{d}D \right). \end{aligned} \qquad (6.67)$$

Table 6.1 Attenuation coefficients of electromagnetic wave A_{prp} [dB km^{-1}] expressed by the rainfall rate R [mm h^{-1}] and constants a_0 and b_0 for various radar wavelengths

Wavelength [cm]	M–P (0°C)	Modified M–P (0°C)	Gunn–East (18°C)
10	0.0007R–0.0009R	0.00082R	0.0003R
5.7	–	–	0.0022$R^{1.17}$
5.5	0.003R–0.004R	0.0031R	–
3.21	0.011$R^{1.15}$	0.013$R^{1.15}$	0.0074$R^{1.31}$
1.24	0.117$R^{1.07}$	0.13$R^{1.07}$	0.12 $R^{1.06}$
0.86	0.27R	0.31R	–
0.62	0.37R–0.50R	0.52R	–

Modified from Wexler and Atlas (1963)

From the Rayleigh approximation in (3.59) and (3.60), $\sigma_s \propto D^6$ and $\sigma_a \propto D^3$. However, the Rayleigh approximation is not very good to compute attenuation for centimeter wavelength radars. Doviak and Zrnić (2006, p. 40) shows σ_t versus D for various wavelengths. If a power-law in the form $\sigma_t = C_\lambda D^n$ is applied for $10^{-4} \leq D \leq 8 \times 10^{-3}$ m, $n = 3.5$, 3.9, and 4.1 fit at 10, 5, and 3 cm wavelengths, respectively as a first approximation. For large drops with $5 \times 10^{-3} \leq D \leq 10^{-2}$ m, the corresponding n values are 4.6, 4.8, and 4.9, respectively. To simplify the discussion, let $n \simeq 4$ (Bringi and Chandrasekar 2001, p. 491). It follows that (6.62) can be expressed as

$$A_{prp} \simeq 4.34 \times 10^3\, C_\lambda \int_0^\infty D^4 N(D)\mathrm{d}D, \tag{6.68}$$

where C_λ is temperature dependent.

The drop size distribution $N(D)$ is contained in (6.68), and it shows that A_{prp} is approximately proportional to the forth moment of the DSD. As we will discuss in Sect. 6.4.4, one of the polarization parameters relates to the forth moment of the DSD, and can be applied for the computation of A_{prp}. However, it is not straightforward to obtain $N(D)$ and A_{prp} simultaneously by single polarization radar. Hence, in a single polarization radar observation, A_{prp} is assumed to be given in an empirical equation with the rainfall rate R [mm h^{-1}]. From (6.9) and (6.30), R is approximately $R \propto D^{3.67}$, if the atmosphere has no vertical motion. Applying this relation into (6.68), A_{prp} is expressed using R and constants a and b as follows (Gunn and East 1954; Hitschfeld and Bordan 1954);

$$A_{prp} = a_0 R_0^b. \tag{6.69}$$

Constants a_0 and b_0 depend on the radar frequency and the atmospheric temperature.[5] The typical values of (6.69) for various radar wavelengths are listed in

[5]It is suggested that b is around $n/3.67$, thus if $n = 4$, b is around $4/3.67 \simeq 1.09$.

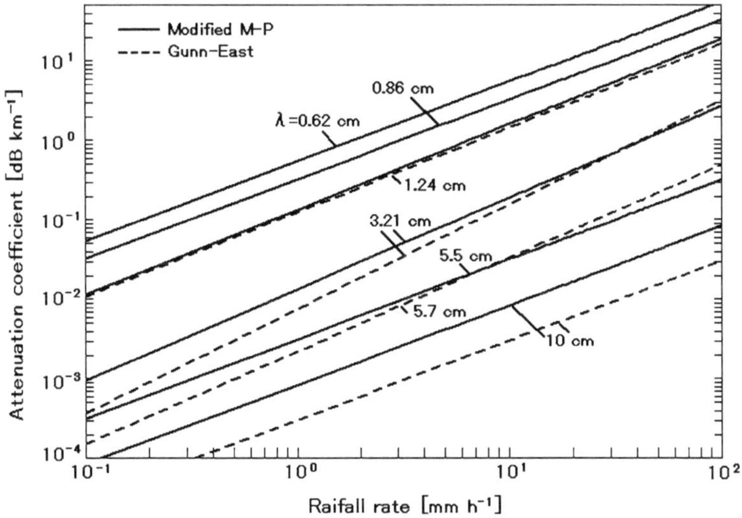

Fig. 6.7 Attenuation coefficients of radio waves due to precipitation calculated for various radar wavelengths

Table 6.1 (Wexler and Atlas 1963). The Modified M–P distribution is sometimes applied for drop size diameter less than 1–1.5 mm where the difference from the exponential distribution becomes large. The Gunn and East (1954) results and the others are in fairly good agreement except at wavelength of 10 cm. The discrepancy in 10 cm is largely attributable to the dissimilarity in temperatures as will be shown later in Table 6.3. The calculated result for the attenuation coefficients of precipitation is shown for different wavelengths in Fig. 6.7.

The R value is derived from Z as discussed in Sect. 6.1.4. It is an iterative attenuation correction scheme, where the measured reflectivity factor Z is used to estimate the attenuation coefficient beginning from a place where Z is not yet substantially affected by the attenuation. The attenuation coefficient A_{prp} is subsequently used to adjust the measurement of Z at next increment of distance from the radar. Thus, even small errors of A_{prp} would grow with increasing distance, causing large errors unless additional information such as a measure of the total path attenuation is available (Weinman et al. 1990; Jameson 1992).

6.4 Polarimetric Radar

All of the rainfall rates which are given in (6.42)–(6.46) represent the typical cases. However, as shown in Fig. 6.3, the variance of R for a specific value of Z is considerably large. Moreover, the DSD varies every moment in actual precipitations. Therefore, it is generally difficult to apply a single Z–R relation to them, or it is impossible to infer R with high precision only by the measurement of radar reflectivity factor Z. Furthermore, a DSD is often approximately expressed by the function which has only two or three parameters. Consequently, it is possible to improve the precision of estimation for precipitation intensity by measuring two or more parameters at the same time with polarimetric radar, which will be discussed in this section. One of the typical techniques proposed for this purpose is dual polarization method. Furthermore, attenuation by precipitation can be estimated more properly not by using the empirical equation of (6.69) but by applying the parameters obtained with the dual polarization method.

6.4.1 Generation of Dual Polarized Wave

Precipitation particles and fog are distinguished by their form, i.e., shape, orientation of the axis of symmetry (typically hydrometeors have an axis of symmetry), and so on. Any information on the form of particles cannot be obtained from backscattered signals of a single polarization radar. On the other hand, the difference of characteristics of backscattered signal due to polarizations directly depends on the shape and orientation of particles. If radar waves are dual polarization waves, i.e., horizontal and vertical polarizations, these characteristics can be obtained. Thus, radar observation with dual polarizations has been introduced as the mean of getting information on the shape of precipitation particles. The conventional meteorological radar transmits and receives a fixed single polarized wave. Meanwhile, polarimetric radar radiates and receives linear polarization waves which are polarized horizontally and vertically. Another method which applies right-hand circular (RHC) and left-hand circular (LHC) polarized waves is also available.

Schematic configurations of polarimetric radar which utilizes dual polarized waves such as horizontally and vertically polarized ones, are demonstrated in Fig. 6.8. Panel (a) shows the case that horizontally and vertically polarized waves are alternately transmitted and received (alternate mode), and (b) horizontally and vertically polarized waves are alternately transmitted, backscattered signals of both polarization are received simultaneously (alternate transmission and simultaneous reception; ATSR mode). In this mode, the linear depolarization ratio, one of the polarization parameters which will be discussed later in Sect. 6.4.2, can be measured. In recent years, it becomes common to transmit and receive both polarized waves simultaneously. The National Weather Service of the U.S.A. uses this mode of operation in their network of polarimetric Doppler weather radars. Panel (c) shows the case of simultaneous transmission and simultaneous reception (STSR mode). In this mode, radar waves of slant linear at 45 degrees are radiated, and backscattered waves of horizontal and vertical polarization are received and processed by two receivers. In the STSR mode, signals at both polarizations are simultaneously sampled which has significant advantage. Meanwhile, the transmitted power of each polarization becomes a half of (a) and (b).

For example, Doviak et al. (2000) and Zrnić et al. (2012, Sect. 2.2.3) discuss the advantage of polarization modes in detail. For example, assuming no propagation effect, in the STSR mode, polarization parameters which will be discussed in Sect. 6.4.2 can be estimated directly, i.e., there is no need to assume the correlation model between successive received echoes, not so for the alternate and ATSR modes. On the other hand, reflectivities of overlaid echoes[6] for even transmitted pulses (second, forth, etc.) are automatically suppressed when using the alternate mode and ATSR mode. As a whole, they concluded that the advantages of the STSR mode are superior to other modes for almost all operational applications expected. A significant problem related to the polarization mode is calibration of differential powers (i.e., Z_{DR}) which will be discussed in next subsection. Several papers, e.g., Hubbert et al. (2003) and Zrnić et al. (2006) discusses this topic.

Backscattering Matrix

As discussed in Sect. 2.1.2, the backscattered wave is, in general, a composition of horizontally and vertically polarized waves of different amplitude and phase. The characteristics of backscattered wave are decided by the distribution, shape, and direction of the main axis of particles, and the permittivity of the particle. If the backscattered wave is composed of horizontally and vertically polarized waves, the backscattered (reception) electric field $\boldsymbol{E}^{\mathrm{b}}$ at the antenna point can be expressed by the incident (transmitted) electric field $\boldsymbol{E}^{\mathrm{i}}$ and the backscattering matrix $[\boldsymbol{S}]$ of linear polarization as follows;

$$\begin{bmatrix} E_{\mathrm{h}} \\ E_{\mathrm{v}} \end{bmatrix}^{\mathrm{b}} = [\boldsymbol{S}] \begin{bmatrix} E_{\mathrm{h}} \\ E_{\mathrm{v}} \end{bmatrix}^{\mathrm{i}} \frac{\mathrm{e}^{-jkr}}{r}, \tag{6.70}$$

where

$$[\boldsymbol{S}] = \begin{bmatrix} s_{\mathrm{hh}} & s_{\mathrm{hv}} \\ s_{\mathrm{vh}} & s_{\mathrm{vv}} \end{bmatrix}, \tag{6.71}$$

and k is the radar wave number ($k = 2\pi/\lambda$), r the distance between radar and scatterer. The subscript h of electric field vector and matrix element means horizontal polarization, while v vertical polarization. The first and the second subscripts show the backscattered and incident electric fields, respectively.

All elements of $[\boldsymbol{S}]$ can be obtained by the radar configuration illustrated in Fig. 6.8b. On the other hand, in the radar configurations Fig. 6.8a and c, only the diagonal elements of $[\boldsymbol{S}]$ can be obtained. To obtain all elements in the alternate mode of (a), the following procedure is needed. The receiver port is not alternately switched, synchronizing with the switching of the transmission port, but switched once per two transmission pulses of horizontally polarized wave, then the

---

[6] See Sect. 8.4.2, the paragraph of suppression of secondary echo.

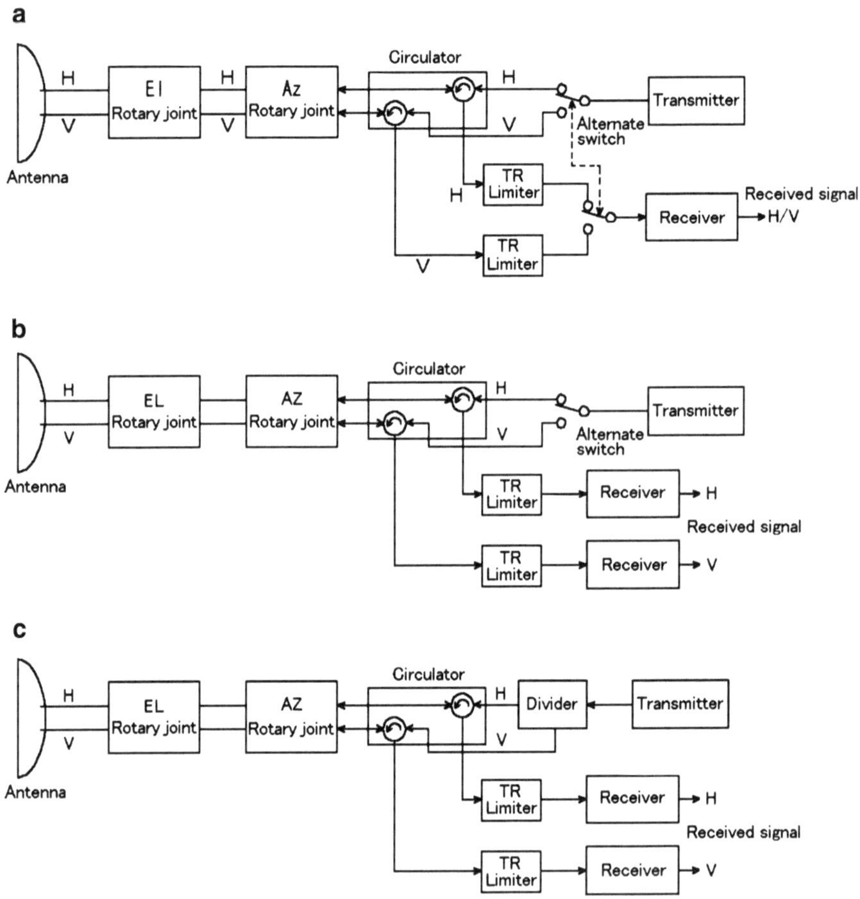

Fig. 6.8 Schematic configurations of polarimetric radars which uses horizontally and vertically polarized waves. (**a**) Alternate H and V transmission and reception (alternate mode), (**b**) alternate transmission and simultaneous reception (ATSR mode), and (**c**) simultaneous transmission and simultaneous reception of H and V signals (STSR mode)

backscattering matrix element s_{hh} appears for the first pulse, and the element s_{vh} appears for the second pulse. In this case, the time period while each element is obtained becomes one quarter of total observation time. To obtain all elements in the STSR mode in (c), the following procedure is needed. Although radar waves are divided into two orthogonal polarization waves for every transmission pulse, interrupt transmission of either polarization wave once of two times. For example, when vertical polarization wave is interrupted, s_{hh} and s_{vh} appear on the H and V ports of the receiver, respectively. By the same way, when horizontal polarization wave is interrupted, s_{vv} and s_{hv} appear on the V and H ports of the receiver, respectively. The period which each element is obtained is a half the total observation time.

If backscattered wave is assumed to be the composition of right-hand and left-hand circular polarized waves, the matrix $[\boldsymbol{S}]$ in (6.71) can be replaced with that of circular polarization wave $[\boldsymbol{S}_\mathrm{c}]$ as

$$[\boldsymbol{S}_\mathrm{c}] = \begin{bmatrix} s_\mathrm{rr} & s_\mathrm{rl} \\ s_\mathrm{lr} & s_\mathrm{ll} \end{bmatrix}, \tag{6.72}$$

where subscript r in the matrix elements shows right-hand circularly polarized wave, and l left-hand circularly polarized wave. Elements of the matrices $[\boldsymbol{S}]$ and $[\boldsymbol{S}_\mathrm{c}]$ are related in the following ways;

$$s_\mathrm{rr} = \frac{1}{2}(s_\mathrm{vv} - s_\mathrm{hh} - j2s_\mathrm{vh}), \tag{6.73}$$

$$s_\mathrm{ll} = \frac{1}{2}(s_\mathrm{vv} - s_\mathrm{hh} + j2s_\mathrm{vh}), \tag{6.74}$$

$$s_\mathrm{rl} = s_\mathrm{lr} = \frac{1}{2}(s_\mathrm{vv} + s_\mathrm{hh}). \tag{6.75}$$

Horizontally and Vertically Polarized Waves

We limit further discussion to horizontally and vertically polarized waves which are commonly applied for meteorological radar observations. The signal voltage backscattered by the nth particle at r_n, v_{ij}, is expressed with the scattering parameter s_{ij} as

$$v_{ij}(\boldsymbol{r}_n) = s_{ij}(n)F(\boldsymbol{r}_n)\,\mathrm{e}^{-j2kr_n}, \tag{6.76}$$

where weighting function $F(\boldsymbol{r}_n)$ is assumed to be identical for horizontally and vertically polarized waves, and that each is purely transmitted, i.e., (6.76) assumes the antenna does not radiate cross-polar fields (not so in practice, and the effect of cross-polar field is discussed by e.g., Hubbert et al. 2010), there is no depolarization along the propagation path, and that the transmitted power in horizontally and vertically polarized waves are equal and alternately transmitted (Doviak and Zrnić 2006, pp. 240–241, errata and supplements). Received signal from the whole scattering volume V_{ij} is the sum from all particles inside, which is given by

$$V_{ij}(r) = \sum_n s_{ij}(n)F(\boldsymbol{r}_n)\,\mathrm{e}^{-j2kr_n}. \tag{6.77}$$

As mentioned in Sect. 5.3.2, since the phase $2kr_n$ is uniformly distributed between $0 - 2\pi$, the mean value of $V_{ij}(r)$ becomes zero.

The second-order moment $\langle V_{ij}V_{kl}^*\rangle$, which is given by the product of $V_{ij}(r)$ and its complex conjugate, where superscript * denotes complex conjugate, is important as the parameter that relates to a wide range of characteristics of precipitation particles. From (6.77), $\langle V_{ij}V_{kl}^*\rangle$ is given by

$$\begin{aligned}[\boldsymbol{C}_{\mathrm{o}}] = \langle V_{ij}V_{kl}^*\rangle &= \left\langle \sum_n \sum_m [s_{ij}(n)s_{kl}^*(m)\mathrm{e}^{-j2k(r_n-r_m)}F(\boldsymbol{r}_n)F^*(\boldsymbol{r}_m)] \right\rangle \\ &= \sum_n \langle [s_{ij}(n)s_{kl}^*(n)] \rangle |F(\boldsymbol{r})|^2 \\ &= \int n(\boldsymbol{r})\langle s_{ij}s_{kl}^*\rangle |F(\boldsymbol{r})|^2 \mathrm{d}V, \end{aligned} \tag{6.78}$$

where $[\boldsymbol{C}_{\mathrm{o}}]$ is a four by four voltage covariance matrix. The ith polarized field are transmitted alternately with the kth polarized field, and echoes received simultaneously in two receivers to obtain all the elements of the voltage covariance matrix. But, because not all signals are received simultaneously, adjustments need to be made to compensate for the loss of signal correlation and phase shifts due to the relative motion of scatterers. In the last equality the summation over n is replaced with the integral over the product of the density $n(\boldsymbol{r})$ of drops at position $\boldsymbol{r}$, the ensemble mean of $s_{ij}s_{kl}$, and the resolution volume weighting function. In this integral form $\langle s_{ij}s_{kl}^*\rangle$ are elements of the backscattering covariance matrix for a drop that represents the ensemble of drops of various diameters, shapes, etc. It might seem strange to assign the same $\langle s_{ij}s_{kl}^*\rangle$ to drops of all sizes and shapes, but it must be realized the size and shape of each scatterer is not known. The best we can do is estimate the expected size and shape of a scatterer for each location (Doviak and Zrnić 2006, pp. 240–241, errata 1st–4th). The expectations in (6.78) are expressed in terms of the distribution of the hydrometeor's properties (i.e., equivalent volume diameter, shape, canting angle, etc.). Thus the general term is

$$\langle s_{ij}s_{kl}^*\rangle = \int p(\boldsymbol{X},\boldsymbol{r})s_{ij}s_{kl}^*\mathrm{d}\boldsymbol{X}, \tag{6.79}$$

where $p(\boldsymbol{X},\boldsymbol{r})$ is the probability density of the scatterer's properties. These properties are represented by a vector $\boldsymbol{X}$ (Doviak and Zrnić 2006, pp. 241–242).

The second-order moment $\langle V_{ij}V_{kl}^*\rangle$ given by (6.78) is generally expressed by a four by four covariance matrix. On the other hand, considering that V_{ij} and V_{ji} are reciprocal,[7] the matrix becomes three by three covariance matrix (Borgeaud et al. 1987). Each matrix element becomes as follows (Doviak and Zrnić 2006, p. 241).

---

[7]In the case of a monostatic radar, the electric field intensity when transmitted by vertically polarized wave, scattered and received by horizontally polarized wave is equal to the electric field intensity when transmitted by horizontally polarized wave, scattered by the same scatterer and received by vertically polarized wave. This principle is called the reciprocity theorem.

$$\begin{matrix} \langle |s_{\rm hh}|^2 \rangle & \langle s_{\rm hv} s_{\rm hh}^* \rangle & \langle s_{\rm vv} s_{\rm hh}^* \rangle \\ \langle s_{\rm hh} s_{\rm hv}^* \rangle & \langle |s_{\rm hv}|^2 \rangle & \langle s_{\rm vv} s_{\rm hv}^* \rangle \\ \langle s_{\rm hh} s_{\rm vv}^* \rangle & \langle s_{\rm hv} s_{\rm vv}^* \rangle & \langle |s_{\rm vv}|^2 \rangle \end{matrix} \tag{6.80}$$

As meteorological radars are usually operated in the wide zenithal angle from horizontal direction to near the zenith, various polarization parameters can be obtained by combining two linearly polarized waves (horizontal and vertical). Meanwhile, the observation area for atmospheric radars is limited to near the zenith, and thus, atmospheric radars generally use single linearly polarized wave for observations. Various characteristics about precipitation particles can be shown by using parameters which are obtained by polarimetric radars as shown in the following sections.

6.4.2 Characteristics of Polarization Parameter

There are two kinds of polarization parameters; one is caused by backscattering of radar wave and the other by propagation. In this section, we will first discuss the relations between the elements of backscattering covariance matrix of backscattering signals and the differential reflectivity, and the linear depolarization ratio. Next, the relation with parameters by propagation such as the specific differential phase will be discussed. Then, methods for estimating precipitation parameters such as the rainfall rate will be discussed.

Polarized Wave and Covariance Matrix Backscattering Signal

Equation (6.78) provides various components of scattered power, e.g., horizontally polarized component of scattered wave for horizontally polarized incident wave can be selectively calculated. From this equation and the discussion on backscattering cross section in Sect. 3.3, radar reflectivity defined by the backscattering cross section in a unit volume is related to each element of backscattering covariance matrix. For example, in the case that both incident and scattered electric field are either horizontal or vertical polarization wave, the following relations are obtained;

$$\eta_{\rm hh} = 4\pi n \langle |s_{\rm hh}|^2 \rangle, \tag{6.81}$$

$$\eta_{\rm vv} = 4\pi n \langle |s_{\rm vv}|^2 \rangle. \tag{6.82}$$

If the backscattering precipitation particle is a complete sphere,

$$\eta = 4\pi n \langle |s|^2 \rangle, \tag{6.83}$$

where $s \equiv s_{\mathrm{hh}} = s_{\mathrm{vv}}$. Substituting (3.69) into (6.81) and (6.82), the radar reflectivity factor Z is expressed using elements of the backscattering covariance matrix as

$$Z_{\mathrm{hh}} = \frac{4\lambda^4 n\langle|s_{\mathrm{hh}}|^2\rangle}{\pi^4|K_{\mathrm{w}}|^2}, \tag{6.84}$$

$$Z_{\mathrm{vv}} = \frac{4\lambda^4 n\langle|s_{\mathrm{vv}}|^2\rangle}{\pi^4|K_{\mathrm{w}}|^2}. \tag{6.85}$$

The radar reflectivity of horizontally polarized scattered wave corresponding to vertically polarized incident wave η_{hv} and that of the vertically polarized scattered wave to horizontally polarized incident wave η_{vh} are respectively given using elements of backscattering covariance matrix as follows;

$$\eta_{\mathrm{hv}} = \eta_{\mathrm{vh}} = 4\pi n\langle|s_{\mathrm{hv}}|^2\rangle. \tag{6.86}$$

Differential Reflectivity Z_{dr}, Z_{DR}

Differential reflectivities for horizontally to vertically polarized waves Z_{dr} and Z_{DR} are expressed using (6.81), (6.82), (6.84), and (6.85) as

$$Z_{\mathrm{dr}} = \frac{\eta_{\mathrm{hh}}}{\eta_{\mathrm{vv}}} = \frac{4\pi n\langle|s_{\mathrm{hh}}|^2\rangle}{4\pi n\langle|s_{\mathrm{vv}}|^2\rangle} = \frac{Z_{\mathrm{hh}}}{Z_{\mathrm{vv}}}, \tag{6.87}$$

$$Z_{\mathrm{DR}}[\mathrm{dB}] = 10\log Z_{\mathrm{dr}} = 10\log\left(\frac{Z_{\mathrm{hh}}}{Z_{\mathrm{vv}}}\right), \tag{6.88}$$

where the lower case subscripts are used to denote the ratio directly (in linear unit), whereas the upper case subscripts are used to express the ratio in decibels.

Parameter Z_{DR} is assumed to depend on the drop shape for horizontally and vertically polarized waves, i.e., the ratio of height and width of drop. Water drop receives upward air resistance when falling down, and generally flatness of the drop increases as the diameter becomes large. Thus, Z_{DR} generally becomes positive. The maximum value observed by ordinary meteorological radar at frequency around 2.8–9.5 GHz is approximately 4 dB.[8] Maximum diameter of raindrop is 6–8 mm. When exceeding this size, drop splits into two or more pieces smaller in size (Komabayashi et al. 1964). On the other hand, in the case of snow and hail, flatness of particle does not change much even if the diameter of particle changes. From these points, the differential reflectivity is considered to be useful for getting information on shape, distinction of particles, and the DSD. This naturally leads to the improvement of the estimation of precipitation intensity (Seliga and Bringi

[8] As will be shown later in Table 6.2, the value for rain and hail becomes approximately to 6 dB.

1976). The precipitation intensity (rainfall rate) estimation with parameters of drop shape will be described later in Sect. 6.4.6.

In the STSR mode, it should be noticed that Z_{dr} biasies is caused by cross coupling of horizontally and vertically polarized waves due to antenna polarization error and so on. If the cross-polar isolation is better than 40 dB, the biasies can be reduced to negligible value (Hubbert et al. 2010).

Linear Depolarization ratio Ldr, LDR

Linear depolarization ratios Ldr and LDR are defined using (6.81), (6.82), and (6.84)–(6.86) as

$$\mathrm{Ldr}_{\mathrm{hv}} = \frac{Z_{\mathrm{hv}}}{Z_{\mathrm{vv}}} = \frac{4\pi n\langle|s_{\mathrm{hv}}|^2\rangle}{4\pi n\langle|s_{\mathrm{vv}}|^2\rangle}, \tag{6.89}$$

$$\mathrm{LDR}_{\mathrm{hv}}[\mathrm{dB}] = 10\log \mathrm{Ldr}_{\mathrm{hv}} = 10\log\left(\frac{Z_{\mathrm{hv}}}{Z_{\mathrm{vv}}}\right), \tag{6.90}$$

and

$$\mathrm{Ldr}_{\mathrm{vh}} = \frac{Z_{\mathrm{vh}}}{Z_{\mathrm{hh}}} = \frac{4\pi n\langle|s_{\mathrm{hv}}|^2\rangle}{4\pi n\langle|s_{\mathrm{hh}}|^2\rangle}, \tag{6.91}$$

$$\mathrm{LDR}_{\mathrm{vh}}[\mathrm{dB}] = 10\log \mathrm{Ldr}_{\mathrm{vh}} = 10\log\left(\frac{Z_{\mathrm{vh}}}{Z_{\mathrm{hh}}}\right). \tag{6.92}$$

where the uses of lower and upper cases are the same as those for Z_{dr} and Z_{DR}, respectively.

The denominator of right-hand side of (6.89) shows the backscattered electric power when vertically polarized wave is transmitted and the same polarized wave is received. On the other hand, the numerator shows the backscattered electric power when vertically polarized wave is transmitted and horizontally polarized wave is received. In this case, the transmission of horizontally polarized wave is ceased. In (6.91), horizontally polarized wave is transmitted. For the transmission and reception of dual polarization radar, LDR shows the ratio of the backscattered electric power which is generated by the orthogonally polarized wave to that which is generated by the same polarized wave. When flat particles fall, LDR depends on the flatness and orientation of the main axis.

The parameter LDR is assumed to be an index for qualitative distinction of shape and direction of ice particles. Thus, it is regarded to be effective for detection of ice and water mixture in the upper layer and for distinction of hail. However the magnitude is as small as at most −20 dB except for inside the bright band. To obtain the correct LDR, it is desirable for observation system to secure a degree of separation of 30–40 dB or more between horizontally and vertically polarized waves. The application of LDR to practical and quantitative measurements is a subject in the future.

Correlation Coefficient at Zero Lag $\rho_{hv}(0)$

The correlation coefficient between horizontally and vertically polarized waves of the received signal is called correlation coefficient at zero lag, and is given using covariance matrix in (6.80) by (e.g., Doviak and Zrnić 2006, p. 242);

$$\rho_{hv}(0) = \frac{|\langle s_{vv} s_{hh}^* \rangle|}{\langle |s_{hh}|^2 \rangle^{1/2} \langle |s_{vv}|^2 \rangle^{1/2}} \,. \tag{6.93}$$

Parameter $\rho_{hv}(0)$ depends on fluctuations of horizontal to vertical length ratio of particles within the scattering volume. As shown in Fig. 3.3, backscattered signal intensity monotonically increases with the diameter of particle in the Rayleigh scattering region. Meanwhile, it significantly changes corresponding to a little change of the diameter in the Mie scattering region. Therefore, if different kinds of particles coexist in the scattering volume, the backscattered signal intensity ratio of horizontally to vertically polarized waves change according to type and spatial distribution of particles, which makes the correlation coefficients between them smaller. Furthermore, if the particle shape is distorted like hail, the signal intensity ratio of horizontally to vertically polarized wave fluctuates when the particle vibrates or rotates, also making the correlation coefficients between them smaller. Consequently, $\rho_{hv}(0)$ is called the parameter that indicates irregular shape and uneven distribution of precipitation particles.

There are other quantities of correlation coefficient between polarized waves, the co-polar elements and cross-polar element s_{hv}, as (Doviak and Zrnić 2006, p. 243)

$$\rho_{h} = \frac{|\langle s_{hv} s_{hh}^* \rangle|}{\langle |s_{hv}|^2 \rangle^{1/2} \langle |s_{hh}|^2 \rangle^{1/2}} \,, \tag{6.94}$$

$$\rho_{v} = \frac{|\langle s_{hv} s_{vv}^* \rangle|}{\langle |s_{hv}|^2 \rangle^{1/2} \langle |s_{vv}|^2 \rangle^{1/2}} \,. \tag{6.95}$$

The practical use of these parameters is also a subject in the future.

Differential Phase Φ_{DP}

The polarization parameter that have been discussed so far is related to the backscattering by precipitation particles within a scattering volume. On the other hand, phase shift is caused in radar wave during propagation. If the phase shifts in round trip between radar and scatterer for horizontally and vertically polarized waves are Φ_{hh} and Φ_{vv}, respectively, Φ_{hh} and Φ_{vv} are almost equal for isotropic scatterers. On the other hand, they become different for anisotropic scatterers. In the case of flat particles such as raindrops in large diameter, phase shift by horizontally polarized wave becomes larger than that of vertically polarized

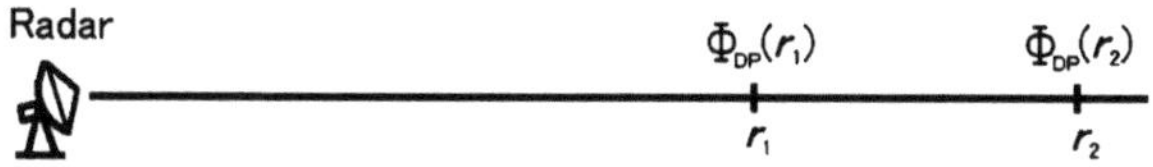

Fig. 6.9 Conceptual diagram of the differential phase between at the distances r_1 and r_2

wave, and accordingly $\Phi_{hh} > \Phi_{vv}$ in a unit distance. The difference between the magnitudes for horizontally and vertically polarized waves is called the differential phase, which is given by

$$\Phi_{DP} = \Phi_{hh} - \Phi_{vv}, \tag{6.96}$$

where Φ_{DP} and $\Phi_{hh,vv}$ are defined in two-way with the unit [deg].

Specific Differential Phase K_{DP}

The change of Φ_{DP} per unit distance is called the specific differential phase K_{DP}, which has been watched in recent years as an important parameter concerned with precipitation intensity (Sachindananda and Zrnić 1985). When radar wave propagates through homogeneous medium like the atmosphere as shown in Fig. 6.9, K_{DP}, which is defined per one-way, is given as the differential phase per unit distance which occurs in round trip between two points r_1 and r_2 as (e.g., Bringi and Chandrasekar 2001, p. 368)

$$K_{DP} = \frac{\Phi_{DP}(r_2) - \Phi_{DP}(r_1)}{2(r_2 - r_1)}, \tag{6.97}$$

where the unit is [deg km^{-1}].

As is clear from the definition of Φ_{hh} and Φ_{vv}, $K_{DP} > 0$ if the horizontally polarized wave is affected more by precipitation particles or flat particles are dominant, while $K_{DP} < 0$ otherwise. The parameter K_{DP}, which is different from Z_{DR} and LDR, is not influenced by the amplitude of received signal, and therefore, its accuracy is not damaged by attenuation of radar waves through the propagation path. By these characteristics, K_{DP} is regarded to be useful as same as Z_{DR} for precipitation intensity estimation and expected for improvement of the estimation accuracy. The technique to obtain precipitation intensity (rainfall rate) based on K_{DP} will be mentioned later in Sect. 6.4.6.

Ranges of polarization parameters and temperature for various hydrometeor species are given in Table 6.2. The ranges are first outlined by Doviak and Zrnić (2006, p. 271) and refined later by Straka et al. (2000) except for the K_{DP} ranges which have been modified to reflect the operation of the classification procedure at the C band (Keenan 2003).

Table 6.2 Ranges of polarization parameters and temperature for various hydrometeor species [from Keenan 2003]

	Z_{hh} [dBZ]	Z_{DR} [dB]	$\lvert\rho_{hv}(0)\rvert$	K_{DP} [deg km^{-1}]	Temperature °C
Drizzle	10 to 25	0.2 to 0.7	> 0.97	0 to 0.06	> -10
Rain	25 to 60	0.5 to 4	> 0.95	0 to 20	> -10
Snow (dry, low density)	−10 to 35	−0.5 to 0.5	> 0.95	−1 to 1	< 0
Snow (dry, high density)	−10 to 35	0 to 1	> 0.95	0 to 0.4	< 0
Snow (wet, melting)	20 to 45	0.5 to 3	0.5 to 0.9	0 to 1	0 to 5
Graupel, dry	20 to 35	−0.5 to 1	> 0.95	0 to 1	< 0
Graupel, wet	30 to 50	−0.5 to 2	> 0.95	0 to 3	−15 to 5
Hail, small $<$ 2 cm wet	50 to 60	−0.5 to 0.5	0.92 to 0.95	−1 to 1	−15 to 20
Hail, large $>$ 2 cm wet	55 to 65	−1 to 0.5	0.90 to 0.92	−1 to 2	−25 to 20
Rain and hail	45 to 80	−1 to 6	> 0.9	0 to 20	−10 to 25

6.4.3 *Shapes of Precipitation Particles and Polarization Parameters*

Shapes of Particles

As previously described, there are certain limits in single-polarization radar observations for estimating intensity and other parameters of precipitation whose DSD changes momentarily. Meanwhile, using polarization parameters available by dual polarization radar, various experiments have been conducted to estimate precipitation intensity and liquid water content by different algorithms that compensate the limits of the single-polarization radar observation.

Observations at low elevation angles are generally emphasized for quantitative radar observations of precipitation. In low elevation angle observations, the most important polarization parameters are differential reflectivity Z_{DR} and specific differential phase K_{DP}. The polarization characteristics of received signal depends on backscattering cross section as the sum of each scatterer. Therefore, it is important to know the relation between drop size and flatness, and that between flatness and dual polarization parameters for estimating precipitation parameters. For the semi-major axis diameter a and the semi-minor axis diameter b of a flat raindrop, some linear approximation which shows the relation between the axis ratio b/a and the diameter of drop D has been derived. For example, Pruppacher and Beard (1970) presented the expression of the axis ratio from a wind-tunnel test as

$$\frac{b}{a} = 1.03 - 62D \qquad 0 \leq D \leq 9 \times 10^{-3}\ [\mathrm{m}]. \tag{6.98}$$

The more accurate polynomial approximation has been presented through a numerical modeling by Beard and Chuang (1987) as

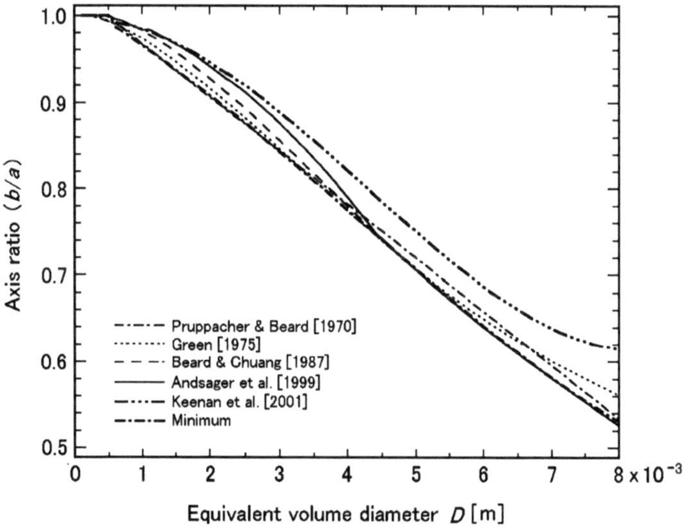

Fig. 6.10 The relation between the axis ratio b/a versus D for various numerical models, where the curve referred to the "minimum" is derived by interpolation of the smallest values from the relations in the literatures [from Park et al. 2005a]

$$\frac{b}{a} = 1.0048 + 5.7 \times 10^{-1}D - 2.628 \times 10^{4}D^{2} + 3.682 \times 10^{6}D^{3} - 1.677 \times 10^{8}D^{4}$$

$$0 \leq D \leq 7 \times 10^{-3}\ [\mathrm{m}], \tag{6.99}$$

where the unit of D is [m]. Figure 6.10 indicates the axis ratio b/a versus D for various numerical models including the results of (6.98) and (6.99), where the curve referred to the "minimum" is derived by interpolation of the smallest values among the axis ratios calculated from the relations in the literatures. These minimum axis ratios are close to those Pruppacher and Beard (1970) for $D < 4 \times 10^{-3}$ m and are close to Beard and Chuang (1987) for $D > 4.4 \times 10^{-3}$ m (Park et al. 2005a).

Relation Between Axis Ratio and Z_{DR}

Polarization parameters which depend on the flatness of raindrop will be discussed below by following Bringi and Chandrasekar (2001, Chap. 7). If the axis ratio shown by either (6.98) or (6.99) is expressed as $b/a \equiv \zeta$, the factor λ_z which depends on flatness of drop is given by

$$\lambda_z = \frac{1+f^2}{f^2}\left(1 - \frac{1}{f}\tan^{-1} f\right), \tag{6.100}$$

where f^2 is expressed as

$$f^2 = \frac{1}{\zeta^2} - 1 = \left(\frac{a}{b}\right)^2 - 1, \tag{6.101}$$

In the Rayleigh–Gans limit, the elements of backscattering matrix s_{hh} and s_{vv} can be expressed as (Bringi and Chandrasekar 2001, p. 382)

$$s_{\text{hh}} = \frac{k^2}{4\pi} \frac{V(\varepsilon_{\text{r}} - 1)}{\left[1 + \frac{1}{2}(1 - \lambda_z)(\varepsilon_{\text{r}} - 1)\right]} \tag{6.102}$$

$$s_{\text{vv}} = \frac{k^2}{4\pi} \frac{V(\varepsilon_{\text{r}} - 1)}{[1 + \lambda_z(\varepsilon_{\text{r}} - 1)]}, \tag{6.103}$$

where k is the radar wave number, $k = 2\pi/\lambda$, V the volume of a drop or $V = (\pi/6)D^3$, and ε_{r} the relative permittivity $(= \varepsilon/\varepsilon_0)$.

From (6.102) and (6.103), it is clear that both s_{hh} and s_{vv} are the function of ζ and D. Hence, s_{hh} and s_{vv} can be expresses as $s_{\text{hh}}(\zeta, D)$ and $s_{\text{vv}}(\zeta, D)$, respectively. Thus, the differential reflectivity Z_{DR} which is given by (6.88) becomes

$$Z_{\text{DR}} = 10\log_{10}\left[\frac{4\pi\langle|s_{\text{hh}}(\zeta, D)|^2\rangle}{4\pi\langle|s_{\text{vv}}(\zeta, D)|^2\rangle}\right]. \tag{6.104}$$

Noting that $Z_{\text{dr}} = 10^{0.1Z_{\text{DR}}}$, the antilogarithm of Z_{DR} leads to

$$Z_{\text{dr}} = \frac{4\pi\langle|s_{\text{hh}}(\zeta, D)|^2\rangle}{4\pi\langle|s_{\text{vv}}(\zeta, D)|^2\rangle}. \tag{6.105}$$

In general, if the DSD is $N(D)$, Z_{dr} is expressed as

$$Z_{\text{dr}} = \frac{\int |s_{\text{hh}}(\zeta, D)|^2 N(D)\text{d}D}{\int |s_{\text{vv}}(\zeta, D)|^2 N(D)\text{d}D}. \tag{6.106}$$

Substituting (6.102) and (6.103) into (6.106), Z_{dr} becomes

$$Z_{\text{dr}} = \frac{\int D^6 N(D)[1 + \lambda_z(\varepsilon_{\text{r}} - 1)]^2 \text{d}D}{\int D^6 N(D)\left[1 + \frac{1}{2}(1 - \lambda_z)(\varepsilon_{\text{r}} - 1)\right]^2 \text{d}D}. \tag{6.107}$$

Substituting the approximation

$$\frac{|1+\lambda_z(\varepsilon_r-1)|^2}{\left|1+\frac{1}{2}(1-\lambda_z)(\varepsilon_r-1)\right|^2} \simeq \frac{1}{\zeta^{7/3}} \tag{6.108}$$

into (6.107), Z_{dr} is expressed as (Bringi and Chandrasekar 2001, pp. 389–390)

$$Z_{dr}^{-1} \simeq \frac{\int \zeta^{7/3} D^6 N(D) dD}{\int D^6 N(D) dD} = \overline{\zeta_Z^{7/3}}, \tag{6.109}$$

where $\overline{\zeta_Z^{7/3}}$ is the mean value of $\zeta^{7/3}$ which is weighted by the radar reflectivity factor.

Relation Between Axis Ratio and K_{DP}

Radar wave propagation through precipitation was first discussed by Oguchi (1983). If the propagation path is uniformly filled with precipitation, the phase delays per unit distance (one-way) in [rad] for horizontally and vertically polarized waves are written as

$$\phi_h = \frac{2\pi}{k} \int N(D) \mathrm{Re}[a_h \cdot f(\zeta, D)] dD, \tag{6.110}$$

$$\phi_v = \frac{2\pi}{k} \int N(D) \mathrm{Re}[a_v \cdot f(\zeta, D)] dD, \tag{6.111}$$

respectively, where k is the radar wave number ($= 2\pi/\lambda$), $\boldsymbol{a}_h$ and $\boldsymbol{a}_v$ are unit vectors of horizontally and vertically polarized waves, respectively, and $\boldsymbol{f}(\zeta, D)$ is the vector amplitude of forwardly scattered wave which is the same as that shown in (3.13). As $\Phi_{hh,vv}$ in (6.96) are defined in two-way path, K_{DP} in (6.97) is given by

$$K_{DP} = \frac{2\pi}{k} \int N(D) \mathrm{Re}[\boldsymbol{a}_h \cdot \boldsymbol{f}(\zeta, D) - \boldsymbol{a}_v \cdot \boldsymbol{f}(\zeta, D)] dD \ \ [\text{rad m}^{-1}]. \tag{6.112}$$

From the Rayleigh–Gans theory, the above equation is replaced to

$$K_{DP} = \frac{\pi k}{12} \int D^3 N(D) \mathrm{Re}[F(\zeta, \varepsilon_r)] dD, \tag{6.113}$$

where $F(\zeta, \varepsilon_r)$ is given by

$$F(\zeta, \varepsilon_r) = \frac{\varepsilon_r - 1}{1 + \frac{1}{2}(1-\lambda_z)(\varepsilon_r - 1)} - \frac{\varepsilon_r - 1}{1 + \lambda_z(\varepsilon_r - 1)}. \tag{6.114}$$

In the Rayleigh scattering, $\mathrm{Re}[F(\zeta,\varepsilon_r)]$ is simplified as

$$\mathrm{Re}[F(\zeta,\varepsilon_r)] \simeq C_k(1-\zeta), \tag{6.115}$$

where C_k is almost constant. For instance, in the frequency band of 3 to 30 GHz, C_k becomes 3.3 to 4.2 for ζ from 1 to 0.5. Substituting (6.115) into (6.113), K_{DP} is expressed as (Bringi and Chandrasekar 2001, p. 387)

$$\begin{aligned} K_{\mathrm{DP}} &= \frac{\pi k C_k}{12}\int D^3(1-\zeta)N(D)\mathrm{d}D \\ &= \left(\frac{\pi}{\lambda}\right)\frac{C_k}{\rho_{\mathrm{w}}}\int\left(\frac{\pi}{6}\right)\rho_{\mathrm{w}}(1-\zeta)N(D)\mathrm{d}D \\ &= \left(\frac{\pi}{\lambda}\right)C_k\frac{W}{\rho_{\mathrm{w}}}\left[1-\frac{\int\zeta D^3N(D)\mathrm{d}D}{\int D^3N(D)\mathrm{d}D}\right], \end{aligned} \tag{6.116}$$

where W is the cloud water content given by (6.18), λ is the radar wavelength, and ρ_{w} is the density of raindrop. Note that MKS units are used throughout. The ratio of integrations in (6.116) means the mass-weighted mean axis ratio $\overline{\zeta}_{\mathrm{m}}$. The value of K_{DP} is commonly expressed in [deg km^{-1}], and thus, if the units of W, ρ_{w}, and λ are [g m^{-3}], [g cm^{-3}], and [m], respectively, K_{DP} becomes (Jameson 1985; Bringi and Chandrasekar 2001, p. 388)

$$K_{\mathrm{DP}} = \left(\frac{180}{\lambda}\right)10^{-3}C_kW(1-\overline{\zeta}_{\mathrm{m}}) \ \ [\mathrm{deg\ km^{-1}}], \tag{6.117}$$

where $C_k = 3.75$. The above equation relates K_{DP} to mass-weighted mean axis ratio $\overline{\zeta}_{\mathrm{m}}$. Furthermore, in (6.117), replacing $\overline{\zeta}_{\mathrm{m}}$ by (6.98) and using the mass-weighted drop diameter D_{m}[m] in (6.22), K_{DP} is expressed as

$$K_{\mathrm{DP}} \simeq 62\left(\frac{180}{\lambda}\right)10^{-3}C_kD_{\mathrm{m}}W \ \ [\mathrm{deg\ km^{-1}}]. \tag{6.118}$$

If W is sufficiently large and λ is small, K_{DP} becomes the magnitude which is practically measurable.

6.4.4 Attenuation Correction Using K_{DP}

As shown in (6.68), the attenuation coefficient due to water droplet A_{prp} is approximately proportional to the forth moment of the DSD. Furthermore, as the definition of W is given by (6.18) and that of D_{m} by (6.22), $D_{\mathrm{m}}W$ in (6.118) is essentially proportional to the forth moment of the DSD. It follows that A_{h} [dB km^{-1}], which is the attenuation coefficient of a horizontally polarized wave due to precipitation and same as A_{prp} in (6.68), is nearly linearly related to K_{DP} for frequencies from 5 GHz to, at least, 19 GHz, and is of the form

$$A_h = \alpha K_{DP}{}^b, \tag{6.119}$$

where $b \simeq 1$. The coefficients α and b depend on the temperature and the radar frequency (Jameson 1992; Bringi and Chandrasekar 2001, pp. 492–493).

Attenuation Correction of Z

If the measured (attenuated) radar reflectivity factor of a horizontally polarized wave at distance r [km] is $Z'_{hh}(r)$ and that of after attenuation correction is $Z_{hh}(r)$, $Z'_{hh}(r)$ is expressed as

$$10\log Z'_{hh}(r) = 10\log Z_{hh}(r) - 2A_h r, \tag{6.120}$$

which can be generalized for an inhomogeneous path as

$$10\log Z'_{hh}(r) = 10\log Z_{hh}(r) - 2\int_0^r A_h(s)\mathrm{d}s. \tag{6.121}$$

Substituting $A_h = \alpha K_{DP}{}^b$ from (6.119) into (6.121), the Z value after attenuation correction becomes

$$10\log Z_{hh}(r) = 10\log Z'_{hh}(r) + 2\int_0^r \alpha K_{DP}{}^b(s)\mathrm{d}s. \tag{6.122}$$

If the linear relation, i.e., $b = 1$ is applied along the propagation path, K_{DP} can be replaced by ϕ_{DP} as

$$10\log Z_{hh}(r) = 10\log Z'_{hh}(r) + \alpha[\Phi_{DP}(r) - \Phi_{DP}(0)]. \tag{6.123}$$

The values α and b in (6.122) for the temperature between 0°C and 30°C are listed on Table 6.3 (Jameson 1992).

Attenuation Correction of Z_{DR} and LDR

From (6.88), the following relation is derived;

$$Z_{DR}[\mathrm{dB}] = 10\log Z_{hh} - 10\log Z_{vv}. \tag{6.124}$$

Thus, if the measured radar reflectivity factor of a vertically polarized wave is $Z'_{vv}(r)$ and that after attenuation correction is $Z_{vv}(r)$, the measured differential reflectivity Z'_{DR} is expressed as

Table 6.3 The values α and b in $A_h = \alpha K_{DP}^b$, β and c in $A_{DR} = \beta K_{DP}^c$, for temperature-average 0–30°C [from Jameson 1992]

Frequency [GHz]	Temperature [°C]	$A_h = \alpha K_{DP}^b$		$A_{DR} = \beta K_{DP}^c$	
		α	b	β	c
	0	2.577×10^{-2}	0.838	4.391×10^{-3}	1.053
2.80	0–30	1.704×10^{-2}	0.836	2.925×10^{-3}	1.052
	30	1.152×10^{-2}	0.834	1.993×10^{-3}	1.051
	0	9.738×10^{-2}	0.959	1.605×10^{-2}	1.177
5.48	0–30	7.268×10^{-2}	0.991	1.331×10^{-2}	1.231
	30	5.339×10^{-2}	1.017	1.080×10^{-2}	1.280
	0	2.480×10^{-1}	0.971	3.321×10^{-2}	1.128
9.34	0–30	2.328×10^{-1}	1.019	3.279×10^{-2}	1.148
	30	2.138×10^{-1}	1.068	3.220×10^{-2}	1.170
	0	4.231×10^{-1}	0.957	5.387×10^{-2}	1.137
13.80	0–30	4.390×10^{-1}	0.975	5.439×10^{-2}	1.131
	30	4.532×10^{-1}	0.994	5.553×10^{-2}	1.127
	0	6.960×10^{-1}	0.995	9.098×10^{-2}	1.181
19.35	0–30	7.218×10^{-1}	0.988	8.802×10^{-2}	1.172
	30	7.593×10^{-1}	0.980	8.681×10^{-2}	1.161

$$\begin{aligned} Z'_{DR} &= 10\log Z'_{hh} - 10\log Z'_{vv} \\ &= 10\log Z_{hh} - 10\log Z_{vv} - 2(A_h - A_v) \\ &= Z_{DR} - 2A_{DR}, \end{aligned} \tag{6.125}$$

where A_v is the attenuation coefficient of a vertically polarized wave due to precipitation and $A_{DR} = A_h - A_v$. As the relation between A_v and the DSD is similar to that of A_h, A_v can be expressed using K_{DP} in the same way. Thus, Z_{DR} at the distance r [km] is finally given in [dB] by the same formula as (6.122) as

$$Z_{DR}(r) = Z'_{DR}(r) + 2\int_0^r \beta K_{DP}^c(s)\mathrm{d}s, \tag{6.126}$$

and as (6.123) in the linear relation of $c = 1$ as

$$Z_{DR}(r) = Z'_{DR}(r) + \beta[\Phi_{DP}(r) - \Phi_{DP}(0)]. \tag{6.127}$$

The values β and c in $A_{DR} = \beta K_{DP}^c$ for the temperature between 0°C and 30°C is also listed on Table 6.3.

As for LDR, one way path contributes to the total attenuation, thus the measured $\mathrm{LDR}'_{\mathrm{hv}}$ and the value after attenuation correction $\mathrm{LDR}_{\mathrm{hv}}$ are in the following relation (Pfeifer et al. 2008).

$$\mathrm{LDR}'_{\mathrm{hv}} = \mathrm{LDR}_{\mathrm{hv}} + A_{\mathrm{DR}} \tag{6.128}$$

$$\mathrm{LDR}_{\mathrm{hv}} = \mathrm{LDR}'_{\mathrm{hv}} - \int_0^r \beta K^c_{\mathrm{DP}}(s)\mathrm{d}s. \tag{6.129}$$

The algorithms to estimate attenuation due to precipitation from K_{DP} or Φ_{DP} have several advantages. These values are derived from phase measurement, and thus they are unaffected by the echo intensity, i.e., absolute calibration error, attenuation, and so on. However, it should be noticed that K_{DP} is relatively noisy at low rain rates. The value of K_{DP} is derived as the range derivative of Φ_{DP} and estimated over a finite path. Moreover, as discussed in the previous subsection, smoothing is intrinsically desired for the practical use of Φ_{DP}. More detailed discussion is done by e.g., Jameson (1992), Bringi and Chandrasekar (2001, Chap. 7), and Park et al. (2005a,b) especially for 9.375 GHz radar.

6.4.5 *Estimates and Variances of Polarization Parameters*

Ryzhkov and Zrnić (1998) have compared theoretically the errors in estimating Φ_{DP} of the STSR mode and the alternate mode and concluded that the received signal of the STSR mode provides for reduced statistical fluctuations. This advantage is more pronounced if Doppler velocity spectrum widths normalized by the Nyquist width $2v_{\mathrm{N}}$ ($= \lambda/2T_{\mathrm{s}}$) is larger than 0.1, values that are typically encountered in severe storms. Bringi and Chandrasekar (2001, Chap. 6) have discussed in detail the variance of the estimates of polarimetric parameters for the ATSR and STSR modes. Based upon their descriptions, the estimates and variances of polarization parameters for the STSR mode are discussed in this subsection, where the received signals for the STSR mode are schematically shown in Fig. 6.11.

Radar Reflectivity Factor and Differential Reflectivity

The estimates of received signal power of the horizontally polarized wave $\hat{P}_{\mathrm{h}}$ and that of the vertically polarized one $\hat{P}_{\mathrm{v}}$ are equal to the respective mean values, i.e., the expected values ($\bar{\ }$), which are given by

$$\hat{P}_{\mathrm{h}} = \overline{P}_{\mathrm{h}} = \frac{1}{M}\sum_{m=1}^{M}|V_{\mathrm{h}}(m)|^2, \tag{6.130}$$

$$\hat{P}_{\mathrm{v}} = \overline{P}_{\mathrm{v}} = \frac{1}{M}\sum_{m=1}^{M}|V_{\mathrm{v}}(m)|^2, \tag{6.131}$$

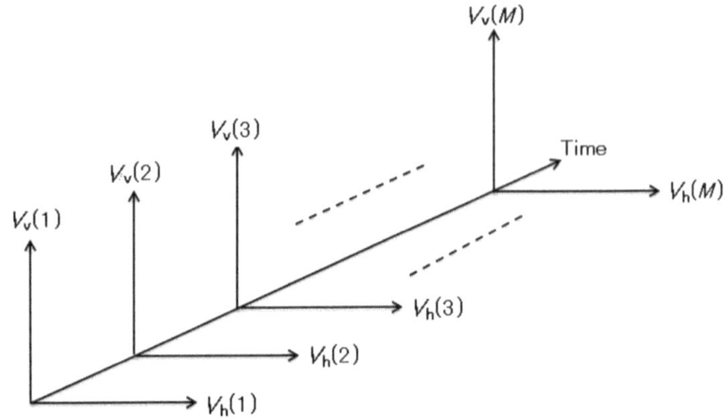

Fig. 6.11 Schematic showing horizontal and vertical components of the received signal $V_h(m)$ and $V_v(m)$ for the pulses of the STSR mode

where M is the number of samples, and V_h and V_v are signal amplitudes of horizontally and vertically polarized waves, respectively. The estimate of differential reflectivity $\hat{Z}_{DR}$ is given by

$$\hat{Z}_{DR}[\mathrm{dB}] = 10\log\left(\frac{\hat{P}_h}{\hat{P}_v}\right). \tag{6.132}$$

The variances of $\hat{P}_h$ and $\hat{P}_v$ are expressed using the number of independent samples M_I and the relations of (5.198) and (5.200) as

$$\mathrm{var}[\hat{P}_h] = \frac{\overline{P}_h^2}{M_I} = \frac{\overline{P}_h^2}{M}\sum_{m=-(M-1)}^{M-1}\left(1-\frac{|m|}{M}\right)\rho_h(m), \tag{6.133}$$

$$\mathrm{var}[\hat{P}_v] = \frac{\overline{P}_v^2}{M_I} = \frac{\overline{P}_v^2}{M}\sum_{m=-(M-1)}^{M-1}\left(1-\frac{|m|}{M}\right)\rho_v(m), \tag{6.134}$$

respectively, where M_I is the number of independent samples, M the total number of samples, and $\rho_{h,v}(m)$ the correlation coefficients of the received power (power sample) of horizontally and vertically polarized waves, respectively. Thus, as discussed in Sect. 5.6.4, $\rho_{h,v}(m)$ are equivalent to $\rho_{S+N}(mT_s) = [\rho_0(m\tau)]^2$ in (5.201). The perturbation approximation for the ratio of two random variables X and Y becomes (Bringi and Chandrasekar 2001, p. 354)

$$\mathrm{var}\left[\frac{X}{Y}\right] \simeq \left(\frac{\overline{X}}{\overline{Y}}\right)^2\left[\frac{\mathrm{var}(X)}{\overline{X}^2} + \frac{\mathrm{var}(Y)}{\overline{Y}^2} - \frac{2\mathrm{cov}(X,Y)}{\overline{XY}}\right], \tag{6.135}$$

where $\overline{X}$ and $\overline{Y}$ are the expected values of X and Y, respectively, and "cov" denotes covariance. Applying the above relation and using (6.133) and (6.134), $\mathrm{var}[\hat{P}_\mathrm{h}/\hat{P}_\mathrm{v}]$ is expressed as

$$\mathrm{var}\left[\frac{\hat{P}_\mathrm{h}}{\hat{P}_\mathrm{v}}\right] = \frac{2}{M}\left(\frac{\hat{P}_\mathrm{h}}{\hat{P}_\mathrm{v}}\right)^2 [1-\rho_\mathrm{hv}(0)^2] \sum_{m=-(M-1)}^{M-1}\left(1-\frac{|m|}{M}\right)\rho_\mathrm{h}(m). \tag{6.136}$$

Taking logarithm of both sides of (6.136) and applying (5.203), the standard deviation SD($\hat{Z}_\mathrm{DR}$) becomes

$$\begin{aligned}\mathrm{SD}[\hat{Z}_\mathrm{DR}]\ [\mathrm{dB}] &= 10\log\left\{1+\left[\frac{2}{M}(1-\rho_\mathrm{hv}(0)^2)\sum_{m=-(M-1)}^{M-1}\left(1-\frac{|m|}{M}\right)\rho_\mathrm{h}(m)\right]^{1/2}\right\} \\ &= 10\log\left\{1+\left[\frac{2}{M}(1-\rho_\mathrm{hv}(0)^2)\frac{1+2\dfrac{S}{N}+\dfrac{(S/N)^2}{2\sigma_{vn}\sqrt{\pi}}}{\left(1+\dfrac{S}{N}\right)^2}\right]^{1/2}\right\},\end{aligned} \tag{6.137}$$

where σ_{vn} is the Doppler velocity spectrum width normalized by the Nyquist width $2v_\mathrm{N}$. Figure 6.12 illustrates $\mathrm{SD}[\hat{Z}_\mathrm{DR}]$ versus M for various σ_{vn} values calculated with (6.137), where the SNR is 10 dB, $\rho_\mathrm{hv}(0)$ values are (a) 0.95 and (b) 0.99.

Correlation Coefficient at Zero Lag Between the Copolar Signals

The correlation coefficient at zero lag between horizontally and vertically polarized signals $\rho_\mathrm{hv}(0)$ is obtained by

$$\hat{\rho}_\mathrm{hv}(0) = \frac{|\hat{R}_\mathrm{hv}(0)|}{(\hat{P}_\mathrm{h}\hat{P}_\mathrm{v})^{1/2}}, \tag{6.138}$$

where $\hat{R}_\mathrm{hv}(0)$ is the covariance at zero lag between V_h and V_v, which is given by

$$\hat{R}_\mathrm{hv}(0) = \frac{1}{M}\sum_{m=1}^{M} V_\mathrm{h}^*(m)V_\mathrm{v}(m), \tag{6.139}$$

where the superscript * denotes complex conjugate.

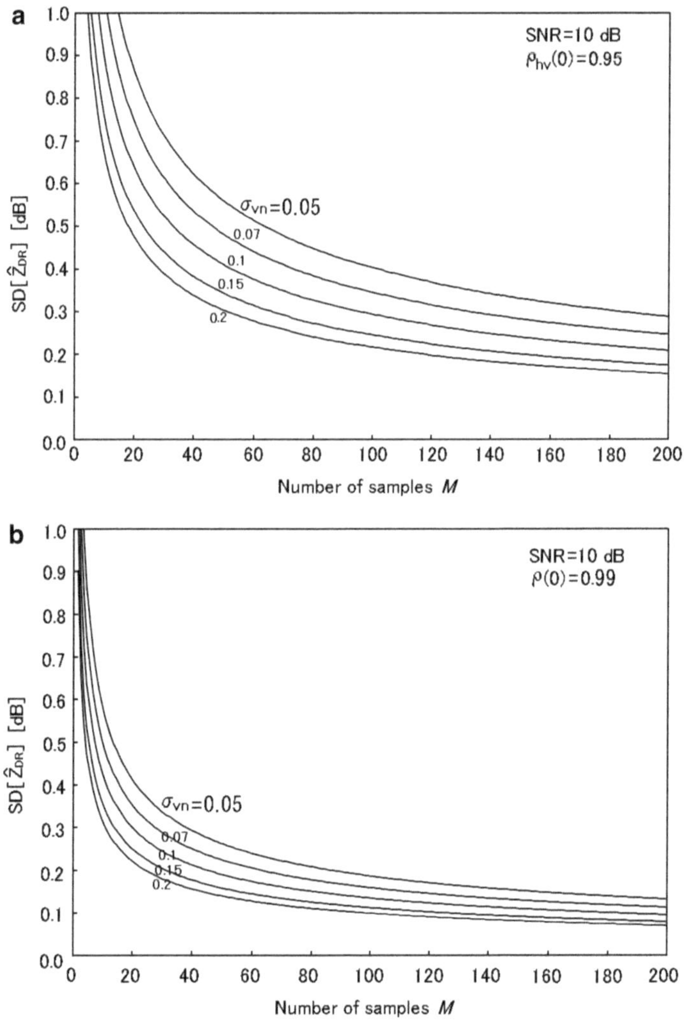

Fig. 6.12 SD$[\hat{Z}_{\mathrm{DR}}]$ versus total number of samples M for various σ_{vn} values calculated with (6.137), where the SNR is 10 dB, $\rho_{\mathrm{hv}}(0)$ values are (**a**) 0.95 and (**b**) 0.99

The presence of noise affects the value of $\rho_{\mathrm{hv}}(0)$. If the correlation coefficient corresponding to observation data in the presence of noise is $\rho'_{\mathrm{hv}}(0)$ and that corrected for the signal-to-noise ratio $\rho_{\mathrm{hv}}(0)$, the value $\rho'_{\mathrm{hv}}(0)$ is expressed as

$$\rho'_{\mathrm{hv}}(0) = \rho_{\mathrm{hv}}(0) \frac{1}{\sqrt{1 + \frac{1}{(S/N)_h}} \sqrt{1 + \frac{1}{(S/N)_v}}}, \tag{6.140}$$

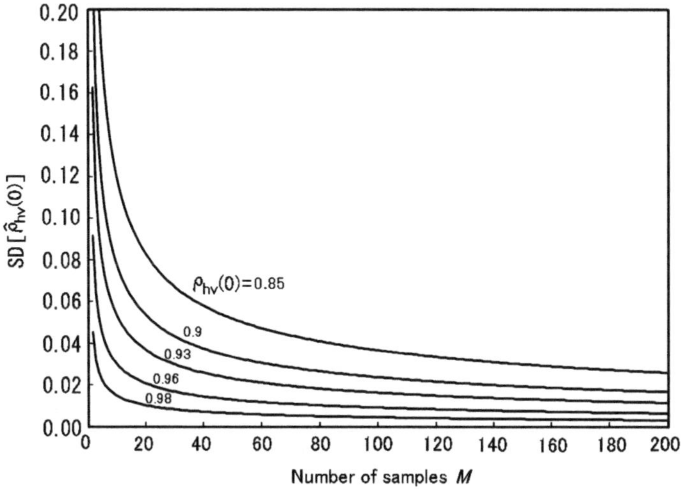

Fig. 6.13 SD[$\hat{\rho}_{\text{hv}}(0)$] for the STSR mode versus total number of samples M for $\rho_{\text{hv}}(0)$ values between 0.85 and 0.98 calculated with (6.142), where the SNR is 10 dB, and σ_{vn} is 0.1

where $(S/N)_h$ and $(S/N)_v$ are the signal-to-noise ratio of horizontally and vertically polarized signals, respectively. With the assumption of $(S/N)_h = (S/N)_v = (S/N)$, we can express $\rho_{\text{hv}}(0)$ as (Shusse et al. 2009)

$$\rho_{\text{hv}}(0) = \rho_{\text{hv}}(0)' \left[1 + \frac{1}{(S/N)}\right] = \rho_{\text{hv}}(0)'[1 + 10^{-(S/N)[\text{dB}]/10}]. \tag{6.141}$$

The variance of $\hat{\rho}_{\text{hv}}(0)$ for the STSR mode can be written as (Bringi and Chandrasekar 2001, p. 364)

$$\text{var}[\hat{\rho}_{\text{hv}}(0)] = \frac{(1 - |\rho_{\text{hv}}(0)|^2)^2}{|\rho_{\text{hv}}(0)|^2} \frac{1}{2M^2} \sum_{m=-(M-1)}^{M-1} (M - |m|)\rho_{\text{h}}(m), \tag{6.142}$$

and the standard deviation is given by $\text{SD}[\hat{\rho}_{\text{hv}}(0)] = \sqrt{\text{var}[\hat{\rho}_{\text{hv}}(0)]}$. Figure 6.13 shows SD[$\hat{\rho}_{\text{hv}}(0)$] versus M for various values of $\rho_{\text{hv}}(0)$ calculated with (6.142) and applying (5.203), where the SNR is 10 dB, and σ_{vn} is 0.1.

Differential Phase

The differential phase of measured received signals between horizontally and vertically polarized waves Ψ_{DP} is given using $\hat{R}_{\text{hv}}(0)$ in (6.139) as

$$\Psi_{\text{DP}} = \arg[\hat{R}_{\text{hv}}(0)], \tag{6.143}$$

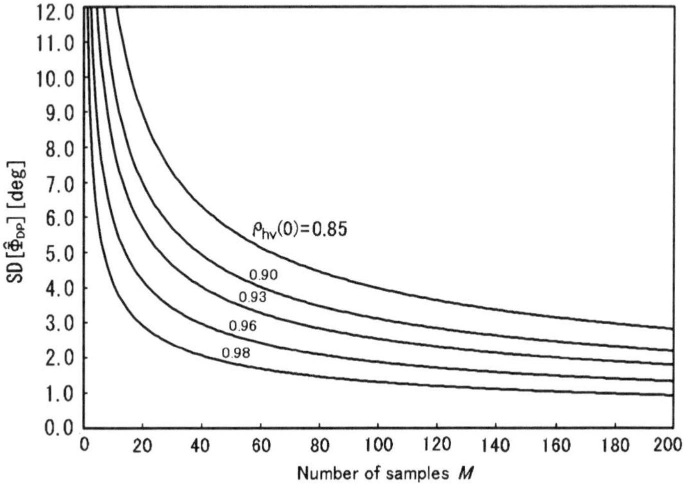

Fig. 6.14 SD[$\hat{\Phi}_{DP}$] versus total number of samples M for $\rho_{hv}(0)$ values between 0.85 and 0.98 calculated with (6.145), where the SNR is 10 dB, and σ_{vn} is 0.1

which is the composition of the differential (propagation) phase Φ_{DP} and the differential scattering phase δ as

$$\Psi_{DP} = \Phi_{DP} + \delta, \tag{6.144}$$

where $\delta = \arg(s_{hh}^* s_{vv})$. The value of Φ_{DP} is range cumulative, and its range profiles are smooth functions of range, whereas δ is dependent on the phase change in each range volume. For scattering from precipitation at a 10-cm wavelength, $\delta \simeq 0$ (Torlaschi and Holt 1993). In the present paragraph, we assume that $\delta = 0$, and estimate the error of Φ_{DP}. The treatment of nonzero δ will be discussed in the following paragraph.

The variance of Φ_{DP} for the STSR mode is given by (Bringi and Chandrasekar 2001, p. 362)

$$\mathrm{var}[\hat{\Phi}_{DP}] = \frac{1 - |\rho_{hv}(0)|^2}{|\rho_{hv}(0)|^2} \frac{1}{2M^2} \sum_{m=-(M-1)}^{M-1} (M - |m|)\rho_h(m), \tag{6.145}$$

and the standard deviation is given by $\mathrm{SD}[\hat{\Phi}_{DP}] = \sqrt{\mathrm{var}[\hat{\Phi}_{DP}]}$. Figure 6.14 shows SD[$\hat{\Phi}_{DP}$] versus total number of samples M for $\rho_{hv}(0)$ values between 0.85 and 0.98 calculated with (6.145), where the SNR is 10 dB, and σ_{vn} is 0.1. Figure 6.15 shows the $\sqrt{M}$ SD[$\hat{\Phi}_{DP}$] values as a function of the normalized Doppler velocity spectrum width σ_{vn} for the STSR mode and the alternate mode (Ryzhkov and Zrnić 1998). Naturally, two results agree well, e.g., if σ_{vn} is 0.1, $\rho_{hv}(0)$ is 0.90, and M is 100, SD[$\hat{\Phi}_{DP}$] becomes almost 3.1°.

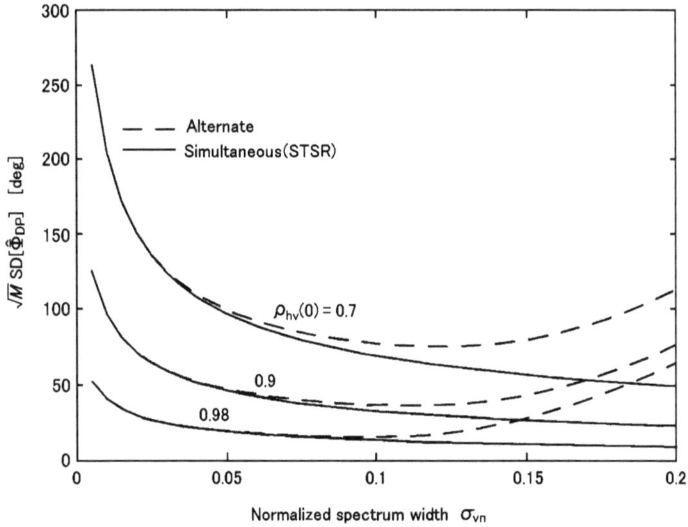

Fig. 6.15 $\sqrt{M}\,\mathrm{SD}[\hat{\Phi}_{DP}]$ as a function of the normalized Doppler velocity spectrum width σ_{vn} for the STSR mode and the alternate mode [from Ryzhkov and Zrnić 1998]

For small σ_{vn} values, the two estimates of the STSR mode and the alternate mode are almost equivalent. Meanwhile, at large spectrum width, the alternate mode is clearly inferior. A sharp increase in the error is observed at a σ_{vn} value around 0.12. At a radar wavelength of 10 cm and a pulse repetition time of 10^{-3} s, $\sigma_{vn} <$ 0.12 corresponds to $\sigma_v < 6\,\mathrm{m\,s^{-1}}$. At higher radar frequencies such as 5- and 3-cm wavelengths, $\sigma_v < 3\,\mathrm{m\,s^{-1}}$ and $< 1.8\,\mathrm{m\,s^{-1}}$, respectively. Note also that $\mathrm{SD}[\hat{\Phi}_{DP}]$ is always lower for long pulse repetition time, i.e., for large σ_{vn} values in the STSR mode (Ryzhkov and Zrnić 1998).

Smoothing the Range Profile of Differential Propagation Phase

If the differential scattering phase δ in (6.144) is nonzero, as occurs in Mie scattering, and if it varies over the range volume, the value of specific differential phase K_{DP} will be in error. Hence, effects due to δ must be eliminated before estimating K_{DP}. Hubbert et al. (1993) and Hubbert and Bringi (1995) discussed in detail how to eliminate the possible effects of δ. The following description is based on their discussion.

Fundamentally, the range profile of Φ_{DP} shows a range-cumulative feature, and so if perturbations are observed in the profile, they can be attributed to δ which is a kind of statistical error. Low pass filtering schemes can be applied to reduce the measurement fluctuations, and to detection, estimation, and elimination of perturbations due to δ. The finite impulse response (FIR) and infinite impulse response (IIR) filters are proposed for these purposes. Figure 6.16a shows magnitude responses of two types of IIR filters; "light" and "heavy" (Hubbert et al. 1993).

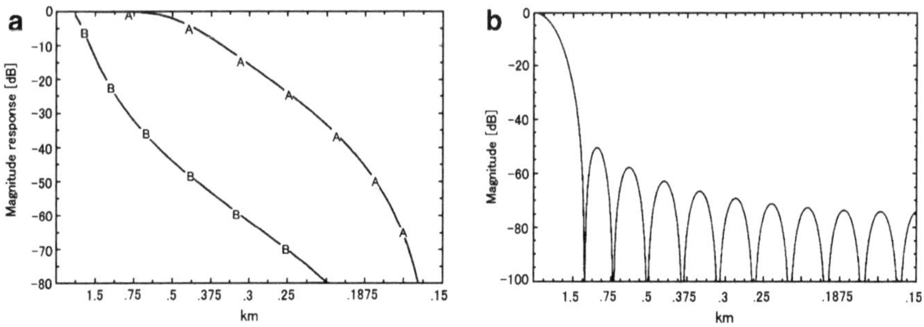

Fig. 6.16 (**a**) Magnitude response of the "light" (*curve A*) and "heavy" (*curve B*) IIR filters. (Hubbert et al. 1993). (**b**) Magnitude response of the filter used in the iterative algorithm [from Hubbert and Bringi 1995]

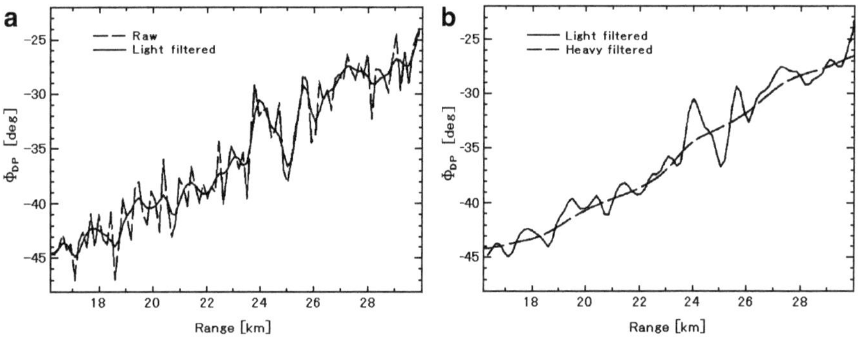

Fig. 6.17 (**a**) Raw (*broken line*) and lightly filtered (*solid line*) Ψ_{DP} profiles using IIR through rain area. (**b**) Filtered (*broken line*) Ψ_{DP} profiles for the data of (**a**) [from Hubbert et al. 1993]

Curve A shows the response of the light filter, whose range variations on the order of 0.5 km and larger are passed, whereas variations of Φ_{DP} on the range scale on the order of 0.25 km and smaller are attenuated by more than 20 dB. Curve B shows that of the heavy filter, where range variations on the order of 1.5 km or less are significantly attenuated. The magnitude response of the filter used in the iterative algorithm is shown in Panel (b) (Hubbert and Bringi 1995).

Figure 6.17 shows the range profile of Ψ_{DP} before and after applying the above-mentioned filters. In Panel (a), perturbations due to δ are observed in raw range profile (broken line) of the range sampling rate of 0.15 km. Applying the light filter (solid line), the gate-to-gate fluctuations are strongly attenuated and range variation on the order of 0.5 km is preserved. Panel (b) shows the light filter output (solid line) and the heavy filter output (broken line) which is applied on the light filter output. In the broken line, the perturbations due to δ are eliminated, and the mean trend of increasing Φ_{DP} with range becomes clear (Hubbert et al. 1993).

Expanding the above concept and selecting proper magnitude response, an iterative range filtering technique can be applied in a wider variety of conditions.

The algorithm has two primary selectable parameters. The first is the cutoff frequency of the filter. When using data with a 0.15 km sampling period, Φ_{DP} estimates are practically averaged over 1–2 km intervals to reduce statistical fluctuations so as to obtain meaningful K_{DP} estimates. Thus, choosing a filter which sharply attenuate fluctuations that occur over 1.5 km or less is appropriate. Figure 6.16b shows an example of the low-pass magnitude response of the FIR filter used in the iterative algorithm. The second is to decide threshold value Ψ_{Th} used to construct the iterated Ψ_{DP} range profile. The algorithm of the iterative filtering is given as follows;

1. Apply observed $\Psi_{DP}(r_i)$ or the result of procedure 3 on the filter and obtain the output $\Psi_{fltDP}(r_i)$.
2. Calculate the value $\Delta\Psi_{DP}(r_i) = |\Psi_{fltDP}(r_i) - \Psi_{DP}(r_i)|$.
3. If $\Delta\Psi_{DP}(r_i) \le \Psi_{Th}$ then adopt $\Psi_{DP}(r_i)|$ as the filter output, if $\Delta\Phi_{DP}(r_i) > \Psi_{Th}$ then adopt $\Psi_{fltDP}(r_i)$ as the filter output, where Ψ_{Th} [deg] is the threshold value.
4. Repeat the above procedures 1 to 3.

The thresholds of 1.25 to 2 times the standard deviation Ψ_{DP} were found to give good results. Furthermore, it was found that repeating the above process around 10 times provides good results (Hubbert and Bringi 1995).

Specific Differential Phase

The specific differential phase K_{DP} after smoothing Φ_{DP} is given by (6.97) and (6.144) as

$$\hat{K}_{DP} = \frac{\Phi_{DP}(r_2) - \Phi_{DP}(r_1)}{2(r_2 - r_1)} \simeq \frac{\hat{\Psi}_{DP}(r_2) - \hat{\Psi}_{DP}(r_1)}{2(r_2 - r_1)}, \tag{6.146}$$

thus the standard deviation of K_{DP} becomes

$$\mathrm{SD}[\hat{K}_{DP}] = \frac{\mathrm{SD}[\hat{\Phi}_{DP}]}{\sqrt{2}_{\Delta r}}, \tag{6.147}$$

where Δr is the range interval $(r_2 - r_1)$.

Assuming that K_{DP} is uniform along the propagation path, it can be estimated as the slope of the cumulative differential propagation phase profile utilizing a least squares approximation with N range samples of $\Phi_{DP}(r_i)$ as (Gorgucci et al. 1999)

$$\hat{K}_{DP} = \frac{\sum_{i=1}^{N} [\Phi_{DP}(r_i) - \overline{\Phi}_{DP}] \left(i - \frac{N+1}{2}\right) \Delta r}{2\sum_{i=1}^{N} \left[\left(i - \frac{N+1}{2}\right) \Delta r\right]^2}. \tag{6.148}$$

The variance of $\hat{K}_{\mathrm{DP}}$ can be given by

$$\mathrm{var}[\hat{K}_{\mathrm{DP}}] = \frac{3}{\Delta r^2} \frac{\mathrm{var}[\hat{\Phi}_{\mathrm{DP}}]}{N(N-1)(N+1)}. \tag{6.149}$$

Thus, using $L = N\Delta r$, the standard deviation of K_{DP} becomes

$$\mathrm{SD}[\hat{K}_{\mathrm{DP}}] = \frac{\mathrm{SD}[\hat{\Phi}_{\mathrm{DP}}]}{L} \sqrt{\frac{3}{N - 1/N}}. \tag{6.150}$$

If $N = 2$, the standard deviation of (6.150) becomes equal to that of (6.147).

Desirable accuracies for K_{DP} are on the order of 0.4 deg km^{-1} for applications such as rainfall rate estimation (Hubbert et al. 1993), and it can be estimated to the accuracy of around 0.3–0.4 deg km^{-1} depending on the path length over which K_{DP} is estimated (Bringi and Chandrasekar 2001, p. 548).

6.4.6 Radar Rainfall Estimation Using Polarization Parameters

The Z–R relation which was discussed in Sect. 6.1.4 is widely used at the present time. Furthermore, as polarimetric parameters Z_{DR} and K_{DP} are deeply related to the DSD, and thus these parameters can be used in various combinations to estimate rainfall rate. Radar rainfall algorithms can be broadly classified into four categories, namely, $R(Z)$ (here Z means Z_{h} in common) which is based on the Z–R relation, $R(Z_{\mathrm{h}}, Z_{\mathrm{DR}})$, $R(K_{\mathrm{DP}})$, and $R(K_{\mathrm{DP}}, Z_{\mathrm{DR}})$, where Z_{h} is in unit of [mm^6 m^{-3}], Z_{DR} is in [dB], and K_{DP} is in [deg km^{-1}].

As of now, most radars operate at single polarization, hence $R(Z)$ is the most popular algorithms to estimate radar rainfall. However, reflectivity-based algorithm requires accurate knowledge of radar performance and is liable to errors in absolute calibration including the attenuation through the propagation path. The differential reflectivity (Z_{DR}) is the value which is obtained by relative power measurement, thus it can be measured accurately without being affected by absolute calibration errors. As it is not possible to measure rainfall rate only using Z_{DR}, any rainfall estimate algorithm which uses Z_{DR} needs to be combined with Z_{h} and or K_{DP}. Algorithms which uses K_{DP} to estimate rainfall rate have several advantages. Since K_{DP} is derived from phase measurements, it is intrinsically unaffected by absolute calibration error and the attenuation through the propagation path. However, it is derived as the range derivative of Φ_{DP}, therefore, there is a trade-off between the accuracy and range resolution of K_{DP}-based rainfall estimates. The advantages and disadvantages of each polarimetric measurement translate into the error structure of algorithms involving Z_{h}, Z_{DR}, and K_{DP} (Bringi and Chandrasekar 2001, p. 537).

Table 6.4 Constants to estimate R [mm h^{-1}] using Z_h and Z_{DR}

Frequency	c_1	a_1	b_1
3 GHz (S-band)	6.7×10^{-3}	0.93	−3.43
5.45 GHz (C-band)	5.8×10^{-3}	0.91	−2.09
10 GHz (X-band)	3.9×10^{-3}	1.07	−5.97

Bringi and Chandrasekar (2001, Chap. 8) discussed in detail the algorithms to estimate rainfall rate using polarization parameters. Teschl et al. (2008) also present rainfall rate estimation algorithms for different raindrop models with 2.8-, 5.635-, and 9.6-GHz polarimetric radars. Maki et al. (2005) also shows the similar algorithms for convective and stratiform rains with 3-cm wavelength polarimetric radar. In the following paragraphs, we will discuss the estimations based upon the description of Bringi and Chandrasekar (2001, Chap. 8).

Estimation by Z_h and Z_{DR}, $R(Z_h, Z_{DR})$

Using antilogarithm expression of the differential reflectivity Z_{DR} which is obtained by (6.88), Z_{dr} $(= 10^{0.1 Z_{DR}})$, and the radar reflectivity factor Z_h which is obtained by the observation of horizontal polarization wave, rainfall rate R [mm h^{-1}] can be given by (Bringi and Chandrasekar 2001, pp. 537–538)

$$R(Z_h, Z_{DR}) = c_1 {Z_h}^{a_1} {Z_{dr}}^{b_1} = c_1 {Z_h}^{a_1} 10^{0.1 b_1 Z_{DR}}, \tag{6.151}$$

where a_1, b_1, and c_1 are constants,[9] and given in Table 6.4.

Estimation by K_{DP}, $R(K_{DP})$

Similar to the discussion of the attenuation by precipitation in (6.69) and (6.119), the relation between R and K_{DP} can be expressed as $a_0 R_0^b = \alpha {K_{DP}}^{\beta}$. Thus the generic form of $R(K_{DP})$ can be given with radar frequency f[GHz] and K_{DP} by (Bringi and Chandrasekar 2001, p. 539)

$$R(K_{DP}) = 129 \left(\frac{K_{DP}}{f} \right)^{b_2}, \tag{6.152}$$

where b_2 is constant, and $b_2 = 0.85$ is generally applied. Equation (6.152) is effective in microwave band of less than around 13 GHz.

[9] Constants $a_1, b_1, c_1, \cdots$ which are used in Sects. 6.4.4 and 6.4.5 are different from those used in Sect. 6.3.3.

Table 6.5 Constants to estimate R [mm h^{-1}] using K_{DP} and Z_{DR}

Frequency	c_3	a_3	b_3
3 GHz (S-band)	90.8	0.93	−1.69
5.45 GHz (C-band)	37.9	0.89	−0.72
10 GHz (X-band)	28.6	0.95	−1.37

Table 6.6 Constants to estimate W [g m^{-3}] using Z_{h} and Z_{DR}

Frequency	c_4	a_4	b_4
3 GHz (S-band)	0.7×10^{-3}	0.89	−4.16
5.45 GHz (C-band)	0.6×10^{-3}	0.85	−2.36
10 GHz (X-band)	0.9×10^{-3}	0.95	−6.18

Estimation by K_{DP} and Z_{DR}, $R(K_{\mathrm{DP}}, Z_{\mathrm{DR}})$

From (6.118), it is shown that K_{DP} is related to W and D_{m}, and the value W can be related to rainfall rate. In addition, from (6.28) D_{m} is proportional to D_0 which is directly related to Z_{DR}. Using the above relation, the generic form of $R(K_{\mathrm{DP}}, Z_{\mathrm{DR}})$ can be given by (Bringi and Chandrasekar 2001, pp. 539–540).

$$R(K_{\mathrm{DP}}, Z_{\mathrm{DR}}) = c_3 {K_{\mathrm{DP}}}^{a_3} {Z_{\mathrm{dr}}}^{b3} = c_3 {K_{\mathrm{DP}}}^{a_3} 10^{0.1 b_3 Z_{\mathrm{DR}}}, \tag{6.153}$$

where a_3, b_3, and c_3 are constants, and given in Table 6.5.

6.4.7 *Estimation of Cloud Water Content*

The cloud water content W is defined using the DSD as (6.18), and W can be also parameterized with either (Z_{h}, Z_{DR}) or K_{DP}, (K_{DP} and Z_{DR}) as similar to rainfall estimation.

Estimation by Z_{h} and Z_{DR}

Using Z_{h} and Z_{DR}, cloud water content W [g m^{-3}] can be given by (Bringi and Chandrasekar 2001, p. 544)

$$W(Z_{\mathrm{h}}, Z_{\mathrm{DR}}) = c_4 Z_{\mathrm{h}}^{a_4} Z_{\mathrm{dr}}^{b_4} = c_4 Z_{\mathrm{h}}^{a_4} 10^{0.1 b_4 Z_{\mathrm{DR}}}, \tag{6.154}$$

where a_4, b_4, and c_4 are constants, which are given in Table 6.6.

Table 6.7 Constants to estimate W [g m^{-3}] using K_{DP} and Z_{DR}

Frequency	c_6	a_6	b_6
3 GHz (S-band)	6.05	0.88	−2.52
5.45 GHz (C-band)	2.32	0.83	−1.11
10 GHz (X-band)	2.13	0.91	−2.19

Estimation by K_{DP}

Using K_{DP} [deg km^{-1}] and radar frequency f[GHz], W [g m^{-3}] can be given by (Bringi and Chandrasekar 2001, p. 544)

$$W(K_{\mathrm{DP}}) = c_5 \left(\frac{K_{\mathrm{DP}}}{f} \right)^{b_5}, \tag{6.155}$$

where b_5 and c_5 are constants, and the values are 0.77 and 3.565, respectively.

Estimation by K_{DP} and Z_{DR}

Using K_{DP} [deg km^{-1}] and Z_{DR}, W [g m^{-3}] can be given by (Bringi and Chandrasekar 2001, p. 545)

$$W(K_{\mathrm{DP}}, Z_{\mathrm{DR}}) = c_6 K_{\mathrm{DP}}^{a_6} Z_{\mathrm{dr}}^{b_6} = c_6 K_{\mathrm{DP}}^{a_6} 10^{0.1 b_6 Z_{\mathrm{DR}}}, \tag{6.156}$$

where a_6, b_6, and c_6 are constants, and given in Table 6.7.

6.4.8 Hydrometeor Classification with Polarization Parameters

Polarimetric parameters can be used to classify hydrometeor types. Although the classification itself is the final purpose, it is also important to improve the estimation quality of rainfall intensity. As for the classification, many methodologies such as statistical discussion model, neural network model, and so on are discussed. In addition, the method of using a fuzzy logic theory has become popular in recent years (e.g., Liu and Chandrasekar 2000; Zrnić et al. 2001; Keenan 2003; May and Keenan 2005; Park et al. 2009). The fuzzy logic classification treats the information which contains ambiguousness in its own boundary and judges the information by modeling as the human being considers and handling it quantitatively. The following discussion is based upon the description of Park et al. (2009).

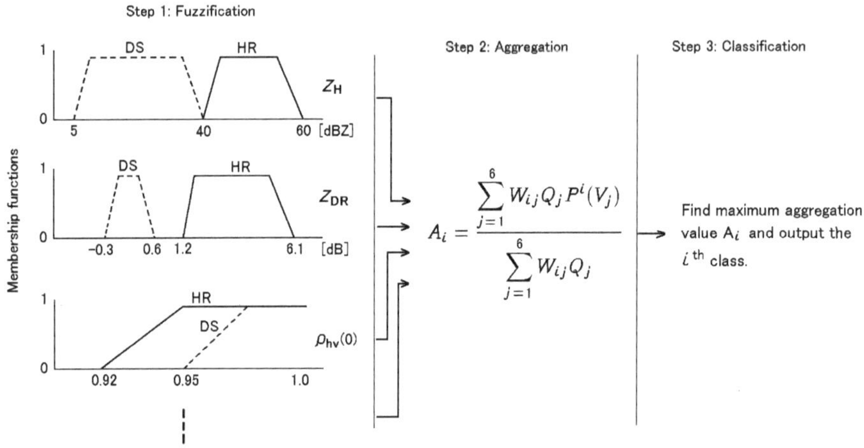

Fig. 6.18 Conceptual flow diagram of the fuzzy logic classification using two example classes, heavy rain (HR) and dry aggregated snow (DS). Labels on the abscissas are based on Park et al. (2009). The Z_{DR} value of HR is for Z_h value of 50 dBZ

Estimation of Hydrometeor Species with Fuzzy Logic Theory

In the fuzzy logic classification, there are generally three steps as shown in Fig. 6.18. The first step is the fuzzification process. From the statistical points, each hydrometeor species, e.g., drizzle, rain, snow, and so on, has a specific range of values for various polarization parameters such as Z_h, Z_{DR}, and so on. Appropriate membership functions are usually applied to characterize these ranges for the fuzzification. The second step is the aggregation process. In this step, the fuzzification results on the first step are aggregated. The final step is the classification to find the most probable radar echo class.

In the first step, the degree, i.e., the probability of an element which belongs to a set (fuzzy set) is expressed by the membership function. If the hydrometeor species are denoted by i (ith class), and the observed parameter is expressed by j (jth variable), the membership function which expresses the probability that the observed element belongs to the ith class and the jth variable is given by $P^i(V_j)$. There are several expression for the membership function such as the beta function (Liu and Chandrasekar 2000) and a Gaussian Bell function (Keenan 2003). Here, we assume $P^i(V_j)$ to have trapezoidal shape with a maximal value of 1 and a minimal value of 0 as shown in Fig. 6.19. The four parameters, $x1$, $x2$, $x3$, and $x4$ for 10 classes to describe trapezoidal functions are given by Park et al. (2009). They give these values for 6 parameters as input, Z_h[dBZ], Z_{DR}[dB], $\rho_{hv}(0)$, $LK_{DP} = 10\log(K_{DP}[\text{deg km}^{-1}])$, SD[$Z_h$][dB], and SD[$\Phi_{DP}$][deg], and output ten classes, ground clutter including that due to anomalous propagation (GG/AP), biological scatterers (BS), dry aggregated snow (DS), wet snow (WS), crystals of various

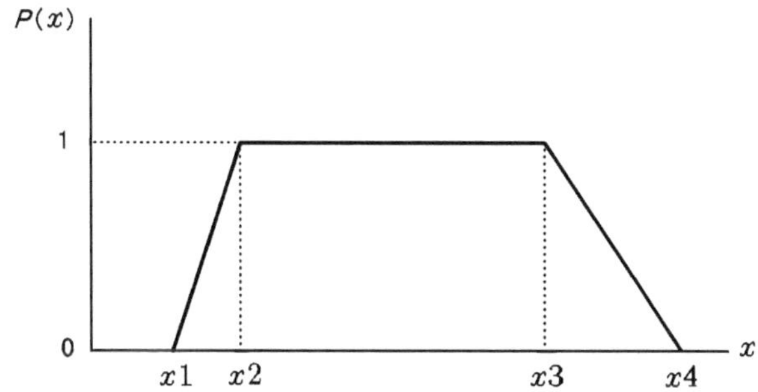

Fig. 6.19 Trapezoidal membership function, where x is an arbitrary polarimetric variable [from Park et al. 2009]

Table 6.8 Trapezoidal function parameters $x1$, $x2$, $x3$, and $x4$ for Z_h[dBZ] [from Park et al. 2009]

x	GC/AP	BS	DS	WS	CR	GR	BD	RA	HR	RH
x_1	15	5	5	25	0	25	20	5	40	45
x_2	20	10	10	30	5	35	25	10	45	50
x_3	70	20	35	40	20	50	45	45	55	75
x_4	80	30	40	50	25	55	50	50	60	80

orientations (CR), graupel (GR), big drops (BD), light and moderate rain (RA), heavy rain (HR), and a mixture of rain and hail (RH). Table 6.8 shows the $x1$ to $x4$ values for the input parameter Z_h.

In the second and final steps, an aggregation value A_i for the ith class of hydrometeor species is found here as

$$A_i = \frac{\sum_{j=1}^{6} W_{ij} Q_j P^i(V_j)}{\sum_{j=1}^{6} W_{ij} Q_j}, \tag{6.157}$$

where W_{ij} is a weight between 0 and 1 assigned to the ith class and the jth variable according to the contribution of each input parameter for various classes, and Q_j is an element of the confidence vector assigned to the jth variable that associates with data quality. The classification decision is based on the maximal aggregation value. The details of W_{ij} and Q_j are also discussed in Park et al. (2009).

The temperature within range volume is important to evaluate the consistency of the classification. Although the temperature cannot be obtained by radar observations directly, in principle, the melting layer can be incorporated from the radar data by measuring the height of the radar bright band. The geometry of the radar beam with respect to the melting layer is illustrated in Fig. 6.20. In the figure, the beam axis is drawn with the thick solid line, where R_t and R_b correspond to geometrical

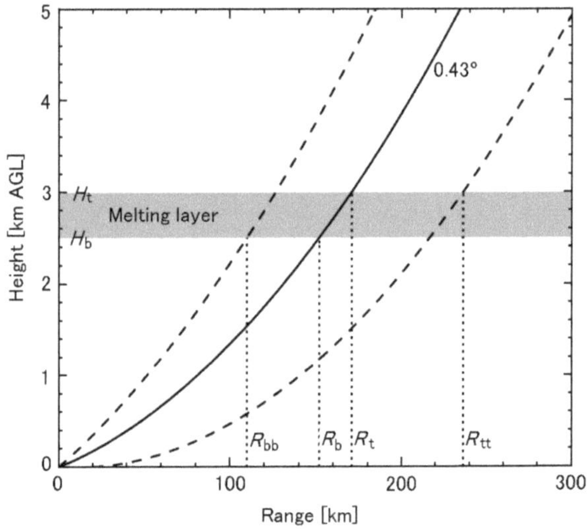

Fig. 6.20 Geometry of the radar beam with respect to the melting layer. The beam axis is drawn with the *thick solid line* whereas the locations of beam extent (3-dB beam width) are indicated with the *dashed curves*. Slant ranges R_{tt} and R_t correspond to the melting layer top height H_t. R_{bb} and R_b correspond to the same layer bottom height H_b [from Park et al. 2009]

projections of the melting layer; top height H_t and bottom H_b, respectively. Dotted lines show the locations of radar beam extent (3-dB beam width), where R_{tt} and R_{bb} correspond to the melting layer H_t and H_b, respectively. Once the boundaries of the melting layer are determined, the fuzzy logic aggregation (6.157) and classification procedure are applied with two restrictions: snow is not allowed below the melting layer bottom and pure rain is prohibited above the melting layer top except for the big-drops category. The following subsets of classes are allowed within five slant range intervals (Park et al. 2009):

$$
\begin{aligned}
&0 < R < R_{bb} && \mathrm{GC/AP, BS, BD, RA, HR, RH;}\\
&R_{bb} < R < R_b && \mathrm{GC/AP, BS, WS, GR, BD, RA, HR, RH;}\\
&R_b < R < R_t && \mathrm{GC/AP, BS, DS, WS, GR, BD, RH;}\\
&R_t < R < R_{tt} && \mathrm{GC/AP, BS, DS, WS, CR, GR, BD, RH;}\\
&R > R_{tt} && \mathrm{DS, CR, GR, RH.}
\end{aligned}
$$

Chapter 7
Radar Observations of the Clear Atmosphere

As seen in Sect. 3.4, refractive index perturbations of half the radar wavelength in scale mostly contribute to atmospheric radar backscattering. Because the turbulent irregularities are advected by the prevailing (background) wind, the component of the wind vector parallel to the radar beam direction causes a frequency shift of the radar carrier frequency by Doppler effect. The baricenter of Doppler spectrum is proportional to the radial component of the mean wind velocity in the radar resolution volume (or scattering volume), the spectral width provides information on the wind variability partly due to turbulence, and the signal power can be related to the intensity of refractive index irregularities. From these primary parameters, vertical profiles of turbulence parameters such as the energy dissipation rate and the vertical eddy diffusivity can be deduced on certain conditions. Combined with the radio acoustic sounding system (RASS), atmospheric radar measurements can also provide vertical profiles of background virtual temperature and water vapor.

In this chapter, the notion of radar detectability is first defined. Standard techniques used to retrieve wind velocity from radar measurements are then described. Next, the methods used for estimating turbulence parameters and other physical parameters obtained from atmospheric radar measurements are discussed. Finally, radar interferometric techniques for improving the spatial resolution are summarized.

7.1 Detectability of Atmospheric Radar Signals

7.1.1 Received Power and Radar Reflectivity

Assuming locally homogeneous isotropic turbulence in the inertial subrange, the so-called refractive index structure constant, C_{n}^2, defined in (3.116), can be related to radar reflectivity η by the relation

S. Fukao and K. Hamazu, *Radar for Meteorological and Atmospheric Observations*, DOI 10.1007/978-4-431-54334-3_7, © Springer Japan 2014

$$C_{\rm n}^2 \simeq \left(\frac{\eta}{0.38}\right)\lambda^{1/3}. \tag{7.1}$$

$C_{\rm n}^2$ is an intrinsic parameter describing the intensity of the refractive index perturbation and thus does not depend on radar frequency. In addition, a relation between η and the received echo power can be established.

The radar equation for Bragg scatterers which is commonly applied for atmospheric radars is (3.40). Thus the expected radar reflectivity is obtained using the expected (or time mean) received power $E[P_r]$ at the distance r from the radar as

$$E[\eta] = \frac{2^{10}(\ln 2)\pi^2 r^2}{P_{\rm t} g^2 \lambda^2 \theta_1^2 c\tau} E[P_{\rm r}]. \tag{7.2}$$

The noise power (expected value or time mean value) $E[P_{\rm n}]$ at the receiver input port is given by (5.36) as

$$E[P_{\rm n}] = kT_{\rm sys}B, \tag{7.3}$$

where k is the Boltzman's constant, B is the receiver bandwidth, and $T_{\rm sys}$ is the system noise temperature given by (5.35). Then by using (7.2) and (7.3), the radar reflectivity is expressed as

$$E[\eta] = \frac{2^{10}(\ln 2)\pi^2 r^2 kT_{\rm sys}B}{P_{\rm t} g^2 \lambda^2 \theta_1^2 c\tau} \frac{E[P_{\rm r}]}{E[P_{\rm n}]}. \tag{7.4}$$

Applying actual radar performance data and measured $E[P_{\rm r}]/E[P_{\rm n}]$ to (7.4), the expected radar reflectivity $E[\eta]$ can be derived which, in turn, provides C_n^2 from (7.1).

7.1.2 Coherent Integration in Atmospheric Radar

The extraction of a radar signal in noise background is generally based on e.g., spectral width of the signal. Furthermore, for atmospheric radars, a criterion is based on the power spectral density (also called power spectrum). It is the reason why signal and noise powers per unit of frequency, i.e., the signal and noise power spectral densities, are important quantities. A typical example of the echo power spectrum obtained from atmospheric radar measurements is plotted against the Doppler shift between $-\delta f$ and $+\delta f$ in Fig. 7.1 (modified from Gage and Balsley 1978). The symbols $S_{\rm S}$ and $S_{\rm n}$ refer to the power spectral density of signal and noise, respectively. Their mean values are noted $\overline{S_{\rm S}}$ and $\overline{S_{\rm n}}$, respectively. $\overline{\Delta S_{\rm n}}$ is the standard deviation of the statistical fluctuations of noise power spectral density around the mean level. The $\pm\delta f$ limits are determined by the transmitted pulse repetition frequency and the number of coherent integrations, i.e., δf [Hz] can be expressed as a function of the number of coherent integrations (i.e., addition of consecutive complex data samples) $M_{\rm coh}$, the total number of points of the periodogram

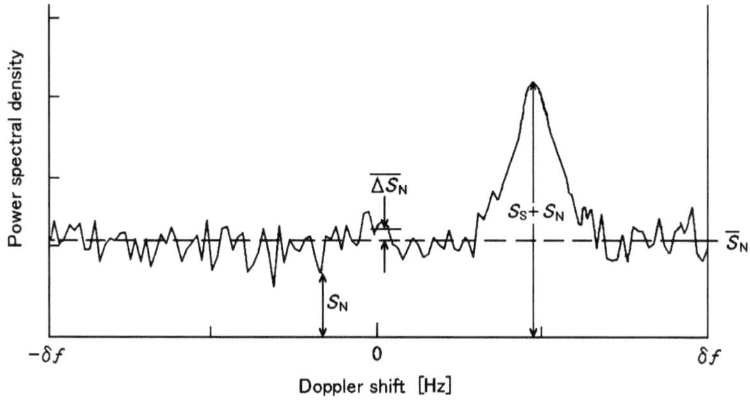

Fig. 7.1 A typical echo power spectrum obtained from atmospheric radar measurements [modified from Gage and Balsley 1978]

(or shortly, fast Fourier transform (FFT) points) M_B, the total number of points of actual signal sample M_s, where $M_s \leq M_B$, and the pulse repetition time T as

$$\delta f = M_s \Delta f = \frac{1}{M_{coh} T} \left(\frac{M_s}{M_B} \right), \tag{7.5}$$

where $\Delta f = 1/(M_{coh} M_B T)$ is the frequency resolution of the power spectral density.

Gage and Balsley (1978) define the detectability D of the received signal in terms of $\overline{S}_S$ (i.e., the spectral peak of the received signal density after averaging) relative to $\overline{\Delta S_n}$ (i.e., the noise fluctuation level after averaging) as

$$D \equiv \frac{\overline{S}_S}{\overline{\Delta S_n}}. \tag{7.6}$$

Then, in terms of the radar equation for distributed scatterers, the detectability criterion may be obtained by using (7.2) and (7.3) as (Gage and Balsley 1978)

$$D = \frac{E[P_r]}{E[P_n]} \left(\frac{M_{coh} M_B}{M_s} \right). \tag{7.7}$$

The terms in the parenthesis of right-hand side of (7.7) express the improvement factor resulting from coherent integration as discussed in Sect. 5.6.2.

Substituting (7.5) into (7.7), D can be expressed as

$$D = \frac{E[P_r]}{E[P_n]} \left(\frac{1}{\delta f T} \right). \tag{7.8}$$

From (7.7) and (7.8), it can be noted that the detectability obtained after M_{coh} coherent integrations and M_B FFT points is equal to the detectability for a

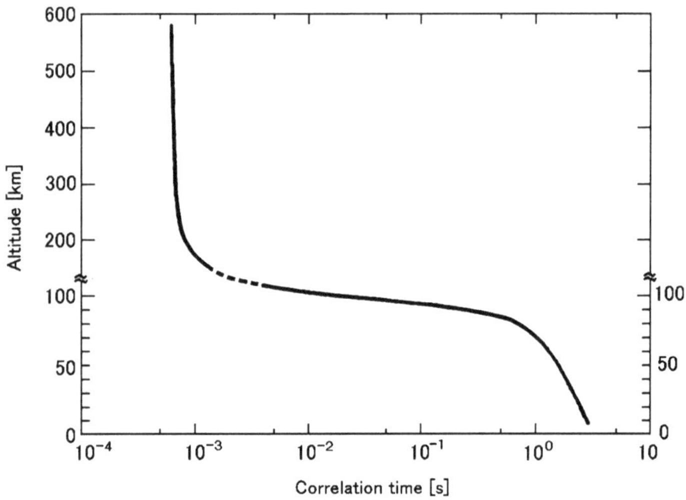

Fig. 7.2 A vertical profile representative of the correlation time expected for received signals from turbulent scattering at 50 MHz [from Kato et al. 1982]

power spectral density of ($M_{\mathrm{coh}} \times M_{\mathrm{B}}$) FFT points without coherent integrations. The reason for this is that the frequency resolution Δf is the same for both cases. However, the memory size necessary for the calculations and the time for FFT calculations significantly differ for the two cases. Filtering by coherent integrations can be used for data compression until frequency aliasing occurs.

7.1.3 Detection of Signal in Noise Background

Figure 7.2 shows a vertical profile representative of the correlation time expected for received signals from turbulent scattering at 50 MHz. The magnitude decreases rapidly for altitudes higher than about 100 km. Below 100 km, the correlation time is estimated to be about 1 s or more. This value is thus much larger than the pulse repetition time for usual atmospheric radars (about 1 ms or even less). Consequently, consecutive samples are highly correlated and the frequency resolution can be selected arbitrarily in the range of correlation time. When applying incoherent integrations of power spectral density (i.e., when adding power spectrum), both signal and noise are added incoherently. In this case, the signal-to-noise ratio (SNR) is not improved but the noisy fluctuations of the power spectral density are reduced, making easier the detection of the peak of power spectral density, especially when the SNR is small. As discussed in Sect. 5.6.3, if the number of incoherent integrations is noted M_{inc}, the standard deviation of the noise fluctuations in the power spectral density is reduced by a factor $1/\sqrt{M_{\mathrm{inc}}}$. Thus, D of (7.8) is improved by a factor $\sqrt{M_{\mathrm{inc}}}$, namely,

$$D = \frac{E[P_{\mathrm{r}}]}{E[P_{\mathrm{n}}]}\left(\frac{\sqrt{M_{\mathrm{inc}}}}{\delta f T}\right). \tag{7.9}$$

If D is equal to three or more, the probability that the noise fluctuations exceed than those of the atmospheric signals is equal to 0.1 % or less. Thus, the level $D = 3$ is generally used as the detection limit.

7.2 Wind Measurements

7.2.1 DBS/VAD Methods

A method used to estimate the wind vector from three (or more) ground-based radar antennas was described in Sect. 4.3.4. The meteorological radar beams can be steered from the horizontal plane to the vertical direction in any azimuth, whereas atmospheric radar beams are generally steered around the zenith direction down to a maximum value of 30° off zenith. Therefore, wind vector can also be measured from a single radar antenna by steering the radar beam in three (or more) directions assuming that the wind field is uniform in the volumes probed by the different radar beams. This method is called the Doppler beam swinging (DBS) method.

Basic Beam Scanning with the DBS Method

The most basic DBS method consists of steering the radar beam in three non coplanar directions. Assuming that the vertical and horizontal wind are uniform, respectively, the east-west (zonal) and the south-north (meridional) winds can be measured using a vertical beam and two oblique beams at zenith angle θ of about 10° to the north and the east directions. As a more robust method in three beam DBS method, the number of beams is sometimes increased to 5, to the south and the west directions in addition to the former three beams as shown in Fig. 7.3. If the vertical and zonal wind components are w and u, respectively, the radial components measured from the eastward beam $V_{r\mathrm{E}}$ and westward beam $V_{r\mathrm{W}}$ can be expressed as follows

$$V_{r\mathrm{E}} = w\cos\theta + u\sin\theta, \tag{7.10}$$

$$V_{r\mathrm{W}} = w\cos\theta - u\sin\theta. \tag{7.11}$$

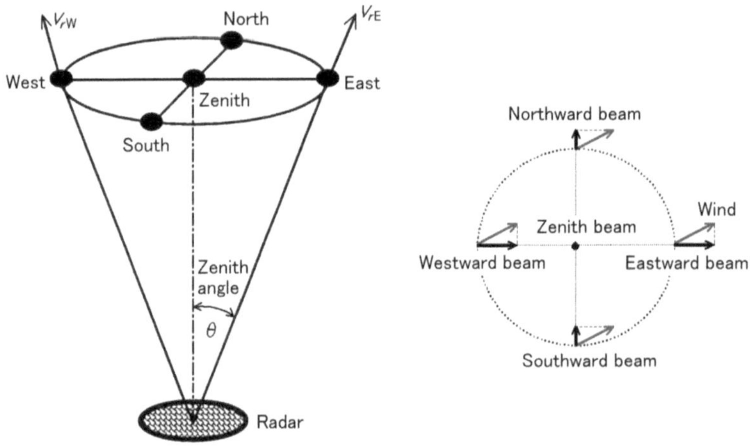

Fig. 7.3 An example of Doppler beam swinging with five beams. *Thick arrows* in the right panel indicate the Doppler velocities of each beam

Subtracting (7.11) from (7.10), one obtains:

$$u = \frac{V_{rE} - V_{rW}}{2\sin\theta}. \tag{7.12}$$

Adding (7.10) and (7.11) provides an estimate of w. Similarly, the meridional wind component v can be obtained by substituting V_{rE} and V_{rW} in (7.12) by the radial wind components V_{rN} and V_{rS} measured from the northward and southward beams, respectively.

Generalization

Let $\mathbf{i}$ be the unit vector along the radar beam direction. The radial component v_d of wind vector $\boldsymbol{v} = (v_x, v_y, v_z)$ is

$$v_d = \boldsymbol{v} \cdot \mathbf{i} = v_x \cos\theta_x + v_y \cos\theta_y + v_z \cos\theta_z, \tag{7.13}$$

where θ_x, θ_y, and θ_z are the angles between the axis defined by the unit vector $\mathbf{i}$ in the radial direction and x, y, and z axes, respectively. The expression (7.13) can be applied for $m = 3$ arbitrary non coplanar directions so that

$$\boldsymbol{v} = \begin{pmatrix} \cos\theta_{x1} & \cos\theta_{y1} & \cos\theta_{z1} \\ \cos\theta_{x2} & \cos\theta_{y2} & \cos\theta_{z2} \\ \cos\theta_{x3} & \cos\theta_{y3} & \cos\theta_{z3} \end{pmatrix}^{-1} \begin{pmatrix} v_{d1} \\ v_{d2} \\ v_{d3} \end{pmatrix}. \tag{7.14}$$

The approach can be generalized to multiple beams ($m > 3$). $\boldsymbol{v}$ can then be estimated by minimizing the residual error ε_v^2 defined from the least square method.

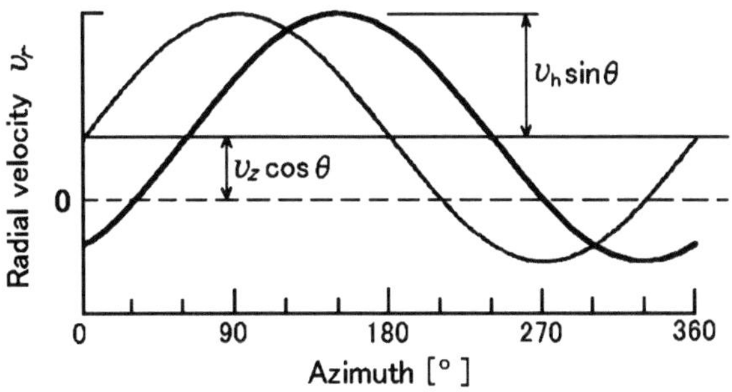

Fig. 7.4 The change of radial velocity according to the antenna rotation in a simplified VAD method

$$\varepsilon_v^2 \equiv \sum_{i=1}^{m} (v_x \cos\theta_{xi} + v_y \cos\theta_{yi} + v_z \cos\theta_{zi} - v_{\mathrm{d}i})^2 . \tag{7.15}$$

The optimal value of $\boldsymbol{v}$ is obtained when the partial derivative of ε_v^2 is 0 for each $\boldsymbol{v}$, namely,

$$\frac{\partial \varepsilon_v^2}{\partial v_j} = 0 \quad (j = x,\ y,\ z). \tag{7.16}$$

It can be shown that $\boldsymbol{v}$ is then given by

$$\boldsymbol{v} = \begin{pmatrix} \sum\cos^2\theta_{xi} & \sum\cos\theta_{xi}\cos\theta_{yi} & \sum\cos\theta_{xi}\cos\theta_{zi} \\ \sum\cos\theta_{yi}\cos\theta_{xi} & \sum\cos^2\theta_{yi} & \sum\cos\theta_{yi}\cos\theta_{zi} \\ \sum\cos\theta_{zi}\cos\theta_{xi} & \sum\cos\theta_{zi}\cos\theta_{yi} & \sum\cos^2\theta_{zi} \end{pmatrix}^{-1} \begin{pmatrix} \sum v_{\mathrm{d}i}\cos\theta_{xi} \\ \sum v_{\mathrm{d}i}\cos\theta_{yi} \\ \sum v_{\mathrm{d}i}\cos\theta_{zi} \end{pmatrix}, \tag{7.17}$$

where $\sum$ denotes $\sum_{i=1}^{m}$.

Simplified VAD Method

A simplified velocity azimuth display (VAD) method is used by atmospheric radars for retrieving the wind vector. Radial velocities are observed by scanning radar beam in azimuth along a circle at a given zenith angle (see Sect. 4.2.2). Using the zenith angle θ instead of the elevation angle θ_e and the wind direction D corresponding to $\delta + \pi$ in expression (4.36), the radial velocity component v_r can be expressed as

$$v_r = v_z \cos\theta - v_{\mathrm{h}} \sin\theta \cos(D - \phi), \tag{7.18}$$

where ϕ is the azimuthal angle. v_r varies with ϕ as indicated by the thick curve in Fig. 7.4. The thin curve shows the result for $D = 270°$ (eastward or westerly wind). From (7.18), the maximum and minimum values $v_{r\max}$, $v_{r\min}$ are given by

$$v_{r\max} = v_z \cos\theta + v_{\mathrm{h}} \sin\theta, \tag{7.19}$$

$$v_{r\min} = v_z \cos\theta - v_{\mathrm{h}} \sin\theta, \tag{7.20}$$

where v_{h} and v_z are the horizontal and vertical wind velocities, respectively. They can be deduced by subtracting/adding (7.20) from/to (7.19) that

$$v_{\mathrm{h}} = \frac{v_{r\max} - v_{r\min}}{2\sin\theta}, \tag{7.21}$$

$$v_z = \frac{v_{r\max} + v_{r\min}}{2\cos\theta}. \tag{7.22}$$

7.2.2 Wind Velocity Measurements from Spaced Antenna Methods

The DBS method previously described is based on a direct measurement of the Doppler velocity in different directions near the zenith with narrow beams (a few degrees). It is considered as an excellent method for wind velocity measurements. However, narrow beams require large antenna diameters. Consequently, the DBS method can only be applied in practice at very high frequency (VHF) and higher frequencies. For lower frequencies, the spaced antenna (SA) method based on correlation methods is generally applied instead of the DBS method. It is based on the measurement of the drift of the diffraction pattern along the ground. The cross correlation and autocorrelation techniques are used to determine drift velocity, and other parameters of moving patterns is examined (Briggs 1984).

The SA method can also constitute an alternative to the DBS method at VHF. The principle of the SA method is based on the measurement of the interference pattern at the ground of signals due to backscattering from refractive index irregularities advected by the wind or backscattering from precipitation, and the horizontal wind velocity component is estimated from the displacement of the diffraction pattern. Contrary to the DBS method, the SA method requires a wide radar beam for measuring the drift of the refractive index irregularities by the wind. The SA method gains advantage from enhanced echoes due to partial reflection near the zenith in the lower atmosphere. It was initially developed for observing the evolution of irregularities in the ionosphere around the altitude of 100 km (Ratcliffe 1956), and was applied later for measuring the horizontal wind velocity in the lower and middle atmosphere (Fraser 1968; Röttger and Vincent 1978). If the radar resolution volume is filled by multiple and randomly distributed refractive index scatterers, backscattered signals interfere each other. As a result, a random diffraction pattern is formed on the ground. The processing method is based on the assumption that the random diffraction pattern on the ground has a spatial correlation that can be described with a Gaussian function. If the scatterers are identical, the distribution is isotropic. The contours of equal three-dimensional spatial autocorrelation of the

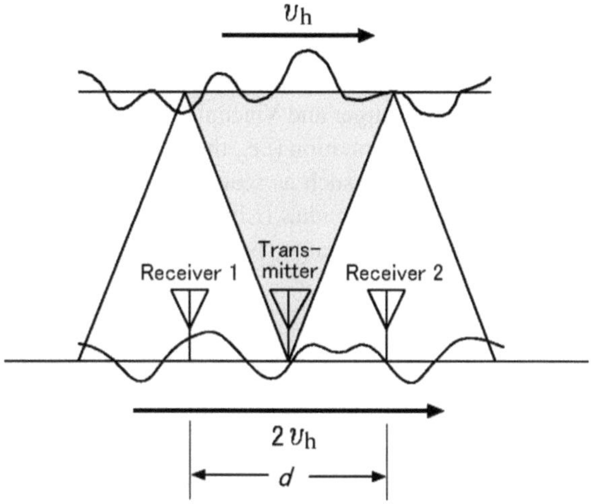

Fig. 7.5 Conceptual diagram of the movement of atmospheric scatterer and the diffraction pattern at the ground in the spaced antenna (SA) method

diffraction pattern at the ground are then given by ellipses using variables ξ, η, and ζ, and constants a, b, and h, i.e.,

$$\rho(\xi,\eta,\zeta) = \exp[-(\xi^2/2a^2 + \eta^2/2b^2 + \zeta^2/2h^2)]. \quad (7.23)$$

It can be shown that autocorrelation functions identical to (7.23) can be observed in any arbitrary planes parallel to scattering layers, i.e., on the ground (Ratcliffe 1956).

The statistical properties of the diffraction pattern at the ground can be determined from the estimation of space-time correlation functions of received signals from several antennas receiving radiation with a wide beam pattern toward zenith and separated by a suitable distance. The correlation values are obtained by calculating the temporal average of the product of complex samples collected from a pair of receiving arrays. If the main factor responsible for temporal fluctuations of the signals is the diffraction pattern advection by the wind (i.e., the scatterers can be considered as almost frozenly advected during the acquisition time), it can be shown that the velocity of displacement of the diffraction pattern at the ground is approximately twice the wind velocity of the scatterers as illustrated in Fig. 7.5. This relation can be intuitively understood since the shadow caused by a screen on a wall moves at twice of speed on the screen when the light source, the screen, and the wall are at the same distance. In the above discussion, it is assumed that the correlation time of the refractive index irregularities (scatterers) advected by the background wind is large enough compared to the acquisition time.

SAD Method

The simplest processing method of the SA method consists in measuring the time lag between signals received between a pair of receiving arrays. This time lag can be estimated by calculating the cross-correlation of time series of samples collected by both antenna arrays. The velocity component along the base line formed by the pair of antennas can then be estimated. This method is called the spaced antenna drift (SAD) method (Röttger and Vincent 1978). It is assumed that the diffraction pattern moves without deformation (i.e., the field pattern moves as the "frozen" or "locally frozen" structures of such as scattering refractive index irregularities move across the radar receiving antennas (Liu et al. 1991)). The two components of horizontal wind velocity can be obtained by applying the method with at least two non coplanar pairs of receiving arrays.

FCA Method

Actually, the frozen advection hypothesis can be flawed mainly due to the random motions of scatterers, and decay and generation of the turbulent irregularities. More sophisticated methods called full correlation analysis (FCA) method or its equivalent in the spectral domain, the full spectral analysis (FSA) method, for example, have been developed. These methods take into account the limited span life of the irregularities (Briggs 1984; Costa and Fougere 1988). Because these methods take account for the anisotropy of the scatterers, information on the characteristics of the scatterers can be an additional product of the methods.

The FCA method assumes that the two-dimensional cross-correlation function ρ of the diffraction patterns received by two ground-based antennas separated by (ξ,η) along two orthogonal baselines, can be expressed as

$$\rho(\xi,\eta,\tau) = \rho(A\xi^2 + B\eta^2 + 2H\xi\eta + K\tau^2), \tag{7.24}$$

where A, B, H, and K are constant, and τ is the time lag. The correlation is 1 if both antennas are located at the same point and tends to 0 for infinite distance separation. In the general case, the shape of (7.24) is not defined. The expression (7.24), however, implies that the contours of equal correlation values are ellipsoids. In this case, the horizontal cross section forms an ellipsoid at any time and only the direction changes. It is thus possible to quantify the anisotropy of the scatterers in the horizontal plane. In particular, the contour for $\rho = 0.5$ is called the correlation ellipsoid or the characteristic ellipsoid.

The second step of the FCA method consists in estimating wind velocity from the interference pattern subjected to random fluctuations. If the interference pattern is advected at a velocity $V = (V_x, V_y)$, expression (7.24) becomes

$$\rho(\xi,\eta,\tau) = \rho[A(\xi - V_x\tau)^2 + B(\eta - V_y\tau)^2 + K\tau^2 + 2H(\xi - V_x\tau)(\eta - V_y\eta)]. \tag{7.25}$$

If we define F and G such that

$$AV_x + HV_y = -F, \tag{7.26}$$

$$BV_y + HV_x = -G. \tag{7.27}$$

Equation (7.25) can be rewritten as

$$\rho(\xi,\eta,\tau) = \rho(A\xi^2 + B\eta^2 + C\tau^2 + 2F\xi\tau + 2G\eta\tau + 2H\xi\eta), \quad (7.28)$$

where $C = AV_x^2 + BV_y^2 + K + 2HV_xV_y$.

Since (7.28) is a quadratic in τ, three unknown variables can be eliminated, from fitting to the observed cross-correlation function. If two sets of independent cross-correlation functions are obtained from three antennas, the six variables A, B, C, F, G, H can be estimated. The horizontal wind velocity components (V_x, V_y) are finally calculated by substituting these values into (7.26) and (7.27).

Another method of analysis was developed by Doviak et al. (1996) and Holloway et al. (1997a), which is based on models of volume scattering and spectrum of randomly distributed refractive index irregularities produced by turbulence. Turbulence parameters and correlation lengths of turbulent irregularities can be derived from these methods. An additional method for measuring wind velocities from another correlation method was also proposed by Holloway et al. (1997b). All these methods only differ slightly by the assumed model. Assessing their validity by intercomparison is still at issue (Hassenpflug et al. 2003).

FSA Method

The Fourier transform of cross-correlation function is the cross-power spectrum. Consequently, the approaches used for the FCA method can also be applied in the frequency domain (Costa and Fougere 1988; Briggs and Vincent 1992) and are called the FSA method. Let us assume that the general cross-correlation function has a Gaussian distribution shape, and let the vectorial separation between one pair of antennas $(i,\ j)$ be $(\xi_{ij},\ \eta_{ij})$, (7.28) can be then expressed as

$$\rho(\xi_{ij},\eta_{ij},\tau) = \exp\{-[A\xi_{ij}^2 + B\eta_{ij}^2 + 2H\xi_{ij}\eta_{ij} + 2(F\xi_{ij} + G\eta_{ij})\tau + C\tau^2]\}. \quad (7.29)$$

Let $W_{ij}(f)$ be the Fourier transform with respect to τ of the cross-correlation function (7.29), W_{ij} the maximum value of $|W_{ij}(f)|$, f_{ij} the e^{-1} half width of the cross-spectrum, and m_{ij} the slope of the cross-spectrum phase. Then,

$$W_{ij} = \left(\frac{\pi}{C}\right)^{1/2} \exp\left\{-\left[A\xi_{ij}^2 + B\eta_{ij}^2 + 2H\xi_{ij}\eta_{ij} - \frac{(F\xi_{ij} + G\eta_{ij})^2}{C}\right]\right\}, \quad (7.30)$$

$$f_{ij} = \frac{C^{1/2}}{\pi}, \quad (7.31)$$

$$m_{ij} = -2\pi\left(\frac{F\xi_{ij}}{C} + \frac{G\eta_{ij}}{C}\right). \quad (7.32)$$

The value of f_{ij} can be accurately estimated from the mean power spectrum obtained by averaging of the power spectrum from three antennas. The estimation can be made by fitting a Gaussian function to the mean power spectrum. In practice, from three antennas forming a triangle, it is possible to estimate W_{ij} and m_{ij}. All unknowns in (7.30)–(7.32) can then be determined (Briggs and Vincent 1992). Substituting these values into (7.26) and (7.27), horizontal wind estimate can finally be obtained.

The SA method only provides the horizontal wind component. In order to measure the vertical wind component, the Doppler shift in the power spectrum measured from vertically oriented beam must be calculated at the same time. With the FSA method, high frequency echoes resulting from sidelobes at low elevation angles and electromagnetic interferences at specific frequencies can be easily removed.

7.3 Turbulence Observations

Basically, turbulence is a chaotic fluid regime producing random and rapid fluctuations of pressure, velocity and other quantities and is not a repeatable process. Turbulent flows are usually composed of eddies or vortices of many different scales. Atmospheric motions can often become turbulent due to instabilities at any scale and atmospheric turbulence occurs most in the planetary boundary layer. However, turbulence can be met at any altitude and despite the static stability of the stratosphere, it is supposed that diffusion caused by small-scale turbulence can play an important role in vertical transport of materials, energy, and momentum. The earliest experimental studies of atmospheric turbulence in the free atmosphere were performed by in situ observations using instrumented balloons and aircrafts (e.g., Lilly et al. 1974). Quantitative estimations of vertical eddy diffusivity and energy dissipation rate by remote sensing techniques became possible owing to the development of atmospheric radars (Weinstock 1981; Hocking 1983).

As discussed in Sect. 3.4, atmospheric radars are sensitive to refractive index perturbations caused by turbulence having the length scale of half the radar wavelength (i.e., the Bragg scale). The Doppler power spectrum peak and its spectrum width can be related to the radial velocity of the refractive index perturbations being advected by the mean wind in the radar resolution volume and to the wind velocity variations caused by turbulence, respectively. The power of the Doppler spectrum is proportional to the intensity of the refractive index perturbations at the Bragg scale. Stronger turbulence mixes more deeply air having vertical gradients of mean potential refractive index, and thus creates refractive index perturbations having larger amplitude. Thus, the power of the Doppler spectrum is indirectly related to the intensity of turbulence as well as the intensity of the gradient of temperature and humidity. Based on these facts and additional information such as background stability, vertical eddy diffusivity and energy dissipation rate can be estimated.

7.3.1 Measurement of Momentum Flux

Atmospheric gravity waves (described in detail in Sect. 12.3) are known to have significant impacts on the general circulation in the middle atmosphere. Through turbulent mixing in the mean circulation, they can transport upward momentum flux and energy from the lower atmosphere and release them in the mesosphere.

The momentum flux is a rate of transfer of momentum across a unit area. The vertical momentum flux (per unit of mass) of horizontal wind perturbations u' is given by the covariance $\overline{u'w'}$ where w' is the vertical wind perturbations. Atmospheric radars (i.e., Bragg scatter radars) are a better tool for estimating these momentum fluxes (Vincent and Reid 1983) because they can better measure backscatter from refractive index perturbations, especially in the upper atmosphere. Momentum fluxes measurement with meteorological radars is also discussed by Doviak and Zrnić (2006, Chap. 10).

Let us consider two radar beams oriented at the same zenith angle but separated by an angle of 180° in azimuth as illustrated in Fig. 7.6a. The corresponding radial wind velocities, V_1 and V_2, can be expressed as

$$V_1 = u_1 \sin\theta + w_1 \cos\theta, \tag{7.33}$$

$$V_2 = -u_2 \sin\theta + w_2 \cos\theta, \tag{7.34}$$

where u_i and w_i ($i = 1, 2$) are horizontal and vertical wind velocities, respectively. If the mean wind velocity $\overline{V_i}$ is the time averaged value during the observation period, wind velocity fluctuation V_i' can be expressed as

$$V_i' = V_i - \overline{V_i}. \tag{7.35}$$

The difference between the mean squared values of V_1' and V_2' (i.e., the variances of radial wind velocities can be expressed as

$$\begin{aligned} \overline{V_1'^2} - \overline{V_2'^2} &= \overline{(u_1' \sin\theta + w_1' \cos\theta)^2} - \overline{(-u_2' \sin\theta + w_2' \cos\theta)^2} \\ &= (\overline{u_1'^2} - \overline{u_2'^2}) \sin^2\theta + (\overline{w_1'^2} - \overline{w_2'^2}) \cos^2\theta \\ &\quad + 2(\overline{u_1'w_1'} + \overline{u_2'w_2'}) \sin\theta \cos\theta, \end{aligned} \tag{7.36}$$

where u_i' and w_i' ($i = 1, 2$) are the horizontal and vertical wind perturbations, respectively. If the statistical values of wind velocity perturbations do not depend on the point of observations as is the case for atmospheric gravity wave propagating over the radar, $\overline{u_i'^2}$, $\overline{w_i'^2}$, and $\overline{u_i'w_i'}$ are identical in both directions. Therefore, the vertical momentum flux can be deduced from (7.36) as (Lhermitte 1968; Vincent and Reid 1983)

$$\overline{u_i'w_i'} = \frac{\overline{V_1'^2} - \overline{V_2'^2}}{2 \sin 2\theta}. \tag{7.37}$$

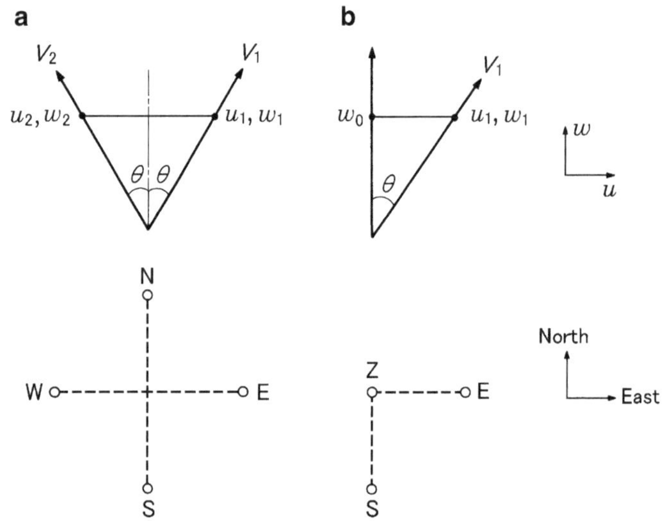

Fig. 7.6 Geometry of the radar beams for (**a**) four- and (**b**) three-beam determinations of momentum flux by MST radars. Side and plan views are given on the *top* and *bottom*, respectively. Beam positions are indicated by N (northward), W (westward), S (southward), E (eastward), and Z (vertical, upward) [from Fukao et al. 1988a]

The momentum flux per unit mass (7.37) can be expressed as that per unit volume by multiplying (7.37) by the atmospheric density. To obtain the vertical momentum flux of the zonal and meridional wind perturbations, two pairs of radar beams are required as shown by Fig. 7.6a.

If a vertical beam is used in addition to an oblique beam as shown in Fig. 7.6b, u' is obtained by the following equation by the estimates of V_1' and w_0',

$$u' = \frac{V_1' - w_0' \cos\theta}{\sin\theta}. \tag{7.38}$$

Using (7.33), u' becomes

$$u' = u_1' + (w_1' - w_0') \cot\theta. \tag{7.39}$$

The perturbations w' are directly measured from the vertical beam. Thus, $w' = w_0'$. As a result, the vertical momentum flux is given by the expression

$$\overline{u'w'} = \overline{u_1' w_0'} + (\overline{w_1' w_0'} - \overline{w_0'^2}) \cot\theta. \tag{7.40}$$

In (7.40), the second term of the right-hand side includes the correlation between vertical wind perturbations from the two beams. Therefore, the momentum flux can be accurately measured from this method if the horizontal wavelength of the atmospheric gravity waves producing the wind perturbations is much larger than the horizontal distance separating the two volumes probed by the radar beams (Fukao et al. 1988b).

7.3.2 *Estimation of the Turbulence Contribution to Spectrum Width*

The radar echo results from backscattered signals from individual scatterers in the radar resolution volume. Each scatterer moves in average with the background wind. Hence, the power spectrum at a given height is the result of Doppler shift due to the mean wind superimposed to the random motions of each scatterer around the mean velocity. As a result, to a first approximation, the Doppler spectrum generally follows a Gaussian shape.[1] The spectrum width reflects the intensity of velocity fluctuations weighted by the intensity of the refractive index perturbations. However, additional non turbulent effects also contribute to the measured spectrum width. Among them, the beam broadening effect is due to the finite width of the radar beam; the shear broadening effect results from the vertical shear of the background wind in the radar resolution volume; transient effects are due to the time variations of the background wind during the acquisition time (e.g., Hocking 1983), which may also contain turbulence of larger scale than the acquisition time. For accurate estimates of turbulence parameters, it is necessary to remove these effects.

Beam Broadening

The radial velocity in the radar resolution volume is given by the velocity component along the main axis of radar beam. However, the radial direction for each scatterer slightly varies because the radar beam is not infinitely narrow. Hence, even if the background wind velocity is constant within the radar resolution volume, the radial velocity in each part of the volume slightly varies as shown in Fig. 7.7 leading to a broadening of the spectrum width. This effect, called beam broadening effect, increases with the width of the radar beam.

The beam broadening effect depends on the mean wind velocity and on the antenna beam pattern. Its correction is based on the hypothesis of a uniform horizontal wind in the radar resolution volume. The procedure consists in calculating the radial components of background wind velocity at each point in the radar resolution volume, and integrating within all the radar resolution volume weighted by the antenna beam pattern. If the radar beamwidth is about several degrees (as is the case for standard atmospheric radars), the spread of spectrum width by beam broadening effect is approximately given by Hocking (1985)

$$\sigma_{\frac{1}{2}\mathrm{B}} = \theta_{\frac{1}{2}} v_{\mathrm{h}}, \tag{7.41}$$

[1]It is not simply the randomness of turbulence that causes roughly the Gaussian shape. Only if the probability distribution of turbulent velocities is described by a Gaussian function, the spectrum also will be Gaussian shaped (Doviak and Zrnić 2006, pp. 115—116).

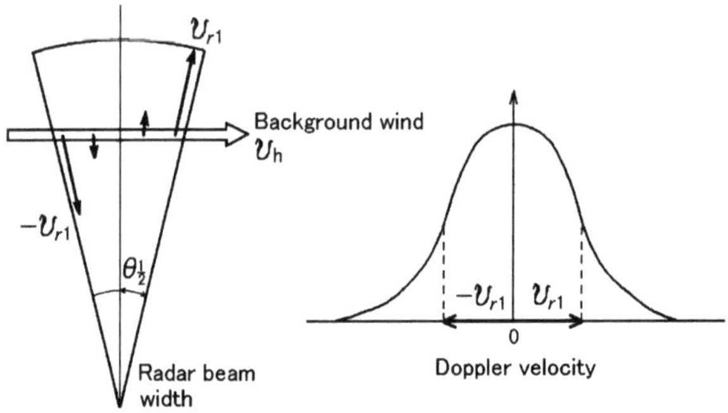

Fig. 7.7 Conceptual diagram of the broadening of the spectrum width due to the antenna beam pattern

where $\sigma_{\frac{1}{2}\mathrm{B}}$ is half-width half power of the Doppler spectrum, $\theta_{\frac{1}{2}}$ is the half power half width of the effective (two way) radar beam, and v_h is the background horizontal wind velocity. For example, if the beamwidth is $4°(\simeq 7.0 \times 10^{-2}\,\mathrm{rad})$ and the horizontal wind velocity is $20\,\mathrm{m\,s^{-1}}$, $\sigma_{\frac{1}{2}\mathrm{B}} \simeq 1.4\,\mathrm{m\,s^{-1}}$. The contribution of beam broadening effect to the measured spectrum width is thus about $2.8\,\mathrm{m\,s^{-1}}$. Practically, beam broadening is the most significant contribution to the spectrum width broadening.

If the radar resolution volume is filled with refractive index perturbations moving with uniform background wind and turbulence, the power spectrum can be expressed in terms of convolution integral of the spectrum by turbulence with that of antenna beam pattern. As these spectrum broadening mechanisms are independent of one another, the square of measured half-width half power of the Doppler spectrum $\sigma_{\frac{1}{2}}$ can be approximated by

$$\sigma^2_{\frac{1}{2}} = \sigma^2_{\frac{1}{2}\mathrm{B}} + \sigma^2_{\frac{1}{2}\mathrm{turb}}, \tag{7.42}$$

where $\sigma^2_{\frac{1}{2}\mathrm{B}}$ and $\sigma^2_{\frac{1}{2}\mathrm{turb}}$ are the second central moments of the two-way antenna power spectrum (half-width) and those of the turbulent power spectrum (half-width) in the radar resolution volume, respectively.

Shear Broadening

Whether the background wind is uniform or not, the radial velocity within the radar resolution volume varies in each part of the volume, and the measured spectrum width becomes wider. Horizontal wind varies with height due to the baroclinicity of the atmosphere (thermal wind) and gravity waves producing a vertical shear of horizontal wind. When the vertical shear is large over the depth of the radar resolution volume, it can be a significant source of the spectrum width broadening.

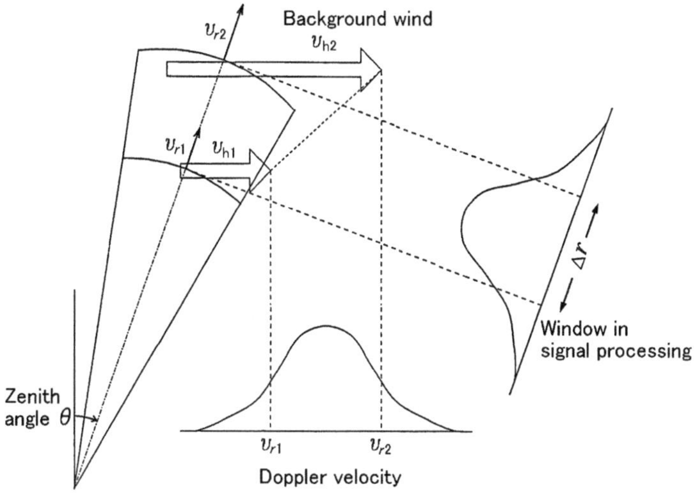

Fig. 7.8 Conceptual diagram of the broadening of the spectrum width due to horizontal shear

This effect is called shear broadening.[2] As for shear broadening, the radial velocity component of background wind contributes as shown in Fig. 7.8. Shear broadening can be ignored when the radar beam is oriented vertically whereas its effect is most pronounced when the radar beam is tilted far away from the zenith.

Let S_w [s^{-1}] be the vertical shear and $\sigma_{\frac{1}{2}\mathrm{S}}$ the contribution of the shear broadening to the measured spectrum width. It is given by

$$\sigma_{\frac{1}{2}\mathrm{S}} = \frac{S_w \sin\theta \Delta r}{2}, \tag{7.43}$$

where θ is the zenith angle and $\Delta r (= c\tau/2$ in Fig. 3.6) is the depth of the scattering resolution volume.

Vertical shear of horizontal wind is mitigated by breaking of atmospheric gravity waves that will be discussed in Sect. 12.3.1, The magnitude of vertical shear for background wind is estimated as follows. Let us note N the Brunt Väisälä frequency[3] of the atmosphere at the height of the turbulence, which is an index

[2]Shear broadening and beam broadening can be lumped together by simply considering the radial velocity shear of the mean wind within the resolution volume (Doviak and Zrnić 2006, Sect. 5.3). If the radial mean velocities along range azimuth and zenith can be measured, the radial velocity shear can be computed under simplifying assumptions, e.g., mean wind shear is uniform.

[3]When a fluid mass in a continuously stratified fluid is raised, the mass density is larger than that of the surrounding fluid. The fluid mass density is smaller when it is lowered. In either case restoring force which is proportional to the vertical displacement works on the fluid mass to return to the original height. A simple harmonic oscillation is induced to the fluid mass, if the fluid is inviscid and not distorted by the mass motion. The frequency of this oscillation is called the Brunt Väisälä frequency. It is proportional to square root of the density change in the vertical direction, and is

of the stability of the stratification and has unit of s^{-1}. The flow is dynamically stable if

$$R_i \equiv \frac{N^2}{|\partial \boldsymbol{u}/\partial z|^2} \geq \frac{1}{4}, \tag{7.44}$$

where R_i is the Richardson number, which defines the dynamical stability of stratified atmosphere constrained to a vertical shear of horizontal wind $|\partial \boldsymbol{u}/\partial z|$. It represents the ratio of potential energy to kinetic energy. If $R_i > 1/4$, shear instability cannot occur because the stratification overcomes the mechanical effects. If the instability develops for $R_i < 1/4$, it is maintained without decaying if $R_i < 1$. If $R_i < 0$, convective instability occurs. The typical value of N in the stratosphere is about $N \simeq 0.02\,\mathrm{s}^{-1}$. It is small compared to $|\partial \boldsymbol{u}/\partial z|$. Therefore, $|\partial \boldsymbol{u}/\partial z| \leq 0.04\,\mathrm{s}^{-1}$. For example, for a zenith angle θ of 10° and $\Delta r = 150\,\mathrm{m}$, expression (7.43) shows that $\sigma_{\frac{1}{2}S}$ is smaller than $0.5\,\mathrm{m\,s}^{-1}$.

Broadening Due to Transient Effects

Incoherent integrations are applied for improving the detectability of the atmospheric echo in the Doppler spectra (see Sects. 5.6.3 and 7.1.3). This procedure increases the total acquisition time τ needed for calculating the atmospheric parameters. It can be an additional source of spectrum width broadening if the acquisition time τ is not significantly smaller than the characteristic time of the wind fluctuations, where fluctuations are caused by; (1) atmospheric waves, e.g., convective rolls, gravity waves, etc., and (2) turbulence, e.g., due to convection, or shear induced breaking of waves. Indeed, if the Doppler shift is slightly different for each Doppler spectrum before the incoherent integration, the width of the composite Doppler spectrum after integration could be larger than the width of a Doppler spectrum without the incoherent integration (Fig. 7.9).[4] The most important effects are obviously produced by atmospheric gravity waves, in particular, those which have an intrinsic frequency close to the Brunt Väisälä frequency N. The contamination is all the more important if acquisition time τ is close to the period T of the gravity wave. Let W be the velocity amplitude of the wave. The variance $\overline{(v'_\tau)^2}$ of radial velocity for an acquisition time τ and a gravity wave of period $T = 2\pi/N$ is (Hocking 1988):

$$\overline{(v'_\tau)^2} = \frac{W^2}{2}\left[1 - \frac{2}{N^2\tau^2}(1 - \cos N\tau)\right]. \tag{7.45}$$

used as a parameter which indicates stability of the stratification. Note that the symbol of N is used to present noise power too in the following chapters as mentioned in the corresponding places.

[4]The possibility that this shift is due to scales of turbulence larger than the acquisition time also remains.

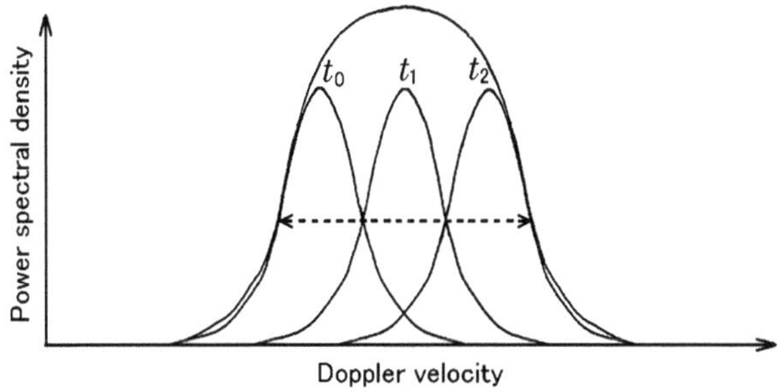

Fig. 7.9 Conceptual diagram of the broadening of spectrum width due to time changes of the wind of gravity waves within the beam. The width of the composite Doppler spectrum obtained after incoherent integrations is indicated by the *dashed line*

The value $\overline{(v'_\tau)^2}$ cannot be directly obtained from radar observations. It is known, however, from the observations that the difference between two successive data of v_r becomes the Gaussian distribution, and the variance $\overline{(v'_{2\tau})^2}$ in the period 2τ can be defined by replacing τ of (7.45) with 2τ. Thus, the squared half-width half power of the Doppler spectrum produced by gravity wave at the Brunt Väisälä frequency becomes

$$\begin{aligned} \sigma^2_{\frac{1}{2}\mathrm{T}} &= 2\ln 2\overline{(v'_\tau)^2} \\ &\simeq 2\ln 2 \frac{1-(2/N^2\tau^2)(1-\cos N\tau)}{1-(1/2N^2\tau^2)(1-\cos 2N\tau)}\overline{(v'_{2\tau})^2} \\ &\simeq 4\overline{(v'_{2\tau})^2}. \end{aligned} \tag{7.46}$$

By extending (7.42), the contribution of turbulence to the spectrum width $\sigma_{\frac{1}{2}\mathrm{turb}}$ is given by subtracting the different broadening effects described in the present subsection (i.e., beam broadening, shear broadening, and transient effects)[5] from the measured spectrum width $\sigma_{\frac{1}{2}}$. It reads

$$\sigma^2_{\frac{1}{2}\mathrm{turb}} = \sigma^2_{\frac{1}{2}} - (\sigma^2_{\frac{1}{2}\mathrm{B}} + \sigma^2_{\frac{1}{2}\mathrm{S}} + \sigma^2_{\frac{1}{2}\mathrm{T}}). \tag{7.47}$$

From (7.47), it is possible to retrieve the value v^2_{rms} which is the variance of fluctuating velocity due to turbulence, true spectrum width defined at -3 dB. From $\exp(-\sigma^2_{\frac{1}{2}\mathrm{turb}}/2v^2_{\mathrm{rms}}) = 1/2$, it is derived as (Hocking 1983, 1985)

[5]In addition, there are following effects which broaden spectrum such as antenna rotation, dispersion of hydrometeor terminal velocities, and hydrometer's oscillation and/or wobbling (Fang and Doviak 2008).

$$v_{\rm rms}^2 = \frac{1}{2\ln 2}\sigma_{\frac{1}{2}{\rm turb}}^2 \simeq 0.72\sigma_{\frac{1}{2}{\rm turb}}^2. \tag{7.48}$$

Hereafter, it will be assumed that the spectrum width has been corrected from the various broadening effects for the estimation of the turbulence parameters.

7.3.3 Estimation of Turbulence Parameters

Inertial Subrange

The time and space domain covered by isotropic and inertial turbulence extends five decades at maximum. Generally, three-dimensional turbulence is generated by convective and dynamical shear instabilities. The turbulence generated is affected by the stability of stratification in the buoyancy subrange (so that the turbulent eddies become flattened and anisotropic). At scales smaller than the buoyancy length scale, the energy "cascades" from the largest eddies to the smallest ones by an inertial (and inviscid) mechanism. This range is called "inertial and isotropic subrange". It has been demonstrated that the one-dimensional (wave number energy) spectrum is proportional to the $-5/3$th power of wave number in this range (see (3.114)) (Ottersten 1969b; Hocking 1985). The smallest eddy is strongly affected by viscosity, and kinetic energy is converted into heat. This is called the viscous subrange or dissipative subrange.

Figure 7.10 shows the height variation of inertial subrange $[l_0, L_{\rm B}]$ (Hocking 1985), where l_0 is the scale for determining the transition region between the viscous and inertial subranges, i.e., the inner scale of turbulence, and $L_{\rm B}$ between the inertial and buoyancy subranges, i.e, the buoyancy length scale. An important scale for analytical expression of l_0 is the Kolmogorov microscale, defined by Tatarskii (1971, p. 53)

$$l_{\rm K} = \left(\frac{\nu^3}{\varepsilon}\right)^{1/4}, \tag{7.49}$$

where ν is the kinematic viscosity (i.e., the dynamical viscosity divided by the fluid density) which has units of $\rm m^2\,s^{-1}$, and ε is the turbulent energy dissipation rate which has units of $\rm W\,kg^{-1}$, then $l_{\rm K}$ has units of meters (Hocking 1985). A widely accepted definition for l_0 was proposed by[6] (Hill 1978; Hill and Clifford 1978):

$$l_0 = 7.4 l_{\rm K}, \tag{7.50}$$

[6]Tatarskii (1971, p. 53) defines $l_0 = l_{\rm K}$ where the dimension of ν is $\rm cm^2\,s^{-1}$, and ε has units of $\rm W\,kg^{-1}$, thus the constant is different from the following discussion.

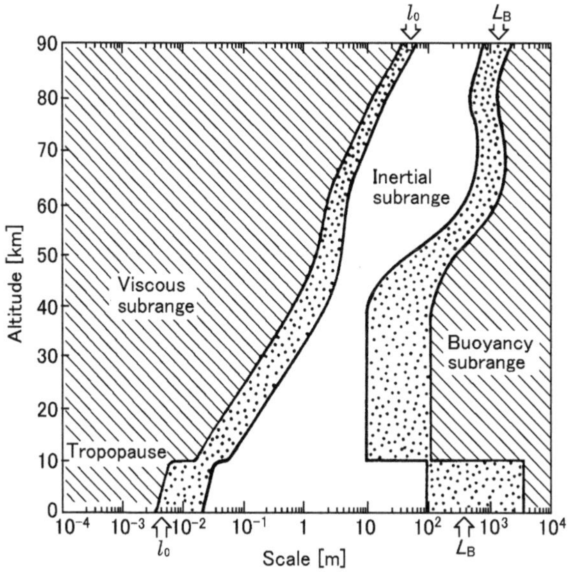

Fig. 7.10 Typical height profile of the inertial subrange. The scales l_0 and L_B are the inner scale of turbulence and the buoyancy length scale, respectively. l_0 and L_B are assumed to vary smoothly between 40 and 60 km. (The region between 30 and 80 km is the most uncertain part of the graph.) The inertial subrange of turbulence strictly only applies for scales somewhat less than L_B and larger than l_0. For a radar wavelength λ, backscatter is only from scales of $\lambda/2$. Thus if a 50-MHz radar is used ($\lambda = 6$ m), then scatter should be possible from isotropic inertial subrange turbulence up to altitudes of about 65–70 km [from Hocking 1985]

where the constant 7.4 is only relevant for air (Hocking 1985). Due to the fact that the ratio of the rates of diffusion of momentum and temperature (i.e., the Prandtl number) is not unity, the inner scale for velocity fluctuations is slightly different, i.e., (e.g., Caughey et al. 1978; Hocking 1985; Wilson 2004)

$$l_0 \simeq 12.8 l_K. \tag{7.51}$$

According to e.g., Weinstock (1978), Crane (1980), and Hocking (1985),

$$L_B = F\varepsilon^{1/2}N^{-3/2}, \tag{7.52}$$

where F is a constant, $F = 2\pi/0.62$ is generally applied, and N is the Brunt Väisälä frequency.

The inertial subrange becomes narrow with height because the Reynolds number Re decreases with height, where the Reynolds number is a dimensionless number which quantifies the relative effect of inertial and viscous forces on fluid. If the travelled length of the fluid is L, and the flow velocity is U, Re is given by $Re = UL/\nu$. The generation and persistence of turbulence depends on the magnitude of Re. In general, the value becomes greater than 10^3. Fluid becomes viscous and turbulence decays, if ν increases and accordingly Re decreases.

Energy Dissipation Rate

In inertial subrange, the one-dimensional spectrum $S_n(\kappa)$ is given using the turbulent energy dissipation rate ε and the wave number of the Bragg scatter κ by Ottersten (1969a); Lilly et al. (1974)

$$S_n(\kappa) = a\varepsilon^{\frac{2}{3}}\kappa^{-\frac{5}{3}}, \tag{7.53}$$

where $a(=1.5)$ is the Kolmogorov constant.

If half the radar wavelength lies in the inertial subrange, observable motions lie in the range from buoyancy length scale to the Bragg scale. Thus, the variance of wind velocity given by (7.48) can be expressed in terms of the integral of the one-dimensional spectrum of (7.53) from buoyancy length scale to the Bragg scale as

$$\frac{3}{2}v_{\mathrm{rms}}^2 = \int_{\kappa_{\mathrm{B}}}^{\kappa_{\mathrm{b}}} a\varepsilon^{\frac{2}{3}}\kappa^{-\frac{5}{3}}\mathrm{d}\kappa, \tag{7.54}$$

where κ_{B} and κ_{b} are wave numbers corresponding to buoyancy length scale $L_{\mathrm{B}} = 2\pi/\kappa_{\mathrm{B}}$ and the Bragg scale $\lambda_{\mathrm{b}} = 2\pi/\kappa_{\mathrm{b}}$, respectively. Solving (7.54) assuming that $\kappa_{\mathrm{B}} \ll \kappa_{\mathrm{b}}$,

$$\varepsilon \simeq a^{-3/2}\kappa_{\mathrm{B}} v_{\mathrm{rms}}^3, \tag{7.55}$$

where, for the lower limit of turbulence,

$$\kappa_{\mathrm{B}} = \frac{N}{v_{\mathrm{rms}}}. \tag{7.56}$$

Substituting (7.56) into (7.55), ε is given by

$$\varepsilon \simeq a^{-3/2} N v_{\mathrm{rms}}^2 = C_{\mathrm{t}} N v_{\mathrm{rms}}^2, \tag{7.57}$$

where C_{t} is a constant. Theoretically, $C_{\mathrm{t}} \simeq a^{-3/2} \simeq 0.5$. However, an experimental value of $C_{\mathrm{t}} \simeq 0.4$ is generally applied (Weinstock 1981) and was confirmed by many estimates made from stratospheric data (Hocking 1983, 1985).

Vertical Eddy Diffusivity

The vertical eddy diffusivity K is defined as

$$K = -\overline{\Theta' w'} \bigg/ \frac{\partial \overline{\Theta}}{\partial z}, \tag{7.58}$$

where z is the altitude, w the vertical wind velocity, and Θ is the potential temperature.[7] The symbols $(\bar{\ })$ and $(')$ denote mean values and perturbations, respectively. The term $\overline{\Theta' w'}$ expresses the vertical heat flux per unit of mass. The potential temperature and the Brunt Väisälä frequency N are related by Lilly et al. (1974)

$$N^2 \equiv \frac{g}{\Theta}\frac{\partial \Theta}{\partial z}, \tag{7.59}$$

where g is the gravitational acceleration. Assuming the stationary state (i.e., the production of turbulent kinetic energy $\frac{\partial \bar{u}}{\partial z}\overline{u'w'}$ is balanced by the buoyancy flux $\frac{g}{\Theta}\overline{\Theta' w'}$ and the energy dissipation rate ε), we obtain

$$\varepsilon = -\frac{\partial \bar{u}}{\partial z}\overline{u'w'} + \frac{g}{\Theta}\overline{\Theta' w'} = \left(1 - \frac{1}{R_{\mathrm{f}}}\right)\frac{g}{\Theta}\overline{\Theta' w'}, \tag{7.60}$$

where u denotes the horizontal wind velocity, and

$$R_{\mathrm{f}} \equiv \frac{g}{\Theta}\overline{\Theta' w'} \bigg/ \frac{\partial \bar{u}}{\partial z}\overline{u'w'} \tag{7.61}$$

is the flux Richardson number. Substituting (7.59)–(7.61) into (7.58), we obtain

$$K = \frac{\beta \varepsilon}{N^2}, \tag{7.62}$$

$$\beta \equiv \frac{R_{\mathrm{f}}}{1 - R_{\mathrm{f}}}. \tag{7.63}$$

The value of β is empirically determined. Assuming $R_{\mathrm{f}} = 1/4$ in the lower stratosphere (Lilly et al. 1974), $\beta \simeq 0.3$.

7.3.4 *Relation Between Refractive Index and Structure Constant for Refractivity Turbulence*

As discussed in Sect. 2.2.2, the refractive index n in the neutral atmosphere depends on pressure p, temperature T, and water vapor pressure e (see (2.84)). The variance $\overline{\delta n^2}$ of the refractive index perturbations due to isotropic turbulence is related to

[7] The potential temperature of an air parcel at pressure p and temperature T is the temperature that would get the air parcel if this parcel is moved adiabatically at a reference pressure p_0 (usually $p_0 = 1{,}000\,\mathrm{hPa}$). Adiabatic motions can be assumed for motions at large scale. Thus, potential temperature is a Lagragian conserved quantity in absence of water phase changes and for non saturated air.

the power spectral density $E_n(\kappa)$ of refractive index perturbations as shown in Sect. 3.4.2;

$$\overline{\delta n^2} = \int_{\kappa_B}^{\infty} E_n(\kappa)\mathrm{d}\kappa, \tag{7.64}$$

where $\kappa_B = 2\pi/L_B$. Rewriting (7.64) with (3.110) for $p = 5/3$ and (3.114),

$$\overline{\delta n^2} = \frac{5}{12} C_n^2 \int_{\kappa_B}^{\infty} \kappa^{-5/3}\mathrm{d}\kappa. \tag{7.65}$$

From (7.65), the refractive index structure constant C_n^2 can be related to the variance of refractive index perturbations by the relation

$$C_n^2 = 5.45 L_B^{-2/3}\overline{\delta n^2}. \tag{7.66}$$

Substituting (7.66) into (3.116), we have

$$\eta = 2.07 L_B^{-2/3}\lambda^{-1/3}\overline{\delta n^2}. \tag{7.67}$$

Tatarskii (1971, p. 76) proposed a relation between C_n^2 and the refractive index gradient $M_n = \mathrm{d}n/\mathrm{d}z$, namely,

$$C_n^2 = \kappa^2 \alpha L_B^{4/3} M_n^2. \tag{7.68}$$

The terms κ and α are constants, $\kappa^2 \simeq 2.8$ and $\alpha \simeq 1.0$. For a dry atmosphere, M_n is derived from (2.84) to (2.86) as[8]

$$M_n = -7.76 \times 10^{-5} \frac{p}{T}\frac{N^2}{g}, \tag{7.69}$$

where p in (7.69) is atmospheric pressure. The substitution of L_B in (7.52) into (7.68) gives

$$C_n^2 \propto \varepsilon^{2/3} M_n^2 N^{-2}. \tag{7.70}$$

If the radar system parameters are known, the radar reflectivity η can be estimated from the measurements of the received power as described in Sect. 7.1.1. Consequently C_n^2 can be estimated from η by (7.1), and ε can be derived from C_n^2 (Hocking 1985; Cohn 1995) when $M_n^2 N^{-2}$ is known.

From (3.116) and (7.70), we obtain

$$\eta \propto \varepsilon^{2/3} M_n^2 N^{-2} \lambda^{-1/3}. \tag{7.71}$$

[8]For a detailed derivation of (7.69), refer to Sect. 7.5.

In (7.69), p/T is proportional to the atmospheric density which slowly varies with height. Thus, in a dry atmosphere, to a first approximation, M_n is proportional to N^2. Applying this relation to (7.71), it is found that η is proportional to N^2. Furthermore, η is proportional to $\lambda^{-1/3}$.

Atmospheric radars operating in VHF band are not only sensitive to isotropic turbulence but also to thin temperature and humidity gradient sheets (Röttger 1980a; Dalaudier et al. 1994; Muschinski and Wode 1998). Partial (or Fresnel) reflection from these quasi laminar and horizontal structures is very often the dominant backscattering mechanism when the radar beam is vertical. When multiple sheets are randomly distributed within the radar resolution volume, the concept of Fresnel scattering was introduced by Gage et al. (1981). For this volume reflection mechanism, the partial reflection coefficient is proportional to the product of M_n^2 and $E_n(\kappa)$, where $E_n(\kappa)$ is the power spectral density vertical displacement at the Bragg scale and κ is the wave number of the Bragg scatter as discussed in Sect. 3.4.2 (VanZandt and Vincent 1983; Gage et al. 1985). Thus

$$|\rho|^2 = CM_n^2E_n(\kappa), \tag{7.72}$$

where C is a constant mainly determined by the radar system parameters. Luce et al. (2007) compared vertical profiles of echo power with the MU radar and profiles of M_n^2 estimated from colocated balloon data. A good agreement was generally found showing that the height profile of echo power is mainly decided by M_n^2 and then N^4 for a dry atmosphere. In the stratosphere where N^2 is approximately constant, M_n decreases in the scale height[9] of atmospheric density (approximately 7.3 km) with altitude.

7.4 Observations of Temperature Profile

7.4.1 Measurement of Atmospheric Temperature with RASS

RASS is the system which measures the virtual atmospheric temperature by the combination of atmospheric radar and acoustic wave source (e.g., Marshall et al. 1972; Matuura et al. 1986; Tsuda et al. 1989a). Very feeble condensations and rarefactions occur in atmospheric density and hence fluctuations of the refractivity index are produced when acoustic waves are radiated in the atmosphere as illustrated in Fig. 7.11. If radio wave of atmospheric radar is radiated to the acoustic wavefront, it is scattered by the produced fluctuations of refractive index. Since sound speed is approximately 10–100 times as high as the atmospheric motion, the frequency spectrum of atmospheric motion is completely separated from that of acoustic wave in received signal. Thus, when the acoustic wave component is taken out

[9]Refer to Sect. 2.2.4 and (2.99).

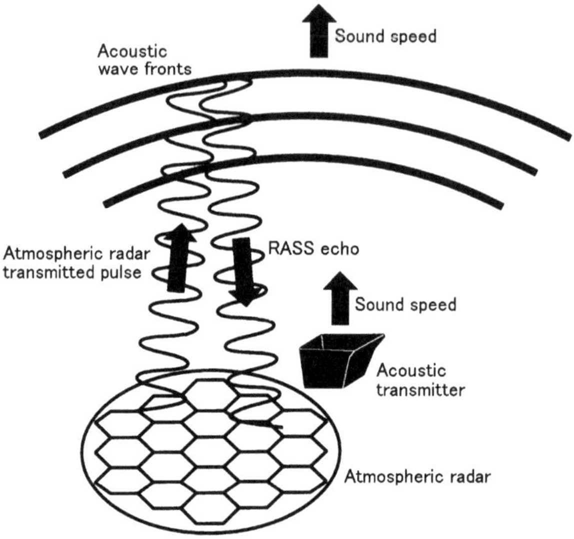

Fig. 7.11 Principle of RASS measurements [from Furumoto 2002]

from the backscattered signal, the sound speed can be deduced from the Doppler shift. Moreover, the atmospheric temperature can be also inferred by using the relationship that the acoustic speed depends on the atmospheric temperature. This measurement can be performed with very high resolution in space and time as compared with radiosonde observation conventionally done from the past.

The apparent sound velocity $\boldsymbol{c}_{\mathrm{a}}$ is given by the sum of the true sound velocity $\boldsymbol{c}_{\mathrm{s}}$ and the background wind velocity $\boldsymbol{v}$ as

$$\boldsymbol{c}_{\mathrm{a}} = \boldsymbol{c}_{\mathrm{s}} + \boldsymbol{v}. \tag{7.73}$$

Thus, the sound velocity which is obtained by the RASS c_{a} becomes

$$c_{\mathrm{a}} = \boldsymbol{c}_{\mathrm{s}} \cdot \mathbf{i} + \boldsymbol{v} \cdot \mathbf{i}, \tag{7.74}$$

where $\mathbf{i}$ is the unit vector in radial direction. The absolute value of apparent sound velocity c_{a} is related to the Doppler frequency f_{da} which is obtained by atmospheric radar as

$$f_{\mathrm{da}} = \frac{2c_{\mathrm{a}}}{\lambda}, \tag{7.75}$$

where λ is the radar wavelength that gives a maximum echo intensity for echoes from the range resolution volume.

When there is no background wind, $c_{\mathrm{a}}(= c_{\mathrm{s}})$ [m s^{-1}] and atmospheric temperature T [K] are related by the following equation.

$$c_{\mathrm{a}} = K_{\mathrm{d}}\sqrt{T}, \tag{7.76}$$

where K_d is a constant which changes with humidity. In dry atmosphere,

$$K_d = \sqrt{\frac{\gamma R}{M}} \simeq 20.047 \quad [\mathrm{m\,s^{-1}\,K^{-1/2}}], \tag{7.77}$$

where γ is the specific heat ratio of ideal gas, and $\gamma \simeq 1.4$ for dry air, R is the gas constant, and M is the mean molecular weight of the atmosphere. Applying f_{da} obtained by atmospheric radar to (7.75), c_a is calculated. Then, by substituting c_a into (7.76), temperature profile $T(z)$ can be inferred at height z. In the actual RASS observation, the radial wind velocity derived from clear air echo observation is subtracted from apparent acoustic velocity obtained with the RASS observation.

Sound speed becomes a little larger in a moist atmosphere than that in a dry atmosphere. In this case, the temperature of moist atmosphere is given by (7.76) as follows;

$$T_v = \left(\frac{c_a}{K_d}\right)^2, \tag{7.78}$$

where T_v is the virtual temperature of the atmosphere. The virtual temperature is the equivalent temperature for a supposed dry atmosphere which has the same atmospheric pressure and volume with the actual moist atmosphere with humidity mixing ratio of q [kg kg $^{-1}$], and is related to actual temperature T by

$$T_v = (1 + 0.608q)T. \tag{7.79}$$

7.4.2 *Change of Refractive Index and Radar Equation for RASS*

Fluctuations of Refractive Index

Electron densities which effect the refractive index n in (2.84) can be ignored in lower atmosphere. In this case, n is given with atmospheric pressure p [hPa], atmospheric temperature T [K], and partial pressure of water vapor e [hPa] by the following equation (Balsley and Gage 1980).

$$n = 1 + 7.76 \times 10^{-5}\frac{p}{T} + 3.73 \times 10^{-1}\frac{e}{T^2}. \tag{7.80}$$

The second and third terms are the contributions of dry and moist atmospheres, respectively. Although the third term is useful for studying moisture profile in the next section, we discard it for simplicity. In dry atmosphere, n can be expressed as

$$n = 1 + 7.76 \times 10^{-5}\frac{p}{T}. \tag{7.81}$$

The change of n caused by fluctuations of the atmospheric pressure $\mathrm{d}p$ and temperature $\mathrm{d}T$ can be given by (7.81) as

$$\mathrm{d}n = 7.76 \times 10^{-5} \left(\frac{\mathrm{d}p}{T} - p\frac{\mathrm{d}T}{T^2} \right). \tag{7.82}$$

Atmospheric pressure changes lead to atmospheric temperature variation in accordance with the adiabatic equation (Tatarskii 1971, p. 74);

$$\frac{\mathrm{d}T}{T} = \frac{\gamma - 1}{\gamma} \frac{\mathrm{d}p}{p}. \tag{7.83}$$

Substituting (7.83) into (7.81), the relation between fluctuations of the refractive index and the temperature can be given by

$$\mathrm{d}n = 7.76 \times 10^{-5} \frac{1}{T\gamma} \mathrm{d}p. \tag{7.84}$$

Acoustic Intensity

Acoustic intensity I_a [W m^{-2}] is the sound power which passes unit area perpendicular to the direction of propagation per second, and given using sound pressure[10] $\overline{p}$ as

$$I_\mathrm{a} = \frac{\overline{p}^2}{\rho_\mathrm{a} c_\mathrm{s}}, \tag{7.85}$$

where ρ_a is atmospheric density [kg m^{-3}], and c_s is true sound velocity [m s^{-1}]. The acoustic intensity of the sound, which is transmitted from the sound source with power P_a and gain g_at, at range r becomes

$$I_\mathrm{a}(r) = \frac{P_\mathrm{a} g_\mathrm{at}}{4\pi r^2}. \tag{7.86}$$

From (7.85) and (7.86), the sound pressure $\overline{p}$ at range r can be expressed as

$$\overline{p}(r) = \sqrt{\frac{\rho_\mathrm{a} c_\mathrm{s} P_\mathrm{a} g_\mathrm{at}}{4\pi r^2}}. \tag{7.87}$$

Applying (7.87) to $\mathrm{d}p$ in (7.84), the rms value of fluctuations of the refractive index $\mathrm{d}n$, $\overline{n}$, becomes

$$\overline{n}(r) = \frac{A_\mathrm{a}}{r}, \tag{7.88}$$

[10] Root mean square (rms) of atmospheric pressure variation caused by passing of sound wave.

where A_a is the constant which is determined by characteristics of the sound source and condition of the atmosphere, and given by

$$A_a = 7.76 \times 10^{-5} \frac{1}{T\gamma} \sqrt{\frac{\rho_a c_s P_a g_{at}}{4\pi}}. \tag{7.89}$$

Radar Equation for RASS

The received power which is backscattered from sound wave surface, P_{ar}, can be given as the detail will be discussed in Appendix D by Marshall et al. (1972); Clifford and Wang (1977)

$$P_{ar} = A_a \frac{4\pi^2 P_t}{\lambda^2 r^2} \left(\frac{\Delta r}{2}\right)^2 \mathrm{sinc}^2 \left[(2k - k_a)\frac{\Delta r}{2}\right], \tag{7.90}$$

where $\mathrm{sinc}x = \sin x/x$. A_a is the constant given by (7.89), P_t and λ the transmitted power and the wavelength of atmospheric radar, respectively, r the range from atmospheric radar to the sound surface, and Δr the range resolution, while k and k_a are the wave numbers of radar wave and acoustic wave, respectively.

7.4.3 Bragg Condition and Background Wind

Bragg Condition

To receive RASS echo, the Bragg condition must be satisfied between the wave number vectors of acoustic wave $\boldsymbol{k}_a$ and of radar wave $\boldsymbol{k}$ as follows.

$$\boldsymbol{k}_a = 2\boldsymbol{k}. \tag{7.91}$$

For this purpose, the directions and the amplitudes of both wave number vectors must satisfy

$$\boldsymbol{k}_a // \boldsymbol{k}, \tag{7.92}$$

$$|\boldsymbol{k}_a| = 2|\boldsymbol{k}|. \tag{7.93}$$

Equation (7.92) shows that the sound wave surface is orthogonal to the direction of radar beam, and (7.93) indicates that the wavelength of acoustic wave is half that of radar wave. For example, frequencies of acoustic wave compatible with atmospheric radars at transmitted frequency of 50 MHz and 1,300 MHz are selected around 110 Hz and 2.86 kHz, respectively. However, atmospheric temperature varies with altitude, and consequently sound speed changes with altitude. Thus, it is generally not easy for a single frequency to satisfy the condition of (7.93) in a wide range of altitude. Therefore, the observations are performed by sweeping frequency of

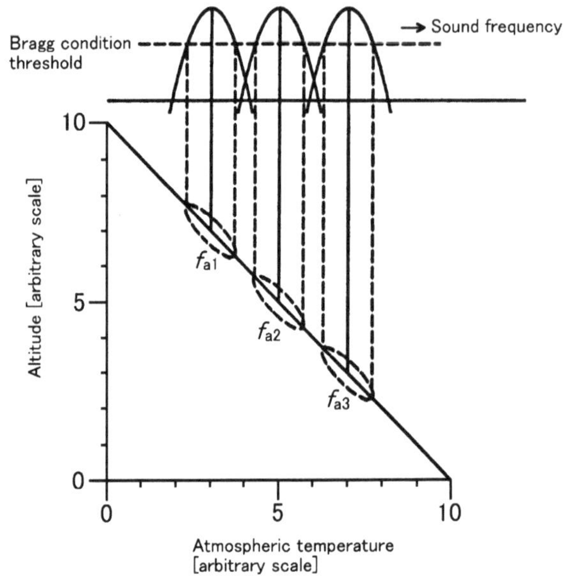

Fig. 7.12 The concept of frequency sweeping for RASS (from Adachi 1996)

acoustic waves in a suitable band where the Bragg condition is almost satisfied (Masuda et al. 1992). Figure 7.12 illustrates the concept of frequency sweeping for RASS. The upper panel shows the frequency sweeping of acoustic waves, and the lower one height profile of atmospheric temperature. For instance, if three discrete frequencies f_{a1}, f_{a2}, and f_{a3} are used for the sound source, temperature measurement becomes possible at the altitudes which satisfy the threshold value of the Bragg condition, i.e., within the ellipses which are shown by broken lines of the same figure (Adachi 1996).

Effects of Background Wind

The second term in the right-hand side of (7.74) is the radial component v_r of background wind velocity. Thus, if background wind velocity is included, (7.78) is corrected to

$$T_{\mathrm{v}} = \left(\frac{c_{\mathrm{a}} - v_r}{K_{\mathrm{d}}}\right)^2. \tag{7.94}$$

Range where the RASS observation is feasible depends considerably on temperature and horizontal wind velocity. Figure 7.13 illustrates loci of acoustic wavefront that propagates from the sound source at the origin every 2 s during 1–29 s after the transmission of the acoustic wave (Masuda 1988). In panel (a) of the figure, it is assumed that sound speed $c_{\mathrm{s}}(z)$ [m s^{-1}] changes linearly with altitude z [m] by $c_{\mathrm{s}}(z) = 331\text{–}2.8 \times 10^{-3} z$ and that the background wind does not

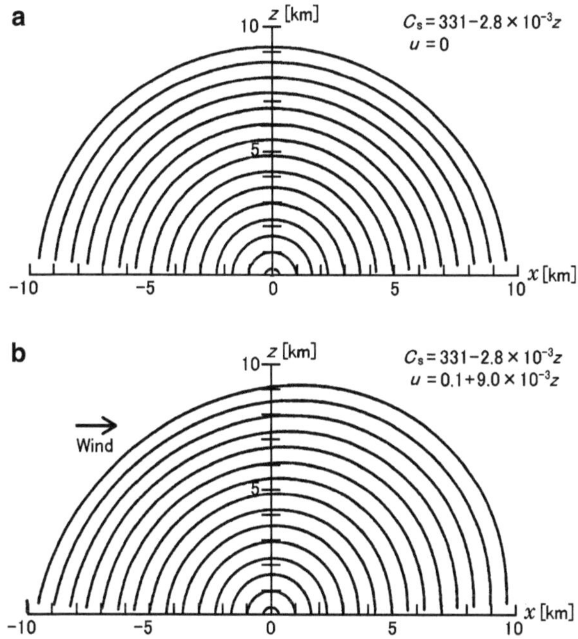

Fig. 7.13 Loci of acoustic wavefront that propagates from the sound source at the origin drawn every 2 s during 1–29 s after the transmission of the acoustic wave. (**a**) represents the windless condition, $dc_s/dz(<0)$ is constant. In (**b**), dc_s/dz is the same as (**a**) but there is a background wind which changes linearly with altitude [from Masuda 1988]

exist, i.e. $u = 0\,\mathrm{m\,s^{-1}}$. Since sound speed decreases with increasing altitude, the loci of acoustic wave fronts have elliptical cross section, which are symmetrical with respect to the altitude direction with the major axis in the horizontal direction. On the other hand, in (b), it is assumed that sound speed is the same as that in (a) while the horizontal wind velocity $u(\mathrm{z})$ [m] changes linearly with altitude by $u(\mathrm{z}) = 0.1 + 9 \times 10^{-3}z$. In this case, the loci of acoustic wave fronts are distorted and the longer axis of the ellipsoid are slanted windward.

Atmospheric temperature, wind direction, and wind velocity generally changes with altitude, and consequently sound speed differs between vertical and horizontal directions. This causes that curvature of acoustic wavefronts differs between vertical and horizontal directions, making it more difficult to point the wave number vector of radar wave perpendicular to the acoustic wavefronts with increasing distance from sound source. This situation limits the range observable with RASS.

Figure 7.14 shows the observable range for RASS simulated by two dimensional ray-tracing of acoustic wavefronts (Adachi 1996). Concentric curves in the figure show the loci of acoustic wavefronts in the interval of 5 s. Shadow indicates observable area for RASS where the radar beam crosses perpendicularly to the acoustic wavefronts, The shadow density changes from high to low with distance

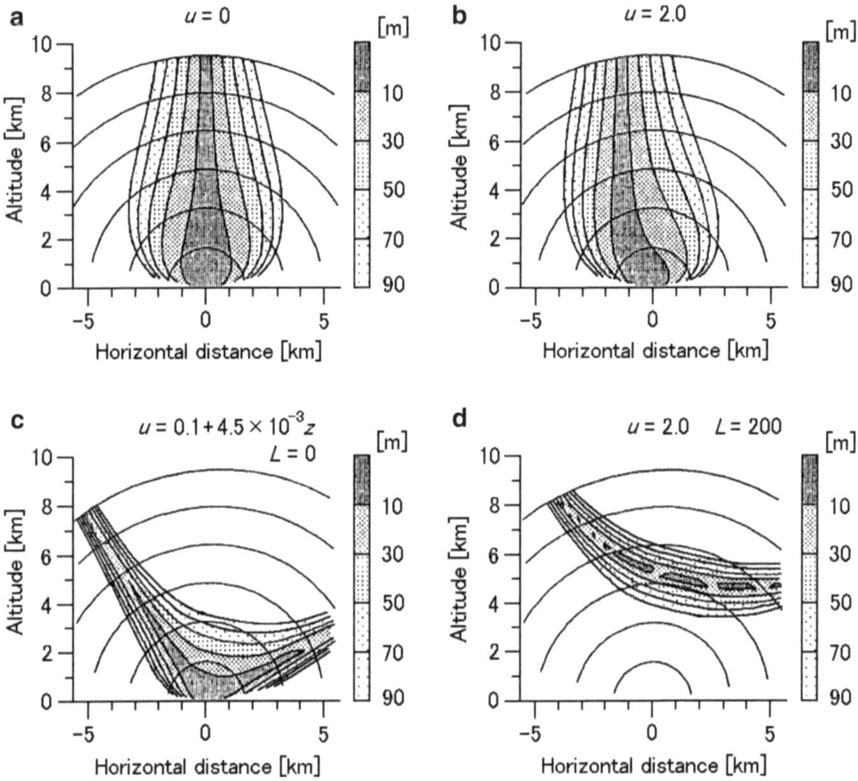

Fig. 7.14 Observable range for RASS simulated by two dimensional ray-tracing of acoustic wavefronts. *Concentric curves* in the figure show the loci of acoustic wavefronts in the interval of 5 s. In each panel, u denotes horizontal wind speed [m s^{-1}], z denotes altitude [m], and L denotes the distance [m] from the center of atmospheric radar antenna to the sound source [from Adachi 1996]

from the point of perpendicularity. In each panel (a)–(d), it is assumed that sound velocity changes with altitude by the same value in Fig. 7.13. In panels (a)–(d), it is assumed that the sound speed linearly varies with altitude as in the case of Fig. 7.13. On the other hand, the following conditions are assumed; (a) horizontal wind velocity is zero, (b) horizontal wind of 2 m s^{-1} blows from left to right, (c) same as Fig. 7.13 (b), i.e., horizontal wind changes linearly with altitude by $u(z) = 0.1 + 4.5 \times 10^{-3}z$, and (d) atmospheric temperature and background wind are the same as those in (c) and placing radar antenna apart from sound source by 200 m.

These figures properly illustrate that the observable altitude for RASS extends higher by scanning radar beam as compared with using fixed beam to the vertical direction, except for the case of no wind. Furthermore, it is effective for expanding observation range of RASS to operate synchronously multiple sound source placed approximately away from radar antenna.

7.5 Estimation of Water Vapor Profiles

The estimation of water vapor profile from atmospheric radar measurements requires to measure the vertical gradient of refractive index (refractive index gradient) M_n and the temperature profile from RASS observations. A major problem is the determination of the sign of M_n because the radar measurements only provide M_n^2. A method for determining the sign must be derived.

Brunt-Väisälä Frequency

The value of Brunt Väisälä frequency N can be derived from the virtual temperature profile estimated from the RASS technique. The square of N in (7.59) can be obtained from the potential temperature Θ or temperature T using the relation:

$$\frac{N^2}{g} = \frac{1}{\Theta}\frac{\partial \Theta}{\partial z} = \frac{1}{T}\left(\frac{\partial T}{\partial z} + \Gamma\right), \tag{7.95}$$

where g is the gravitational acceleration, z is altitude, Γ is the dry adiabatic lapse rate given by $\Gamma = g/C_\mathrm{p} \simeq 9.80$ [K km^{-1}], where C_p is the specific heat capacity at constant pressure ($\simeq 1{,}004$ [J K^{-1} kg^{-1}]). Θ can be expressed in terms of T and p through the relation

$$\Theta = T\left(\frac{p_0}{p}\right)^{\frac{R_\mathrm{sp}}{C_\mathrm{p}}}, \tag{7.96}$$

where $p_0 = 1{,}000$ hPa, and R_sp is the specific gas constant for dry air ($\simeq 287$ [J K^{-1} kg^{-1}]). Thus $R_\mathrm{sp}/C_\mathrm{p} \simeq 0.286$.

Mixing Ratio

The mixing ratio q is defined as the mass of vapor divided by the mass of dry air, and is related to p and partial pressure of water vapor e through the expression

$$q = 0.622\frac{e}{p-e} \simeq 0.622\frac{e}{p}. \tag{7.97}$$

From (7.96) and (7.97), it is possible to express the refractive index n of (7.80) in terms of Θ and q. It gives

$$n = 1 + 7.76\times 10^{-5}\frac{p}{\Theta}\left(\frac{p_0}{p}\right)^{\frac{R_\mathrm{sp}}{C_\mathrm{p}}}\left[1 + 7800\frac{q}{\Theta}\left(\frac{p_0}{p}\right)^{\frac{R_\mathrm{sp}}{C_\mathrm{p}}}\right]. \tag{7.98}$$

Vertical Gradient of Refractive Index

The refractive index gradient M_n is by definition $\mathrm{d}n/\mathrm{d}z$ (see Sect. 7.3.4). From (7.95) to (7.97), we obtain (Ottersten 1969c)

$$M_n = -7.76 \times 10^{-5} \frac{p}{T} \left(\frac{N^2}{g} + 15600 \frac{q}{T} \frac{N^2}{g} - \frac{7800}{T} \frac{\mathrm{d}q}{\mathrm{d}z} \right). \tag{7.99}$$

Expression (7.99) is a first-order differential equation in $q(z)$ and thus can be rewritten as

$$\frac{\mathrm{d}q}{\mathrm{d}z} + A(z)q = B(z), \tag{7.100}$$

where

$$A(z) = -\frac{2N^2}{g}, \tag{7.101}$$

$$B(z) = 1.65 \frac{T^2}{p} M_n + \frac{1}{7800} \left(\frac{\mathrm{d}T}{\mathrm{d}z} + \Gamma \right). \tag{7.102}$$

Assuming boundary conditions in (7.100), i.e., $q(z_0) = q_0$ and $\Theta(z_0) = \Theta_0$ for $z = z_0$, the height profile of mixing ratio $q(z)$ can be expressed as

$$q(z) = \Theta^2 \left[\int_{z_0}^{z} B(z)\, \Theta^{-2} \mathrm{d}z + \frac{q_0}{\Theta_0^2} \right]. \tag{7.103}$$

Because η is proportional to M_n^2 as shown by (7.71), the sign of M_n (appearing in (7.102)) is impossible to determine from the sole radar observations. Methods based on complementary balloon observations and consisting in estimating the correlation function between M_n and N^2 for example have been proposed. Recently, a method for estimating precipitable water from the propagation delay of global positioning system (GPS) radio waves has been tested and validated (e.g., Gossard et al. 1999; Tsuda et al. 2001). That is to integrate water vapor profile in altitude which is estimated by radar observation and calculate precipitable water, and to correct the sign of M_n appropriately so as not to differ with the precipitable water obtained by the GPS. The virtual potential temperature Θ in (7.103) can be calculated from the virtual temperature retrieved from RASS observations. Since the difference between virtual potential temperature and potential temperature is only several K in the lower troposphere where water vapor has the highest density, a maximal error of about 1 % in Θ is expected. The error in mixing ratio by this assumption is at most 1–2 % (Furumoto et al. 2003). The profiles of the relative humidity measured by radar and that by radiosondes will be shown in Sect. 12.4.3.

7.6 Radar Interferometry Techniques

Radar interferometry (RI) techniques have been applied soon after the development of atmospheric radars. More recently, radar imaging techniques based on the use of multiple receivers and/or multiple frequencies have also emerged for improving the spatial resolution of atmospheric radars and the rejection of outliers and ground clutters. For these objectives, more and more sophisticated signal processing methods are being applied.

As already explained in Sect. 3.1, the range resolution of a pulse radar is defined by the width of the transmitted pulse. Therefore, the narrower the pulse width be, the better the range resolution will be. However, the transmission of a narrow pulse requires a wide frequency bandwidth which is not always available due to both technical and frequency allocation limitations. In addition, reducing the pulse width implies a decrease of the transmitted power and thus of the SNR. In standard mode, the range resolution of UHF/VHF atmospheric radars is generally about 75–600 m with various pulse compression techniques.

The angular resolution is defined by the radar beamwidth which essentially depends on the diameter of the antenna array at a given radar wavelength. Thus, the angular resolution can be improved as much as the antenna array dimension can be increased. However, physical space limitations for installation of the antenna arrays cannot always be avoided especially at VHF and lower frequencies. As a result, radar beams cannot be typically narrower than a few degrees. Radar Interferometry techniques can constitute alternatives to large antenna arrays and narrow pulses at the expenses of hypotheses on the atmospheric scatterers or other limitations (e.g., Fukao and Palmer 1991). Recent works aim at reducing the effects of these limitations and hypotheses through further advanced signal processing.

7.6.1 SDI and FDI Techniques

The spatial domain interferometry (SDI) technique was developed for determining the direction of arrival (DOA) of signals backscattered from isolated atmospheric scatterers at a high angular resolution in a wide radar beamwidth. The method is based on the use of pairs of receiving antennas separated by a suitable distance and on the measurement of the difference of phases between the received signals. These phase differences are related to the angular positions of the scatterers (Palmer et al. 1991; Larsen et al. 1992).

If the whole radar resolution volume is filled with atmospheric scatterers, the angular resolution is obviously governed by the radar beamwidth. For a radar wavelength λ and a diameter D_a of the antenna array, the radar beamwidth is proportional to λ/D_a as will be discussed in Sect. 8.2.1. For an isolated scatterer in the radar resolution volume, the angular resolution rather depends on the size of the scatterer. In the limit case of a discrete scatterer, the angular resolution become

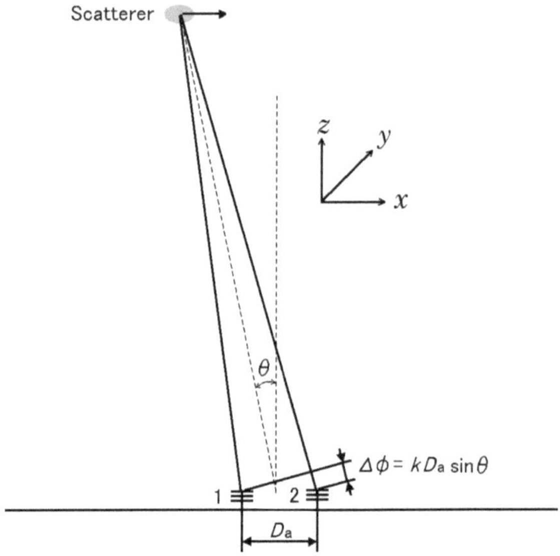

Fig. 7.15 Conceptual diagram of the SDI technique. Antennas 1 and 2 are horizontally separated by a distance D_a and receive signals backscattered from a scatterer advected by the horizontal wind at a zenith angle θ [from Fukao and Palmer 1991]

infinite and the DOA can be determined from simple geometrical considerations. Basically, the DOA is estimated by measuring the phase difference between received signals from different receiving antennas. In practice, three non coplanar receiving antennas are used. For ease of explanation, we describe the technique with two antennas. In Fig. 7.15, antennas horizontally separated by a distance D_a receive signals backscattered from a scatterer advected by the horizontal wind at a zenith angle θ. Obviously, it is assumed that the scatterer is located in both radar beams. The signals are received by antenna 2 with a time delay with respect to the signals received by antenna 1. This time delay corresponds to a phase difference given by

$$\Delta\phi = \phi_1 - \phi_2 = kD_a \sin\theta, \tag{7.104}$$

where $k = 2\pi/\lambda$ is the radar wave number. Relation (7.104) is valid if one assumes that the scatterer is in the far field (i.e., one assumes the Fraunhofer approximation) so that the wavefront of the backscattered signal is plane (as illustrated in Fig. 7.15). Even for a small range difference $D_a \sin\theta$, a significant phase difference can occur depending on θ and the carrier frequency of the radar.

The typical shape of Fourier transforms of complex signals received by antennas 1 and 2 is shown in Fig. 7.16. The radial components of horizontal wind velocity $v\sin\theta$ (where v is the horizontal wind velocity) from the two antennas appear as peaks in the modulus of the complex spectrum (upper panel of Fig. 7.16). Because the distances between the scatterer and both antennas are approximately the same in the far field hypothesis, the signal powers received by the two antennas are also almost the same. The lower panel shows the corresponding phase of the complex spectrum. The phases of the signals are surrounded by circles. The phase difference and finally the angular location of the scatterer can be deduced from (7.104).

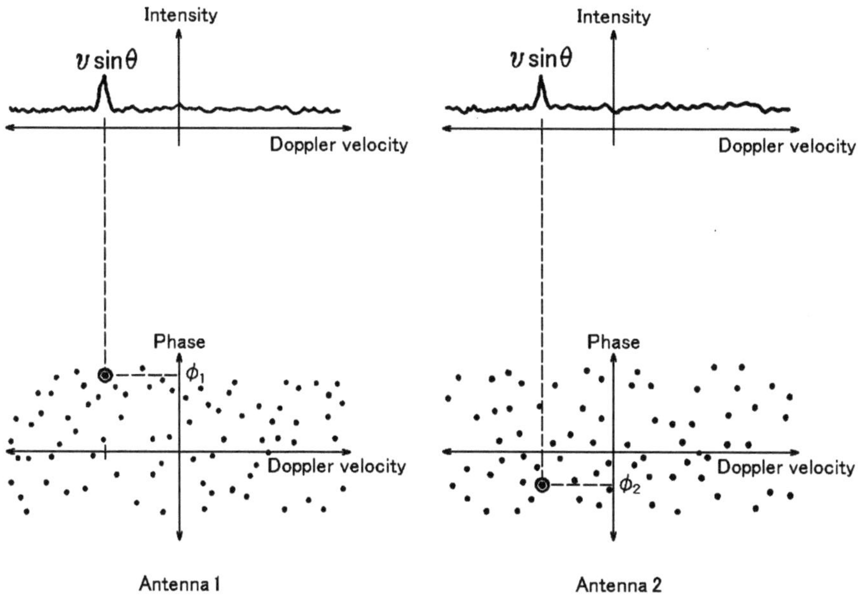

Fig. 7.16 Typical shape of the Fourier transforms of complex signals received by antennas 1 and 2 [from Fukao and Palmer 1991]

Similarly, the range of a single scattering layer within the radar volume can be estimated by measuring the phase difference between received signals when using two distinct carrier frequencies. The technique, called (dual) frequency domain interferometry (FDI) technique, is thus based on the transmission, pulse to pulse, of two closely separated frequencies (Kudeki and Stitt 1987). The radar beam is generally oriented overhead in accordance with the horizontal distribution of the atmospheric layers. As illustrated by Fig. 7.17, the phase difference $\Delta\phi$ varies according to the position of the scattering layer in the radar volume. By appropriately choosing the frequencies f_1 and f_2 of the two consecutive transmitted pulses S1 and S2, i.e., if $\Delta f = f_2 - f_1 < 1/\tau$ where τ is the pulse width, the phase difference of the received signals does not exceed 2π so that ambiguity in distance is avoided. Thus, in practice, the relative position of a single scattering layer in the radar volume, z_l, can be estimated by measuring the phase difference and is given, as a first approximation, by

$$\Delta\phi = \phi_1 - \phi_2 = 2\Delta k z_l, \tag{7.105}$$

where Δk is the wave number difference of the transmitted signals. Equation (7.105) is the equivalent expression of (7.104) for the SDI. In addition, the depth of the scattering layer can be estimated from the signal coherence or normalized covariance of the received signals by assuming a Gaussian distribution of the single scattering layer (e.g., Franke 1990).

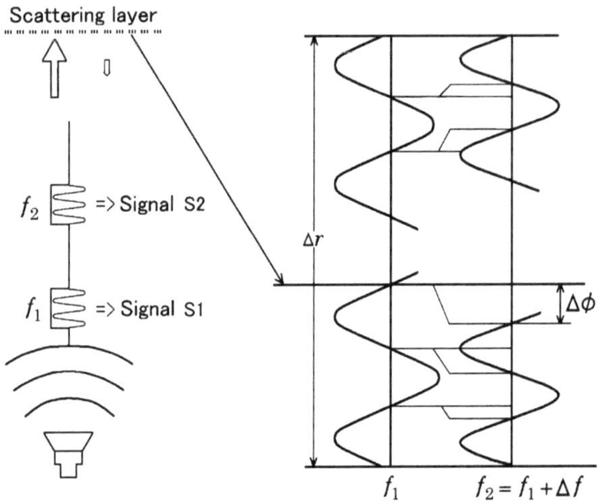

Fig. 7.17 Conceptual diagram of the FDI technique

The SDI and dual FDI techniques are now obsolete for determining the distribution of the scatterers in the radar volume. These techniques only provide apparent features consistent with the restrictive hypotheses necessary for their application. Indeed, these techniques are based on two antennas or two frequencies and thus information that can be obtained is limited. Radar imaging techniques that will be described in Sect. 7.6.2 are now being developed not only for improving the spatial resolution of atmospheric radars but also for suppressing ground clutter and moving clutters (such as aircrafts, birds, and insects).

Wind Estimation

Let us consider the zenith angle θ of the scatterer (see Fig. 7.15) to be small. For a scatterer moving at the radial velocity v_r and two antennas arranged along a baseline as shown by Fig. 7.18, the phase difference ϕ can be rewritten as

$$\phi = kD_\mathrm{a}\cos(\alpha - \delta)\sin\theta \simeq kD_\mathrm{a}\cos(\alpha - \delta)\theta, \tag{7.106}$$

where α and δ are azimuthal angles of the antenna baseline and of the wind velocity, respectively. In addition, the radial velocity v_r is given by

$$v_r = v_\mathrm{h}\sin\theta + w\cos\theta \simeq v_\mathrm{h}\theta + w, \tag{7.107}$$

where v_h and w are the horizontal and vertical wind velocities, respectively. Substituting (7.107) into (7.106), ϕ can be expressed in terms of v_r as (Palmer et al. 1991)

Fig. 7.18 Loci of two antennas applied for the wind estimation by the SDI technique [from Fukao and Palmer 1991]

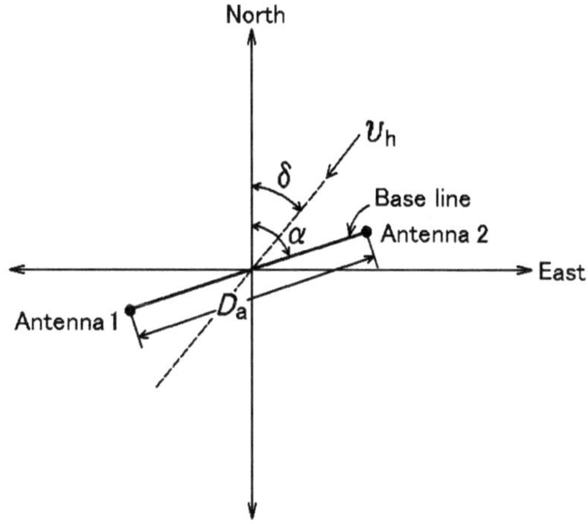

$$\phi(v_r) = \left[\frac{kD_\mathrm{a}\cos(\alpha-\delta)}{v_\mathrm{h}}\right] v_\mathrm{r} - \left[\frac{wkD_\mathrm{a}\cos(\alpha-\delta)}{v_\mathrm{h}}\right]. \tag{7.108}$$

The relation (7.108) does not only apply to a single scatterer but also to distributed scatterers as is the case in the atmosphere. The phase of received signals depends on the position of the scatterers. Since the wind velocity is assumed to be uniform in a narrow radar beam of several degrees, the radial velocity v_r varies continuously as the zenith angle of the scatterer varies from positive to negative values. For example, the value of v_r at the zenith angle $+\theta$ is equal to the value at the zenith angle $-\theta$ but with an opposite sign (i.e., $-v_r$). Consequently, ϕ is a linear function of v_r as revealed by the expression (7.108). This property was verified from observations performed with several antennas and the calculation of cross spectrum[11] (Farley et al. 1981; Röttger et al. 1990b). Thus, (7.108) can be rewritten as $\phi(v_r) = mv_r + b$, where the slope m and the intercept b are given by

$$m = \frac{kD_\mathrm{a}\cos(\alpha-\delta)}{v_\mathrm{h}}, \tag{7.109}$$

$$b = \frac{-wkD_\mathrm{a}\cos(\alpha-\delta)}{v_\mathrm{h}}, \tag{7.110}$$

respectively. Applying $u' = \sin\delta/v_\mathrm{h}$ and $v' = \cos\delta/v_\mathrm{h}$, (7.109) becomes

$$m = kD_\mathrm{a}(u'\sin\alpha + v'\cos\alpha). \tag{7.111}$$

If two sets of independent m and α values are estimated from two pairs of antenna arrays, u' and v' can be obtained. Let m and α for antennas 1 and 2 be $(m_{12},\ \alpha_{12})$ and $(m_{13},\ \alpha_{13})$ for antennas 1 and 3. The parameters u' and v' are then given by

[11] Power spectrum is the Fourier transform of autocorrelation function, whereas cross spectrum is defined by the Fourier transform of the cross correlation function.

$$u' = \frac{1}{kD_a\Delta}[m_{12}\cos\alpha_{13} - m_{13}\cos\alpha_{12}], \quad (7.112)$$

$$v' = \frac{1}{kD_a\Delta}[m_{13}\sin\alpha_{12} - m_{12}\sin\alpha_{13}], \quad (7.113)$$

respectively, where $\Delta = \sin\alpha_{12}\cos\alpha_{13} - \sin\alpha_{13}\cos\alpha_{12}$. Thus, from (7.112) and (7.113), the horizontal wind velocity v_h and its azimuth δ are

$$v_h = \frac{1}{\sqrt{(u')^2 + (v')^2}}, \quad (7.114)$$

$$\delta = \arctan\left(\frac{u'}{v'}\right). \quad (7.115)$$

The vertical wind velocity component can be either estimated from (7.110) and v_h obtained by (7.114), or from (7.108) for $\phi = 0$ such as

$$w = v_r|_{\phi=0} = -\frac{b_{12}}{m_{12}}. \quad (7.116)$$

At VHF (~50 MHz), if the scattering layer is slightly tilted, the vertical wind velocity measured from vertically pointing radar beams can be biased due to the contamination by the horizontal wind. The advantage of the SDI is to provide an unbiased measurement of the vertical wind velocity because there is no component other than the true vertical velocity when the phase between two sets of antenna is zero, i.e., the radar beam is correctly oriented vertically.

7.6.2 *Radar Imaging Techniques*

Coherent Radar Imaging

Coherent radar imaging (CRI) is basically an extension of the SDI and refers to an interferometric technique based on the application of a spatial filter to complex signals received from multiple antenna subarrays with wide and vertically pointing beam patterns. This interferometric technique was originally developed in the field of radio astronomy and acoustics. The filtering operation consists in multiplying the received samples by complex coefficients so that the radar beam can be steered in any desired direction (within the wide beam pattern). The so-called beamforming technique is the most basic approach: signals are simply phase-shifted and added so that a beam is formed in a desired direction. The processing corresponds to a fast Fourier transform of the received signals (as seen in the following paragraph). It can be made off-line by software on time series of received signals and is called the post beam steering (PBS) processing method (Röttger and Ierkic 1985; Van Baelen et al. 1991). It can be shown that phase compensation can be applied after statistical processing of the signals. The method is then called the post statistic steering (PSS) processing method (Kudeki and Woodman 1990; Palmer et al. 1990, 1993). With PBS, time series of complex signals must be recorded whereas cross correlation functions and/or complex cross spectra of signals only need to be recorded with PSS, saving calculation amounts and memory capacity. These techniques need one receiver for each antenna subarrays. However, Hélal et al. (2001) proposed the sequential post beam steering (SPBS) method based on multiple antenna subarrays but a single receiver, the signals being received sequentially at a high time rate, assuming that the signal is stationary during the acquisition time.

The Fourier-based method was first applied to the observation of ionospheric instabilities in the equatorial zone by Kudeki and Sürücü (1991). It is a classical filter bank processing method for estimating the signal intensity (also called "brightness" in radio-astronomy) in a direction determined by the wave number vector $\boldsymbol{k}$. As $\boldsymbol{k}$ can be selected in software, it is possible to focus to any direction in the transmitted radar beam so that the horizontal distribution (map) of signal intensity can be obtained. The resolution of the map ultimately depends on the total length of the receiver baselines. Let $[W_{\mathrm{f}}]$ be the weighting vector which expresses Fourier-based filtering operation. It is expressed with the steering vector $[\mathrm{e}(\boldsymbol{k})]$ as follows:

$$[W_{\mathrm{f}}(\boldsymbol{k})] = [e(\boldsymbol{k})] = [\exp(j\boldsymbol{kd}_1)\ \exp(j\boldsymbol{kd}_2)\ \cdots\ \exp(j\boldsymbol{kd}_N)]^{\mathrm{T}}, \tag{7.117}$$

where $\boldsymbol{d}_i$ is the vector indicating the center of the receiving array i, $\boldsymbol{kd}_i$ means the inner product $\boldsymbol{k}\cdot\boldsymbol{d}_i$, and the upper subscript T denotes transposition operator. Let $[v(t)]$ be the signal vector from the N receivers, the scalar output of the filter in the direction of $\boldsymbol{k}$, $y(\boldsymbol{k},t)$, can be expressed as

$$y(\boldsymbol{k},t) = [W_{\mathrm{f}}(\boldsymbol{k})]^{\dagger}\cdot[v(t)], \tag{7.118}$$

where the upper subscript † denotes the Hermitian operator. The variance $B_{\mathrm{f}}(\boldsymbol{k}) = \langle y(\boldsymbol{k},t)\cdot y^{*}(\boldsymbol{k},t)\rangle$ expresses the brightness in the direction $\boldsymbol{k}$ and is given by

$$B_{\mathrm{f}}(\boldsymbol{k}) = [W_{\mathrm{f}}(\boldsymbol{k})]^{\dagger}\cdot[R]\cdot[W_{\mathrm{f}}(\boldsymbol{k})], \tag{7.119}$$

where $[R] = \langle [v(t)]\cdot[v(t)]^{\dagger}\rangle$ is the covariance matrix of dimension N. This matrix is Hermitian, i.e., $[R] = [R]^{\dagger}$.

Palmer et al. (1998) estimated the autocorrelation function of $y(\boldsymbol{k},t)$, $R_y(\boldsymbol{k},\tau) = \langle y(\boldsymbol{k},t+\tau)\cdot y^{*}(\boldsymbol{k},t)\rangle$, where τ means time lag, assuming a stationary signal in a wide sense. $B_{\mathrm{f}}(\boldsymbol{k})$ is $R_y(\boldsymbol{k},0)$. In matrix notations, it can be expressed as

$$R_y(\boldsymbol{k},\tau) = [W_{\mathrm{f}}(\boldsymbol{k})]^{\dagger}\cdot[R(\tau)]\cdot[W_{\mathrm{f}}(\boldsymbol{k})], \tag{7.120}$$

where $[R(\tau)] = \langle [v(t+\tau)] \cdot [v(t)]^{\dagger} \rangle$. The Fourier transform of $R_y(\boldsymbol{k}, \tau)$ is an estimation of the brightness for each Doppler frequency bin in the direction $\boldsymbol{k}$. This approach is particularly useful for imaging the clear air and precipitation echoes separated when they are Doppler-sorted.

Capon Imaging

The Fourier-based (or periodogram) method has well-known characteristics but has also limited performances. The Fourier filter is just constrained to minimize the background noise contribution and to maintain the gain of the filter to 1, without phase distortion, in the desired direction. The side lobes can be the cause of contaminations from intense sources outside the desired direction (outliers, clutter, etc.). Capon's method (Capon 1969) is also a filter bank method but is defined by a constrained optimization problem. The filter is now designed to be "as selective as possible" for the input consisting of the studied data themselves. It does not only maintain the gain to 1 in the selected direction but also minimizes the brightness in the other directions. For this reason, Capon's method is also called the "minimum variance" method. This constrained optimization problem is mathematically described as follows. The quantity

$$[W_{\mathrm{C}}(\boldsymbol{k})]^{\dagger} \cdot [R] \cdot [W_{\mathrm{C}}(\boldsymbol{k})] \tag{7.121}$$

should be minimum everywhere except in the desired direction where the filter passes the signals, without attenuation and phase distortion, i.e.:

$$[e(\boldsymbol{k}_d)]^{\dagger} \cdot [W_{\mathrm{C}}(\boldsymbol{k}_d)] = 1. \tag{7.122}$$

This minimizing problem can be solved using Lagrange methods (e.g., Palmer et al. 1998). It can be deduced that the weighting vector of Capon imaging $W_{\mathrm{C}}(\boldsymbol{k})$ and the brightness $B_{\mathrm{C}}(\boldsymbol{k})$ can be expressed as

$$[W_{\mathrm{C}}(\boldsymbol{k})] = \frac{[R]^{-1} \cdot [e(\boldsymbol{k})]}{[e(\boldsymbol{k})]^{\dagger} \cdot [R]^{-1} \cdot [e(\boldsymbol{k})]}, \tag{7.123}$$

$$B_{\mathrm{C}}(\boldsymbol{k}) = \frac{1}{[e(\boldsymbol{k})]^{\dagger} \cdot [R]^{-1} \cdot [e(\boldsymbol{k})]}, \tag{7.124}$$

respectively. As shown by (7.117), the weighting vector of the Fourier-based method does not depend on the signal input. On the other hand, the weighting vector of Capon's method depends on the covariance matrix of the signals (7.123) and thus on the SNR. Capon's method is therefore an adaptive (non parametric) filter bank method. Since the output power of the filter is minimized, possible interfering signals can be minimized (at the cost of higher sidelobes in directions of weak sources). Interfering signals from known directions can be totally suppressed by adding the constraint of a null in the direction of the interfering signal. Figure 7.19

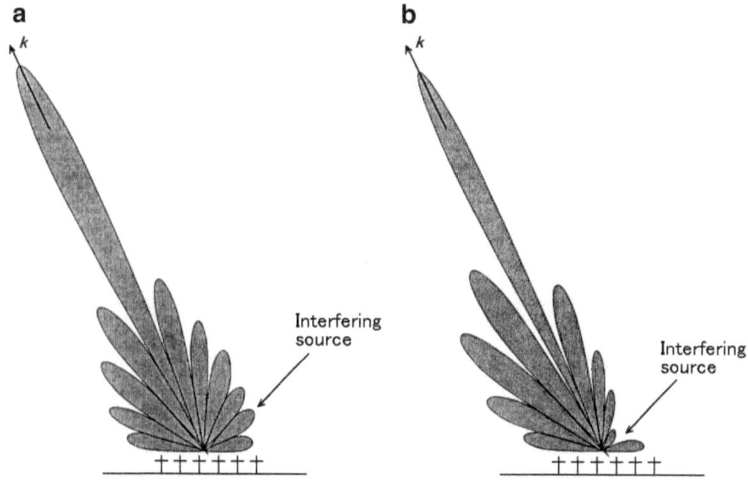

Fig. 7.19 Examples of antenna beam patterns using (**a**) Fourier-based and (**b**) Capon's method. In (**b**), a null is imposed in the direction of the interference [from Palmer et al. 1998]

shows examples of antenna beam patterns using Fourier-based method (a) and Capon's method (b). In (b), a null is imposed in the direction of the interference (Palmer et al. 1998). In addition, the angular resolution obtained from Capon's method is better than that obtained from Fourier-based method, and Capon's method is more robust than autoregressive methods such as the maximum entropy method (MEM) applied in spectral analysis. Indeed, it can be shown that Capon estimator is an average of the MEM estimators (Nickel 1988). Capon's method is most effective when the SNR is high. For very low SNR, the performances of Capon estimator tend to the performances of the Fourier estimator.

The methods used in the spatial domain can be directly transposed into the frequency domain for range imaging with N frequencies steering pulse to pulse (Palmer et al. 1999; Luce et al. 2001b). The technique is either called the range imaging (RIM) or the frequency domain interferometric imaging (FII). Let z be the range within the radar resolution volume. The range brightness for Fourier-based and Capon's methods can be directly deduced from (7.119) and (7.124):

$$B_{\mathrm{f}}(z) = [e(z)]^{\dagger} \cdot [R] \cdot [e(z)], \tag{7.125}$$

$$B_{\mathrm{C}}(z) = \left([e(z)]^{\dagger} \cdot [R]^{-1} \cdot [e(z)]\right)^{-1}, \tag{7.126}$$

respectively. (7.125) and (7.126) are the expressions equivalent to (7.119) and (7.124) for 1D (range) imaging. Capon RIM/FII improves by a factor 5–10 the range resolution if the maximum frequency difference $\Delta f_{\max} = 1/\tau$ and if SNR exceeds 0 dB. For SNR < 0, the performances are degraded but at least a factor 3 of improvement can be obtained in the lower stratosphere where SNR often ranges below 0 dB (e.g., Luce et al. 2007).

CRI with Fourier-based and Capon's methods can also be used for retrieving spectral parameters (mainly wind velocity) at high resolution within the radar resolution volume (e.g., Yu and Brown 2004; Cheong et al. 2004) and in particular the horizontal and vertical winds. By applying (7.120) with W_{f} or W_{C}, Doppler spectra can be calculated from the reconstructed time series in the desired directions. However, some challenging problems still need to be resolved due to the intermittency or inhomogeneities of the scatterers distribution within the radar volume (e.g., Cheong et al. 2004).

Both CRI and RIM/FII can be applied together for a 3D radar imaging (e.g., Yu and Palmer 2001; Hassenpflug et al. 2008). The steering vector for both angular and range imaging $[e(\boldsymbol{k},z)]$ has the form:

$$\begin{aligned}
[e(\boldsymbol{k},z)] = [\; & \mathrm{e}^{j(2k_1 z-\boldsymbol{k}_1\boldsymbol{D}_1)} \quad \mathrm{e}^{j(2k_1 z-\boldsymbol{k}_1\boldsymbol{D}_2)} \quad \cdots \quad \mathrm{e}^{j(2k_1 z-\boldsymbol{k}_1\boldsymbol{D}_N)} \\
& \mathrm{e}^{j(2k_2 z-\boldsymbol{k}_2\boldsymbol{D}_1)} \quad \mathrm{e}^{j(2k_2 z-\boldsymbol{k}_2\boldsymbol{D}_2)} \quad \cdots \quad \mathrm{e}^{j(2k_2 z-\boldsymbol{k}_2\boldsymbol{D}_N)} \\
& \vdots \\
& \mathrm{e}^{j(2k_M z-\boldsymbol{k}_M\boldsymbol{D}_1)} \quad \mathrm{e}^{j(2k_M z-\boldsymbol{k}_M\boldsymbol{D}_2)} \quad \cdots \quad \mathrm{e}^{j(2k_M z-\boldsymbol{k}_M\boldsymbol{D}_N)}]^{\mathrm{T}},
\end{aligned} \tag{7.127}$$

where the number of frequencies and antenna subarrays is M and N, respectively. The covariance matrix has the dimension $(MN) \times (MN)$.

The filter bank processing methods have degraded performances for low SNRs. Recently, some methods using the difference of time correlation between the signal and the noise have been developed for improving the performances of these methods. Le et al. (2010) used a multilag (ML) correlation technique developed for meteorological radars (e.g., Melnikov 2006). The aforementioned methods use the covariance, i.e., the correlation function at zero lag which is affected by noise. The noise effects should be removed or, in practice, strongly attenuated by using the covariance for time lags 1 ($= T_{\mathrm{s}}$) and 2 ($= 2T_{\mathrm{s}}$), or higher. By assuming a Gaussian spectrum of the covariance function, and a Gaussian spectrum of the backscattered signal derived the following expression of the ML brightness (Le et al. 2010),

$$B_{\mathrm{ML}}(\boldsymbol{k}) \simeq \left(\frac{|[W]^{\dagger}[R(T_{\mathrm{s}})][W]|^4}{|[W]^{\dagger}[R(2T_{\mathrm{s}})][W]|} \right)^{1/3}. \tag{7.128}$$

where the weights $[W]$ are given by Fourier-based or Capon's method.

Extremely High Resolution Using SVD Methods

The eigensystem approach (Bienvenu and Kopp 1983) uses a decomposition of the covariance of dimension $(N \times N)$ in singular values. It is called the singular value decomposition (SVD) method. The decomposition leads to two orthogonal vector sub-spaces, associated with the noise and the signal, respectively. This decomposition can be made by determining the eigenvalues and eigenvectors of the covariance matrix. This approach has been developed for the detection of discrete sources assuming that their number is smaller than the number of receivers. Their location is determined using the multiple signal classification (MUSIC) algorithm (Schmidt 1986). Hélal et al. (2001) and Luce et al. (2001b) first applied this method to atmosphere radar observations for angular and range imaging, respectively.

Let us assume a covariance matrix $[R] = \langle [v(t)] \cdot [v(t)]^{\dagger} \rangle$ of dimension $N \times N$ and $M < N$ sources. Since $[R]$ is Hermitian, the eigenvalues $\lambda_1, \lambda_2, \cdots, \lambda_N$ are non-negative real values, and the eigenvectors $[q_i]$ $(\forall i \in [1, N])$ are orthogonal, respectively.

When sorting the eigenvalues λ_1 to λ_N in the decreasing order, the first M eigenvalues are associated with the signal subspace and the eigenvectors $[q_1]$ to $[q_M]$ constitute a basis of the signal sub-space. The $N-M$ eigenvalues correspond to the noise sub-space and the eigenvectors $[q_{M+1}]$ to $[q_N]$ constitute the basis of the noise sub-space. λ_{M+1} to λ_N are equal to noise power. Consequently, the number of sources M can be estimated, in principle, from the distribution of λ_1 to λ_N.

The noise and signal sub-spaces are orthogonal. For localizing the sources, we use the fact that each $[e(\boldsymbol{k}_n)]$ $(\forall n \in [1, M])$ is orthogonal to vector basis of the noise sub-space, i.e.,

$$[q_l]^{\dagger} \cdot [e(\boldsymbol{k}_n)] = 0, \quad \forall n \in [1, M], \quad \forall l \in [M+1, N] \tag{7.129}$$

is verified. For the MUSIC algorithm, we apply

$$B_{\mathrm{M}}(\boldsymbol{k}) = \left(\sum_{l=M+1}^{N} \left| [q_l]^{\dagger} \cdot [e(\boldsymbol{k})] \right|^2 \right)^{-1} = \left([e(\boldsymbol{k})]^{\dagger} \cdot [Q] \cdot [Q]^{\dagger} \cdot [e(\boldsymbol{k})] \right)^{-1}, \tag{7.130}$$

where $[Q]$ is a $(N \times (N-M))$ matrix formed by the eigenvector $[q_{M+1}]$ to $[q_N]$ of the noise sub-space.

Maxima of (7.130) provide the spatial location of the discrete sources. If (1) the number of (discrete) sources is known, (2) the noise output is equal to 1, and (3) SNR tends to infinity, it can be shown that the brightness distribution provided by Capon's method tends to the one provided by the MUSIC algorithm (Nickel 1988). However, the MUSIC algorithm and, more generally, approaches based on the SVD methods, fail to image distributed or correlated sources. For this reason, Capon's method is the most popular method for imaging with atmospheric radars.

Observation results using radar imaging techniques applied on various atmospheric radars are shown in Sect. 12.5.1.

Chapter 8
Overview of Radar

8.1 Brief Discussion on Two Types of Radar

In Chap. 4, we discussed coherent-type radars whose transmitted pulse phase is so stable that phase information of successively received pulses can be precisely utilized. There are two groups in coherent radars regarding the modulation system of transmitted signals, i.e., frequency-modulated continuous waves (FMCW) radars and pulse radars mentioned in Sect. 3.1.1.

8.1.1 FMCW Radar

It is a radar which changes transmitted frequency linearly in a fixed period. When a scattered signal is received, the frequency of transmitted signal is already changed, thus the distance to the scatterer can be estimated by counting the frequency difference between transmitted and received signals, i.e., the beat frequency. FMCW radars are operated with lower peak transmitted power compared with pulse radars, and with the higher range resolution.

An FMCW radar changes the transmitted frequency following a certain function in a constant time interval T_{M}. If the transmitted frequency is linearly changed from f_0 to $f_0 + F$ as shown in the thick line in the upper panel of Fig. 8.1, the returned echo from the scatterer at the distance r is received with a delay time of $\Delta t = 2r/c$, and mixed with the transmission signal. The beats for the received echo appears at the frequency $f_{\mathrm{r}} = F\Delta t/T_{\mathrm{M}}$ as shown in the thick line of the lower panel of the same figure. Thus, the distance from radar to scatterer can be known by

$$r = \frac{T_{\mathrm{M}} c f_{\mathrm{r}}}{2F}. \tag{8.1}$$

For observations of weather and the atmosphere, the typical magnitudes of F and T_{M} are around 10 MHz and 10^{-3} s, respectively.

S. Fukao and K. Hamazu, *Radar for Meteorological and Atmospheric Observations*,
DOI 10.1007/978-4-431-54334-3_8, © Springer Japan 2014

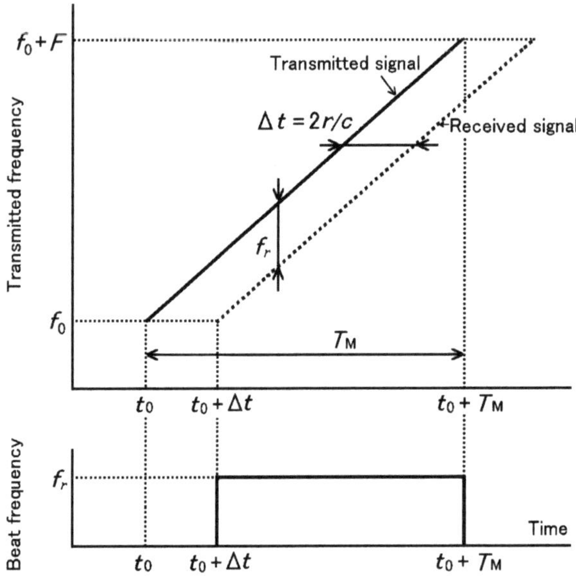

Fig. 8.1 Frequency-time relationship in FMCW radar

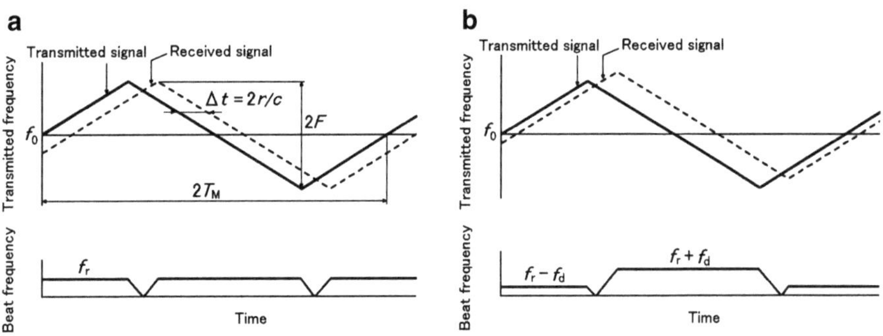

Fig. 8.2 Doppler effect in FMCW radar for (**a**) stationary scatterer and (**b**) moving scatterer

Doppler Frequency

In operational radars, the transmitted frequency is usually changed in a triangular waveform as shown in Fig. 8.2 to avoid modulation frequency from becoming discontinuous. In the case of stationary scatterer, the received echo appears as the broken line in the upper panel of the same figure (a), and the beats appear as shown in the lower panel of the same figure. On the other hand, when the scatterer moves, the received echo appears as the broken line of the upper panel in Fig. 8.2b, and the beats appear as shown in the lower panel of the same figure. If the scatterer is approaching the radar and the transmitted frequency increases, the beat frequency $f_{b(up)}$ becomes the magnitude that the frequency change f_r for the stationary scatterer is reduced by the Doppler shift f_d. On the other hand, when the transmitted frequency decreases, the beat frequency $f_{b(down)}$ becomes the sum of f_r and f_d. Thus,

$$f_{b(up)} = f_r - f_d, \tag{8.2}$$

$$f_{b(down)} = f_r + f_d, \tag{8.3}$$

where f_d is the same value with the Doppler frequency which was discussed in Sect. 4.1.1. From (8.2) and (8.3), the frequency change which is related to distance to the scatterer f_r becomes

$$f_r = \frac{f_{b(up)} + f_{b(down)}}{2}. \tag{8.4}$$

In the same way, Doppler frequency f_d is given by

$$f_d = \frac{f_{b(down)} - f_{b(up)}}{2}. \tag{8.5}$$

Points to be Remarked in Actual Observations

Both the frequency change f_r and the Doppler shift f_d are contained in the beat frequency of the FMCW radar output. Thus, it is necessary to estimate the influence of the Doppler shift which would make worse the accuracy of range measurement. However, in general, the Doppler shift caused by weather or the atmosphere scatterer is small compared with f_r, and can be ignored. For example, if the radial velocity of a scatterer is observed as $10\,\mathrm{m\,s^{-1}}$ with the radar of frequency modulation bandwidth $F = 10\,\mathrm{MHz}$, the time interval $T_M = 2.5\,\mathrm{ms}$, and the radar wavelength $\lambda = 10^{-1}\,\mathrm{m}$, f_d becomes 200 Hz from (4.11). The distance change (or error) Δr caused by the f_d becomes $\Delta r = 7.5\,\mathrm{m}$ from (8.1). This value is small enough to be ignored in ranging scatterers.

If more than one scatterer exist in the same observation range, multiple signals with different frequencies whose number is equal to the number of scatterers are obtained. Therefore, it is necessary to separate these frequencies by narrow filter banks and so on to determine each frequency, i.e., distance to each scatterer. For an FMCW radar, transmitted power is radiated via the antenna without interruption, hence the sensitivity of receiver is deteriorated by the leakage of transmitted power which overlaps received signal. Therefore, the transmitting antenna and receiving antennas are often separated from each other so as to isolate the received signal from the leakage of transmitted power.

One of the advantages of the FMCW radar is its high range resolution. Range resolution of pulse radars is generally determined by the transmitted pulse width, whereas it is determined by the value of f_r in the FMCW radar. As f_r is determined by the frequency modulation band width F, the range resolution becomes precise if

Table 8.1 Typical meteorological radars and atmospheric radars

Band	Typical radar	Frequency	Application
Atmospheric radar			
VHF	MST radar	50 MHz	Middle and upper atmosphere
	ST radar	50 MHz	Stratosphere and troposphere
UHF	ST radar	400 MHz	Stratosphere and troposphere
	BL radar	900 MHz	Boundary layer
L	Lower troposphere radar	1.35 GHz	Lower troposphere and boundary layer
Meteorological radar			
S	Weather radar	2.7–3.0 GHz	Wide area weather
C	Weather radar	5.3 GHz, 5.6 GHz	Middle area weather
X	Weather radar	9.5 GHz	Local area weather
Ku	TRMM precipitation radar	13.8 GHz	Satellite-borne, precipitation
Ka	Cloud and fog radar	35 GHz	Cloud and fog
W	Cloud radar	95 GHz	Cloud

F is expanded wider. If the scatterer is single, the range resolution of FMCW radars generally becomes about one digit higher than that of pulse radars. Furthermore, because the FMCW radar uses a continuous wave as the transmitted waveform, the signal to noise ratio (SNR) of the desired level can be realized by smaller transmitted peak power than that of pulse radar which requires high voltage modulation circuitry. Thus, FMCW radars are used in collision preventive radars installed on automobiles, altimeters on board aircraft and satellite, and so on.

8.1.2 Pulse Radar

It is the radar which successively radiates short transmitted pulses separated by long intervals during which echoes are received. The distance to a scatterer can be known by measuring the time between the pulse transmission and the reception of the echo. Radars that measure the phases of the transmitted radio frequency (rf) pulse and received echoes are called coherent pulse radars, whereas radars which cannot are called non-coherent pulse radars. There are two groups of coherent pulse radars: one does not perform phase modulation to pulse, and the other does. The phase modulation is widely used by the radars which implement pulse compression techniques as will be discussed in Sect. 8.3.4.

The basic configuration of a pulse radar is shown in Fig. 3.1. The typical transmitted frequency bands of meteorological radar and atmospheric radar are shown in Table 8.1. The transmitter and the receiver of a coherent pulse radar, which was mentioned to explain the principle of Doppler radar in Sect. 4.1.1, will be discussed again in detail in Sect. 8.3. A pulse radar intermittently radiates rf pulses of constant width and typically with constant spacing between pulses.

As shown in Fig. 5.6, the carrier signal is modulated with fixed pulse duration and (pulse) interval, and the transmitted pulses are radiated via the antenna which has directive power density gain. Generally, the transmission and the reception of radar signals are performed with a common antenna. The path control of the transmitted pulse and received signals are performed with the duplexer (sometimes called a Transmit/Receive switch or, for short, a T/R switch). The duplexer for small peak transmitted power is usually a semiconductor diode, whereas for high power switching it is generally composed of a waveguide circulator. The circulator[1] connects transmitter, antenna, and receiver, which lets the transmitted pulse pass from transmitter to antenna, and lets the received signal pass from antenna to receiver. In the receiver, the received signal is amplified. Various (sometimes very sophisticated) signal processing is performed to obtain desired signals.

Pulse radars usually consist of the master oscillator and power amplifier (MOPA) system, which can control the phase of the transmitted pulse produced by modulating an rf amplified source signal as will be discussed in Sect. 8.3. Digitalization of the received signal is generally performed to convert the IF (intermediate frequency) signal into a digital signal in the stage of phase detection.

In the remaining sections of this chapter, attention will be focused on the pulse radar and its signal processing methods.

8.1.3 Echo Power

Echo power is obtained by measuring receiver output power[2] and then it is referred to the input to obtain the echo power at the input to the receiver. Echo power from a scatterer at a certain distance can be calculated by the radar equation which was discussed in Sects. 3.3 and 3.4. Echo power from isolated scatterer such as aircraft is obtained by (3.8), and it is inversely proportional to the 4th power of the distance between radar and scatterer. On the other hand, echo power from discretely distributed scatterers such as precipitation, clouds and fog are in inversely proportional to the 2nd power of the distance and given by (3.40). If scattering occurs from continuously distributed scatterers as the Bragg scatters, the echo power is obtained by (3.40) and (3.116). In the case of partial reflection, it is

[1]A circulator blocks leakage of the transmitted pulse from entering into the receiver. In practice, around 40 dB suppression of the leaking power is feasible. A TR limiter is typically also inserted in front of the receiver to prevent the receiver from breaking down by the leakage of transmitted power.

[2]Actually, it is often obtained as the 0th moment of power spectrum as previously discussed in (5.98).

obtained by (3.117). The echo power from these distributed scatterers or from partial reflection scatterers is also in inversely proportional to the 2nd power of the distance between radar and scatterers.

Assuming that the echo power from scatterers obtained by actual observation is identical to the result of the above mentioned radar equation, physical parameters which contribute to the echo power is determined. These physical parameters include the radar cross section σ for isolated scatterer, radar reflectivity factor Z, radar reflectivity η for distributed scatterers, and reflection coefficient $|\rho|$ for partial reflection scatters. To estimate these physical parameters, it is necessary to have an absolute calibration of signal power both from transmitter to antenna and from antenna to the receiver. It is noted, however, as will be mentioned in Sect. 8.2.5, the absolute calibration for large scale atmospheric radar is not always feasible because of its system configuration.

Because electromagnetic waves are attenuated when propagating through the atmosphere, it is necessary when using the radar equation to take the attenuation into consideration in the calculation of echo power. Attenuation includes both absorption and scattering as discussed in Sects. 3.3.3 and 3.3.4. The attenuation is mainly caused by oxygen, water vapor, cloud, and precipitation, as discussed in Sects. 6.3.2 and 6.3.3. Note that the values of specific attenuation estimated in these sections are one way attenuations per unit distance, and the total loss in dB is the product of the specific attenuation and the round trip distance. Thus the received echo power in dB is reduced by the total atmospheric loss factor in dB.

8.1.4 Scanning Methods

The meteorological phenomena like precipitation and clouds which are observed with meteorological radar, and atmospheric turbulence and various atmospheric phenomena which are studied by atmospheric radar, have three dimensional structure. Therefore, it is necessary to scan the radar beam to receive scattering signals in three-dimensional space in order to understand structure and dynamics of scatterers. The frequency bands which are applied in meteorological radar and atmospheric radar differ from each other because of the difference of the scattering mechanism of precipitation particles and refractive index perturbations. Thus, radar characteristics such as transmitted power, beamwidth, and the antenna's physical size differ for meteorological and atmospheric radars. Furthermore, there is a significant difference in SNR for echoes received by meteorological and atmospheric radars, and thus data acquisition time of these radars differs much. Because of the above differences in observations, there are differences in the methods of beam scanning and the volume of space being scanned by the beams when using meteorological and atmospheric radars.

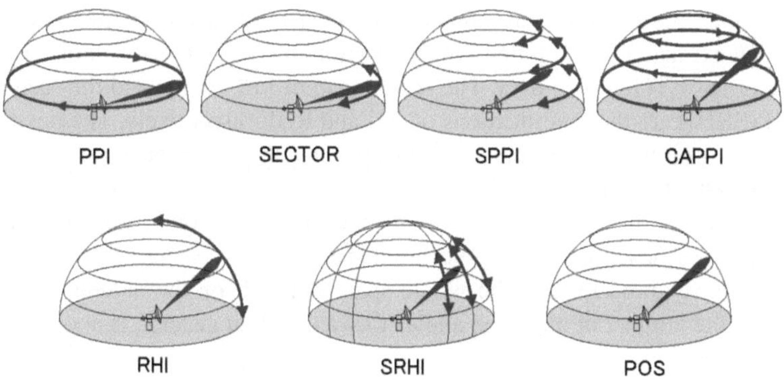

Fig. 8.3 Antenna scan mode and scan area in the space

Scanning with Meteorological Radar

The antenna diameter of ordinary meteorological radar is one to several meters, and accordingly its radar beamwidth is around 0.3–2°. Thus, as for meteorological radars, it is possible to scan the whole three dimensional space considering its physical size and mechanical configuration. In addition, the beamwidth is narrow enough to suppress the interference from ground and sea scatterers (clutter) when the beam is directed at low elevation angles. Furthermore, there are following reasons that meteorological radars do not perform coherent integration (as for coherent integration, see Sect. 5.6.2). Atmospheric radars are mostly pointed in the vertical direction or near vertical and Doppler velocities are relatively small, whereas the mean Doppler velocities measured with meteorological radars easily span the entire unambiguous velocity interval. Moreover, the weather signal spectrum typically occupies an appreciable portion of the unambiguous velocity interval (i.e., $-v_N$ to $+v_N$). Thus, in general the coherent time of meteorological radar echoes are relatively short,[3] and coherent integration would not only reduce noise power but also signal power.

Because meteorological radar antenna sizes are relatively small, these radars generally scan their antennas mechanically, and their beams are pointed to specific directions or areas in the space as shown in Fig. 8.3. The most basic scan method is the volume scan, which rotates antenna around azimuthal direction continuously with changing elevation angle in sequence. A polar coordinates indication of data in azimuth and range at a specific elevation angle is called the plan position indicator (PPI). On the other hand, data extraction and indication at a constant altitude from the observation data of each elevation scan is called the constant altitude PPI (CAPPI). The vertical cross sectional view extracted from the three dimensional

[3]As discussed in Sect. 5.6.1, the correlation time is in inverse proportion to the spectrum width.

data is called the range height indicator (RHI). The radar observation data which have been extracted from limited ranges in azimuthal and elevation angles is called the sector (SECTOR) scan data. The sector scan of PPI (SPPI) and the sector scan of RHI (SRHI) are the modifications of PPI and RHI, respectively. The positioning data or the pointing (POS) data are obtained by the observation in the specific direction. The data of PPI, RHI, and SECTOR scan can be obtained not only by volume scan but also by scanning in specific elevation angle, specific azimuthal angle, and specific angle range of elevation and azimuth, respectively.

The time required to scan the whole space depends on the antenna revolution speed and the number of elevation angles to scan, and, in general, is about 3–5 min.

Scanning with Atmospheric Radar

Transmitted frequency of atmospheric radar is around 50 MHz to 1 GHz.[4] To realize the radar beamwidth narrower than several degrees, the apertures are required to be several meters for 1-GHz radars and around 100 m for 50 MHz radars. Thus, it is not easy for the antenna of atmospheric radar to mechanically scan in three dimensions from physical and mechanical aspects. Furthermore, as the radar beamwidth of atmospheric radar is wider compared with that of meteorological radar, influence by ground and sea clutter cannot always be avoided when the beam is scanned at low elevation angles near the ground surface.

Generally, scattered signals received by meteorological radars are considerably stronger than the noise level, whereas that of atmospheric radars are often weaker than the noise level. Therefore, for the atmosphere radar, detection of received signals by coherent integration becomes indispensable. It takes approximately for 10 s to about 1 min for atmospheric radar to observe in one direction, and thus, it is restricted by the observation time to scan the whole space.

Because of the above reasons, atmospheric radar observations are generally conducted changing the direction of radar beam by electronic means to observe the space over the radar. For the most basic method, the radar beam is directed to the zenith, then sequentially directed to the north, south, east, and west at fixed zenith angles to a maximum zenith angle of about 10–15°. The electronic scanning of the radar beam is performed by a phased array antenna, which controls the phases of the wave transmitted from each element of the array. The phased array method has many advantages, especially, it is possible to change the radar beam direction instantly, and the space can be observed with more than one beam at the same time.

[4]Considering the portability, the higher frequency of 3.05 GHz is sometimes utilized. Refer to Sect. 10.4.1 regarding this point.

8.2 Radar Antenna

The antenna is the instrument which efficiently radiates high power electromagnetic waves generated by the transmitter, and efficiently receives the echoes from scatterers.

The shape and direction of the beam containing the radiated electromagnetic field is determined by the electromagnetic field distribution in the antenna aperture. It is the vector sum of electromagnetic fields that correspond to the individual electric currents which flow on each antenna element. Parabolic reflector antennas and planar array antennas which will be discussed in Sects. 8.2.2 and 8.2.4, respectively can radiate intense rf power density by concentrating the radiation in a narrow beam. Figure 8.4 shows the theoretically computed pattern (the angular distribution of calculated rf power density) of the MU radar antenna (see Fig. 10.4); the area which rf power density is concentrated is the main lobe and surrounding small peaks are sidelobes.

8.2.1 Radar Antenna Parameters

The same antenna is generally used for transmission and reception of radar signals. Hereafter, we assume that the antenna is applied both for transmission and reception, and discuss mainly the parabolic antenna which is generally used for meteorological radar and a planar array antenna which is often used for atmospheric radar.

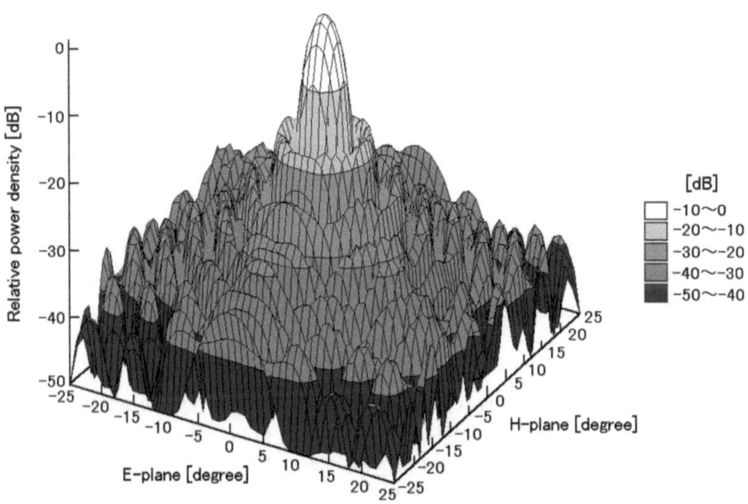

Fig. 8.4 The angular distribution of calculated rf power density for the MU radar antenna, which is shown in Fig. 10.4

Near and Far Fields

The area which is the nearest to the antenna aperture is called near field, and the adjoining area is called the Fresnel region, beyond which is called the Fraunhofer region or far field.[5] The concentration of rf power density, i.e., radar beam is gradually made as electromagnetic wave leaves the antenna aperture, and the final conically shaped beam is formed in the far field. In the near field the power density is quasi uniform with range because the beam has a relatively fixed cross section with range, whereas in the far field power density decreases with range because this is where the beam has a fixed angular width, but a cross section increasing with range. The distance to the boundary between the Fresnel region and the far field r_F is related to the antenna diameter D_a and radar wavelength λ as $r_{F1} = D_a^2/\lambda$ or $r_{F2} = 2D_a^2/\lambda$. For example, r_{F2} is approximately 600 m for a meteorological radar transmitting at a frequency of 5.6 GHz and using an antenna aperture of 4 m diameter. The accepted values of the minimum distance at which the far field rf power density distribution pattern (radiation pattern) measurements may properly be made varies between r_{F1} and r_{F2}. The ratios of antenna gains at the distance r_{F1} and r_{F2} to those at the infinite distance are 0.94 and 0.99, respectively. There is a little difference between them; but the r_{F2} criterion is to be favored (Silver 1949, p. 199). However, if space limitations do not allow measurements at ranges r_{F2} or larger, then a distance of r_{F1} may be tolerated (Silver 1949, p. 575).

Antenna Gain and Beamwidth

Antenna gain gives the relative level of electromagnetic power density in a specific direction, typically along the beam axis.

As discussed in Sects. 2.1.4 and 3.2.1 directivity g_D is a function of the effective antenna aperture A_e and radar wavelength, and given as shown in (2.83) by (Skolnik, 1970, p. 9.5)

$$g_D = \frac{4\pi\eta_a A}{\lambda^2} = \frac{4\pi A_e}{\lambda^2}, \tag{8.6}$$

where the physical antenna aperture A by $A_e = \eta_a A$, where η_a is called the antenna aperture efficiency.[6] The effective aperture area of the antenna represents a smaller, uniformly illuminated aperture having the same gain as that of actual, nonuniformly

[5] Antenna engineers usual call this region the far field, whereas optical physicists call it the Fraunhofer region (Skolnik 2001, p. 545).

[6] The antenna aperture efficiency depends on the nature of the current illumination across the antenna aperture. If the illumination is uniform, the efficiency becomes maximum, i.e., $\eta_a = 1$, but the radiation pattern has a relatively high peak-sidelobe level. On the other hand, if the illumination is the maximum at the center of the aperture and tapers off in amplitude towards the edges, the peak-sidelobe level becomes lower but the antenna is less efficient than if it were to be uniformly illuminated (Skolnik 2001, p. 544).

illuminated aperture, i.e., $\eta_a < 1$ (Skolnik 1990, p. 6.4). A crude approximation to the directivity for the definition in (2.77) is expressed as (Skolnik, 1970, p. 9.3)

$$g_D = \frac{4\pi}{\vartheta_B \varphi_B}, \tag{8.7}$$

where ϑ_B and φ_B are radar's one-way half-power beamwidths in radians in the E- and H-planes. On the other hand, as discussed in Sect. 3.3.2, the radiation pattern (i.e., beam shape) is well represented by a Gaussian function.[7] Then with beamwidths radians, g_D is given by $g_D = 16\ln 2/(\vartheta_B \varphi_B)$ (Doviak and Zrnić 2006, p. 478). The difference this equation and (8.7) is due to the difference in modeling the beam shape. If the beamwidths are given in degrees, (8.7) is given by (Skolnik, 1970, p. 9.5)

$$g_D \simeq \frac{4100}{\vartheta_B \varphi_B}. \tag{8.8}$$

For practical antenna, the total radiated power is less than the power applied to the antenna port because of losses. The ratio of the power radiated to the power applied to the antenna port is the efficiency of the antenna, $\eta_l (\leq 1)$, thus the antenna power gain (or simply gain) g is related with directivity by

$$g = \eta_l g_D. \tag{8.9}$$

Skolnik (1970, p. 9.5) gives typically as

$$g \simeq \frac{32000}{\vartheta_B \varphi_B}, \tag{8.10}$$

where the beamwidths are in degrees. If ϑ_B and φ_B are $1°$, g in (8.10) is approximately 3.2×10^4 ($\simeq 45\,\text{dB}$). It should be noticed that the coefficients in the numerator of (8.10) depends on the efficiency of the antenna η_l. The relation $g \simeq 26000/(\vartheta_B \varphi_B)$ is also recommended as an excellent approximation for general use (Stutzman 1998; Skolnik 2001, p. 541).

The angular width at which electromagnetic power density of antenna radiation pattern is reduced to the half of the peak value, i.e., 3 dB down from the maximum value, is called a half-power beamwidth, and generally the beamwidth is defined by this value. In the case of a symmetrically illuminated circular aperture, the antenna beamwidths in the E- and H-planes become equal, and are functions of antenna diameter D_a and radar wavelength λ. The radar beamwidth in radians is approximately given by (Silver 1949, p. 195; Doviak and Zrnić 2006, p. 34)

$$\vartheta_B = \varphi_B = \frac{1.27\lambda}{D_a} \qquad [\text{rad}]. \tag{8.11}$$

[7] The radar equation (3.40) is also derived analytically on the assumption that the radiation pattern is Gaussian shaped.

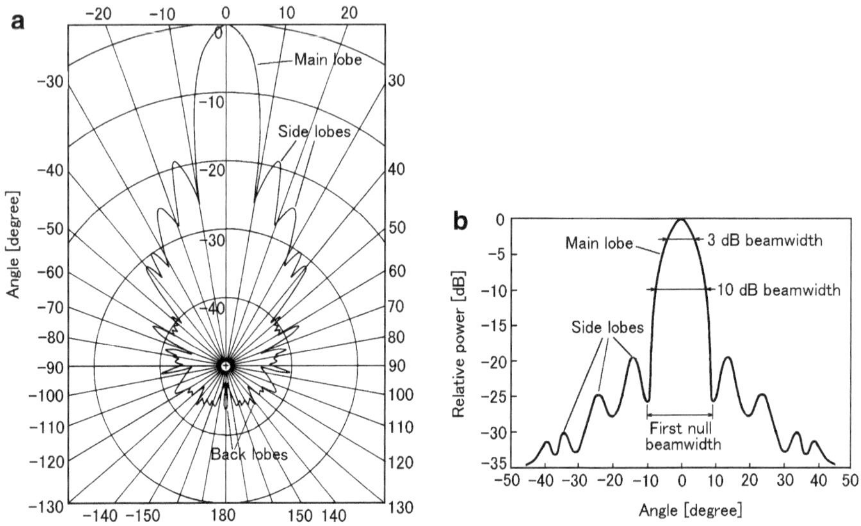

Fig. 8.5 Radiation pattern and beamwidth of a circular aperture antenna. (**a**) Polar coordinate system and (**b**) rectangular coordinate system [from Ulaby et al. 1981, p. 98]

This value strongly depends on η_a. For η_a equal to 1.0, 0.75, and 0.56, the coefficients in the numerator of (8.11) become 1.02, 1.27, and 1.47, respectively (Silver 1949, p. 195).

The beamwidths at the levels where the rf power density reduced by 3 or 10 dB from the peak value are called the 3-dB(half-power) or 10-dB beamwidths, respectively. For a circular aperture antenna, these relations are illustrated in the polar coordinate system (a) and the rectangular coordinate system (b) in Fig. 8.5.

8.2.2 Parabolic Antenna

One of the techniques to realize the antenna characteristics of high antenna gain and sharply narrowed beam is to utilize the similar reflectors as those in optical telescopes. A parabolic antenna is widely adopted for meteorological radar. Electromagnetic waves which are radiated from the feed horn installed at the focal point of the parabola, as shown in Fig. 8.6a, are transformed by the parabolic surface into a plane wave, and then radiated propagating along the beam axis. It should be noted that in the range beyond r_{F2} (i.e., in the far field), the phase fronts of the radiated waves are spherical, not planar. The same figure (b) shows the parabolic antenna with 4 m in diameter which is actually used with a dual polarization radar.

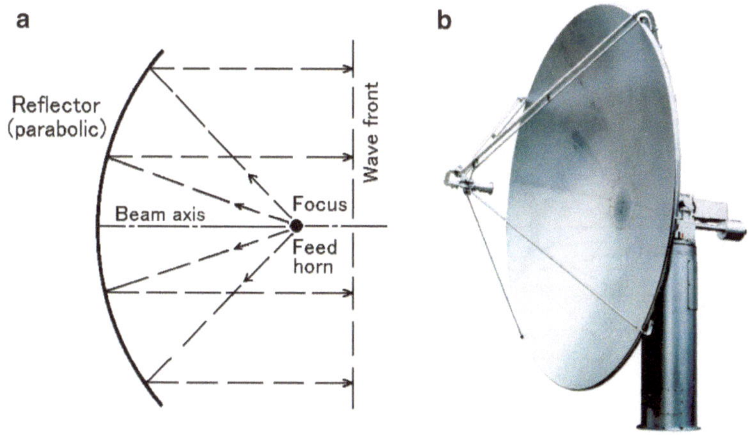

Fig. 8.6 Fundamental parabolic antenna. (**a**) Radiation path and (**b**) an actual parabolic antenna, having a 4 m diameter aperture, used with a polarimetric radar (note the pair of waveguides attached to the spars supporting the feed horn; one guide contains the power radiated with H polarization and the other with V polarization [courtesy of Mitsubishi Electric Corporation]

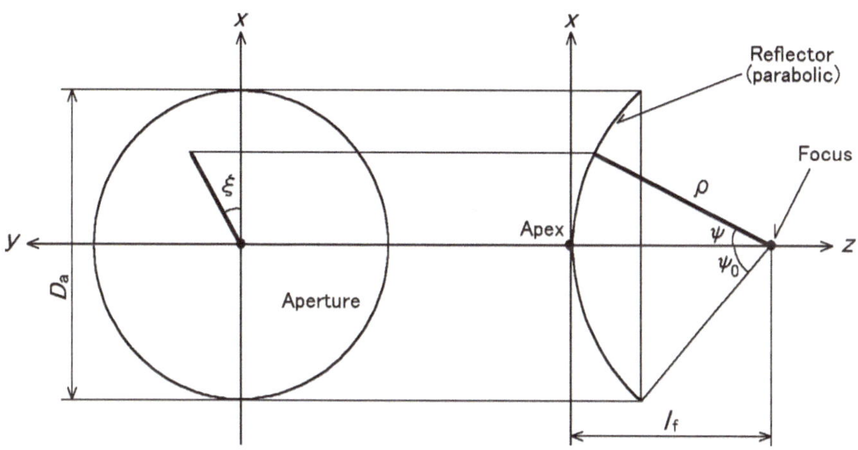

Fig. 8.7 Coordinate system of a parabolic antenna

In the rectangular coordinate system (x, y, z) of Fig. 8.7 where the origin is located at the apex of the paraboloid, the focal length l_f is related to the paraboloid as follows (Skolnik 1990, p. 6.12):

$$z = \frac{x^2 + y^2}{4l_f}. \tag{8.12}$$

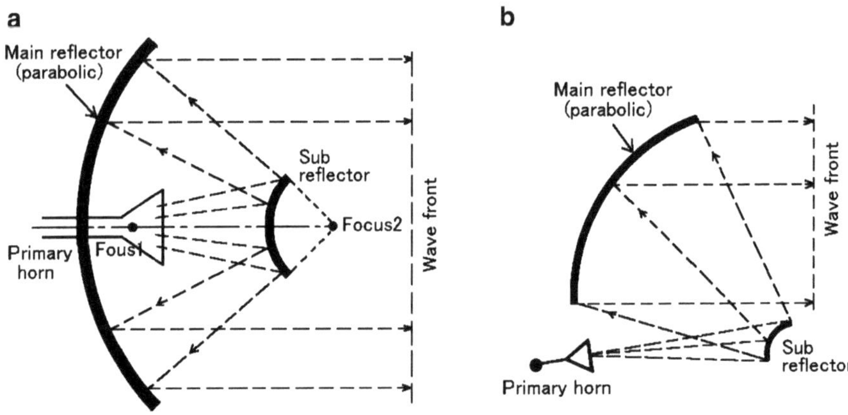

Fig. 8.8 Reflector types of (**a**) Cassegrain antenna and (**b**) offset Cassegrain antenna [modified from Skolnik 1990, p. 6.25]

Whereas in the polar coordinate system (ρ, ψ, ξ) where the origin is the focal point, the above relation is expressed as[8]

$$\rho = \frac{2l_f}{1+\cos\psi} = l_f \sec^2\frac{\psi}{2}. \tag{8.13}$$

$$\tan\frac{\psi_0}{2} = \frac{D_a}{4l_f}. \tag{8.14}$$

The aperture angle, i.e., the angle to see the edge of aperture from the focal point, is given by $2\psi_0$. Between the aperture angle, l_f, and antenna diameter D_a, the following relationship is held. The ratio l_f/D_a of most reflectors are chosen to have a focus length between 0.25 and 0.5 times the diameter (Skolnik 1990, p. 6.12). In the case of parabolic antenna of Fig. 8.6b, the ratio l_f/D_a is approximately 0.4.

The Cassegrain antenna is an application of the parabolic antenna. It is composed of the main parabolic reflector which is rotationally symmetric with respect to the main axis, the hyperboloid which is sub-reflector also rotationally symmetric, and a primary horn as illustrated in Fig. 8.8a. It has a variety of advantages. One of them is that it is possible to extend the focal length, and thus curvature of the main reflector

---

[8]For the derivation of (8.13), the following relationship between the two coordinate systems

$$x = \rho\sin\psi\cos\xi,\quad y = \rho\sin\psi\sin\xi,\quad z = l_f - \rho\cos\psi$$

are applied. Substituting these equations into (8.12), the following equation is obtained;

$$l_f - \rho\cos\psi = \frac{\rho^2\sin^2\psi}{4l_f}.$$

Transforming this equation, the relation (8.13) can be obtained.

can be significantly reduced which in turn reduces cross-polar radiation. The other advantage is that the primary horn can be installed near the main reflector, and thus waveguide can be shortened and waveguide loss is reduced. On the other hand, because the sub-reflector is placed along the beam axis, a part of power reflected from the main reflector is blocked (this phenomenon is called blocking) and results in a reduction of antenna power gain and an increase of the level of sidelobes. Therefore, the Cassegrain antenna is generally applied in the condition that the influence of blocking is comparatively small as the aperture of the main reflector is about 80–100 times or more larger than the radar wavelength, and the radar beamwidth is narrower than about 1°. When using the offset Cassegrain antenna which is illustrated in Fig. 8.8b, the sub-reflector can be made offset from the front of the main reflector so that the blocking can be reduced.

The CSU-CHILL radar, in Greeley, CO, the U.S.A, is a dual-polarized, dual-wavelength (S- and X-band) radar of 8.5 m reflector. The beamwidths of the S- and X-band radars are 1° and 0.3°, respectively. The reflector was at first a center-fed parabolic one, and is recently replaced with a dual-offset Gregorian parabolic reflector. The Gregorian antenna is similar to the Cassegrain antenna except that the sub-reflector is concave in shape. The antenna aperture efficiency of the Cassegrain type is in the order of 65–70 %, whereas that over 70 % can be achieved with the Gregorian design. The antenna performance is greatly improved, e.g., the ground clutter is greatly reduced because of rapid falloff in the sidelobe levels with increasing elevation angle, and the linear depolarization ratio (LDR) system limit is shown to be −40 dB or better[9] (Bringi et al. 2011).

8.2.3 Radome

Scanning of meteorological radar antennas is made by, in general, mechanically steering, as discussed in Sect. 8.1.4. Thus, the radar antenna is usually covered with radome (a compound word of radar dome), which protects the antenna reflector and mechanical elements from heavy wind and rain to keep scanning of the antenna stable. On the other hand, the antennas for atmospheric radar usually do not have mechanically steered beams, but beams that are scanned electronically. Furthermore, atmospheric radar antennas, many of which are significantly larger than meteorological radar antennas, are often installed near the ground. Thus, most atmospheric radars do not require a radome. However, some small atmospheric radar are housed in a radome, these are typically those radar sited in heavy snowfall areas so as to avoid snowfall on the antenna surface.

In general, the radome has a spherical structure which connects panels of polyurethane by pinching together with fiber-reinforced plastic (FRP). Radomes typically have a power transmissivity of 90 % or more, and that the beam deflect

[9]In general, the value of well-designed radars is around −35 dB or better.

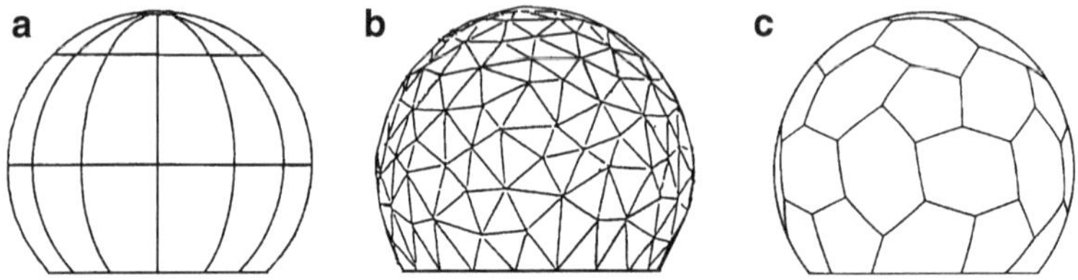

Fig. 8.9 Various types of radome. (**a**) Solid laminate, (**b**) metal space frame, and (**c**) sandwich

angle, the angle which the main beam deflects when penetrating the radome, is smaller than 0.1°. Representative types of spherical radome are solid laminate, metal space frame, and sandwich type as illustrated in Fig. 8.9.

Solid Laminate Radome

Vertically divided segments of solid fibergrass are used to construct a spherical radome. Typically, the radome of about 7 m in diameter encloses a parabolic antenna with an aperture diameter of 3–4 m. Because of its structure, the connecting lines are made necessarily straight in both vertical and horizontal directions. The difference in the vertical and horizontal structure causes the attenuation to differ for vertically and horizontally polarized waves. Thus in general, the radome of this type is not used for polarimetric radars but used for single polarized wave radars.

Metal Space Frame Radome

Metal rods are connected to construct triangular frames of various sizes and shapes that compose the spherical dome, and panels are stuck to the frames. It has good electrical characteristics over a wide frequency band from 0.5 to 100 GHz, and the scattering characteristic becomes approximately homogeneous for both horizontally and vertically polarized plane waves. Thus, it is possible to use this type of radome for dual polarization radar.

Sandwich of Laminates

Multi-layer (e.g., 3-layer) panels of hexagonal shape and of various sizes and orientation are connected together to construct spherical radome. Impedance matching is often employed to decrease the insertion loss to a particular frequency band. The radome of this type is generally applied for radars of 10 or more GHz. The scattering characteristic becomes almost homogeneous for horizontally and vertically polarized waves, and the electrical characteristics are the best among the

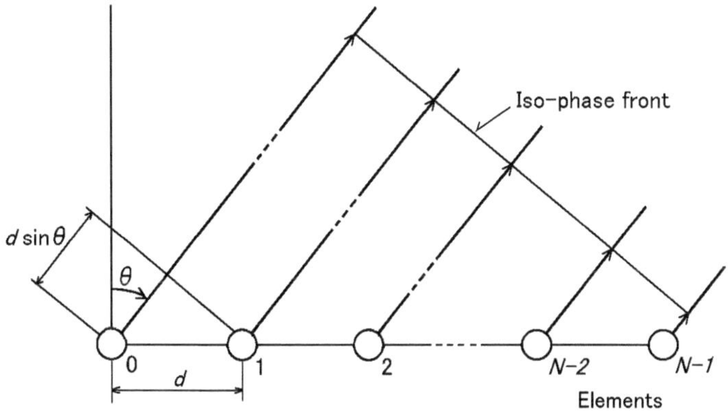

Fig. 8.10 Linear array antenna which arranges N pieces of half-wave length dipole (element) in alignment

three types of radome. For the airport weather Doppler radars in Japan (5.3-GHz), a sandwich type radome with 11 m diameter is applied for the parabolic antenna of 7 m in diameter.

8.2.4 Array Antenna

An array antenna consists of multiple radiating elements regularly or sometimes irregularly arranged. If the phase of the electromagnetic wave supplied to each individual antenna elements is controlled, the antenna is called a phased array antenna. A phased array antenna can change the direction of the beam electrically, and thus it is possible to scan the radar beam with a stationary antenna. It is common to use the phased array antenna for atmospheric radars.

Linear Array Antenna

Suppose a linear array antenna with N half wavelength dipoles (elements) arranged in a straight line as shown in Fig. 8.10. If the distribution of electric current on each element is identical, the radiation pattern of each element (element pattern) is expressed by the same function.[10] Therefore, the radiation pattern of array antenna can be given by adding phase difference corresponding to antenna element positions to the current distribution on each element. In the same figure, if the phase difference

[10]Strictly speaking, the current distribution on each element is not identical because cross-couplings between elements differ according to positions of them. However, when the overall pattern of array antenna is considered, the difference of this current distribution can be ignored.

between the successive elements in the Fraunhofer region is δ, and the electric current which flows on the mth element is I_m, the radiation pattern of linear array antenna $F(\theta)$ can be expressed as

$$F(\theta) = g(\theta) \sum_{m=0}^{N-1} I_m e^{jm\delta}, \tag{8.15}$$

where θ is the offset angle from the broadside, which is perpendicular to the base line of linear array antenna, $g(\theta)$ is the element pattern, and $m\delta$ is the phase of electric current which flows on the mth element. The phase difference between the successive elements δ is given by

$$\delta = kd \sin\theta, \tag{8.16}$$

where d is the distance between successive elements, and $k = 2\pi/\lambda$ is the wave number (λ : wavelength). From (8.15), the radiation pattern $E_a(\theta)$ without the element pattern is given by

$$E_a(\theta) = \sum_{m=0}^{N-1} I_m e^{jm\delta}. \tag{8.17}$$

This pattern is called the array factor. If $I_m = 1$, the right-hand side of (8.17) is given by the sum of the geometric series with the first term 1, with successive term ratio $e^{j\delta}$, and term number N. Thus, $E_a(\theta)$ becomes

$$E_a(\theta) = e^{j\delta(N-1)/2} \frac{\sin\left(\frac{N\delta}{2}\right)}{\sin\left(\frac{\delta}{2}\right)}. \tag{8.18}$$

Hereafter, we assume, for simplicity, the array factor excluding phase term $e^{j\delta(N-1)/2}$. If $N = 20$, the array factor can be shown as Fig. 8.11. It is clear from the same figure that the antenna radiation pattern becomes a periodic function of δ in the linear array antenna whose elements are arranged in the regular interval. In (8.16), $|\sin\theta| \leq 1$, and accordingly δ varies as follows:

$$-kd \leq \delta \leq kd \quad \text{that is} \quad -\frac{2\pi d}{\lambda} \leq \delta \leq \frac{2\pi d}{\lambda}. \tag{8.19}$$

If the distance between the adjoining elements is half wavelength, the range of $|\delta| \leq \pi$ corresponds to the real space $|\sin\theta|$ as shown in Fig. 8.11, and the radiation pattern outside this range does not appear in the real space. The radiation whose amplitude becomes the maximum at $\delta = 0$ is the main lobe, and other radiations with smaller amplitude are sidelobes. The region where the radiation pattern appears in space is called the visible range, and the outside of the region is called the invisible range. In the region where $\delta \geq 2\pi$, i.e., $d \sin\theta \geq \lambda$, a large radiation which is generated in the direction $\delta = \pm 2s\pi$(s: integer) enters the visible range.

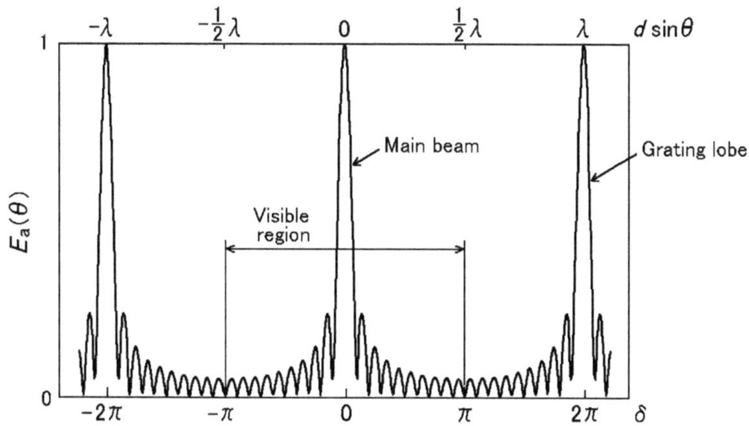

Fig. 8.11 Array factor 20 elements linear array antenna. Distance between elements is half-wave length

This large radiation is called grating lobe. The grating lobe for $d = \lambda$ appears in the direction of $\theta = \pi/2$ (horizon line), and it approaches the broadside when d becomes larger than λ. If the distance between thc clements becomes larger, cross-coupling among each element decreases, and radar beamwidth becomes narrower. The area which one element covers (occupation area) is determined by the array interval and the arrangement of array antenna, whereas the gain expected for one element is given by (2.83). For example, in the case of regular triangle array (see Fig. 8.13b), 0.7λ (= 4.5 m) on a side, the gain derived from the occupation area S_e is $4\pi S_e/\lambda^2 = 5.3$ (= 7.2 dB), and thus the element whose gain is greater than this value is used. Within this limitation, the distance between elements are made as long as possible in the region where grating lobe does not appear in the visible range (Hansen 2009, Chap. 2).

Beam Scanning with a Linear Array Antenna

The main beam of array antenna can be formed to arbitrary directions by adjusting phases of individual element. If each element is excited so that the contribution from all elements become in-phase (or cophase) in the direction of $\theta = \theta_0$ in the Fraunhofer region, the phase of mth element ($m = 0, 1, 2, \ldots$) becomes

$$\phi_m = -mkd \sin\theta_0. \tag{8.20}$$

The phase difference between adjoining element in this case is

$$\delta = kd(\sin\theta - \sin\theta_0). \tag{8.21}$$

The array factor is expressed as

$$E_{\mathrm{a}}(\theta)=\frac{\sin\left[\frac{Nkd}{2}(\sin\theta-\sin\theta_0)\right]}{\sin\left[\frac{kd}{2}(\sin\theta-\sin\theta_0)\right]}. \tag{8.22}$$

If the position of the main beam changes, that of the grating lobe accordingly moves. For avoiding occurrence of grating lobe by cophase arrays in all directions, the following condition

$$|\delta|=|kd(\sin\theta-\sin\theta_0)|\leq 2\pi \tag{8.23}$$

is required, i.e.,

$$\frac{d}{\lambda}<\frac{1}{1+|\sin\theta_0|}. \tag{8.24}$$

Practically, the effect of beamwidth θ_{B} must be considered. If the maximum scan angle of the main beam is $\theta_{0\,\mathrm{max}}$, the requirement in (8.24) is modified as follows:

$$\frac{d}{\lambda}<\frac{1}{1+|\sin\theta_{0\,\mathrm{max}}|+\sin\theta_{\mathrm{B}}}. \tag{8.25}$$

It is also necessary to consider that the beamwidth of array antenna changes corresponding to the change of scan angle, and it becomes $\theta_{\mathrm{B}}(\theta_0)=\theta_{\mathrm{B}}(0)/\cos\theta_0$, where $\theta_{\mathrm{B}}(0)$ is the beamwidth in the scan angle $\theta=0$.

As previously discussed, the synthetic radiating pattern of array antenna $F(\theta)$ becomes the product of the array factor E_{a} and the element pattern $g(\theta)$. If the scan angle of the main beam is θ_0, $F(\theta)$ becomes[11]

$$F(\theta)=g(\theta)\frac{\sin\left[\frac{Nkd}{2}(\sin\theta-\sin\theta_0)\right]}{\sin\left[\frac{kd}{2}(\sin\theta-\sin\theta_0)\right]}. \tag{8.26}$$

Rectangle Planar Array Antenna

Consider a rectangular planar array antenna with M rows and N columns in the x and y directions, respectively, as shown in Fig. 8.12. If the amplitude of radiation pattern

[11] From the same reason as that of the previous footnote (10), $g(\theta)$ at the outer edge element differs from that of the central element. However, in the case of planar array antenna composed of several hundred elements can be assumed that all $g(\theta)$ values are the same as that of the central one (Fukao et al. 1986).

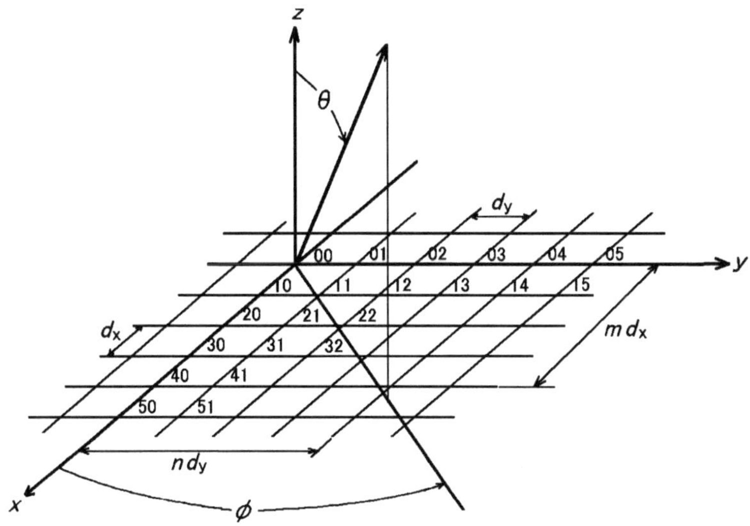

Fig. 8.12 Alignment of rectangle planar array antenna of M row N column for x and y directions, respectively

of the element at the point of intersection between the mth row and nth column is I_{mn}, the array factor can be expressed as

$$E_a(\theta,\phi) = \sum_{m=0}^{M-1}\sum_{n=0}^{N-1} I_{mn} e^{j(m\delta_x + n\delta_y)}. \tag{8.27}$$

In the above equation, the phases of the (m,n)th element are

$$\delta_x = kd_x \sin\theta\cos\phi, \tag{8.28}$$

$$\delta_y = kd_y \sin\theta\sin\phi, \tag{8.29}$$

where k is the wave number, d_x and d_y are the distances between elements in the x and y directions, respectively. If the radiation I_{mn} from the (m,n)th element in a planar array antenna is made as

$$I_{mn} = I_m I_n, \tag{8.30}$$

Eq. (8.27) becomes

$$E_a(\theta,\phi) = \sum_{m=0}^{M-1} I_m e^{jm\delta_x} \sum_{n=0}^{N-1} I_n e^{jn\delta_y}. \tag{8.31}$$

The above equation means that the array factor of rectangular planar array antenna can be expressed as the product of that of linear array antennas (IEICE 2008, Chap. 8).

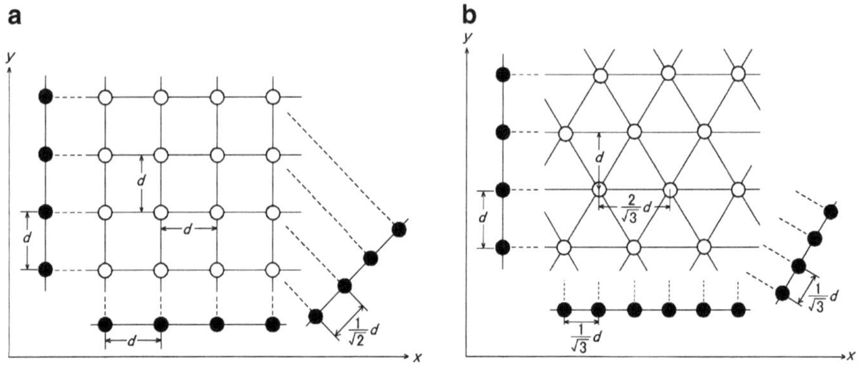

Fig. 8.13 Alignment of elements. (**a**) Quadrate arrangement and (**b**) triangular arrangement, where *white circle* shows actual position of element and *black circle* shows equivalent position of that in the corresponding direction of planar array antenna

Alignment of Elements

In designing actual phased array antennas, it is desirable that grating lobes do not appear and moreover the occupation area by an element is as wide as possible with the least number of elements.[12] In Fig. 8.13, white circle shows actual position of element and black one the equivalent position to that in the corresponding direction in planar array antenna. For the square arrangement in the same figure (a), the equivalent interval between each element becomes maximum along the x- and y-axes. For example, if the maximum scan angle of the main beam is 40° from the zenith with the maximum distance between elements d which produces no grating lobe, $d \simeq 0.61\lambda$ is obtained from (8.24) where λ is the wavelength. On the other hand, if element is arranged at each apex of triangle as shown in the same figure (b), which is called a triangular arrangement, the equivalent distance between elements becomes maximum along the y axis. The distance d in this case which produces no grating lobe should also be $d \simeq 0.61\lambda$. The actual distance between elements in this case is longer than this distance or about 0.70λ as seen in (b). It means that the actual distance between elements can be made wider by approximately 15 % in the alignment of (b) than that of (a).

Shape of Array Antenna

The antenna of atmospheric radar is, in general, the array antenna composed of numerous elements. The most representative configurations are the square array

[12] In reducing number of elements, both the expense of power consumption and cost for manufacturing elements can be suppressed. On the other hand, it should be noticed that total output power of active phased array antenna is proportional to the number of element.

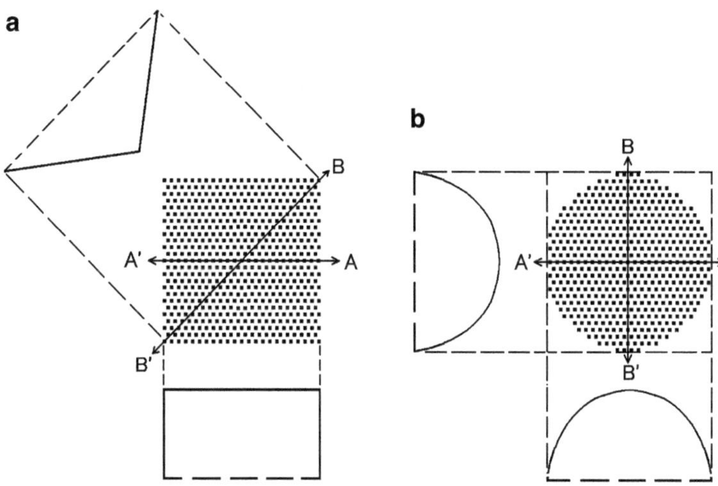

Fig. 8.14 Configuration of array antenna and integrated current distributions, i.e., summations of current distribution of individual element projected on each axis along A-A' and B-B'. (**a**) Square array antenna and (**b**) circular array antenna, where *black dots* show elements

shown in Fig. 8.14a and the circular array in the same figure (b). The black dots in each figure show the positions of elements schematically. The integrated current distributions, summations of current distribution of individual element projected on each axis along A-A' and B-B', are also shown in each figure. The integrated current distributions in the respective directions are not the same in the square array antenna of (a), since the distance between each element and the number of elements which contribute to the integrated current distributions along A-A' and B-B' are different. As a result, the antenna radiation patterns become different between both directions. In other words, main-beam width along A-A' becomes narrower but the current distribution becomes discontinuous at the edges of array. The sidelobes of radiation pattern which correspond to higher frequency spectral components becomes large. On the other hand, the sidelobe levels are improved but the main-beam width along B-B becomes wider than that along A-A'. In the case of circular array antenna of (b), the current distribution in all directions which include A-A' and B-B' becomes approximately identical, and therefore, the radiation pattern also becomes almost the same in all directions. Consequently, omitting elements at the four corners of square array antenna or using circular array antenna are effective to improve the antenna pattern. When giving priority to the radiation pattern of the main beam and the rotation symmetry of sidelobes, circular array is often chosen. On the other hand, when having much interest to the easiness in array alignment, square array is usually preferred.

Array Antenna in General Shape

Equation (8.31) can be applied to the expression of the radiation pattern for square array antenna. However, the same equation cannot be applied to the general array antennas in alignment at irregular intervals which include circular array antenna. Hence, the actual antenna radiation pattern is, in general, obtained by numerical calculation, considering actual alignment of individual element. Generally, the array factor for arbitrary three-dimensional array antenna with N elements can be expressed as follows. If the position vector and phase of exciting current on nth element are $\boldsymbol{d}_n = (x_n,\ y_n,\ z_n)$ and ϕ_n, respectively, the array factor E_{a} in the direction of unit vector $\boldsymbol{r} = (x,\ y,\ z)$ can be expressed as

$$E_{\mathrm{a}} = \sum_{n=1}^{N} e^{j\phi_n} e^{jk\boldsymbol{d}_n \cdot \boldsymbol{r}}. \tag{8.32}$$

If the unit vector in the direction of beam scanning is $\boldsymbol{r}_b = (x_b,\ y_b,\ z_b)$, the phase angle of radar beam can be given by

$$\phi_n = -k\boldsymbol{d}_n \cdot \boldsymbol{r}_b. \tag{8.33}$$

Phase Shifter and Feeding System

As previously described, the main beam of phased array antenna can be electrically steered by changing the relative phase of each element. By giving the relative phase shown in (8.20) to each phase shifter, the main beam is pointed to the direction θ_0 from the broadside direction. The phase of the transmitted pulse on the mth element is delayed by $kmd \sin\theta_0$ from that on the 1st element. Hence, by giving this delay to the phase of each element, the main beam can be steered to the direction θ_0.

There are roughly two ways of alignment of phase shifters for steering the radar beam of planar phased array antenna. One way is to attach one phase shifter to each row and column of elements, and thus $M+N$ phase shifters are needed in total for M rows and N columns. In this case, scanning is limited in the planes perpendicular to the directions of the row (x) and column (y). The other way is to attach phase shifter to all elements one by one. In this case, $M \times N$ phase shifters are needed in total for the same planar array antenna, however, for the sake the main beam can be steered to arbitrary directions within the range where the main beam is allowed to move. One of the typical examples of feeder system for planar phased array antenna is to divide one transmitter output power to whole elements as shown in Fig. 8.15a. In the past, the above system was often adopted with high power transmitter tube. In recent years, the transmitting and receiving module which houses transmitter amplifier and phase shifter is commonly adopted. In the system which uses these modules, one transmitter amplifier supplies its output power to one or several elements as shown in the same figure (b). After the phase-control for each element is applied, each

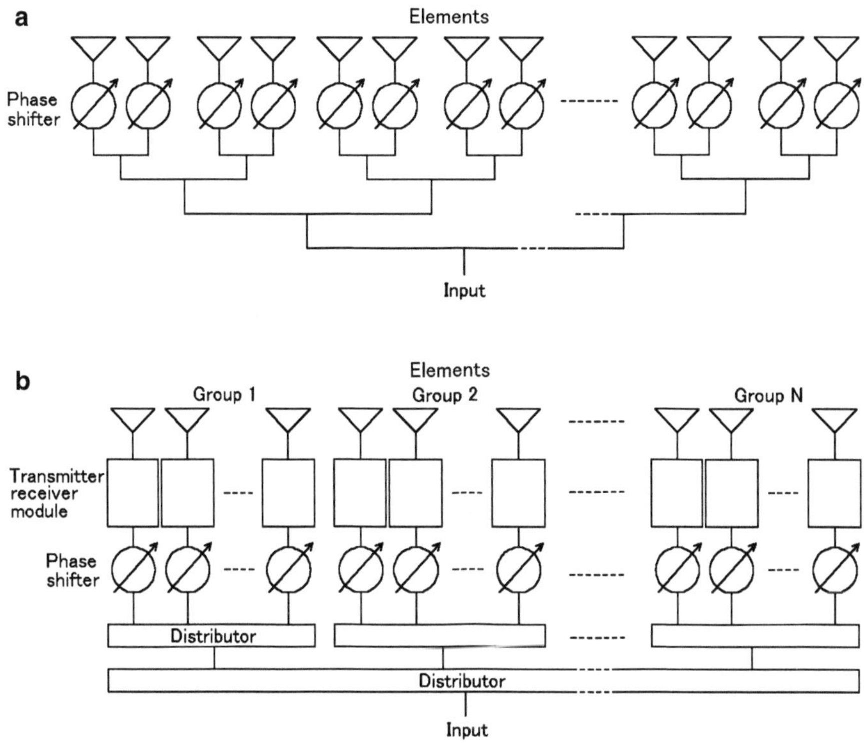

Fig. 8.15 Feeder system of planar phased array antenna. (**a**) Passive phased array. One transmitter output power is divided to whole elements. (**b**) Active phased array. Phase shifter, transmitter, and receiver of the set is connected to one or several elements

element radiates amplified pulse. The whole transmitted powers from individual elements are composed to high power in space. The radar system of this type is called active phased array radar.

Example of Planar Array Antenna

Table 8.2 shows examples of typical planar array antenna for atmospheric radar. The antenna power gain [dB] and the beamwidth [deg] of each antenna which is calculated by (8.10) and (8.11), respectively, are also listed. As one of the actual planar phased array antenna, the antenna for the Equatorial Atmospheric Radar (EAR) installed in Indonesia by Kyoto University in Japan is shown in Fig. 8.16. Its antenna field is almost circular plane in diameter 110 m with 560 elements of Yagi, and the radar frequency is 47 MHz (Fukao et al. 2003). The feeder system of the EAR is composed as shown in Fig. 8.15b, and the phase shifter is equipped to each element.

Table 8.2 Characteristics of typical planar array antenna of atmospheric radar

Frequency [MHz]	Aperture	Antenna power gain	Beamwidth
50	Diameter 110 m	33 dB	4.0°
400	11 m × 11 m	31 dB	5.0°
1,300	4 m × 4 m	33 dB	4.2°

Fig. 8.16 Planar phased array antenna of the Equatorial Atmospheric Radar (EAR) which is installed in Indonesia by Kyoto University in Japan. Five hundred and sixty set of elements are composed of Yagi antenna. Transmitter and receiver module with phase shifter is equipped to each element [from Fukao et al. 2003]

Multi-beam Antenna with Lens

Multi-beam antenna can form multiple beams for different radiation directions by use of lens. The Luneburg lens is one example of this kind of antenna. A Luneburg lens can collect a plane wave which is incident on the surface of spherical dielectric to the focus of the opposite surface by varying the dielectric constant according to the distance from the center of the spherical dielectric. As the dielectric is a sphere, each incident wave arriving from different direction has its focus on the surface of the sphere as shown in Fig. 8.17a. Therefore, if the transmitter–receiver is placed on the focal position corresponding to the radiation direction and operated, the Luneburg lens can perform the function of multi-beam antenna. The palaboric antenna has single focus thus it is necessary to move entire antenna to change the radiation direction, whereas Luneburg lens can change the beam direction by changing the direction of primary horn.

In recent years, several Luneburg lens antennas which configure low power phased array antenna for atmospheric radar are performed by switching the input and output ports of each antenna primary radiator, which is shown in Fig. 8.17b. The antenna is composed of seven Luneburg lenses, each Luneburg lens has five primary horns corresponding to zenith, 15° from zenith to north, south, east, and

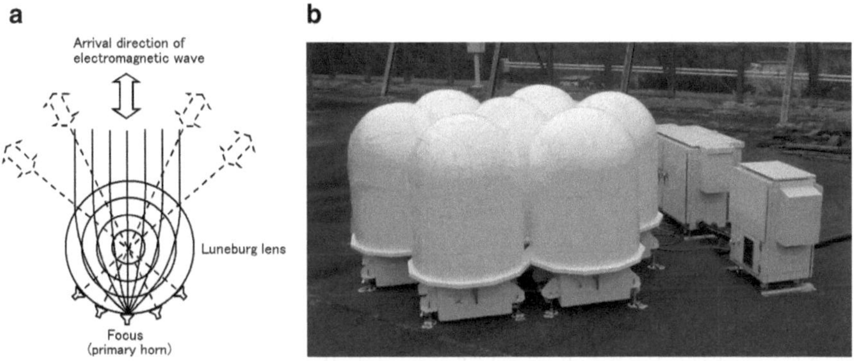

Fig. 8.17 (**a**) Principle of Luneburg lens and (**b**) phased array antenna of atmospheric radar which is composed of seven Luneburg lenses [courtesy of Kyoto University]

west, respectively. High-speed electronic scanning is performed by changing the phase between the switch and the primary radiator according to the beam direction at the same time.

8.2.5 *Measurement of Antenna Radiation Pattern*

Designed pattern of the antenna radiation is validated by actual measurement. It is known that the antenna patterns are reciprocal or identical for transmission and reception. Therefore, the antenna radiation pattern is generally obtained, for simplicity of measurement, in mode for reception.

In general, small antennas of diameter of less than several meters such as those applied for meteorological radar[13] can be easily scanned mechanically. The actual measurement is conducted as follows: a test transmission signal is radiated through the well-calibrated transmitting antenna, and received by the antenna under measurement which is located in the far field. The main lobe and possibly the first few sidelobes can be measured accurately from the intensity change of echoes received by changing sequentially azimuth and elevation angles. The test field should be selected so that the received signal is not contaminated with interference, scattered signals from stationary targets and so on.

On the other hand, the above-mentioned method is not applicable for large aperture antennas, such as that used for large scale atmospheric radar. The atmospheric radar generally steers its beam only to the limited directions near the zenith.

[13]As for meteorological radar antennas, the aperture is, in general, less than a few meters. The antenna used for WSR-88D, which will be mentioned in Sect. 9.2.2, is comparatively large or 8.5 m in diameter.

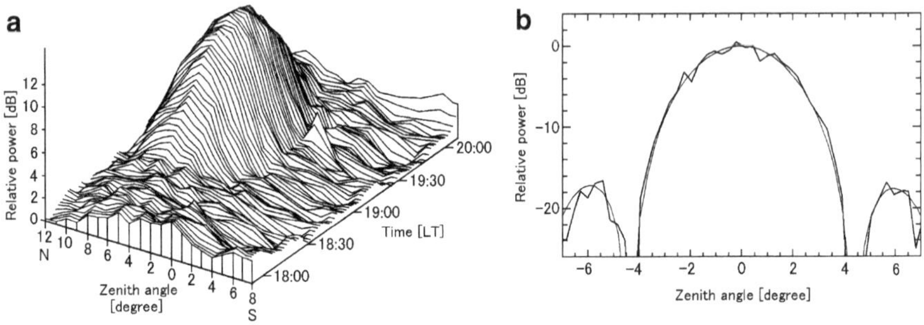

Fig. 8.18 Antenna radiation patterns. (**a**) Received noise power of Cassiopeia A by the SOUSY radar at 53.5 MHz in Germany (Czechowsky et al. 1984). (**b**) Radiation pattern of the MU radar obtained using the reflection from the moon surface. *Thick lines* show measured data, and the *thin lines* the result of fitting [from Fukao et al. 1985c]

In general, there are two methods to measure the radiation pattern of large aperture antenna. The one method is to use the astronomical bodies such as discrete radio sources, solar radio waves, or reflection from the moon surface or orbiting metallic spheres. The other is to use the flight objects such as aircrafts, artificial satellites or balloons. The position of the radio stars, the moon or metallic spheres can be precisely computed, and the reproducibility of measurements is good. As for the discrete radio source with high brightness, Cassiopeia A (the celestial declination 58.58°N) is well known at high and middle latitudes. On the other hand, Cygnus A (the celestial declination 40.73°N) is well known at low latitudes. However, it is not easy to measure sidelobes with this method because there is limitation in direction of scan angles of the measured antenna, and the SNR is low.

Figure 8.18a shows the received noise power of Cassiopeia measured with the SOUSY radar[14] in Germany which was obtained while scanning radar beam in the north–south direction near the zenith (Czechowsky et al. 1984). The radiation pattern in the east–west direction is determined by using the rotation of the Earth. The maximum rf power is observed in the north approximately 7°, which shows that the direction of the main beam is correct. Analyzing the radiation pattern, the antenna radiation pattern of the main beam in the north–south cross section, approximate east–west cross section and the half power beamwidth can be obtained. Figure 8.18b shows the antenna radiation pattern of the MU radar[15] obtained by using the reflection from the moon surface (Fukao et al. 1985c). The measurement is conducted by directing the main beam to the meridian direction of the moon, and the antenna radiation pattern of approximately east–west cross section is obtained

[14]The MST radar which was installed in Lindau, Germany in 1974. The radar frequency is 53.5 MHz, and the beamwidth is 5°.

[15]The pattern of the three dimensional power distribution of the MU radar is shown in Fig. 8.4, which corresponds to the thin line of Fig. 8.17b. The details on the MU radar will be discussed in Sect. 10.2.3.

by using the rotation of the Earth. The measured value (the heavy line) of the main beam direction, the half power beamwidth, and the first sidelobe level agree well with the theoretical value (the thin line). The fluctuation of the measurement is caused by turbulence (the scintillation) in the ionosphere.

The measurement which utilizes orbiting satellites has the advantage in performing the same measurement repeatedly down to low elevation angles for an extended period, although the initial manufacturing cost is high. In this method, the receiver is on board a satellite, and receives the signal power from the ground-base radar. The antenna radiation pattern is determined along the orbital plane. A calibration signal, which is separated 50 kHz from that of radar signal for the MU radar, is transmitted from another antenna that is calibrated beforehand so as to eliminate the effects due to posture changes of the satellite antenna, and intensity fluctuation of received signal which depends on the distance between satellite and radar. The satellite receiver separates the two signals received. The radiation pattern of the measured antenna relative to the calibration antenna can be obtained from the intensity ratio (Fukao et al. 1985c). The MU radar of Kyoto University and the 404 MHz band radar[16] which composes the wind profiler network in the central part of the United States of America are measured in the above-mentioned method at their early stages of operation.

In the case of active phased array radar, each element is excited by individual transmitter module, and thus the above mentioned reciprocity cannot be applied. Therefore, the antenna radiation patterns both for transmission and reception should be measured. For this purpose, the method to utilize the surface reflection from the moon and artificial satellite in transmission mode is effective.

8.3 Transmitters and Receivers

The transmitter–receiver system consists of the transmitter which generates and radiates the rf high power pulses via the antenna, and the receiver which receives the rf backscattered signals, converts them to the IF signals, and performs phase detection for analyses in the follow-on processor. The peak power of transmitted pulse varies from several hundreds watt (W) to several hundreds kW according to uses, the scale of the radar system, and so on. Some large-scale atmospheric radars have the peak transmitted power of several MW, whereas the minimum power that the receiver can detect is the order of 10^{-11} mW, which is determined by the receiver noise, bandwidth, and circuit temperature as discussed in Sect. 5.1. The dynamic range of typical receiver is approximately 80 dB, and the received signal power is roughly between 10^{-3} mW (-30 dBm) and 10^{-11} mW (-110 dBm).

---

[16] Refer to Sect. 10.3.1 for the details.

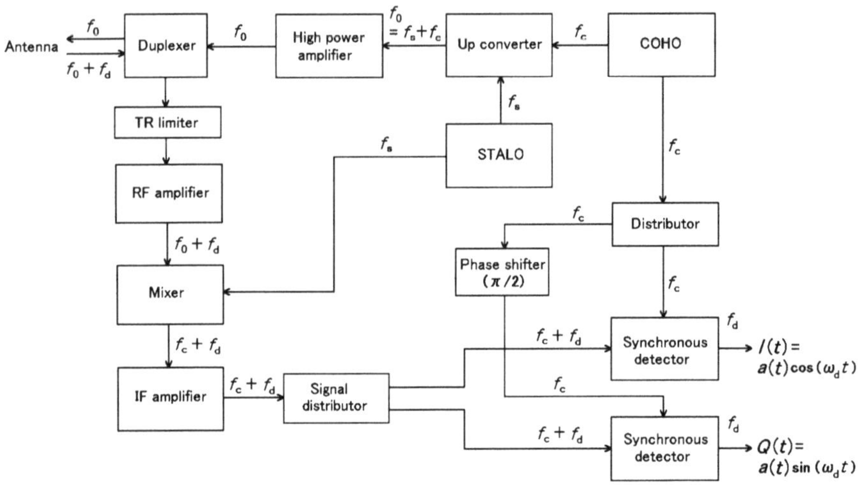

Fig. 8.19 Schematic configuration of a transmitter–receiver system of coherent pulse radar which is conceptually shown in Fig. 5.6. $a(t)$ denotes the amplitude of detector output

Figure 8.19 illustrates the schematic configuration of a transmitter–receiver system of coherent pulse radar which is conceptually shown in Fig. 5.6. The IF signal of frequency f_c which is generated by the coherent oscillator (COHO) and the local frequency (LO) signal of frequency f_s which is produced by the stabilized local oscillator (STALO) are mixed, up-converted to the rf carrier frequency $f_0 = f_c + f_s$, which is amplified by the high power amplifier to the high power transmission pulse. In general, $f_c \ll f_s$. For example, f_c is commonly chosen around 30 MHz for 5-GHz band weather radar, and around 5 MHz for 50-MHz atmospheric radar. The type of the modulation and amplification system of a coherent pulse radar which amplifies the source signal step by step is called MOPA. The frequency of received signal from moving target is affected by the Doppler shift of f_d, and shifted to $f_0 + f_d$. Applying frequency conversion and phase detection as discussed in Sect. 5.2.2, two components of received signal, I and Q signals, are obtained. Various radar parameters are derived from them.

8.3.1 Transmitter

Primary parameters which define the performance of the transmitter for a pulse radar is transmitted frequency f_0, transmitted power P_t, transmitted pulse width τ, and pulse repetition frequency (PRF) f_p. Pulse repetition time is given by $T = 1/f_p$, and the transmitter's duty cycle R_d becomes

$$R_d = \frac{\tau}{T}. \tag{8.34}$$

Transmitted power P_t generally refers to the peak power. The time average transmitted power $\overline{P}_t$ and P_t are related as follows;

$$\overline{P}_t = R_d P_t. \tag{8.35}$$

For example, if $\tau = 1\,\mu s$ and $T_s = 1$ ms, R_d becomes 0.001. The typical value of R_d in meteorological radar is around 0.0005–0.005. On the other hand, for atmospheric radars using pulse compression which will be discussed in Sect. 8.3.4, the value of τ usually becomes long compared with that of meteorological radar.[17] Thus the duty cycle, R_d, of atmospheric radar is higher than that of meteorological radar, and R_d can be as large as 0.2.

To perform Doppler velocity measurements with pulse radar, it is indispensable to keep coherency between transmitted pulses. Coherent transmitters typically adopt the MOPA system (see Sect. 8.3.2). The basic method to generate the rf pulse and radiate it from the antenna is common between meteorological radar and atmospheric radar. For meteorological radar, a parabolic antenna is commonly used to radiate the high peak power pulse, and a large transmitter tube such as klystron is generally used as the final high power amplifier. On the other hand, atmosphere radar commonly uses a planar array antenna, and its high power transmitted pulse is synthesized in space by combining incremental electromagnetic power radiated from each of the antenna elements. Furthermore, the number of atmospheric radars which use active phased array antennas is increasing, and each antenna element of the phased array is being integrated into one module containing transmitter and receiver.

8.3.2 *Transmitter Used for Meteorological Doppler Radar*

Transmitter of the MOPA Type

In the past, magnetron, a self-excited oscillator was widely applied as the transmitter tube. However, because the magnetron is an incoherent oscillator and cannot generate coherent transmitted pulses (i.e., the phase of a transmitted pulse is not correlated with the phase of any other pulse transmitted by the magnetron). Moreover, because the sinusoidal oscillations within the transmitted pulse are not uniform, the magnetron also transmits signals outside the nominal bandwidth of the rf pulse. Thus transmitters of the MOPA system became common in recent years. In the background, to cope with the worldwide wireless local area network (LAN) demand, specifically in the frequency band of meteorological radar, the development and practical application of a technology, which reduces frequency allocation and suppresses interference to other radio stations is desired; the MOPA

[17] In recent years, meteorological radars which adopt pulse compression technique appear in practical use.

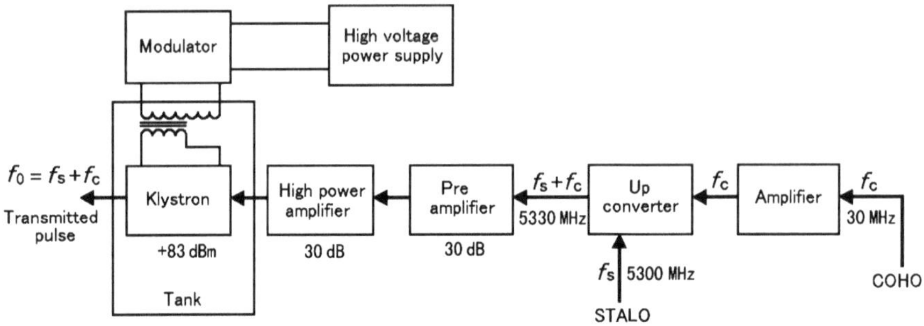

Fig. 8.20 Configuration of the transmitter which adopts the MOPA system with a klystron; 5.3-GHz band and a peak transmitted power of 200 kW [adapted from Hamazu et al. 2002]

technology achieves the goal of reducing interference. Typical examples which use the MOPA system to obtain coherent transmitted pulses having peak powers of several kW or more are klystrons, traveling wave tubes (TWT), and high power transistors. The peak transmitted power of traveling wave tubes is several kW, and that of high power transistors is several 10–100 W, and those output powers are generally low compared with that of klystrons of about a few hundred kW and up to a couple of megawatts. However, pulse compression techniques were developed in recent years to obtain high resolution and long observation range as conventional radar, but with less peak power. Figure 8.20 shows an example of the configuration of a transmitter which adopts the MOPA system using a high peak power klystron (Hamazu et al. 2002). Klystrons can obtain high peak power, high efficiency, and high gain, and several times or more longer lifetime compared with that of magnetrons. Moreover, in klystron, the adjustment of output power and frequency is easy and the configuration of peripheral circuits is simple, thus the use of klystrons is spreading as the transmitter tube of the MOPA system. In recent years, the development of high power gyroklystrons is also advancing, mainly in millimeter wave radar.

Klystron

The operation of the klystron is based on the principle that velocity modulated electron beam produces amplification of pulse power. A basic configuration of the klystron is shown in Fig. 8.21. It is an amplifier tube and is composed of the electron gun which radiates electron beam, the cavity to amplify microwave, the collector which catches electron beam, the output window which pull out amplified microwave, and so on. Also, a focusing coil (solenoid) is installed to surrounds klystron main unit. The electron beam radiated from electron gun of diode structure is focused by the magnetic field which is produced by the focusing coil, and it

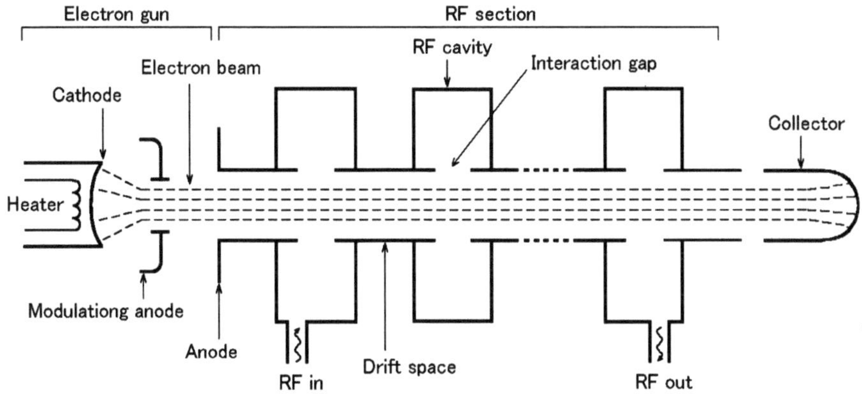

Fig. 8.21 Basic configuration of a klystron [adapted from Skolnik 2001, p. 694]

goes along the axial direction of the klystron to the collector. To simplify the description, we assume that the cavities are composed of one resonance cavity and one output cavity. When electron beam reaches resonance cavity, a part of electron is accelerated by the electromagnetic field of microwave which is added to the input port and a part is decelerated. As a result, the electron beam which is stored in the drift space (in the drift tube) receives bunching action. It means that the electron which is worked on deceleration reaches the output cavity with low velocity, and that worked on acceleration reaches with high velocity. If choosing the length of the drift space suitably, it is possible to make those electrons arrive at the output cavity at the same time, and the density of the electron beam can be maximized. Output voltage is generated in the output cavity according to the density of the passed electron current. In other words, klystron transforms the velocity modulation of electron beam to the density modulation, and pulls out amplified microwave signal from the resonated output cavity. The actual klystron has several cavities, and the electron beam passes these resonance cavities. In this process, the electron beam receives bunching action repeatedly, thus the kinetic energy of electron increases, and the bunching action itself becomes large (Skolnik 2001, pp. 694–695).

Traveling Wave Tube (TWT)

Like the klystron, the TWT is also a kind of amplifier tube. If phase velocity of electromagnetic waves and velocity of electron beam are arranged to be almost equal and come close each other, a strong interaction arises and electronic kinetic energy is transferred into the energy of electromagnetic waves. The TWT applies this principle. Figure 8.22 shows the basic configuration of a TWT. A TWT is composed of an electron gun which generates focused electron beam, a slow-wave circuit which decreases the speed of electromagnetic wave, and a collector which

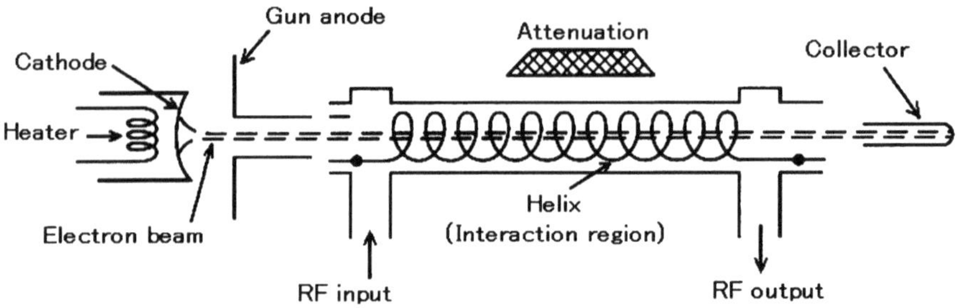

Fig. 8.22 Basic configuration of a traveling wave tube (TWT) [from Skolnik 2001, p. 699]

removes the electron beam from the inside of the tube. A helix is used for the slow-wave structure[18] (Skolnik 2001, pp. 698–699).

The electron beam (direct current) passes through the slow-wave structure. The velocity of propagation of the inputted rf signal is slowed down by the slow-wave structure so that it is nearly equal to the velocity of the electron beam. The synchronism between the electromagnetic wave (inputted rf signal) propagating along the slow-wave structure and the electron beam propagating inside the slow-wave structure results in a cumulative interaction which transfers direct current energy from the electron beam to increase the energy of the electromagnetic wave (Skolnik 2001, p. 699). Since the direction of the electromagnetic wave field reverses for every half-wavelength, the region which accelerates the speed of electrons and the region which decreases the speed of electrons are separated by half a wavelength. Thus, accelerated electrons overtake the decelerated electrons causing the electron density to vary or bunch in space and time as the electrons flow along the axis of the tube. The bunching action amplifies the electromagnetic wave.

A combination of solid-state (semiconductor) amplifiers and a TWT is also often designed to configure microwave power module. Although solid-state amplifiers and a TWT can be applied independently, the combination can create a more compact transmitter than applying either device alone. As the gain of TWT is a function of the physical length of the tube, if it can be minimized, the TWT can be made smaller. Thus for example, instead of requiring a 50 dB gain TWT, a solid-state driver of 25 dB or so of gain (where power levels are reasonably low) and a TWT of 25 dB or so of gain where the power levels are high can be used. As the solid-state device can generate reasonable powers at wide bandwidths and the TWT is inherently wideband, the microwave power module is also capable of wide bandwidths.

[18]The helix TWT is limited to voltages of about 10 kV and a peak transmitted power of a few kilowatts. For higher transmitted power radar, other types of slow-wave structures, e.g., the ring-bar circuit and the coupled-cavity circuit, have been employed (Skolnik 2001, p. 699).

Magnetron

In older meteorological radars which used only intensity data, it was not necessary to obtain phase information. For that case, it is generally accepted practice to adopt the magnetron as the transmitter tube which is relatively low cost and can get high power easily. The magnetron has a cylindrical anode and a cathode which is in the center axis, and applies magnetic field axially. Klystrons generate bunches of electrons along a straight beam, whereas the magnetron's resonant cavities, uniformly spaced in the cylindrical anode, generate several radially extended bunches of electrons streaming from the cylindrical cathode to the anode. Because of the strong axial magnetic field, these radially extended electron bunches sweep by the resonant cavities and amplify the electromagnetic fields within these cavities.

Coherent Radar Obtained with a Magnetron

The frequency of magnetron oscillation is random (i.e., not representative of the average frequency across the pulse) at the beginning of the pulse. Thus the phase is measured near the center of the pulse where it is most representative of the carrier frequency. Although the phase is derived from an I, Q sample (see Sect. 5.2.2) at the center of the transmitted pulse, the phase is also random (i.e., it is not coherent from pulse to pulse), but it is the phase of a relatively constant frequency across the rf pulse. Hence, when assembling a pulse Doppler radar using a magnetron, it is indispensable to measure the phase of the transmitted pulse and compare that to the phase of the echo sample. In the past, it was general practice to couple the transmitted pulse at much reduced power and then pass this lower power rf pulse to the receiver and synchronize the phase of the coherent oscillator (i.e., COHO) in the receiver to that of transmitted pulse in analog circuitry so as to obtain coherency and to perform phase detection. This method is often used at relatively low radar frequencies such as in the 5.3/5.6-GHz and 9.5-GHz bands to build a low cost pulse Doppler radar. However, using this method, it becomes difficult at higher frequencies and for narrower pulse widths to sample a phase of a stable and representative frequency within transmitted pulse width. Thus, the new method (the digital phase reference method) as shown in Fig. 8.23 has been developed wherein the initial phase of each transmitted pulse is recorded (upper panel), and it is used as a reference phase for real time digital processing (lower panel) (Hamazu et al. 2003). Although the noisy portion of the transmitted spectrum is relatively high compared to that of the pulse transmitted using the MOPA, coherent processing can be performed in real-time. This technique is useful for high frequency Doppler radar in the millimeter wave band.

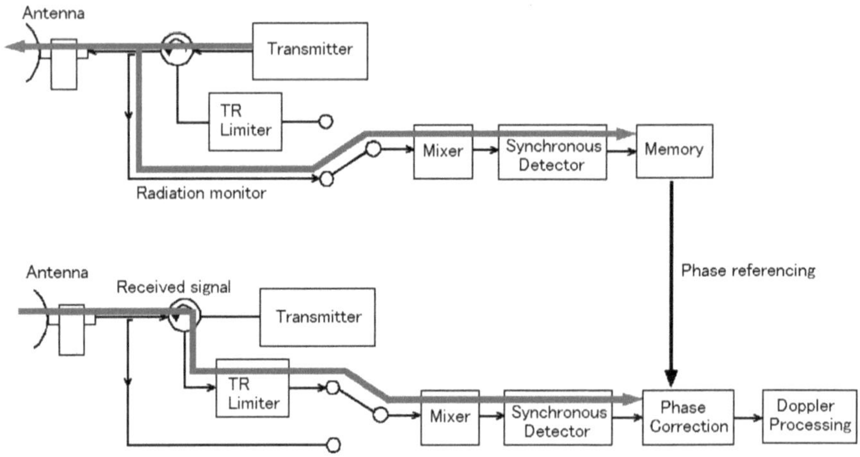

Fig. 8.23 Coherent processing of a magnetron radar with digital phase referencing

Double Pulse Technique to Increase the Nyquist Velocity

As discussed in Sect. 4.1.2, the Nyquist velocity $v_N = \lambda f_p/4$, where λ is the wavelength and f_p is the pulse repetition frequency. Generally, short wavelength (i.e., less than 3 cm) radars are chosen to detect clouds and thus v_N is relatively small. Thus to increase v_N of short wavelength radars, f_p can be increased.

However, some higher rf oscillators such as the magnetron of 3 cm wavelength cannot increase their pulse repetition frequencies easily because of the restrictions of the maximum duty cycle. Then by shortening the pulse width and increasing f_p, v_N can be increased without impacting the duty cycle. Although this solution would allow increase of v_N without violating the duty cycle limitation, it comes at the cost of decreased detection capability. The loss of detection performance results because a wider receiver bandwidth is required to received the shorter width pulses. But a wider receiver bandwidth increases the noise power and thus decreases the per pulse SNR. Furthermore, decreasing the inter-pulse spacing of a uniform train of pulses comes at the cost of multiple trip echoes being overlaid.

A solution that maintains detection performance without losing detection capability, and limits the number of overlaid echoes, is one in which closely spaced pairs of pulses, separated by long intervals, are transmitted as seen in Fig. 8.24. The upper panel of the figure shows the ordinary method to transmit pulses with a constant repetition rate. The lower panel shows a pulse series which is applied for this method, where the pair pulses having a short inter-pulse interval are transmitted between long pulse intervals. This solution keeps the duty cycle fixed while increasing v_N. In this case, the Doppler velocity measurement by the autocovariance

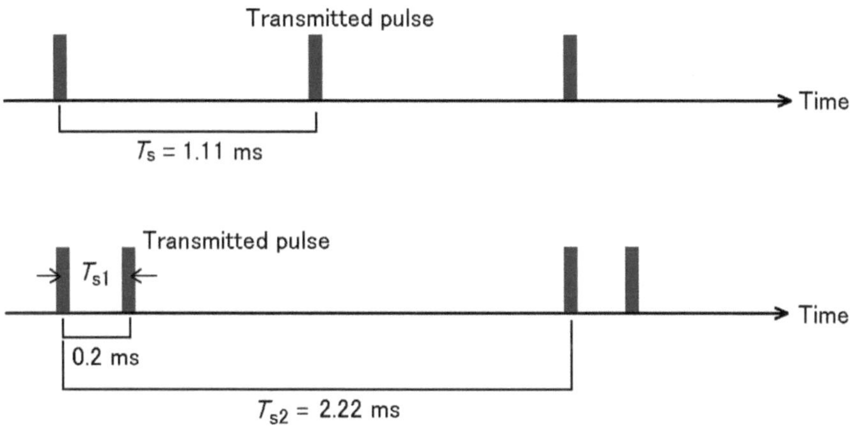

Fig. 8.24 Single pulse transmission (*upper panel*) and double-pulse transmission where pair pulses of short pulse interval are transmitted in a long pulse interval (*lower panel*)

Table 8.3 Peak transmitted power of typical meteorological radars

Frequency [GHz]	Peak transmitted power [kW]
2.8	500–800 (MOPA, magnetron)
5.3	200–300 (ditto)
9.5	30–100 (ditto)
35	30–50 (MOPA), 50–150 (magnetron)
95	2–3 (MOPA)

processing which was discussed in Sect. 5.5.4 is determined by the two pulses with short interval, and the value of v_N is five times expanded compared with that of upper panel. However, the spectrum analyses such as the FFT can not be performed with this method. If weather extends beyond the range $cT_s/2$, where T_s is the sample time interval, echoes only from the two closely spaced pulses will be overlaid (i.e., assuming the interval between pairs of pulses is sufficiently long). Although accurate measurements would be limited because of the overlaid echoes, the pair of overlaid echo samples are not correlated because they come from different ranges. Thus Doppler velocity estimation is not biased, but estimate variance will increase.

Peak Transmitted Power of Meteorological Radar

Meteorological radar generally transmits a uniform sequence of high power pulses. In addition, pulse compression transmitters which adopt relatively low transmitted peak power has been developed recently. The values of typical frequency band and typical peak power of meteorological radars are shown in Table 8.3.

Table 8.4 Breakdown power of rectangular waveguide

Frequency [GHz]	Waveguide	a × b [mm^2]	Breakdown power [kW]
2.8	WRJ-3	72.1 × 34.0	2,450
5.3	WRJ-5	47.6 × 22.2	1,270
9.5	WRJ-9	28.5 × 12.6	446
35	WRJ-320	7.11 × 3.56	30.1
95	WRJ-900	2.54 × 1.27	3.77

Characters a and b denote long side and short side, respectively, of the rectangular waveguide (inside dimension)
Withstand voltage of air at the normal temperature and 1 atm is $E = 1,500\,\mathrm{V\,mm^{-1}}$ where safety ratio is 2

Breakdown Power of Waveguides

When using small size waveguide in high power, e.g., for millimeter wave pulse radar, there are some problems such as avoiding arcing in the waveguide due to a high transmitted power as well as losses in the waveguides. So as to keep withstand voltage high, the atmospheric pressure in a waveguide is raised. The breakdown power (peak power) of rectangular waveguide can be given by the following equation (Marcuvitz 1965, p. 61).

$$P = \frac{E^2}{4 \times 120\pi} \sqrt{1 - \left(\frac{\lambda}{2a}\right)^2}\, ab, \tag{8.36}$$

where E is the [$\mathrm{V\,mm^{-1}}$] of air at the normal temperature and 1 atm, λ is the wavelength [mm], and a and b are the rectangular waveguide sizes of long side and short side, respectively. The higher the frequency, the smaller the waveguide size, and the withstand voltage declines. The values of breakdown power at safety ratio 2 for various sizes of rectangular waveguide are shown in Table 8.4. For example, the breakdown power for 35-GHz band waveguide becomes around 30 kW. Thus, to realize a pulse radar having 100 kW of peak transmitted power in this frequency band, it is necessary to increase breakdown power by pressurization.

Solid State Power Amplifier

As will be discussed in the next subsection, power-transistors are adopted for transmitter modules of atmospheric radars. In recent years, meteorological radars have begun using power-transistors for the transmitter. The solid state power amplifier (SSPA) is a MOPA system, and its operation is the same as that of klystron as shown in Fig. 8.20. However, the operational voltage of the SSPA is only a fraction of that needed by klystrons and magnetrons. Although the peak output power of the SSPA is quite low, it can be operated at high duty cycle up to several tens of a percent. Thus applying the pulse compression technique (Sect. 8.3.4) with

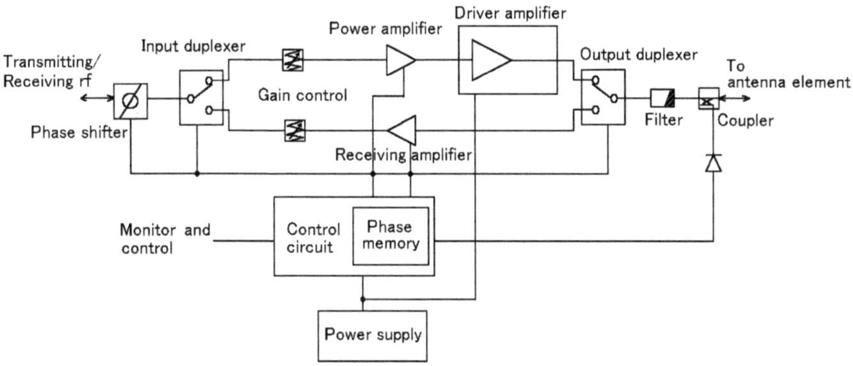

Fig. 8.25 Configuration of the transmitter and receiver module for 50-MHz band atmospheric radar [from Fukao et al. 2003]

pulse width of several tens μs to several hundred μs, a SSPA radar can achieve almost the same observation performance as that of a klystron radar. Moreover, the output power spectrum of the SSPA in linear amplification region is pure and its spurious radiation is better suppressed. Thus, a SSPA radar can minimize interferences to neighboring radars in the crowded area of the same frequency band. At 3-cm wavelength, SSPA radars of peak output power of around 200 W whose transmitter module is composed with gallium nitride (GaN) is in practical use (Wada and Uyeda 2011). 5-cm wavelength SSPA radars are also under development.

8.3.3 Transmitter of Atmospheric Radar

In the atmospheric radar which adopts phased array antenna, active phased array radar is becoming general, where elements and integrated transmitter–receiver modules are combined to generate the transmitted pulse. A transmitter and receiver module is composed of a power amplifier, receiver, phase shifter, TR switch, control circuit, and related peripherals. The example of the transmitter and receiver module for 50-MHz band atmospheric radar is shown in Fig. 8.25. As previously discussed, there is more than one method to connect the array elements to a transmitter–receiver module. For example, one method is for the transmitter–receiver module to be connected to one antenna element, and the other method is for one transmitter–receiver module to supply rf power to several antenna elements. The method of connection is determined according to the desired function of radar beam control.

In the typical example of large-sized atmosphere radar in the 50-MHz band (to be discussed later in Chap. 10), the output rf power of each transmitter–receiver module is around 200 W to 2 kW, and the active phased array antenna is composed of about 500 antenna elements. In this case, the combined output rf power becomes around

100 kW to 1 MW. In the case of 1,300-MHz band radar for boundary layer and lower troposphere observation, whose transmitted frequency is one of the highest in atmospheric radar, the rf power generated by one transmitter module is divided and supplied to several elements. As the observation range of this type of radar is relatively low, the output rf power of one transmitter module is reduced to about 200 W, and the combined output rf power is about 2 kW. Using about ten transmitter modules, each module supplies rf power for two to four elements.

8.3.4 Pulse Compression

The methods to improve SNR of pulse radar are to increase peak transmitted power, expand transmitted pulse width and so on. However, there is limitation in transmitter devices to increase peak rf power. Furthermore, expansion of pulse width degrades range resolution (altitude resolution). Pulse compression technique is introduced with the atmospheric radar as the method to solve this problem and improve SNR. The basic concept of pulse compression is to apply appropriate phase modulation on a long transmitted pulse and then demodulate and decode the received signal to obtain an equivalent echo sample having a much higher peak power than that transmitted, and having significantly better range resolution than that of the long transmitted pulse if it were not coded. In this technique, all the rf power which is equivalently contained in the pulse can be centered in a short time period without degrading range resolution and exceeding peak transmitted power limitations.

In meteorological radar, it is not so difficult to obtain desired SNR compared with atmospheric radar, and high power transmission tube is sometimes easily produced. Hence, in the past, pulse compression techniques were not generally applied to meteorological radar. Meanwhile, there are significant advantages in converting high peak power and short pulse to low peak power and long pulse, e.g., lower peak power makes the design of transmitters easier since the components used to construct the circuits must be able to prevent breakdown. Furthermore, lower peak power transmission is preferable to avoid contaminating adjacent radars with transmitted waves. Thus in recent years meteorological radars which adopt pulse compression technique also attract attention from the point of radio wave allocation plan.

Basic Concept of Pulse Compression Technique

To compress a long pulse into a short one in the receiver, the easiest method is to modulate a waveform that is uncorrelated in time but known at the receiver. A cross-correlation operation at the receiver will compress the longer received waveform into a shorter one. This is due to the autocorrelation properties of the transmitted waveform, which is maximum at zero-lag and almost zero at lags other than zero.

As shown in (5.84), the autocorrelation function of the received signal $v(t)$ is given by $R(\tau) = E[v(t)v^*(t+\tau)] = [v(0)v^*(\tau)]$. The value $R(\tau)$ becomes maximum for $\tau = 0$ (where $R(0)$ is equal to the power).

The alignment of the phase in frequency domain for the time domain signal $v(t)$ can be performed by multiplying the frequency spectrum $V(f)$ of $v(t)$ and the complex conjugate $V^*(f)$. Taking into consideration that the frequency spectrum of $v^*(-t)$ becomes $V^*(f)$, the product $V(f)V^*(f)$ in frequency domain is equivalent to the convolution $v(t) * v^*(-t)$ in time domain as from (5.79). Furthermore, calculation of $v(t) * v^*(-t)$ is nothing other to obtaining the autocorrelation of $v(t)$. Moreover, it is equivalent to the matched filter which was previously discussed in Sect. 5.3. From the above procedure, it is possible to say that the pulse compression is obtained by implementing the following procedure.

1. Generate a modulated long pulse width signal as a transmitted pulse where the modulation signal is a quasi-random signal that has a very narrow autocorrelation function. In this stage, the frequency spectrum of the generated signal is spread and the phase is not aligned.
2. In the receiving process, perform demodulation by applying cross-correlation between the detected received signal and the modulation signal which was adopted to generate transmitted pulse.

As general classification, there are two typical methods in the pulse compression, one is the frequency modulation method and the other is the code modulation method. The former is the method which performs frequency modulation of the transmitted pulse. The latter is performed by modulating with a discrete code the phase and amplitude of the rf pulse, applying discrete coding, where the coding function is selected that the autocorrelation function is analogous to the delta function. There are some kinds of coding depending on the selection of code series. Some typical methods are discussed in the following paragraphs.

Pulse Compression by Linear Frequency Modulation (FM Chirping)

Figure 8.26 presents this method. The frequency modulation of the transmitted rf pulse is shown in (a), and the concept of demodulation of the received signal is shown in (b). As shown in the upper panel of (a), linear frequency modulation of bandwidth Δf is applied on the transmitted signal of pulse width T. As a result, the carrier frequency of the transmitted pulse becomes as shown in the lower panel of (a). When receiving signals, the received signal is applied to a circuit which has the delay time characteristic corresponding to the frequency as shown in the upper panel of (b), large delay occurs for the lower frequency components on the first half of the transmitted pulse, and small delay occurs for the higher frequency components on the last half of the transmitted pulse. Therefore, when the received signal passes the demodulation circuit, the frequency components of the signal which are distributed in the pulse concentrates to the narrow time zone at the center of the waveform, and

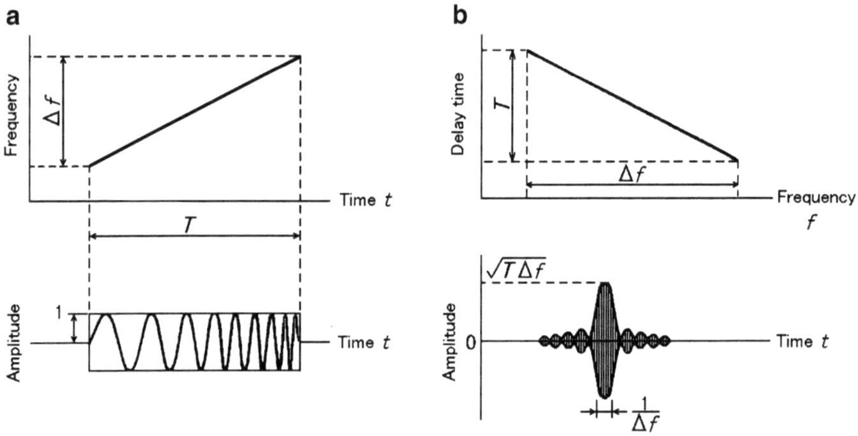

Fig. 8.26 Principal of the pulse compression by linear frequency modulation (FM chirping). (**a**) Frequency modulation of transmitted rf pulse, and (**b**) demodulation of received signal

it becomes the steep pulse as shown the lower panel of (b). The envelope curve $f(t)$ of the waveform after the demodulation circuit becomes

$$f(t) = \sqrt{T\Delta f}\,\frac{\sin \pi\Delta f t}{\pi\Delta f t}, \tag{8.37}$$

where $\sqrt{T\Delta f}$ is the peak amplitude (voltage) of demodulation, and $1/\Delta f(\equiv \tau_p)$ is the demodulated (final) pulse width. The product of transmitted pulse width T and bandwidth Δf, $T\Delta f$, is called the pulse compression ratio. In pulse compression; this ratio, $T\Delta f = T/\tau_p$, expresses the effective increase of transmitted peak power.

Pulse Compression by Phase Coding

The transmitted rf pulse can be modulated by phase coding. In this case, the transmitted pulse is divided in N time slots and the phase of the transmitted pulse is changed at the beginning of each time slot. The binary phase coding which can perform phase modulation digitally by 180° is the most practical. In this method, the phase of $+(0°)$ or $-(180°)$ (which is equivalent to changing the sign of the signal) is applied to the transmitted pulse at each of N slots. In the reception, to decode the signal, the cross-correlation is performed between the detector output signal and phase modulated code series. The range resolution and the amplitude of the decoded signal becomes $1/N$ and N times, respectively compared with original signal, and almost all the rf power in the pulse is transferred to one slot. At this time, the pulse compression ratio is N, and the rf power of the signal becomes N^2 times larger. On the other hand, as the noise has no correlation to each other, the noise power of N samples become N times that of the original noise power. Thus, SNR is improved N times.

In a N bits binary phase code series, if the unit amplitude signal series is expressed as

$$(X_1, X_2, \cdots, X_n) \quad \text{where } X_i = 1 \text{ or } -1, \tag{8.38}$$

the pulse compression waveform can be digitally modulated and the autocorrelation function $\Omega(k)$ can be given by

$$\Omega(k) = \sum_{i=1}^{N-|k|} X_i X_{i+|k|} \quad k = -(N-1) \sim N-1. \tag{8.39}$$

The above function is an even function which is symmetric around $k = 0$. Generally, the autocorrelation function of binary phase code series does not become a single pulse of width N because there are residual signals, called range sidelobes, before and after the peak signal. Typical phase code series which suppress sidelobes to the utmost are shown below.

(1) Barker codes

For the phase code series given by (8.38), the autocorrelation function (8.39) becomes N for time lag 0, and 0 or ± 1 for the other lags. In this code series (8.38), $\Omega(k)$ which is given in (8.39) satisfies

$$|\Omega(k)| \leq 1 \tag{8.40}$$

for $k \neq 0$. The phase code series shown in Table 8.5 has been discovered for $N = 2, 3, 4, 5, 7, 11, 13$, and two kinds of phase code series exist for $N = 2$ and 4. In the above table, code element $+$ indicates 0 phase, and $-$ indicates a phase of π radians. The longest bit length of the phase code series which adapts between 1 and 6084 is 13 bits. Figure 8.27 shows the autocorrelation function, i.e., decoded output for $N = 7$, 11, and 13 in the upper, middle, and lower panels, respectively. The characteristics of Barker codes that the range sidelobes become 0 or ± 1 are shown in these panels. If the conversion that are shown in (8.41)–(8.43) are applied on the phase code series in Table 8.5, another Barker code series is generated.[19]

$$X_{i1} = (-1)^i X_i, \tag{8.41}$$

$$X_{i2} = (-1)^{i+1} X_i, \tag{8.42}$$

$$X_{i3} = -X_i. \tag{8.43}$$

(2) Complementary codes

The complementary codes are a pair phase code series of equal length for which the autocorrelation function of each code series has a sign opposite of the

[19]In the case of $N = 2$, only the conversion of $X_1 = 1$ is approved.

Table 8.5 Barker codes (from Skolnik 1990, p. 351)

Code length N	Code elements	Sidelobe level [dB]
2	+−, ++	−6.0
3	++−	−9.5
4	++−+, +++−	−12.0
5	+++−+	−14.0
7	+++−−+−	−16.9
11	+++−−−+−−+−	−20.8
13	+++++−−++−+−+	−22.3

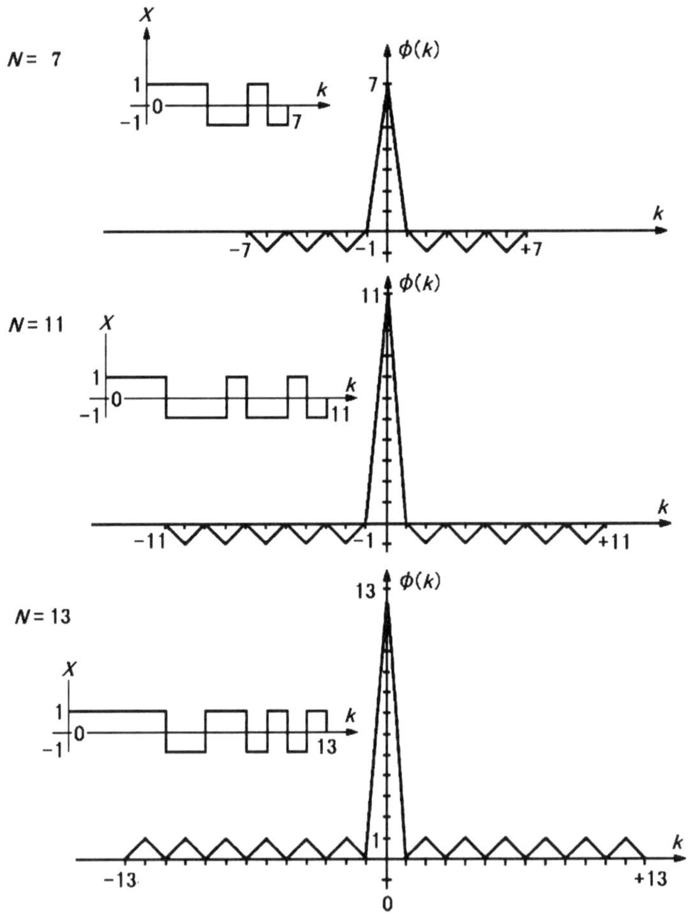

Fig. 8.27 Autocorrelation function of Barker codes of lengths $N = 7, 11, 13$

other (complementary code) in the sidelobe region. Thus, the range sidelobes are erased by adding the outputs of the autocorrelation of both code series. Because the autocorrelation functions of the two phase code series $(X_1, X_2, \cdots, X_N)$ and $(Y_1, Y_2, \cdots, Y_N)$ of length N are

$$\Omega_X(k) = \sum_{i=1}^{N-|k|} X_i X_{i+|k|} \tag{8.44}$$

$$\Omega_Y(k) = \sum_{i=1}^{N-|k|} Y_i Y_{i+|k|}, \tag{8.45}$$

respectively, they are a pair of complementary codes if the following relation is satisfied,

$$\Omega_X(k) + \Omega_Y(k) = \begin{cases} 0 & \text{for } k \neq 0 \\ 2N & \text{for } k = 0, \end{cases} \tag{8.46}$$

where $k = -(N-1) \sim N-1$. Equations (8.44) and (8.45) show the case for k of non-negative integer. Those autocorrelation functions are symmetric for positive and negative values each other, thus (8.46) is approved even for $k < 0$. As evident from (8.46), the summation of the above pair code series has no range sidelobe.

Figure 8.28 shows an example of a 32 bit complementary codes autocorrelation functions, and the summation of the respective autocorrelation functions (Woodman 1980). If the code length of X and Y is N bits, $X+Y$ and $X+\overline{Y}$, which are composed by combining those two codes, become $2N$ bits complementary codes, where $\overline{Y}$ means the code having signs that are inverted from those of Y. Thus, the length of complementary code has no upper limitation. It is known that complementary code series of bit lengths such as $N = 2^n$ and 10×2^n can exist, where n is integer.

Applying this code series on actual pulse radar, pulses are transmitted with a phase modulation by phase code series X and Y alternately. In reception of radar signals, the received signal for each phase code series is decoded independently. After the decoding, the received signals obtained for each code series are added. This code series is applicable for the signals that have a correlation time sufficiently longer than the pulse repetition time and successive received signals of equal to or more than two pulses are coherent. For example, this code series is suited for received signals of atmospheric radar observations of troposphere and stratosphere where the correlation time is sufficiently long.

(3) Pseudo-Barker code

This code is similar to Barker codes but the maximum value of range sidelobe level allowed to be ± 2. The code $N = 28$ is in practical use.

Table 8.6 Code sequence of Spano codes (from Spano and Ghebrebrhan 1996b)

Code length N	Code name	Code elements
4	C_1	$+---$
	C_2	$++-+$
8	C_1	$-++-----$
	C_2	$--++-+-+$
	C_3	$----+--+$
	C_4	$-+-+++--$

If sampled complex data series for kth transmitted pulse are $\{v_1^k, v_2^k, \cdots, v_N^k, \cdots, v_R^k\}$, and the decoded complex data series are $\{w_1^k, w_2^k, \cdots, w_N^k, \cdots, w_R^k\}$, then w_j^k is given by

$$w_j^k = \begin{cases} \sum_{m=1}^{j} c_{m+N-j}^k v_m^k & 1 \le j \le N-1 \\ \sum_{m=1}^{N} c_m^k v_{m-N+j}^k & N \le j \le R, \end{cases} \tag{8.47}$$

where $\{w_1^k, w_2^k, \cdots, w_{N-1}^k\}$ are signals in the truncated range.

The final decoded signal w_j in range gate G_j can be obtained by the summation of w_j^k. In the truncated range, as the received signals from subpulses which are necessary for the decoding are partially missing, thus complete decoding (correlation processing) is not possible. In general, matrix operation is required to perform full decoding for the phase modulated signal by complementary code modulation. On the other hand, when using the Spano codes, the matrix is diagonalized and particular decoding processes for the truncated range is not necessary (Spano and Ghebrebrhan 1996a). Thus, decoding becomes possible regardless of the truncated range and the untruncated range and observed data can be obtained.

Let the phase modulation of 0° and 180° be simply expressed by + and −, respectively. It is known that if the signals are phase modulated with the Spano codes as shown in Table 8.6, and transmitted as the sequence shown below, good decoding performance can be achieved (Spano and Ghebrebrhan 1996b).

$$\text{case of 4 bit } C_1,\ C_2,\ C_2',\ C_1',\ C_2,\ C_1,\ C_1',\ C_2' \tag{8.48}$$

$$\begin{aligned} \text{case of 8 bit } & C_1,\ C_2,\ \overline{C_2}',\ \overline{C_1}',\ C_2,\ C_1,\ \overline{C_1}',\ \overline{C_2}', \\ & C_3,\ C_4,\ \overline{C_4}',\ \overline{C_3}',\ C_4,\ C_3,\ \overline{C_3}',\ \overline{C_4}'\ , \end{aligned} \tag{8.49}$$

where C' and $\overline{C}$ show the reverse bit sequence and sign reversal of C, respectively.

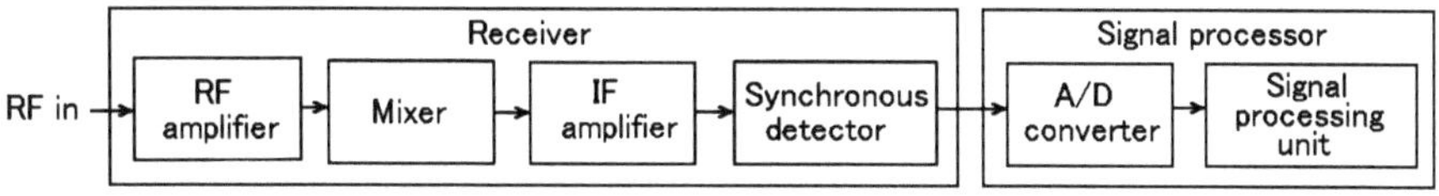

Fig. 8.31 Basic configuration of rf signal reception and processing

8.3.5 *Receiver*

The fundamental function of a radar receiver is receiving echoes from scatterers, amplifying them, and applying appropriate filter to detect the desired signals. The scope of receiver has been changed according to the progress of technology. In the past, the receiver amplified the rf signal, translated the signal spectrum to an intermediated carrier frequency (IF), and finally obtain video signal by detection. On the other hand in recent years, the IF signal is usually converted to digital numbers by an analog to digital (A/D) converters and the mixing to baseband is done digitally. In this section, we define the scope of receiver between the stage of rf signal reception and phase detection as shown in Fig. 8.31. The following signal processing including A/D conversion is described in Sect. 8.4.

Radio Frequency Amplifier

It was mentioned in Sect. 5.1 that the suppression of background noise which is generated inside receiver is important for the receiver at GHz bands.[20] If the received signal from scatterers is weak, it is buried in noise and cannot be detected. Hence, it is possible to say that the optimal design of receiver is to maximize SNR.

The most important parameter at the rf stage is background noise in the receiver. Therefore, the first stage of rf signal reception and processing uses a low noise amplifier (LNA). The role of low noise amplifier is to amplify the extremely weak rf echoes and externally generated noise without decreasing the signal to noise ratio. In the past, the parametric amplifier of the cooling type or the non-cooling type which has excellent low noise characteristic was usually employed. However, it is generally complex and expensive. On the other hand, in recent years various types of transistor amplifier became available at the radar frequency band for use in receiver front-ends. As a typical receiver front-end, the high electric mobility transistor (HEMT) which used III-V family compound semiconductor, such as centered on gallium arsenic, is well known. Furthermore, the monolithic microwave integrated circuit (MMIC), which is composed of the HEMT and other devices such as matched circuit for input and output, bias circuit, and so on a common

[20] As for MHz band receivers such as that of atmospheric radar, space background noise is generally much higher than receiver internal noise.

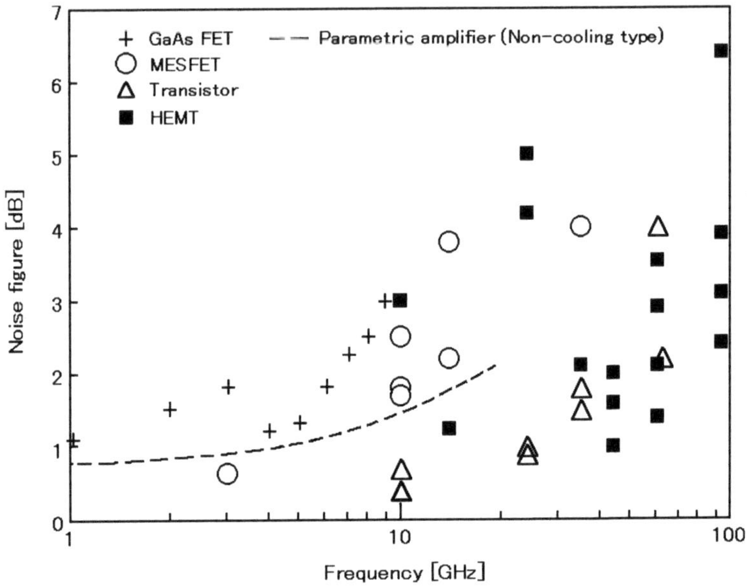

Fig. 8.32 Characteristics of typical low noise amplifiers. The *broken line* is from Skolnik (1981, p. 351)

semiconductor board, has been developed and is widely used. The Gallium Arsenide field effect transistor (GaAs FET) or metal-semiconductor FET (MESFET) is used mainly for high performance microwave applications and in semiconductor rf amplifiers. The GaAs FET is also applied for the low noise amplifier of the receiver of satellite communication earth station. The characteristics of the typical low noise amplifiers are shown in Fig. 8.32.

Receiver Principle

As shown in Figs. 8.19 and 8.31, a typical radar receiver is based on the super-heterodyne principle. The rf received signal, which contains the signal at the carrier frequency signal, is amplified in the first stage, where filtering is applied in a comparatively wide frequency band. The filtered and amplified signal is then mixed with the LO signal which is generated by the STALO. If the frequency of the LO is f_s and the receiver input signal is centered on f_1, two mixer output signals of frequencies $f_1 - f_s$ and $f_1 + f_s$ are generated.[21] The signal of frequency $f_1 - f_s$ is selected by filtering, and is called IF received signal which is then amplified

[21] As shown in Fig. 8.19, f_1 is the summation of the COHO frequency f_c, LO frequency f_s, and the Doppler shift f_d.

before synchronous detection. Major advantages of superheterodyne principle are as follows:

1. Amplification of signals in IF band is lower in cost and more stable in operation than in rf band.
2. The ratio of the pass band and IF is much larger than that of the pass band and rf (carrier frequency). Thus the filter design and its operation are easier than those in rf band.
3. Even if the transmitted frequency changes, the receiver operates only by changing the frequency of the local oscillator. There is no need to change the IF and latter stage circuits.

Although other receiving methods are used (e.g., the direct detection or homodyne receiver Doviak and Zrnić 2006, p. 30 and Skolnik 2001, p. 738; the superregenerative receiver Skolnik 1990, p. 3.3), it can be said that almost all radar receiver adopts the superheterodyne method.

There are from one to several steps to convert received signal to final IF frequency signal of several MHz to around 100 MHz. Using multiple frequency conversion steps, the rejection performance of interference due to image signal reception can be improved. In meteorological radar, receivers of double superheterodyne principle that includes two frequency conversion steps is generally adopted. On the other hand, in atmospheric radar, as the transmitted frequency is lower compared with that of meteorological radar, the single superheterodyne principle is in general.

Optimal Design of Receiver

Significant parameters in IF band which determine the characteristics of receiver are the center frequency, the band width of the band pass filter, the gain of the amplifier, the dynamic range, and so on. The filter which maximizes SNR in the receiver, where external noise and Johnson noise are added, is called matched filter as was mentioned in Sect. 5.2.1. The matched filter delays input signal and reverses the time axes of it as shown in (5.30). If received signal is applied to a filter of finite bandwidth, loss of received signal occurs compared with the filter of infinite bandwidth. The loss in the case of applying rectangular pulse to the matched filter is around 1.8 dB (Nathanson and Smith 1972).

As is difficult to compose the perfect matched filter in practice, an approximate alternative is actually employed. The degradation of SNR, due to the loss by employing alternative filter instead of the perfect matched filter, depends on the characteristic of the input signal waveform and that of alternative filter. If the filter band width is B_{f} and the pulse width of the signal is τ, the loss due to filter mismatching is determined by the product between B_{f} and τ ($B_{\mathrm{f}}\,\tau$ product). The $B_{\mathrm{f}}\,\tau$ products that minimize the degradation of SNR for various input signals and various filter characteristics are shown in Table 8.7. The 6-dB width (B_6), the 3-dB width, and all power in the table mean the band width that the power is decreased by 6 dB, by 3 dB, and the bandwidth to pass all power, respectively. The loss [dB] in the table

Table 8.7 Efficiency of nonmatched filters compared with matched filters (adapted from Skolnik 1990, p. 3.21)

Input signal	Filter	Optimum ($B_6\tau$) 6-dB width	3-dB width	All power	Loss in SNR [dB]
Gaussian pulse	Gaussian	0.88	0.44	0.50	0
Gaussian pulse	Rectangular	1.05	0.74	0.79	0.51
Rectangular pulse	Gaussian	1.05	0.74	0.70	0.51
Rectangular pulse	5 cascaded single-tuned stages	0.97	0.67	0.76	0.51
Rectangular	2 cascaded single-tuned stages	0.95	0.61	0.75	0.56
Rectangular pulse	Single tuned	0.70	0.40	0.63	0.88
Rectangular pulse	Rectangular	1.37	1.37	1.37	0.85

means the degradation of SNR due to the bandwidth for B_6. For example, if the input signal of Gaussian waveform is applied to Gaussian filter, the degradation of SNR does not occur. On the other hand, if the input signal of rectangular waveform is applied to the same filter, the degradation of SNR, i.e., the loss due to filter mismatch, becomes about 0.5 dB. The SNR for various input signal waveforms become maximum if the B_6 value is the reciprocal of around τ, i.e., $B_6\tau \simeq 1$.

Actually, input signal is not a complete rectangular wave. Hence, considering the other fluctuation factors, too, the receiver band width is designed to $B_6\tau \simeq 1.5 - 2$, and is a bit wider than the optimum value. Also, as shown in the above table, the loss due to the degradation caused by the mismatch of filter is around 0.5–0.9 dB. Typically, the value of 0.5 dB is employed as the loss by the mismatch of filter. Thus, in the computation of actual radar equation, the total loss of 2.3 dB due to the limitation of filter bandwidth and filter mismatch should be taken into account.

Usually, the carrier frequency of pulse radar is modulated with the pulse. Hence, the above mentioned alternative filter is applied on both I and Q signals after phase detection is performed.

Phase Detection Circuit

The procedure to extract the information of amplitude and phase from the received signal is performed by phase detection that was mentioned in Sect. 5.2.2. The received signal which is converted to IF with superheterodyne receiver is applied to two synchronous phase detectors as shown in Fig. 5.6. The output signal of the COHO is divided to two ports, the one is directly applied to the phase detector and produces I signal, and the other is applied to the other phase detector after phase shift of 90° and produces Q signal. The procedure to treat I and Q signals in analog form and convert them to digital form is called analog IQ detection method. On the other hand, the procedure to convert IF received signals into digital signals directly is developed in recent years. This procedure is called digital IQ detection method. In

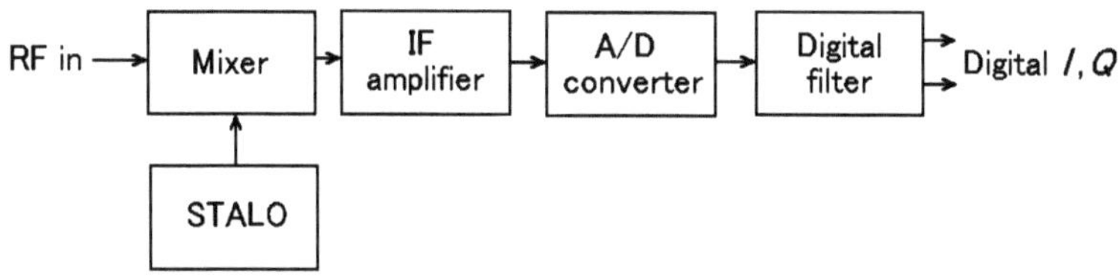

Fig. 8.33 Signal sequence of digital phase detection

this case, the mixer output IF signal is amplified in the IF amplifier, converted from analog to digital (A/D), and applied to digital filter to produce digital I and Q signals as shown in Fig. 8.33. In analog IQ detection method, it is necessary to adjust the offset level of I and Q signals and to calibrate those amplitudes and phases. On the other hand, in digital IQ detection method, the above mentioned adjustments can be treated easily in digital signal process or software algorithm, thus the precision of IQ detection is markedly improved.

8.4 Digital Signal Processing System

In early days, signal processing was performed on the analog signals. Nowadays, with the development of semiconductor devices, signal processing is performed on the digital signals. Today, with the development of high speed CPU processing, various algorithms which were performed with special purpose processors can be treated in real time with the software using the CPU for general use. In this section, we will overview the flow of radar signal processing and discuss the basic processing algorithms.

8.4.1 Signal Processing

In meteorological radar, the estimation of echo intensity, Doppler velocity, Doppler velocity spectrum width, and so on are performed using received echoes from each resolution volume at (r, θ, ϕ) in Fig. 3.6. On the other hand in atmospheric radar, the main process is to obtain the height profile of the above mentioned data with multiple antenna scans around zenith. Although the details of the processing such as removal of unwanted signals are different between in meteorological radar and atmospheric radar, basic technique, such as spectral analysis which is adopted on the estimation of Doppler velocity and velocity spectrum width, is common.

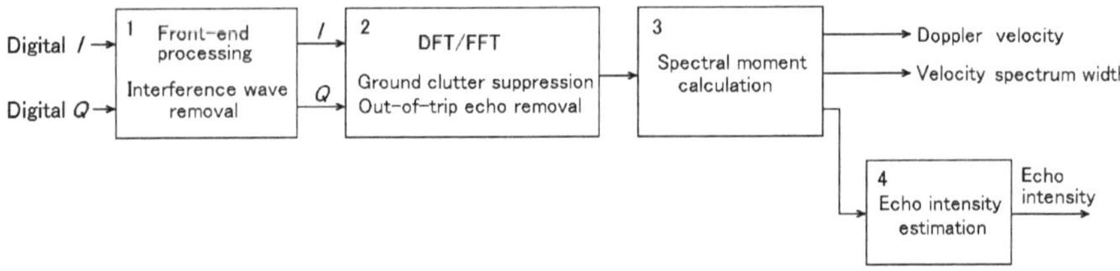

Fig. 8.34 Signal processing sequence of typical meteorological radar

Signal Processing with a Meteorological Radar

Typical flow of the signal processing in meteorological radar is shown in Fig. 8.34. The process is composed of the following steps.

1. The front-end processing using the digitized I and Q signals to remove the interference waves from neighboring radars. In general, rf filters are used to remove interference waves. Even if they remain through the filtering, they are removed in this step.
2. Spectral analysis, where the time series signal samples are transformed to the spectral domain using a discrete Fourier transform (DFT) or a fast Fourier transform (FFT). The processing of ground clutter suppression and out-of-trip echoes removal (see Sect. 8.4.2) are done in this step.
3. Spectral moment calculation (see Sect. 5.4.3) and/or inverse DFT (IDFT) or inverse FFT (IFFT) if necessary, i.e., the frequency domain signal samples are transformed back to the time series signal samples, and the autocovariance processing is then performed to estimate spectral moments (see Sects. 5.5.4 and 5.5.5).
4. The processing to estimate the echo intensity from the radar reflectivity factor using the results of the former step.

Here, we assume a 5.6-GHz meteorological radar having a 150 km observing range and obtaining data every 300 m and every 1.0° azimuth. The rough amount of data which is generated by this radar for 360° of azimuth coverage at each elevation angle can be determined as follows. The number of cells per scan becomes $(150\,\text{km}/0.3\,\text{km}) \times 360 = 180000$. In real time processing, if the antenna rotating speed is 2 rpm (revolutions per minute), the spectral analysis and other calculations are performed for $180000/30 = 6000$, i.e., 6,000 cells/1 s. In above observed range, the minimum pulse repetition time is $(2 \times 150\,\text{km})/3 \times 10^5\,\text{km} = 1.0\,\text{ms}$, where the out-of-trip echoes are removed by the method such as will be discussed in Sect. 8.4.2. In the case of an antenna rotating at 2 rpm, the number of the transmitted pulses for every azimuthal division (every 1.0°) becomes $30\,\text{s}/(360 \times 1.0 \times 10^{-3}\,\text{s}) \simeq 83$. Thus, a DFT of 80 points or more or an FFT of 64 points can be performed. The maximum measurable Doppler velocity, i.e., the Nyquist limit v_N, is given by (4.12). Substituting the wavelength of 5.6-GHz radar $\lambda =$

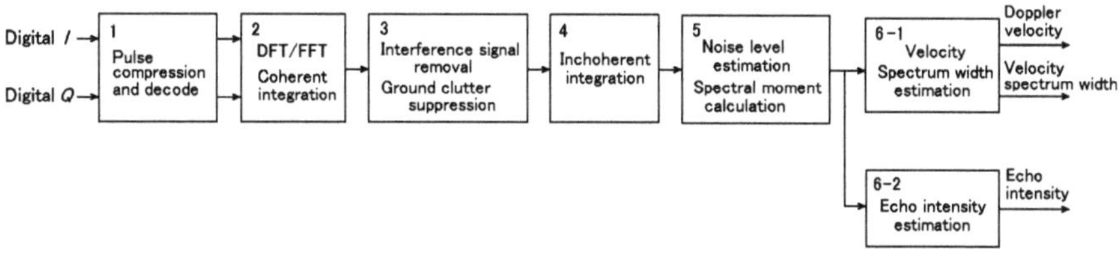

Fig. 8.35 Signal processing sequence of typical atmospheric radar

5.36×10^{-2} m, $v_N = 5.36 \times 10^{-2}\,\mathrm{m}/(4 \times 1.0 \times 10^{-3}\,\mathrm{s}) \simeq 13.4\,\mathrm{m\,s^{-1}}$ is obtained. If spectral processing is performed with a 64 point FFT, the velocity resolution becomes $13.4\,\mathrm{m\,s^{-1}} \times 2/64 = 0.418\,\mathrm{m\,s^{-1}}$.

To perform the spectrum moment estimation, e.g., a 64 points FFT for 14 bit data is executed once for 6,000 cells per 1 s. In the case of the autocovariance estimation, an IFFT of the same amount should be added to transform the data to the time series signal samples again.[22] Today, these processing becomes executable sufficiently with single or parallel processor.[23]. As will be mentioned in Sect. 8.4.3, in recent years, the A/D conversion bit length of 12–14 bits is often adopted to obtain a dynamic range of 60–70 dB or more.

Signal Processing with an Atmospheric Radar

Basic flow of the signal processing in an atmospheric radar is shown in Fig. 8.35. The typical process is mainly composed of the following steps.

1. The processing of decoding for coded received signal in pulse compression (see Sect. 8.3.4).
2. The DFT or FFT and the coherent integration, i.e., the processing to integrate the signal data within the coherency time of the signal, and to suppress receiver noise (see Sect. 5.6.2).
3. The removal of interference signal and ground clutter suppression based on the power spectrum processing (see Sect. 8.4.2).
4. The incoherent integration, i.e., averaging multiple spectra from the same location (see Sect. 5.6.3).
5. Estimation of noise level and calculation of the moments of the signal power spectrum.
6. Estimation of mean Doppler velocity, velocity spectrum width, and echo intensity along the beam.

---

[22] If the signal processing including ground clutter suppression and autocovariance estimation is done only in the time domain as shown in Fig. 5.16, the FFT and IFFT are not needed.

[23] It should be noted that if the out-of-trip echoes removal is done in the frequency domain, further FFT and IFFT combination is needed as will be shown in Sect. 8.4.2.

The processing after the data digitization is usually performed by the software using the general processor like the one used in meteorological radar. The amount of the data which is generated with atmospheric radar can be calculated as follows. First, A/D conversion of bit length of 12–14 bits is often adopted (i.e., the same as that for meteorological radar). The intended height range is different according to the purpose of the atmospheric radar. For example, if the observation range is 15 km in the vertical direction, the minimum pulse repetition time becomes 10^{-4} s. In case the vertical height resolution is 150 m, the number of resolution cells is $15\,\text{km}/0.15\,\text{km} = 100$ for each beam direction. If the antenna is scanned in five directions, the total number of data cells becomes 500 needed to obtain one vertical sounding of wind, turbulence, and reflectivity estimates. The coherent integration is performed for each cell according to the correlation time of the signal. Then, the data of the required FFT points are collected. If 128 points FFT is performed without coherent integration, it is necessary to perform the spectrum analyses for 500 data cell of the bit length of 12–14 in every $12.8\,\text{ms} \times 5 = 64\,\text{ms}$. In the spectral analysis, FFT processing, coherent integration, incoherent integration, and so on are performed. Coherent integration and incoherent integration are arithmetic addition, and the needed processing time for those operations is negligible compared with that needed for the FFT and the IFFT. In the atmospheric radar, the above mentioned processing steps are generally performed by a single processor.

8.4.2 *Removal of Unwanted Signal*

In the reception of backscattered signal, unwanted signals, such as the out-of-trip echo, ground clutter associated with mountains or buildings, echoes from aircraft, and interference from neighboring radars, sometimes contaminate desired signals such as precipitation echoes and atmospheric echoes. Various techniques to remove these unwanted signals which have been developed are discussed in this section.

Reduction of the Effects from Out-of-Trip Echoes

The maximum observable Doppler velocity in pulse radar is proportional to the PRF($= f_p$) and the radar wavelength, whereas the PRF is inversely proportional to the unambiguous range. Hence, it is desirable to fix the PRF to the value that does not cause range ambiguities. If the observation of 150 km range is performed with the PRT of 1.0 ms, the first trip echoes come from ranges between 0 and 150 km. However, if there are storms beyond 150 km, the second trip echoes come from ranges between 150 and 300 km, and the third trip echoes are often seen in wide spread severe storm situations. Echoes from those storms can overlap echoes from nearby storms as shown in Fig. 8.36.

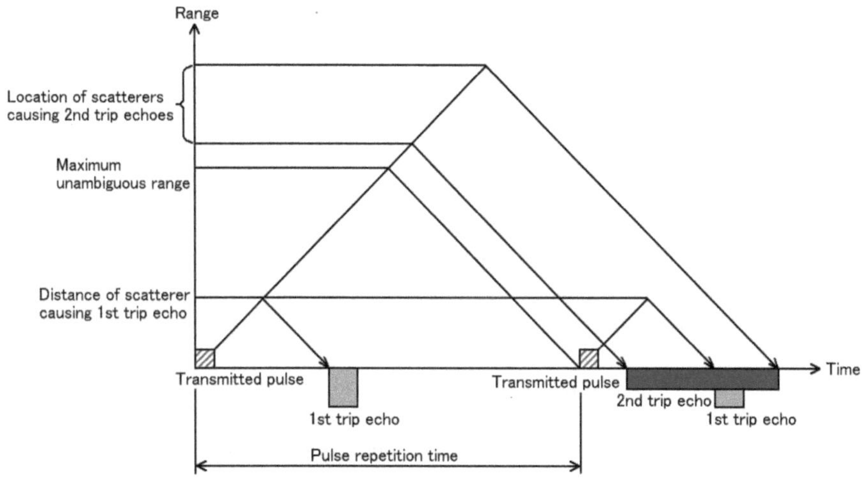

Fig. 8.36 Conceptual diagrams of received first trip echo overlapped by the second trip echo

To suppress out-of-trip echoes, two approaches are common. One relies on variable pulse repetition time (PRT $= 1/f_\mathrm{p}$), e.g., the staggered PRF that spreads echoes in the time domain (e.g., Zrnić and Mahapatra 1985). This approach is based on the fact that the location of the first trip echo is invariable even if the PRF changes, whereas the location of the second trip echo varies according to the PRF. Thus, by carefully selecting the PRFs, range obscuration from distant storms can be minimized over a specified region (e.g., Crocker 1988). The hybrid multi PRF (HMP) method developed by Yamauchi et al. (2006, 2007) which applies multi PRF and spatial continuity of Doppler velocities, is effective for out-of-trip echo removal and automatic dealiasing.

The other approach uses the frequency domain to separate overlaid echoes (e.g., Laird 1981; Siggia 1983). In this approach, the phase coding scheme alters the spectra of the overlaid signals, enables one to estimate the spectral moments of each with appropriate processing techniques. Let the received signal for the mth transmitted pulse be V_m, and the first trip echo, the second trip echo, and the total system noise be V_{m1}, V_{m2}, and N_m, respectively. Thus, the value V_m is expressed as

$$V_m = V_{m1} + V_{m2} + N_m \,. \tag{8.50}$$

If the mth pulse is transmitted with the initial phase shifted Φ_m from the reference phase, and if the received signal is compensated to negate the initial phase shift, the received signal becomes

$$V_m \mathrm{e}^{-j\Phi_m} = V_{m1}\mathrm{e}^{-j\Phi_m} + V_{m2}\mathrm{e}^{-j\Phi_m} + N_m\mathrm{e}^{-j\Phi_m}, \tag{8.51}$$

where V_{m2} in the second term of the right-hand side of above equation is the scatterer signal caused by the $(m-1)$th transmitted pulse. Thus, its original phase difference to the reference phase is Φ_{m-1}, and it becomes to $\Phi_m - \Phi_{m-1}$ nevertheless after above phase compensation. If the signal of the reference phase is expressed with the suffix "s", $V_{m1} = V_{m1s}e^{j\Phi_m}$, $V_{m2} = V_{m2s}e^{j\Phi_{m-1}}$, and (8.51) is expressed as

$$V_m e^{-j\Phi_m} = V_{m1s} + V_{m2s} e^{-j(\Phi_m - \Phi_{m-1})} + N_m e^{-j\Phi_m}. \tag{8.52}$$

If the series of Φ_m is selected as $\Phi_m - \Phi_{m-1}$ becomes random between $[0, 2\pi]$, the second term of the right-hand side of (8.52) is assumed to be random phase noise. Sachidananda and Zrnić (1986) proposed several phase-coding techniques. Doviak and Zrnić (2006, p. 169) shows the conceptual diagram of the rejection of overlaid signals using the random phase method. By applying the FFT to the received signal series after the phase compensation, only the first trip echo spectrum remains intact whereas the second trip echo spectral power is spread quasi uniformly between $\pm v_N$, where v_N is the Nyquist velocity. Thus the spectral spread of the second trip power increases the apparent noise, and the SNR for the first trip echo becomes degraded. To avoid the degradation of the SNR, the second trip echo spectrum is first cohered and then suppressed by notch filtering. Next, the IFFT is applied to the portion of the spectrum remaining after filtering to obtain the time series of the first trip echoes to which autocovariance processing techniques can be applied to extract the spectral moments.

The basic concept of the frequency domain approach is the same for various phase coding schemes; however, the relative merits of the phase codes differ and can be gauged by the accuracy of estimating the spectral moments. Sachidananda and Zrnić (1999) applied not a random phase code but a periodic phase code, the SZ-code, which generally outperforms the random code and its expansion is adopted to the upgrade of the NEXRAD.

Ground Clutter Suppression

In general, ground clutter has a long correlation time, which means that its spectrum width is very narrow, and its mean Doppler velocity is zero (Doviak and Zrnić 2006, p. 202). In the past, most meteorological radars have been built using the approach of a selectable but fixed notch-width infinite impulse response (IIR) filter, a high-pass filter with a pass band Doppler velocity larger than e.g., 1 m s^{-1}. Some radar systems have used clutter maps and the decision to apply the IIR filter depends on the strength of the clutter. If the spectral processing of Doppler signal, i.e., the FFT, is employed, the Doppler velocity spectrum of ground clutter can easily be recognized and hence removed prior to Doppler moment estimation which was discussed in Sect. 5.4.3. The advantage of a spectral approach is that ground clutter filtering can be made adaptive by modeling the clutter and weather echo spectra in the frequency domain. The disadvantage of this approach is that meteorological echoes of zero Doppler are suppressed along with ground clutter. One of the techniques to avoid the above disadvantage is to remove clutter points and replace those points by assuming Gaussian meteorological spectrum, e.g., the Gaussian model adaptive processing (GMAP) developed by Siggia and Passarelli (2004).

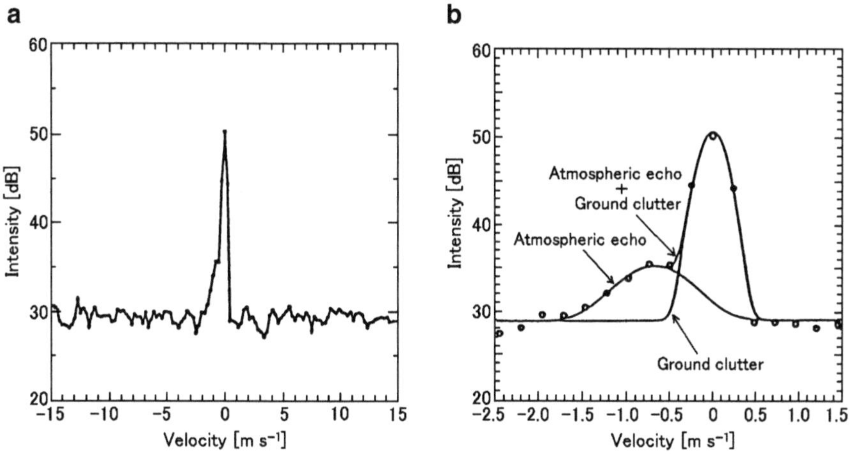

Fig. 8.37 Doppler spectrum of atmospheric radar received signal. (**a**) Echo including ground clutter. (**b**) Expanded view of the spectrum about zero Doppler velocity. Two Gaussian curves who have the peaks of zero Doppler and $-0.7\,\mathrm{m\,s^{-1}}$ are the ground clutter and the atmospheric echo, respectively. They are fitted by the most likelihood method [from Law et al. 1994]

As for the atmospheric radar observations, it is also indispensable to suppress ground clutter to estimate the moments of an atmospheric echo accurately. As with the discussion of the meteorological radar, a simple and easy way to eliminate ground clutter in real time is to assume that zero Doppler spectral lines and those nearby lines are caused by ground clutter, and then to suppress these coefficients and replace them by zero. In atmosphere radar, unwanted signal elimination based on the spectral analysis in the frequency domain has been performed since the earliest days of radar development. In addition, the spectral estimation methods which are based on prediction theory, such as the maximum likelihood method (MLM), the maximum entropy method (MEM), and so on, are also generally adopted (Ogura and Yoshida 1981). If the line of sight Doppler velocity is small, these methods are not always reliable. Figure 8.37a shows an example in which the Doppler velocity of atmospheric echo is small and its spectrum peak and that of ground clutter are adjacent to each other. It is observed with the 402 MHz atmospheric radar which composes the U.S. wind profiler network.[24] The velocity spectrum about zero Doppler is shown in an expanded view in Panel (b). In Panel (b), two Gaussian shaped spectral models are fitted to the data using the maximum likelihood method; one (the clutter spectrum) has a peak at zero Doppler, and a second (the atmospheric echo spectrum) has a peak at $-0.7\,\mathrm{m\,s^{-1}}$. Eliminating the spectrum of the former curve as the ground clutter, the remaining spectral coefficients are associated with atmospheric echoes (Law et al. 1994).

[24]Refer to Sect. 10.3.1.

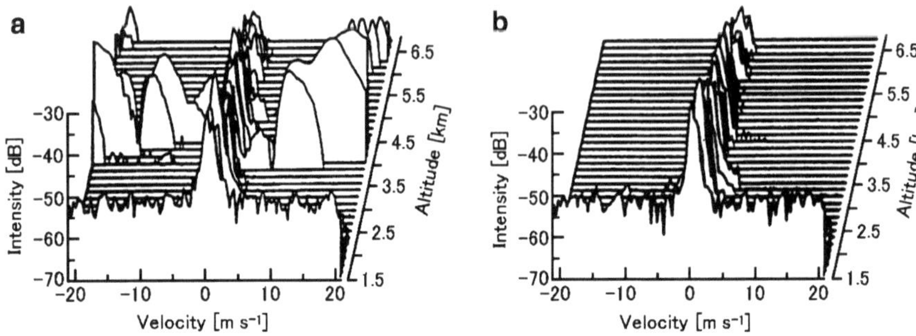

Fig. 8.38 Doppler velocity spectrum of atmospheric radar echo including aircraft echo when the aircraft flies passing over the radar observation area. (**a**) During the passing of aircraft and (**b**) after passing

Moreover, it can be shown that the narrow ground clutter spectrum, which is caused by fluctuations in the air refractivity along the ray path and movement of ground scatterers (i.e., tree limbs, leaves, grass, autos, etc.), can spill over the entire frequency range if the time window for every DFT is shorter than the fading clutter correlation time. It sometimes becomes a few tens of dB stronger than the atmospheric echo. An effective estimation method for such particular cases is already practicable (Sato and Woodman 1982).

Removal of Clutter from Moving Scatterers

The Doppler velocities of clutter from moving scatterers such as vehicles, rail traffic, ships, aircrafts, and birds typically are not zero; thus this clutter cannot be suppressed by zero Doppler filtering. Moving isolated scatterers (Table 3.1) can be detected using a point target filter. This filter uses multiple range gates of data to identify clutter associated with a moving isolated scatterer. The point target filter looks for targets with reflectivity much higher than the neighboring range gates. This types of filter can remove clutter from moving isolated hard scatterers, but weather data in the same range gates are lost (Cho 2009). A more modern approach filters data in the two-dimensional, Doppler velocity spectrum vs. range. The two-dimensional filter utilizes the characteristics that moving clutter spectral signals tend to be spectrally compact and discontinuous in range, whereas meteorological and atmospheric received signals are continuous (e.g., Sasaoka 2003; Meymaris 2007; Cho 2009). This technique is applicable to both meteorological radar and atmospheric radar.

Figure 8.38 shows the Doppler velocity spectra of atmospheric echoes including echoes from aircraft passing through the radar observational volume. Panel (a) shows the data during the passing of aircraft, and (b) shows that of after passing, respectively. The radar beam is directed to the zenith, and about 1 min of Doppler velocity spectra are averaged. In Panel (a), the strong echo which is received at the height above 4 km or higher is caused by aircraft, where the altitude of the aircraft is only an apparent altitude because the echoes from the aircraft are received through antenna sidelobes; that is, the apparent altitude is the range to the aircraft. The Doppler velocity spectrum which is intermingled with aircraft echo can be easily discriminated and eliminated with e.g., above-mentioned two-dimensional filter because it does not continue both in time and in altitude.

The atmospheric radar of the frequency of 400 MHz or higher usually receives the backscattered signal from migrating birds (e.c., Wilczak et al. 1995). The birds are assumed to be small ones that fly on the low-level jet under 2–4 km height, in clear night of the spring and the autumn. As a bird of heavy water content has a large radar cross section, it is easily observed with radar. The 404 MHz band radar of the U.S. wind profiler network is estimated to detect birds of the density of $3.6 \times 10^{-8}\,\mathrm{m}^{-3}$ at the height of 2 km (Wilczak et al. 1995). When the echo intensity of migrating bird is stronger than meteorological or atmospheric echo, the radar sometimes estimates not the movement of meteorological or atmospheric echo but the travelling speed of the birds. Several algorithms to eliminate migrating bird echo are developed by e.g., Pekour and Coulter (1999) and Sasaoka (2003) for the atmospheric radar, and by e.g., Zhang et al. (2005) and Liu et al. (2005) for the meteorological radar.

8.4.3 Analog to Digital Conversion

Quantization noise is caused in the process of sampling analog signals and converting them to digital numbers. Thus, the main parameters which determine the performance of analog to digital (A/D) conversion are the dynamic range and the quantization noise of the signal.

Dynamic Range

If V_{op} is the absolute value of the maximum input voltage to the A/D converter, the full scale voltage range (span) between $-V_{\mathrm{op}}$ and V_{op} is $2V_{\mathrm{op}}$, and N is the A/D resolution in bits, then the resolution Δ of the A/D converter, equal to the least significant bit (LSB) voltage, is given by

$$\Delta = \frac{2V_{\mathrm{op}}}{2^N - 1}. \tag{8.53}$$

For example, if $V_{\mathrm{op}} = 5\,\mathrm{V}$, A/D resolution is 10 bits, the A/D converter's voltage resolution is $\Delta = 2 \times 5\,\mathrm{V}/(2^{10} - 1) \simeq 9.8 \times 10^{-3}\,\mathrm{V}$.

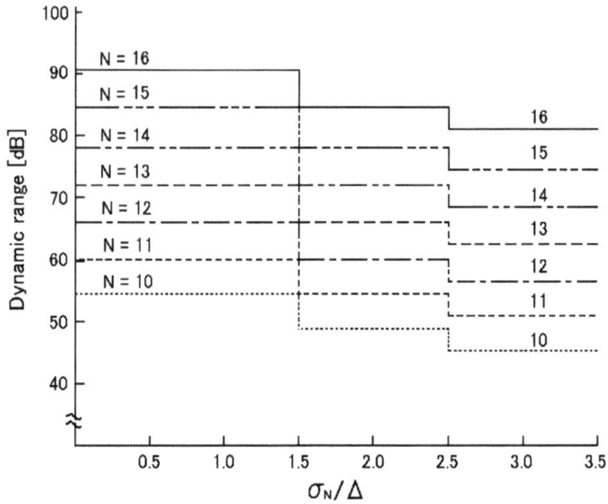

Fig. 8.39 Dynamic range of A/D conversion for bit lengths $N = 10$–16

The maximum dynamic range, D_{rmax}, is obtained if noise is not included in the input signal, and the value is expressed as

$$D_{\text{rmax}} = 20\log\frac{V_{\text{op}}}{\Delta} \quad [\text{dB}]. \tag{8.54}$$

Substituting (8.53) into (8.54), $D_{\text{r}_{\text{max}}}$ becomes

$$D_{\text{rmax}} = 20\log\frac{2^N - 1}{2} \quad [\text{dB}]. \tag{8.55}$$

Even though noise is contained in the input signal, if the noise level σ_{n} is low enough to satisfy $\sigma_{\text{n}} < 3\Delta/2$, the rms value of the quantized noise becomes Δ. Hence, the dynamic range is equal to the value given by (8.55). If noise increases over the above level and $(n+1/2)\Delta \leq \sigma_{\text{n}} < (n+3/2)\Delta$, the rms value of noise becomes $(n+1)\Delta$, where n is the positive integer. Although signal levels below noise power are detectable with appropriate signal processing such as coherent integration (see Sect. 5.6.2), if the dynamic range is defined between the output signal power and the quantization noise flower, the dynamic range of signal decreases to

$$D_{\text{r}} = 20\log\frac{V_{\text{op}}}{(n+1)\Delta} = 20\log\frac{2^N - 1}{2(n+1)} \quad [\text{dB}]. \tag{8.56}$$

Figure 8.39 shows the dynamic range of A/D conversion for bit lengths $N = 10$–16, where the abscissa axis expresses the noise level normalized by the minimum bit Δ.

Quantization Noise

Concerning the quantization unit of the A/D conversion Δ which is given by (8.53), the probability density function $p(x)$ for uniformly distributed input signal x is

$$p(x) = \frac{1}{\Delta} \quad \text{where} \quad -\frac{\Delta}{2} \leq x \leq \frac{\Delta}{2}, \tag{8.57}$$

and the noise power generated by the quantization error, $\overline{x^2}$, which is the square mean value of x becomes

$$\overline{x^2} = \sigma_{\mathrm{q}}^2 = \int_{-\infty}^{\infty} x^2 p(x)\mathrm{d}x = \left[\frac{x^3}{3\Delta}\right]_{-\Delta/2}^{\Delta/2} = \frac{\Delta^2}{12}. \tag{8.58}$$

Degradation of SNR Due to A/D Conversion

If the A/D conversion input signal and noise power are S_{i} and N_{i}, respectively, the input SNR is $\mathrm{SNR_i} = S_{\mathrm{i}}/N_{\mathrm{i}}$. Here, the input noise amplitude is σ_{n}, thus,

$$N_{\mathrm{i}} = \sigma_{\mathrm{n}}^2, \tag{8.59}$$

and the A/D conversion output noise power N_{o} is given by the summation of input noise power and quantization noise power as

$$N_{\mathrm{o}} = \sigma_{\mathrm{n}}^2 + \sigma_{\mathrm{q}}^2, \tag{8.60}$$

On the contrary, the signal power does not change before and after the A/D conversion. Thus, the signal power after the A/D conversion S_{o} is

$$S_{\mathrm{o}} = S_{\mathrm{i}}. \tag{8.61}$$

As a result, using (8.59)–(8.61), the change of the SNR through the A/D conversion, i.e., the degradation of SNR, I_{SN}, is given by

$$I_{\mathrm{SN}} = \frac{S_{\mathrm{o}}/N_{\mathrm{o}}}{S_{\mathrm{i}}/N_{\mathrm{i}}} = \frac{S_{\mathrm{o}}}{S_{\mathrm{i}}}\frac{N_{\mathrm{i}}}{N_{\mathrm{o}}} = \frac{\sigma_{\mathrm{n}}^2}{\sigma_{\mathrm{n}}^2 + \sigma_{\mathrm{q}}^2}. \tag{8.62}$$

Substituting (8.58) into (8.62), then

$$I_{\mathrm{SN}} = \frac{1}{1 + \frac{1}{12}\left(\frac{\Delta}{\sigma_{\mathrm{n}}}\right)^2}. \tag{8.63}$$

The change of I_{SN} against $\sigma_{\mathrm{n}}/\Delta$ is shown in Fig. 8.40. From the figure, it is evident that if the quantization unit is suppressed within around 1.7 times of the A/D conversion input noise, i.e., $\sigma_{\mathrm{n}}/\Delta > 0.6$, it is found that the degradation of SNR due to A/D conversion is 1 dB or less.

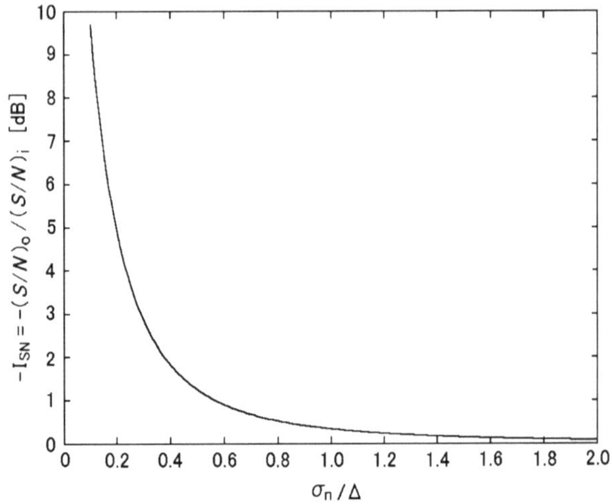

Fig. 8.40 Change of I_{SN} against σ_n/Δ

8.4.4 Spectral Analysis

The basics of the signal processing for meteorological radar and atmospheric radar are efficiently finding the power spectrum from *I* and *Q* signal series as discussed in Sect. 5.4 and getting echo intensity (or radar reflectivity factor), Doppler velocity, and velocity spectrum width. The main operations of the spectral transform of the discrete *I* and *Q* signals, which are obtained by the A/D conversion, are the DFT and the IDFT as were described in Sect. 5.5.1. Major portion of the processing time performed for received signal is spent for the DFT and IDFT.

FFT

The FFT which was devised by Cooley and Tukey (1965) is the technique to perform DFT at high speed. High-speed computing apparatus such as the parallel processor and the array processor were built to perform the FFT of data efficiently processing in real time. In recent years, it can be realized readily by the software processing owing to the performance improvement of computer. The basic algorithm of the FFT is to divide the DFT points *N*, where the number of points *N* is a power of two, one after another to the small radix, radix-2, and perform the DFT by the small radix. The FFT algorithms generally fall into two classes, i.e., two different algorithms that perform the same operation to find the DFT.

One is called the decimation-in-time (DIT) FFT algorithm which is the technique to divide the sequence according to the number of sample in the time axis. The DIT radix-2 FFT recursively partitions a DFT into half-length DFTs of the even-indexed and odd-indexed time samples. The other one is to divide the sequence according to the number of samples along the frequency axis, and is called the decimation-in-frequency (DIF) FFT algorithm. The DIF radix-2 FFT partitions the DFT computation into even-indexed and odd-indexed outputs, which can each be computed by shorter-length DFTs of different combination of input samples. Both differ in the order of the calculation only and there is no difference in calculation time. Incidentally, there are the same two techniques in the IFFT which correspond to that of the FFT, respectively.

Reduction of Calculation by the FFT

The FFT operation which is based on above mentioned algorithms is discussed in Appendix B. In an M points DFT, $M(M-1)$ complex calculations are required, i.e., $4M(M-1)$ real number product operations and the same real number addition are needed. On the other hand, an M points FFT is implemented, $\log_2 M$ recursive operations are required; these recursive operations are each composed of $M/2$ butterfly operations,[25] i.e., $2\log_2 M$ real number product operations, and $3M\log 2M$ real number additions. Thus, a rough estimate of the number of operations for the DFT and the FFT can be compared based only on the number of product operations which is more time-consuming than addition. Table 8.8 shows the comparison of the number of product operations and additions in the DFT and in the FFT (Ziemer et al. 1998). It is evident from the table that the larger is M, the smaller is the per cent of time needed for the FFT compared to the DFT.

8.4.5 *Window Function*

In actual signal processing of the DFT which treats discrete data, spectral distortion due to the finite length of data is unavoidable. To mitigate the distortion, an adequate sampling window is usually applied, where a window function is a mathematical function that goes to zero smoothly yet sufficiently rapidly at the edges of the chosen interval. If the length M of discrete time series $v(m)$ (M is the number of points in

---

[25] If the length M in (5.125) is $M=2$, the DFT of the time signal series $v(m)$ becomes

$$Z(k) = \sum_{m=0}^{1} v(m)W_2^{mk} \quad k = 0, 1,$$

where $W_2^1 = \mathrm{e}^{-j2\pi/2} = -1$. Thus, the above equation is expressed for $k=0$ and $k=1$ as

$$Z(0) = v(0) + v(1) \qquad\qquad Z(1) = v(0) - v(1),$$

respectively. These equations are the operations of the radix-2 FFT algorithm, and the operation flow is analogous to butterfly, and called the butterfly operation.

Table 8.8 Comparison of the number of times for product operation and addition in the DFT and in the FFT (from Ziemer et al. 1998, p. 503)

FFT points	DFT	FFT	Ratio
2	8	4	2
4	48	16	3
8	224	48	5
16	960	128	8
32	3,968	320	12
64	16,128	768	21
128	65,024	1,792	36
256	261,120	4,096	64
512	1,046,528	9,216	114
1,024	4,190,208	20,480	205
2,048	16,769,024	45,056	372
4,096	67,092,480	98,304	683

Table 8.9 Various windows applied for spectral analysis (from e.g., Ziemer et al. 1998, p. 522)

Window type	Window function[a]
Rectangular	1
Hanning	$0.5-0.5\cos\left(\frac{2\pi m}{M}\right)$
Hamming	$0.54-0.46\cos\left(\frac{2\pi m}{M}\right)$
Blackman	$0.42-0.5\cos\left(\frac{2\pi m}{M}\right)+0.08\cos\left(\frac{4\pi m}{M}\right)$

[a] For $0 \le m \le M-1$, otherwise 0

the DFT), is equal to the fundamental period of $v(m)$, a discontinuity will not occur at the boundary of the data. However, in general, radar received signals are not periodic. As a result of the DFT, high frequency components which are not included in the original signal are sometimes generated. The window function is applied to constrain the rapid change of the received signal. The window, i.e., the weighting function, heavily weights toward zero the samples at the beginning and end of the sampling interval. Thus, the relatively smooth transition of sample power to zero at the window edges reduces the spilling of spectral power into velocity bins that would not have echo power if wider windows (i.e., longer dwell times) were used.

Typical window functions applied to the spectral analysis is listed in Table 8.9. As shown in Fig. 8.41, window function has the characteristic which the both sides of the sample section converge on around 0. Window functions except for the rectangular window suppress the discontinuity of the signal at the window boundary.

Rectangular Window

A function that is constant inside the interval and zero elsewhere is called a rectangular window. This window function does not eliminate the discontinuity of the signal at the window boundaries and thus spills the largest amount of power into Doppler velocity intervals outside the intrinsic spectrum of the atmospheric echoes.

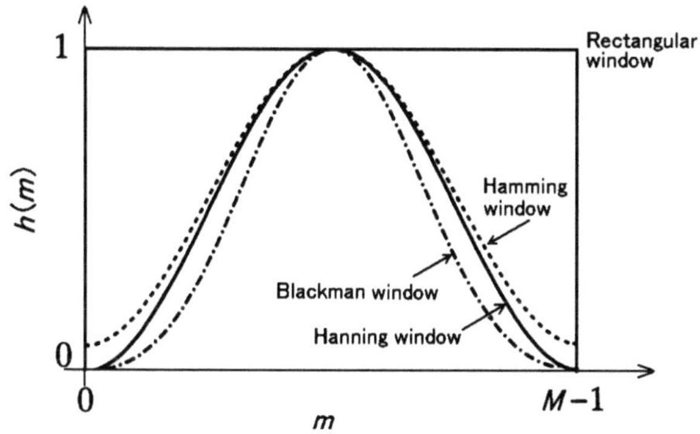

Fig. 8.41 Typical window functions which are applied for spectral analysis

If the window width is T_W, the first minimum value is generated at the frequency of $\pm 1/T_W$ [Hz]. The frequency resolution in the center of the window is excellent. However, the waveform deformation is easily caused at the window edges, and the sidelobe levels are the highest of any of the window functions.

Hanning Window

This window function is a raised cosine waveform whose median value is 1. The frequency resolution of the weather signal spectrum is worse than that of the rectangular window, but the spectral sidelobes are comparatively small. Thus, it is better suited to detect, in the spectral domain, weak signals that are nearby strong signals. On the other hand, the signal components near the window edges are attenuated and the spectral power associated with these components is not well represented in the spectral domain.

Hamming Window

This window function is the sum of a rectangle and a Hanning window, and has amended above mentioned fault of the Hanning window. The characteristic is near that of the Hanning window, and is suppressing the sidelobe around the mainlobe in addition to utilize the signal in the both edges of the window. The frequency resolution of the main component is rather meager, however it is suited for detecting low power spectrum.

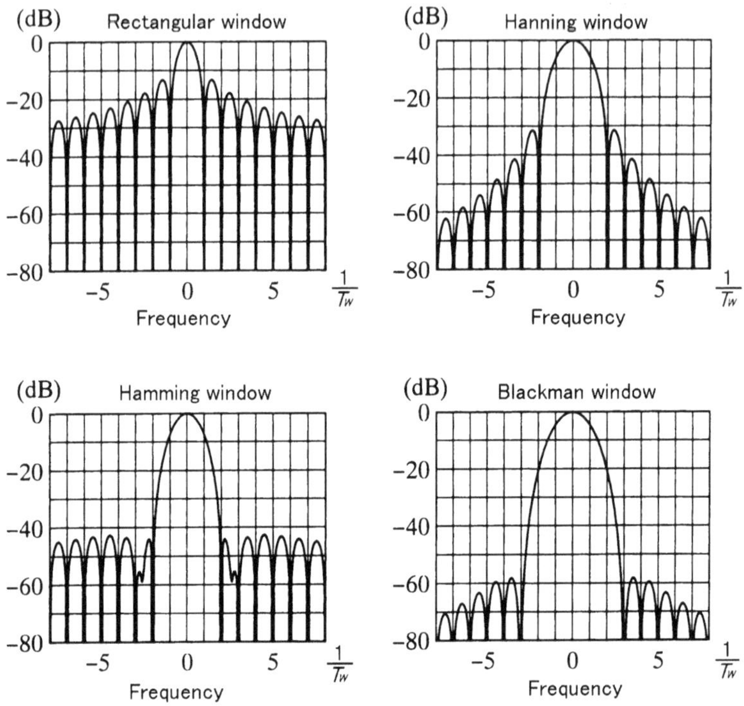

Fig. 8.42 Spectrum of typical window functions which are applied for spectral analysis

Blackman Window

This window function is quite similar to Hann and Hamming windows, and has the characteristic that the sidelobe levels are significantly lower. Thus, it is useful when strong and weak signals need to be resolved in the spectral domain, but it has poorer spectral resolution.

The spectrum of typical window functions is shown in Fig. 8.42. The unit of the abscissa scale is the frequency component $1/T_{\mathrm{W}}$.

The Effect of the Window

Generally, in radar signal processing, the ratio of the maximum power spectral component and that of the minimum one becomes to be about several tens of dB, and the higher frequency resolution is more desirable. Therefore, the Hanning window and the Hamming window are often adopted. Here, the effect of window functions on the power spectrum is examined. Now, the frequency signal series (the spectrum)

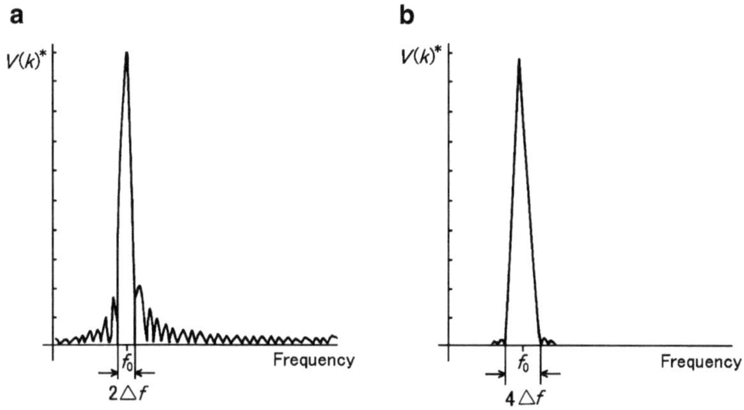

Fig. 8.43 Conceptual diagram to show the effect of the window functions. The convolutions of a line spectrum and (**a**) rectangular window and (**b**) Hanning window, respectively. In both panels, the envelopes are shown, where the center frequency is f_0, and $\Delta f = 1/T_W$. In each panel, the *vertical axis* is an arbitrary linear scale

$V(k)$ with length M which is the DFT of the discrete time signal series $v(m)$ is given by (5.125), and the reverse transform is given by (5.126). Applying these equations, the spectrum $V(k)^*$ which is the product of $v(m)$ and the window function $h(m)$ becomes

$$V(k)^* = \sum_{m=0}^{M-1}[v(m)h(m)]W_M^{mk} = \sum_{m=0}^{M-1}\left[\frac{1}{M}\sum_{n=0}^{M-1}H(n)W_M^{-mn}\right]v(m)W_M^{mk}$$
$$= \frac{1}{M}\sum_{n=o}^{M-1}H(n)\sum_{m=0}^{M-1}v(m)W_M^{(k-n)m} = \frac{1}{M}\sum_{n=0}^{M-1}H(n)V(k-n), \qquad (8.64)$$

where $H(k)$ is the spectrum of window function $h(m)$. Equation (8.64) expresses that $V(k)^*$ is given by the convolution of the spectrum $V(k)$ and $H(k)$. It means that the frequency series signal of the window function output is regarded as the spectrum of the input time series signal, $V(k)$, smoothed with the filter $H(k)$.

If the input signal $v(t)$ is a continuous wave expressed by $v(t) = a\cos(2\pi f_0 t)$, and the window width is a integer times the period of the wave, $V(k)$, the DFT of $v(t)$, becomes single line spectrum at frequency f_0. However, if the sampled signal has a period that is not an integer times the window width, as usually occurs in actual radar observations, spill over of the wave's power into frequency intervals outside that at f_0 will occur. Figure 8.43 shows the conceptual diagram of $V(k)^*$ of (8.61), where $v(t)$ is sampled periodic, thus $V(k)$ is the line spectrum, and $H(k)$s of (a) rectangular window shown in Fig. 8.42 top left panel and (b) Hanning window shown in the same figure top right panel, respectively. In both panels, the vertical axis is arbitrary scale, and the envelopes are shown, where the center frequency is f_0, and $\Delta f = 1/T_W$. It should be noted that the sidelobe almost disappear by applying the Hanning window. On the other hand, the width of the peak value (mainlobe) of Hanning window expands twice Rectangular window.

8.4.6 Parameters for the DFT

The sampling frequency f_s for the DFT and the FFT is determined according to the sampling theorem that $f_s \geq 2W_B$, where W_B is the band width of the signal (assumed low-pass). If the point number of samples used in the DFT or the FFT is M, and the window width is T_W [s], f_s becomes

$$f_s = \frac{M}{T_W} \geq 2W_B \quad [\text{Hz}], \tag{8.65}$$

and the frequency resolution Δf is

$$\Delta f = \frac{1}{T_W} = \frac{f_s}{M} \quad [\text{Hz}]. \tag{8.66}$$

From (8.65) and (8.66), Δf and W_B are in the following relation.

$$\Delta f \geq \frac{2W_B}{M} \quad [\text{Hz}] \tag{8.67}$$

Example of Parameter Selection

If the FFT of $M = 64$ points is performed in a 5.6-GHz band radar whose pulse repetition frequency (PRF) is 1 kHz (the PRF is the sampling frequency f_s), the maximum signal band width W_B is $f_s/2 = 500\,\text{Hz}$, and the frequency resolution becomes $\Delta f = f_s/64 = 15.6\,\text{Hz}$. As the radar wavelength is $\lambda = (3 \times 10^8)/(5.6 \times 10^9) = 5.36 \times 10^{-2}\,\text{m}$, the resolution of Doppler velocity becomes $\Delta v = \lambda \times f_s/2 = (5.36 \times 10^{-2} \times 15.6)/2 = 0.418\,\text{m s}^{-1}$ and the maximum measurable Doppler velocity is $\pm 0.418 \times 64/2 \simeq \pm 13.4\,\text{m s}^{-1}$.

To sample the signal, whose maximum frequency component W_B is 10 kHz, without frequency aliasing, it is necessary to set the sampling frequency to $f_s \geq 2W_B = 20\,\text{kHz}$. If the FFT is performed for the above signal with frequency resolution of 50 Hz, as $20 \times 10^3/50 = 400$, thus 512 points sample data are needed, where 512 is the minimum exponential of 2 that exceed 400. At this time, the sample window width is required to be $T_W = 512/(20 \times 10^3) = 25.6\,\text{ms}$. If the correlation time of the phenomenon is longer than this value, the processing is performed sufficiently.

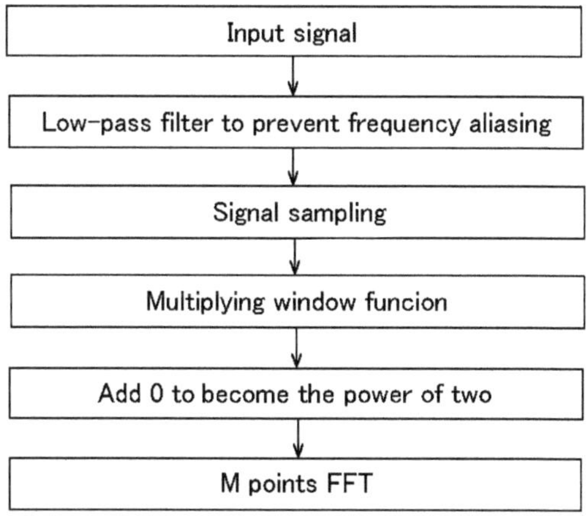

Fig. 8.44 Procedure to perform spectrum analysis by the FFT

The spectral analysis in radar signal processing is performed by the procedure which is shown in Fig. 8.44. First, the input signal $v(t)$ is limited with the low-pass filter to prevent frequency aliasing, and M discrete signals of $v(0)$, $v(1)$, $\cdots$, v(M-1) are obtained. Next, the window function $h(m)$ is multiplied to $v(m)$ where m is the integer and $0 \leq m \leq M-1$. Furthermore, the DFT of $v(m)h(m)$ which is gotten by the above is calculated. The data number M is not always equal to the power of two which is needed for the FFT. At that case, 0 is added to the data series until M becomes the power of two. Finally, the DFT is obtained using the FFT.

Chapter 9
Practical Meteorological Radars

As forementioned in Sect. 1.2, by means of theoretical and experimental research through the 1940s, it became apparent that the backscattered signal, so called weather clutter, is caused from precipitation. The detection of precipitation accelerated the development of meteorological radar. The pulsed Doppler radar was developed in World War II to detect moving targets such as aircraft easily from among ground clutter and sea clutter. It was applied for the meteorological radar observation in United Kingdom for the first time in 1953 (Barratt and Browne 1953), and became the present meteorological Doppler radar through various research and development efforts.

9.1 Meteorological Radars of Various Frequency Bands

Generally, diameters of the precipitation scatterers (i.e., raindrops, snowflakes, and hailstones) range from about 100 μm to several mm. On the other hand, water particles in clouds and fogs are smaller than 100 μm in general. To observe precipitation and clouds, centimetric waves and millimeter waves are used. The former is adopted for the observation of precipitation, and the later for clouds observation. From the point of view of radio wave administration, actual frequencies developed for meteorological radar are confined to the six bands[1] listed in Table 9.1. Microwave radars of the wavelength between 3 and 10 cm are applied for the observation of raindrops, hailstones and so on. Millimeter wave radars of about 3–9 mm wavelength are generally applied for the observation of clouds and fogs. Except for a part of airborne meteorological radars, pulse radars are generally adopted. In fact, all examples which are described in this chapter are pulse radars.

---

[1]Here, 5.3 and 5.6 GHz, so-called C band, are assumed to be one kind all together.

S. Fukao and K. Hamazu, *Radar for Meteorological and Atmospheric Observations*, DOI 10.1007/978-4-431-54334-3_9,

Table 9.1 Frequency band of meteorological radar

Frequency (GHz)	Wavelength (cm)	Frequency band (GHz)		Maximum observable range
2.8	10.7	2.7–3.0	S	200 km or more
5.3[a]	5.7	5.25–5.35	C	Around 200 km
5.6	5.4	5.60–5.65	Ditto	Ditto
9.5	3.2	9.3–9.7	X	Around 60 km
13.8	2.2	13.8	Ku	Satellite-borne (TRMM)
35	0.86	34.5–35.5	Ka	Around 30 km
95	0.32	94.5–95.5	W	Around 10 km

[a] 5.6 GHz is generally used. In Japan, 5.3 GHz band is used due to the internal radio wave conditions. The band limits may be changed according to the radio wave allocation plan

2.8-GHz Band Radar

Radars of 2.8-GHz band are mainly used for the observation of precipitation in wider area of about 200–500 km radius. In 2.8-GHz band, wave attenuation due to precipitation is small. Hence, radars of this frequency band are employed as the meteorological radars in many relatively low latitude countries having episodes of intense precipitation. The Mt. Fuji radar which had been operated between 1964 and 1999 in Japan adopted the same frequency band. Its observation range reached 800 km to observe the approach of typhoons.

5.3/5.6-GHz Band Radar

Radars of 5.3-GHz band or 5.6-GHz band are also used as the radar for precipitation observation. Generally, the observation range of these radars is about 200 km. The WSR-74 which was developed in the U.S.A. used 5.6-GHz band. The electromagnetic wave attenuation due to precipitation in this band is relatively small but greater than that of 2.8-GHz band. Hence, the radars of this frequency band are used in many middle latitude countries and in Europe. In Japan, too, operational meteorological Doppler radars and the Radar Raingauge adopt the frequency of 5.3-GHz band. Furthermore, these frequency band radars are installed at the airport in the U.S.A., Japan, and some countries for the terminal Doppler weather radar to detect wind shears such as microburst and gust front. Because of the difference of radio wave administration, 5.3-GHz band is generally employed in Japan, and 5.6-GHz band is employed in other many countries. However, there is no essential difference between both bands in actual operation.

9.5-GHz Band Radar

In this frequency band, the attenuation of electromagnetic wave due to precipitation increases, thus it is unsuitable for long range observations. However, the radar is comparatively small. Furthermore, it is employed not only as a meteorological radar but also for great number of shipboard radars. Hence, it is possible to compose the radar at lower cost compared with other frequency band, and the radar becomes popular in research use, too. Also, it is often used for the weather observations in the limited area, in less precipitation and snowy districts, and higher latitude districts. In Japan, there are examples to employ it for the sewerage drainage control, for the snowfall observation for the cleaning snow from the road, and so on. The observation range of about 60–80 km radius is general in this frequency band.

In recent years, it becomes clear that some kind of polarimetric parameters are effective to compensate the attenuation of electromagnetic wave due to precipitation. Using the advantage, it is expected to play a role to overcome the weakness caused by the electromagnetic wave attenuation due to precipitation. Thus, polarimetric radar equipped with to transmit horizontally and vertically polarized waves is expected to increase the meteorological utility of these shorter wavelength radars.

35/95-GHz Band Radar

In these frequency bands, the attenuation of electromagnetic wave due to precipitation is large. Hence, the radars of these frequency bands are unsuitable to observe precipitation. On the other hand, it shows utility for the observations of cloud and fog whose particle sizes are far smaller than those associated with precipitation. As for the means to observe cloud and fog, optical sensors such as the visibility meter, lidar, and the ceilometer are commonly known. However, it is difficult to optically observe inside clouds and fog because of the attenuation of the optical wave. Therefore, millimeter wavelength radar measurement is the best technique to observe motions inside clouds and fog which are three-dimensionally spread. The diameters of the minute drops inside clouds and fog are about several tens μm. In the case of microwave meteorological radar, the backscattering cross sections of the tiny drops inside clouds and fog are tens of dB smaller compared to that associated with precipitation particles. Millimeter radars such as the Atmospheric Research Measurement program (ARM) cloud radars (Moran et al. 1998) and Japan's fog radar (see Sect. 9.3.1) (Hamazu et al. 2003) are about 10–15 dB more sensitive than typical 10 cm wavelength meteorological radar. Although their sensitivities are not enough for long range observations, they can be applicable for short range observations of, e.g., around ten to a few-tens kilometer. Moreover, the millimeter radars have other attributes such as high angular resolution and their greater immunity to ground clutter contamination. As a result, the millimeter band is mostly used for the observation of clouds and fog. As shown in Fig. 6.5, attenuation due to atmospheric molecules is quite severe at millimeter wavelengths but is relatively small in 35-GHz band and 95-GHz band compared with adjacent frequencies. Thus, radars that operate in these frequency bands are extensively used for the research of cloud physics.

9.2 Precipitation Observation Radar

9.2.1 2.8-GHz Band Radars

Radars of this frequency band which are suited to observe precipitation over a wide area have been applied in the U.S.A. and many countries from the early days. Many of meteorological radars which the World Meteorological Organization (WMO) has distributed to developing countries operate at 2.8-GHz band. The WSR-57 developed in the U.S.A. had been the principal meteorological radar operated by the US National Weather Service (NWS) for decades.

Research for toward the development of Doppler weather radars for observation of mesoscale systems such as hurricanes and squall lines, started in the U.S.A. in early 1960s. The program to develop the next generation weather radar (NEXRAD) began during late 1970s by the multiagencies of the NWS, the US Federal Aviation Administration (FAA), and the US Air Force Weather Service (AFWS).

9.2.2 NEXRAD: WSR-88D

The WS-88D radar is designed to significantly enhance tornado warnings, improve the detection and measurement of damaging winds, severe turbulence, wind shear and hailstorms, and more accurately delineate areas that are threatened by severe weather. The first deployment the radar began in 1990, and had been completed by the end of 1990s. The radars cover contiguous U.S., Alaska, Hawaii, Guam, and Puerto Rico. They operate in the band 2.7–3.0 GHz. Operational frequencies are chosen for each site to eliminate or minimize interference from nearby weather radars, or other radars. The distribution of the WSR-88D radars and the appearance of the radar tower are shown in Fig. 9.1.

Functions and Performance

The WSR-88D is a fully coherent radar and has a parabolic dish antenna of 8.5 m diameter generating a pencil beam of width less than 1°. The antenna is protected from environmental elements by being housed in a spherical radome of 12 m diameter. The transmitted peak power is 475 kW, and two transmitted pulse widths of 1.57 μs (short pulse) and 4.57 μs (long pulse) are selectable. The radar data are processed with computer and the products are serviced for various meteorological and hydrological organizations.

Under legacy resolution, it has provided reflectivity data at 1 km by 1° to 460 km range and Doppler data at 0.25 km by 1° to a range of 230 km. The data processing software has been modified so that data can be obtained with higher resolution for both reflectivity and velocity, and to longer ranges for Doppler velocity, i.e.,

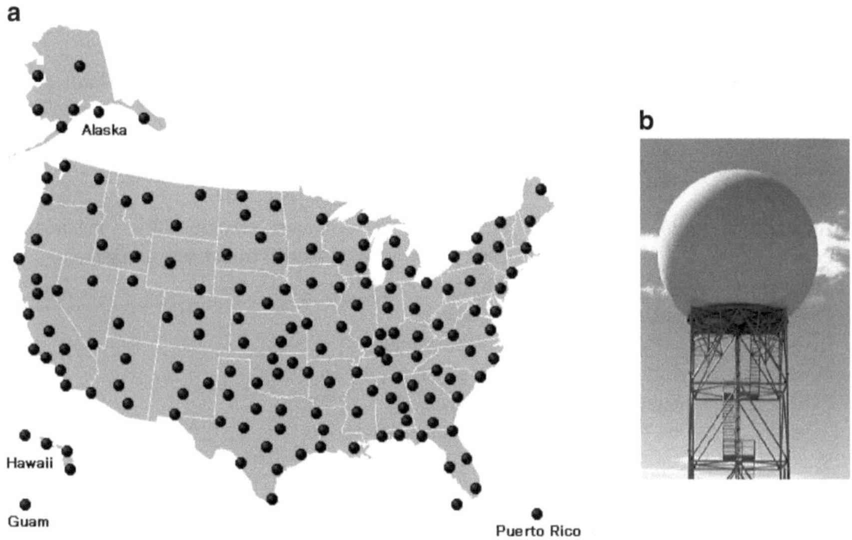

Fig. 9.1 (**a**) Distribution of the WSR-88D radars in the contiguous U.S., Alaska, Hawaii, Guam, and Puerto Rico [http://radar.weather.gov/] and (**b**) the appearance of radar tower [http://www.roc.noaa.gov/WSR88D/Engineering/NEXRADTechInfo.aspx]

reflectivity and Doppler data every half a degree and 250 m, and Doppler velocity to 300 km is now available.[2]

The dynamic range of received signal is 95 dB, and the minimum detectable signal for signal-to-noise ratio (SNR) of 0 dB at range 50 km is −8 dBZ. As the radar is equipped with such a wide dynamic range and high sensitivity, in addition to the precipitation observation of the wide area, it can observe weak backscattering phenomena which are accompanied with no precipitation at close range. The radar is also equipped with a five pole infinite impulse response (IIR) digital filter which can adjust the notch width with software to suppress ground clutter (Heiss et al. 1990). As a result, the filter performs the ground clutter suppression to about 54 dB for fixed target. Furthermore, the radar has been updated under open radar data acquisition (ORDA) program (e.g., Ice et al. 2004, 2005). The ORDA uses a particular types of spectral clutter filter, the Gaussian model adaptive processing (GMAP) filter developed by Siggia and Passarelli (2004). The Main Characteristics And Parameters of the radar are listed on Table 9.2.

[2]The updated resolution is at the expense of data variance. Because the radar scans at a relatively high angular rate (i.e., about 3 rotations per minute (RPM)), the actual angular resolution was about 1.4° before the upgrade. With the upgrade the angular resolution is about 1.0° (see Fig. 7.25 of Doviak and Zrnić 2006, p. 196). So although the data are outputted every half a degree, the actual resolution is about 1.0°. This is still better than the legacy resolution of about 1.4°. Thus the increase in Doppler range is due to the fact that the angular resolution increased from 1.4° to 1.0°. The 300 km range is that to which large tornadoes can be resolved.

Table 9.2 WSR-88D radar system characteristics

Radar range	Reflectivity, 460 km
	Doppler velocity, 300 km (upgraded from 230 km)
Transmitted frequency	2.7–3.0 GHz
Antenna type	Parabolic dish, diameter 8.534 m
Antenna gain	45.8 dB at 2.85 GHz (midband)
Beam width	0.96° at 2.7 GHz, 0.88° at 3.0 GHz
Antenna scan speed	Azimuth 30° s^{-1} maximum, elevation 30° s^{-1} maximum
Polarization	Dual polarization (upgraded from linear horizontal)
Transmitter	Klystron
Peak transmitted power	750 kW
RF duty cycle	0.002 maximum
Pulse width	1.57 μs (short), 4.57 μs (long)
Pulse repetition frequency	Short pulse 320–1,300 Hz
	Long pulse 320–450 Hz
System noise power	−113 dBm
Noise figure	3 dB
Receiver dynamic range	95 dB maximum including STC and AGC
Intermediate frequency	57.6 MHz
Sampling rate	600 kHz

Modified from Doviak and Zrnić (2006, p. 47) and Mahapatra (1999, pp. 248–249)

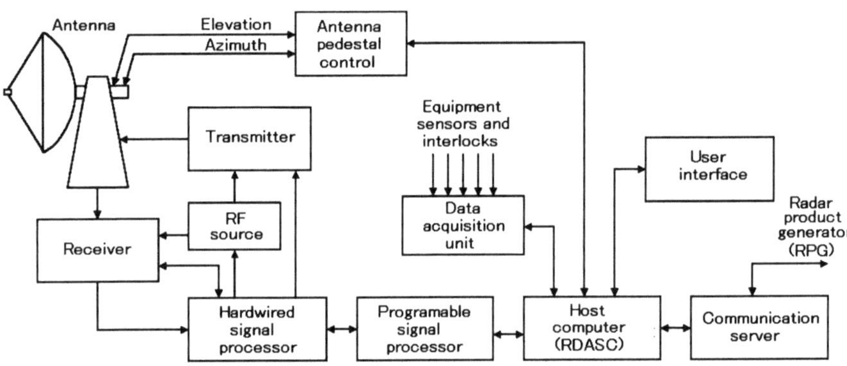

Fig. 9.2 The radar data acquisition (RDA) subsystem architecture of the WSR-88D

Configuration

The WSR-88D is configured by three blocks, the radar data acquisition (RDA) unit, the radar product generator (RPG), and the principal user processor (PUP). As shown in Fig. 9.2, the RDA unit is composed of transmitter, antenna, receiver, signal processor, and other basic components. The RPG is composed of computer which processes radar data obtained by the RDA unit, and generates pictures for display. The PUP is composed of communication controller and display unit.

Table 9.3 Major radar products of the WSR-88D

Radar reflectivity factor	Mean radial velocity
Composite reflectivity	Velocity azimuth display (VAD) wind profile
Layer composite reflectivity	Wind shear
Echo top	Turbulence
Vertically integrated liquid (VIL) water	
One-hour rainfall accumulation	
Three-hour rainfall accumulation	
Storm total rainfall accumulation	
Hourly digital rainfall array	

Radar Products

The WSR-88D produces wind profile based on the VAD method and many high-order radar products which meet the use of various organization in addition to the echo intensity (radar reflectivity factor), the Doppler velocity, and the Doppler velocity spectrum width. A lot of products relate to the precipitation parameter. In addition, the other products which relates to aeronautical meteorology such as the echo top and the wind shear which is gotten from the Doppler radar are also produced. The main radar products are listed on Table 9.3.

Enhancement to Dual H, V Polarization

The upgrade to polarimetric radar, which adds vertical polarization to the current horizontally polarized radar waves, provides more accurate observations. This has been accomplished in 2012. In addition to the three radar moments now collected, three dual polarization parameters will also be corrected. They are the differential reflectivity (Z_{DR}), the correlation coefficient (ρ_{hv}), and the differential phase (Φ_{DP}). The quantitative precipitation estimate algorithm will become possible with polarimetric radar. It is expected to improve the quantitative precipitation estimates, to focus on aviation weather hazards such as convection and outflow boundaries, winter weather, melting layer (or bright band), freezing rain, and so on.

9.2.3 5.6-GHz Band Radar: The Terminal Doppler Weather Radar

General

The Terminal Doppler Weather Radar (TDWR), for the detection of low level wind shear that occurs around airport, is installed in major airports in the U.S.A. Low level wind shear, which occurs at the height of less than about 1,600 ft (about 500 m), is

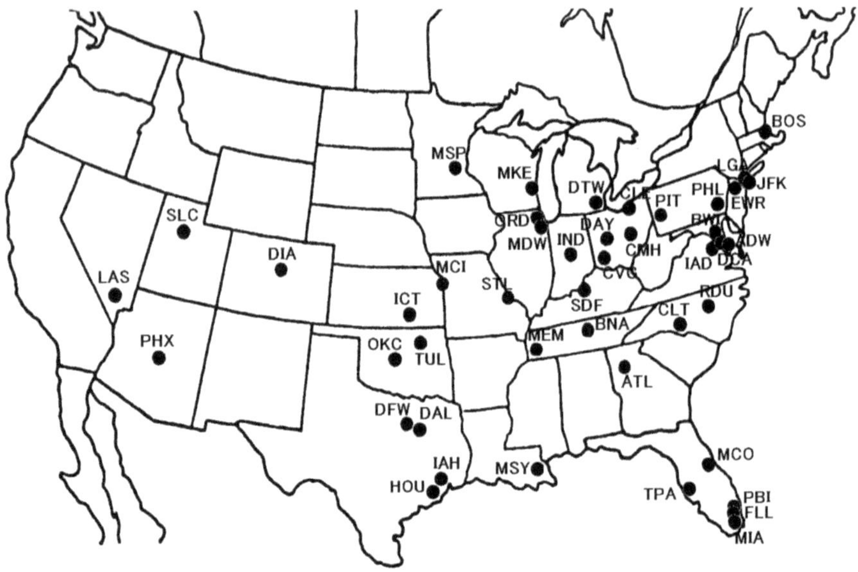

Fig. 9.3 Location of the TDWRs across the U.S. (modified from Mahapatra, 1999, p.280)

an extremely dangerous phenomenon and is difficult to avoid for aircraft in taking off and landing. Therefore, the research for the mechanism of this phenomenon was carried forward in the second half of the 1970s in the U.S.A.

Simultaneously with the initiation of the WSR-88D, the FAA began the work of automatic turbulence and storm cell identification and tracing algorithm. Development of the low-level wind shear detection algorithm started in 1980s (McCarthy et al. 1980; Laird and Evans 1982; McCarthy and Wilson 1985). Fujita (1985, Chaps. 1, 2) describes his long evolutionary discovery of the microburst, a phenomenon often generated by a convective storm. As a result, the conclusion was found out that the radar which can detect dangerous sign before the aircraft encounters to wind shear in the wide area around the airport is desirable. In late 1980s to the first half of the 1990s, microwave Doppler radars were widely introduced as the TDWR and installed in about 50 main airport in the U.S.A. (Fujita and McCarthy 1990) for the detection of down bursts and the protection of life and property. The location of the TDWRd is shown in Fig. 9.3.

Feature and Performance

The primary mission of the TDWR is to detect thunderstorm-related wind shears and microbursts that are potentially hazardous to aircraft during landing and takeoff operations, e.g., within three nautical miles on final approach and within two nautical miles on departure. The TDWR can effectively observe microburst outflows.

Table 9.4 Basic specifications of the Terminal Doppler Weather Radar (TDWR) (Mahapatra 1999, p. 277) and the Doppler Radar for Airport Weather (DRAW)

	TDWR	DRAW
Radar range	Reflectivity 460 km	Reflectivity 240 km
	Doppler velocity 89 km	Doppler velocity 120 km
Transmitted frequency	5.60–5.65 GHz	5.25–5.35 GHz
Antenna diameter	7.5 m	7 m
Antenna gain	50 dB	48 dB
Beam width	Less than 0.55°	0.7°
Polarization		Linear horizontal
Antenna scan speed		Azimuth 30° s^{-1} maximum
		Elevation 15° s^{-1} maximum
Pulse repetition frequency	2,000 Hz (maximum)	2,000 Hz (maximum)
Transmitter	Klystron	Klystron
Peak transmitted power	250 kW	200 kW
Average transmitted power	550 W	500 W
Pulse width	1.1 μs (165 m)	1 μs (150 m)
Minimum receive power		−110 dBm
Noise figure	2.3 dB	3 dB
Receiver dynamic range	129 dB	120 dB
Data digitization		14 bits
Doppler velocity resolution		0.5 m s^{-1} (64 points FFT)
		1.0 m s^{-1} (32 points FFT)

To facilitate the measurement of wind shear at such low altitudes, efficient ground clutter rejection is a primary design feature of the TDWR (Michelson et al. 1990). Further, in its scan cycles spanning in 5 min, the lowest elevation level is visited once every minute to conform to the microburst detection criterion.

The TDWR is a coherent Doppler radar operating in the 5 cm wavelength band, and has a parabolic dish antenna of 7.5 m diameter, with −3 dB beam width of 0.55°. The antenna is protected by a spherical sandwich radome of 11 m diameter. The transmitted peak power is 250 kW and the average power is 550 W. It has the pulse width of 1.1 μs, and the maximum pulse repetition frequency is 2,000 Hz. The range of Doppler velocity observations of the TDWR is 89 km, which is enough to cover the terminal area and an appreciable distance beyond. The limited surveillance range may cause the ambiguity problem. To deal effectively with the problem, the TDWR is capable of selecting its operating pulse repetition frequency (PRF) to mitigate the overlay over specified regions. In each scan cycle it performs a low-PRF surveillance. Data from this can be used to predict the amount of obscuration as a function of PRF, and the value of PRF can then be chosen adaptively and automatically that minimizes the obscuration effect over the selected area (Crocker 1988; Michelson et al. 1990). Table 9.4 lists some of the basic parameters of the TDWR with that of the Doppler Radar for Airport Weather (DRAW) in Japan that will be mentioned in the next subsection.

Detection of Low Level Wind Shear

Some of the convective precipitation clouds such as cumulonimbus cause rapid down drafts, one of the main origins of low level wind shear. The down draft becomes radial airflow when it collides with the ground, and spreads radially to the horizontal direction. The one being especially small with strong outflows is called a microburst. The boundary of the rapidly moving cold air outflow and ambient air is called the gust front. Fujita (1985, p. 8) defined microburst as follows: a small downburst with its outburst, damaging winds extending only 4 km (2.5 miles) or less. In spite of its small horizontal scale, an intense microburst could induce damaging winds as high as $75\,\mathrm{m\,s^{-1}}$ (168 mph). The division of microbursts into two classes, "dry microburst" and "wet microburst", has been recognized (e.g., Rodi et al. 1983; Wolfson 1983). Wakimoto (1985) used the following definition for them.

Dry microburst: A microburst that is accompanied by little or no rain between the onset and the end of the high winds. This type of microburst is usually associated with virga from altocumuli or shallow, high-based cumulonimbi.

Wet microburst: A microburst that is accompanied by heavy rain between the onset and the end of high winds. This type of microburst is usually associated with intense precipitation shafts from thunderstorms.

As the Doppler velocity data obtained by a Doppler radar are radial components of velocity vectors of scatterers, it is necessary to operate at least two Doppler radars simultaneously to obtain three components of wind vectors (east-west, south-north, and vertical wind) as discussed in Sect. 4.3. However, it is possible to extract two-dimensional patterns of the flow, which are characteristics of downbursts and gust fronts by one Doppler radar if gradient data of radial and/or azimuthal Doppler velocities at low elevation angle are obtained (i.e., Uyeda and Zrnić 1986).

Conceptual appearances of a microburst and a gust front and the algorithms to detect them by a Doppler radar are shown in Fig. 9.4. Panel (a) shows the schematic vertical cross sections of a microburst (left) (adapted from Mahapatra 1999, p. 108) and a gust front (right) (adapted from Droegemeier and Wilhelmson 1987), respectively. Panel (b) shows the two-dimensional (azimuthal sweep) wind pattern of the microburst (left) and the gust front (right), respectively. Panel (c) shows the shear segments of the microburst (left) and the gust front (right) which are extracted from the radial convergence and azimuth shear obtained by the Doppler observation.

In the case of the microburst, the algorithm first searches for the touchdown and outflow of air mass followed by precipitation with plan position indicator (PPI) scan at the lower elevation angle. Then the algorithm extracts the range interval where the velocity increases from a negative direction (toward the radar) to positive (away from the radar) and monotonically increases for each radial of data at various azimuths. Finally, the algorithm combines the segments at various contiguous azimuths to form a quasi-ellipse as a microburst area. On the other hand,

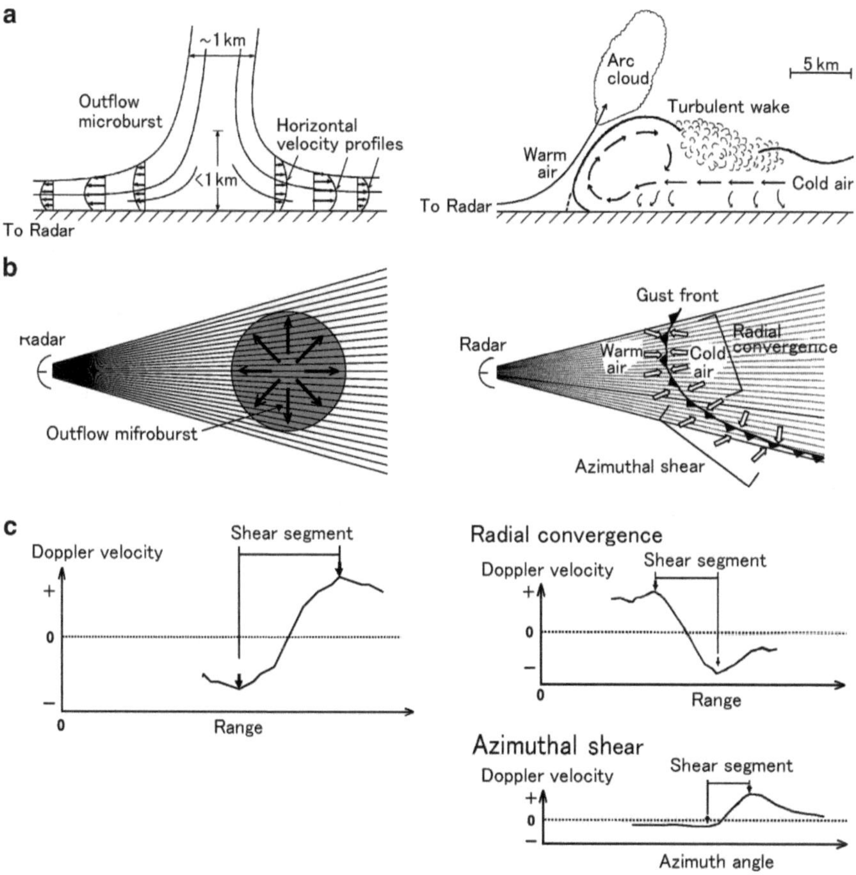

Fig. 9.4 Conceptual appearances of a microburst and a gust front and the algorithms to detect them: (**a**) vertical cross sections of a microburst (*left*) [adapted from Mahapatra 1999, p. 108] and a gust front (*right*) [adapted from Droegemeier and Wilhelmson 1987], (**b**) two-dimensional (azimuthal seep) wind pattern of the microburst (*left*) and the gust front (*right*), and (**c**) shear segments of the microburst (*left*) and the gust front (*right*) which are extracted from the radial convergence and azimuthal shear

gust front detection is performed in two fields, radial convergence and azimuthal shear. First, in the field of radial convergence, the convergent area and wind shift area where the Doppler velocity in the radial direction changes from positive to negative is identified. Second, azimuthal shear area where the wind blows near perpendicular to the radar line of sight direction is identified. Finally, the radial convergent area and the azimuthal shear are combined.

Fig. 9.5 The photo of the DRAW installed at Kansai international airport, in Japan [from Hamazu et al. 2000a]

9.2.4 5.3-GHz Band Radar: The Doppler Radar for Airport Weather in Japan

General

The development research of the similar system with the TDWR is carried forward by the Japan Meteorological Agency (JMA) and so on in the first half of the 1990s, and was practicalized as the DRAW of 5.3-GHz band (Hamazu et al. 2000a,b). This radar was first installed at the Kansai International Airport (KIX) and has been installed at 9 major airports in Japan. The photo of the radar is shown in Fig. 9.5. The system configuration of the DRAW is shown in Fig. 9.6. The basic part of the radar is composed of coherent pulse radar which is shown in Fig. 8.19. The data which is observed by the radar is delivered to the data display site through the optical fiber cable, and various processing results such as detected low level wind shear are displayed in real time. With dual configuration of the main equipment, it is prepared that the outage time of the system becomes minimum when accidental failure occurs.

Feature and Specification

The basic specifications of the DRAW is also listed on Table 9.4. The radar has a parabolic dish antenna of 7 m diameter, with the gain of 48 dB and the beam width of 0.7°. It is protected from severe weather conditions by being housed in a sandwich type spherical radome of 11 m diameter. The transmitter is composed of the master oscillator and power amplifier (MOPA) with a klystron to generate coherent transmitting pulse of a peak power of 200 kW and 1 μs pulse width.

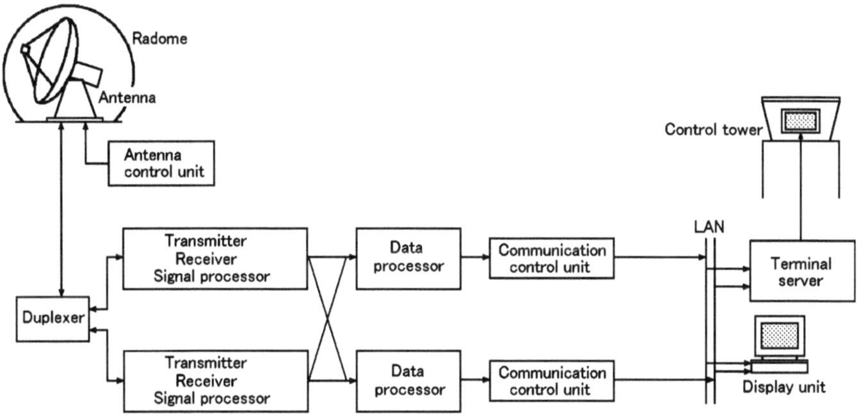

Fig. 9.6 The system configuration of the DRAW. Main equipment except for the antenna and related units of the radar are dualized

The receiver is equipped with a low noise amplifier on the front, and operates coherently with the transmitter. The receiver has a high dynamic range of 120 dB or more including the sensitivity time control (STC) and the automatic gain control (AGC). The received signal is applied synchronous detection to generate in-phase (I) and quadrature phase (Q) signals. The I and Q signals are converted to 14 bits digital signals, and applied to obtain the Doppler velocity, velocity spectrum width, and radar reflectivity factor. The data processor performs the low-level wind shear detection, and the various radar pictures are distributed for the aid of air traffic control and other use.

9.2.5 5.3-GHz Band Meteorological Radars in Japan

General

In Japan, the 5.3-GHz band radar, which is suited from the point that the attenuation of electromagnetic wave due to precipitation is comparatively small, became the radar of choice. Furthermore, because Japan is a mountainous country, radars need to be sited much closer to each other than in flat regions. Thus the attenuation at long ranges is not a significant issue and therefore a shorter wavelength radar can be used effectively. Now, 20 systems of the 5.3-GHz meteorological radar of the most basic configuration, cover whole country. These radars are operated by the JMA.

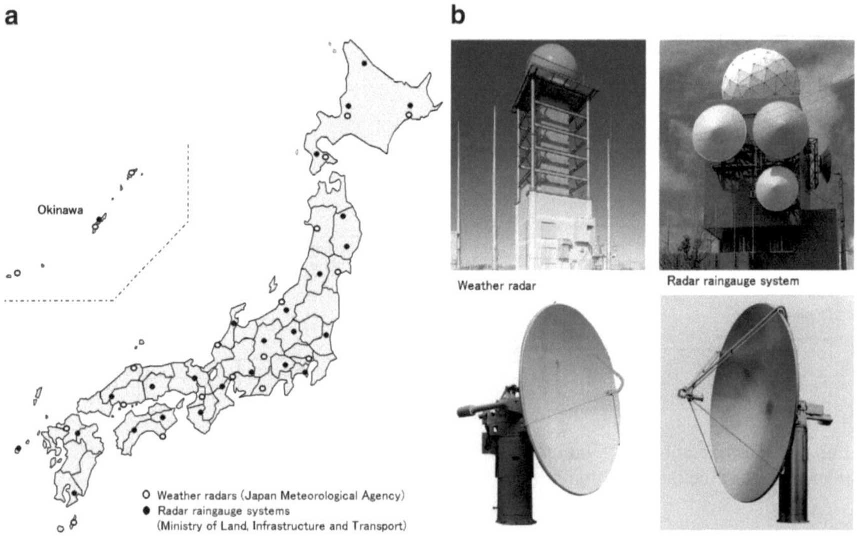

Fig. 9.7 (**a**) Location and coverage of the operational meteorological radars and the radar-raingauges in Japan. (**b**) Appearance of the radar site. *Left*: Hakodate meteorological radar at Mt. Yokotsu and its antenna. *Right*: Miyama radar raingauge and its antenna (same as that on Fig. 8.6b) housed in the spherical radome at the top of the building

Configuration

Almost all operational meteorological radars are equipped with the Doppler capability. These radars typically adopt a klystron for the MOPA transmitter. The I and Q signals are obtained from the received signal with phase measurement which is described in Sect. 5.2.2, and the echo intensity, Doppler velocity, and Doppler velocity spectrum width are produced. The basic configuration of the transmitter and receiver is as shown in Fig. 8.19. The process following to the intermediate frequency (IF) stage is performed by the Radar Observation data Processing System (ROPS), where the digital signal processing system is adopted. The radar and its associated processing system is the operational meteorological radar of the JMA.

The present placement of the operational meteorological radars and the radar raingauges of the same radar-band which will be mentioned in the following subsection is shown in Fig. 9.7a, and the examples of the radar appearance are shown in the same figure (b).

Feature and Performance

The main specifications of the operational meteorological radar are shown in Table 9.5. The receiver sensitivity, i.e., calculated SNR of the radar is shown in Fig. 9.8, where the received signal power is calculated based on Table 9.5 and using (3.73), where the noise power is given by (5.36). The SNR is given for various

Table 9.5 Basic specifications of the operational meteorological radar of the Japan Meteorological Agency

Radar range	Reflectivity 400 km, Doppler velocity 150 km
Transmitted frequency	5.25–5.37 GHz
Antenna type	Parabolic dish, diameter 4 m
Antenna gain	42 dB
Beam width	1.0° (half power width)
Polarization	Linear horizontal
Antenna scan speed	Azimuth 36° s^{-1} maximum, elevation 15° s^{-1} maximum
Pulse repetition frequency	1,500 Hz maximum
Transmitter	Klystron
Peak transmitted power	250 kW
Pulse width	1 μs (short), 2.5 μs (long)
Minimum receive power	−110 dBm
Noise figure	3 dB
Receiver dynamic range	80 dB
Data digitization	14 bits
Range bin spacing	1.67 μs (250 m)

magnitude of the equivalent radar reflectivity factor[3] Z_e. Here, the system loss[4] is 7.0 dB and the loss due to the atmosphere[5] is 0.01 dB km^{-1}. Also, it is assumed that there is no precipitation between the radar and the scatterer.

Radar Information

The output of the radar is sent to the radar data processing center system at the Meteorological Agency central office through the dedicated network. The date which are sent from the meteorological radars in whole country to the center system are synthesized with the Automated Meteorological Data Acquisition System (AMeDAS). The AMeDAS is the weather observation system which covers whole of country, and its rainfall rate measurement[6] is performed at about 1,300 places which are approximately arranged in the regular intervals of about 17 km × 17 km. The synthesized data is returned to the regional meteorological observatories.

---

[3]Like Z which was described in Sect. 3.3.5, the logarithmic expression of $Z_e[\mathrm{mm}^6\mathrm{m}^{-3}]$ is given by $10\mathrm{Log}(Z_e)$ in dBZ. If using (6.42) as the relation between the rainfall rate and the radar reflectivity factor, the rainfall rates for $Z_e = 30$ and 40 dBZ correspond to 3 and 12 mm h^{-1}, respectively.

[4]The transmission loss of 2.5 dB from the transmitter output port to the antenna including radome loss and the reception loss of 4.5 dB from the radome to the receiver head (low noise amplifier; LNA) are counted for the radar equation as the typical values.

[5]The value is the summation of the specific attenuation due to the oxygen which is shown in Fig. 6.5a and that of the water vapor which is shown in the same figure (b).

[6]The measurements of the atmospheric temperature, the wind direction, the wind velocity, and the hours of sunshine are performed at 840 places in the intervals of about of about 21 km × 21 km.

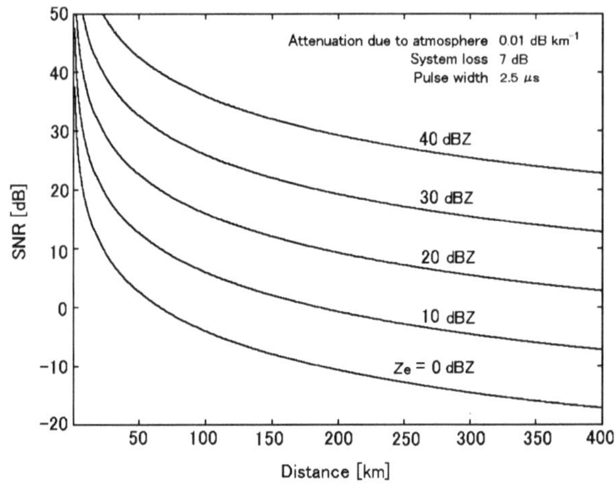

Fig. 9.8 The SNR of the operational meteorological radar, where the received signal power is calculated based on Table 9.5. The SNR is given for various magnitude of the equivalent radar reflectivity factor. Here, the system loss of 7.0 dB, the transmission loss of 2.5 dB and the reception loss of 4.5 dB, and the loss due to the atmosphere of 0.01 dB km^{-1} per one way are assumed. It is also assumed there is no precipitation between the radar and the scatterers

In addition, it is serviced for each authority and the general use outside, too. The meteorological radar can observe the scatterers in the space with high resolution as the two dimensional or three dimensional echoes. However, because of the influence of the disorder of the wind due to the changing of falling precipitation particle, the landform, the building, and so on, the calculated precipitation rate may indicate a bit different values with that of the in situ AMeDAS data. Therefore, the Meteorological Agency creates the radar-AMeDAS analysis rainfall rate (hereinafter, analysis rainfall rate) which is based on the radar data compensated with the AMeDAS data. The analysis rainfall rate is the 1 h rainfall rate, it was created at first in each 5 km mesh of the latitude of roughly 0.05° and the longitude of roughly 0.0625°. From April 2001, the data has been created in each 2.5 km mesh.

9.2.6 *Radar Raingauge*

General

In Japan, the radar raingauge whose transmitters frequency is identical with that of the operational meteorological Doppler radar of the JMA is utilized by the Ministry of Land, Infrastructure and Transport (MLIT) of Japan and the observation network covers the whole country. This system is composed of the automated radar equipment and the analysis processor which processes the observed data. The radars are installed at 26 mountaintop radar sites of good prospect and operated by remote control. Some MLIT radars are dual polarized and others operate with single polarization, and observe only echo intensity. However, some of them are equipped for Doppler observation. The observed result is inputted to the analysis processor through the dedicated microwave line every 5 min after time averaging. Each analysis processor converts reflectivity data to the rainfall rate. The output data of more than one analysis processor of neighboring radars are transmitted to the central synthetic processor.

Data Utilization

The central synthetic processor synthesizes the data from all the radars to generate a wide area rainfall map. The synthetic processing to choose the data of more than one radar overlapped area is automatically performed based on the Digital National Information, each radar height, and each radar elevation angle. Furthermore, the compensation of echo power attenuation due to mountain blocking and the rainfall rate correction with ground raingauge are automatically performed.

The data of the JMA Doppler radar is mainly utilized for rainfall nowcasting and forecasting, whereas the data of the MLIT radar raingauge is applied to calculate the area rainfall[7] of the river basin, and mainly applied for the river management and so on. The data are also serviced for local governments use and general use. Moreover, the use of the integrated radar data of the JMA Doppler radar and the MLIT radar raingauge are progressing.

9.2.7 9.5-GHz Band Radars

Many operating radars of 9.5-GHz band use magnetron for the transmitter tube. Today, radars of the same frequency band of klystron transmitter is in practical use. Moreover, as the technology of high power transistor progresses in recent years, solid state power amplifier (SSPA) radars which adopt high power amplifier of semiconductor for the transmitter have been advancing.

9.4-GHz Magnetron Radar

The 9.4-GHz mobile radar of Japan National Research Institute for Earth Science and Disaster Prevention (NIED) is a polarimetric Doppler radar which uses

[7]The spot rainfall rate means the rainfall rate in the corresponding spot whereas the area rainfall rate means the average rainfall rate of the wide area of the river basin and so on.

Fig. 9.9 Appearance of the 9.4-GHz radar and the satellite communication equipment for data transmission (*right-hand side*) and the 35/95 GHz radar (*left-hand side*) of NIED. The 35/95 GHz radar which will be mentioned in Sect. 9.3.2 [from Iwanami et al. 2001]

magnetron (Iwanami et al. 2001). The 9.4-GHz radar and the satellite communication equipment for data transmission are housed on a vehicle as shown in the right-hand side of Fig. 9.9. The schematic drawing of the radar system is shown in Fig. 9.10, and its main characteristics are listed on Table 9.6. Radar reflectivity factor Z, Doppler velocity v, and Doppler velocity spectrum width σ_v can be obtained with single polarization Doppler observation. Furthermore, Z_{DR}, ρ_{hv}, Φ_{DP}, and K_{DP} can be obtained with polarimetric Doppler observations with waves of horizontal and vertical polarization. Using the observation data, rainfall intensity, water content, and the information about the nature, shape, and size of scatterer can be estimated. The radar on the vehicle and the base station are connected with the satellite communication channel of 384 kbps, thus observed data can be transmitted to the base station in real time.

Polarimetric observation with horizontally polarized (H) and vertically polarized (V) waves are performed simultaneously by the method shown in Fig. 6.8c. High power rf pulses are generated by one magnetron. A part of the generated pulse (burst signal) is stored in the signal processor for the coherent processing of received signals as shown in Fig. 8.23. The generated pulse is divided to two transmitted pulses, H and V, of the same phase and equal magnitude. They pass the two-channel azimuth rotary-joint and each elevation rotary-joint, and radiated through the same antenna. The radiated wave could be slant linear at 45°. The backscattered signals from scatterers are received by the H and V ports of the same antenna, and applied phase measurement separately by two receivers for H and V waves.

The obtained reflectivity and phase data of H and V waves are processed by the signal processor with time sharing. Polarimetric parameters are estimated using these data by the data processor.

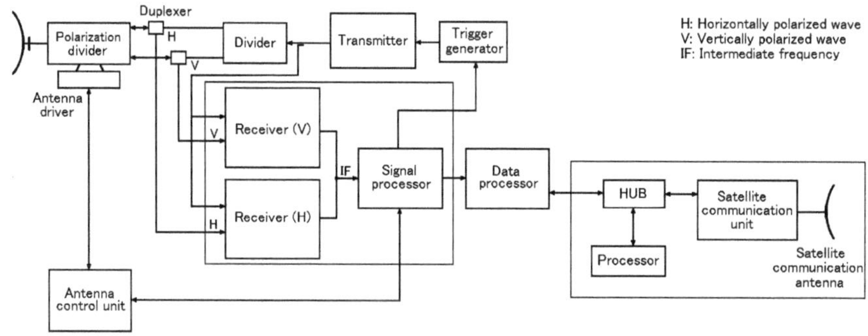

Fig. 9.10 Schematic drawing of NIED 9.4-GHz radar and the satellite communication equipment for data transmission

Table 9.6 Basic specifications of the NIED 9.4-GHz radar [from Iwanami et al. 2001]

Radar range	80 km
Transmitted frequency	9.357 GHz
Antenna type	Parabolic dish, diameter 2.1 m
Antenna gain	42 dB
Beam width	1.3° (half power width)
Polarization	Linear horizontal and vertical
Beam directions	360° azimuth, −2° to +92° elevation
Antenna scan speed	Azimuth 36° s^{-1} maximum Elevation 18° s^{-1} maximum
Pulse repetition frequency	1,800 Hz maximum
Transmitter	Magnetron
Peak transmitted power	50 kW
Pulse width	0.5 μs
Minimum receive power	−109 dBm
Noise figure	2.3 dB
Doppler processing	FFT, autocovariance method
Radar products	Radar reflectivity factor Z [dBZ] Doppler velocity v [m s^{-1}] Velocity spectrum width σ_v [m s^{-1}] Differential reflectivity Z_{DR} [dB] Correlation coefficient ρ_{hv} Differential phase Φ_{DP} [°] Specific differential phase K_{DP} [° km^{-1}]

9.7-GHz SSPA Radar

9.7-GHz SSPA dual polarization radars have been developed and become practical use. Table 9.7 shows the basic specifications of the 3-cm wavelength SSPA dual polarization radar of the MLIT. The peak transmitted power of each port of horizontal and vertical channel is 200 W. To obtain the similar observation

Table 9.7 Basic specifications of the 3-cm wavelength SSPA dual polarization radar [from Wada and Uyeda 2011]

Observation range	80 km or more
Frequency	9.7–9.8 GHz
Pulse width	1 μs, 32 μs
Peak transmitted power	200 W (horizontal polarization) 200 W (vertical polarization)
Receiver dynamic range	95 dB
Radome diameter	4.5 m
Antenna diameter	2.2 m
Antenna gain	41 dB
Beam width	1.2 deg
Range resolution	150 m
Elimination of ground clutter	Coherent MTI system
Output data	Received signal power (P_h, P_v), v, σ_v, ρ_{hv}, Φ_{DP}

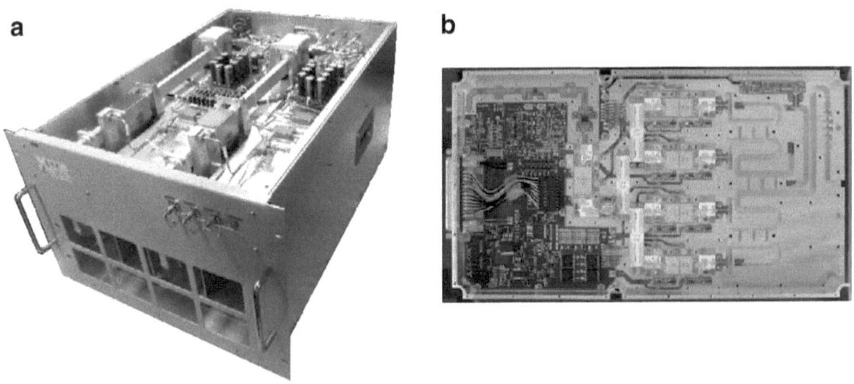

Fig. 9.11 Appearances of (**a**) the 9.7-GHz band SSPA transmitter containing dual polarization channels and (**b**) the SSPA of 200 W peak transmitted power housed in (**a**) [Courtesy of Mitsubishi Electric Corporation]

performance as a magnetron radar and a klystron radar, the pulse compression is performed by giving linear frequency modulation to the long transmitted pulse of 32 μs so as to achieve the range resolution of 150 m. The long pulse duration becomes the truncated range, thus the observation in this range cannot be done. To solve this problem and to enable full coverage, the observations by short pulse of 1 μs for short range and that by long pulse for long range are performed alternately. To avoid contamination of echoes for short pulse and long pulse observations, the transmitted frequencies of short and long pulses are detuned (Wada and Uyeda 2011).

SSPA radars have various advantages such as low operational voltage in the transmitter and suppression performance of interferences. One of the striking features is that the scale of the transmitter becomes small. Figure 9.11 shows the appearance of (a) the 9.7-GHz band SSPA transmitter containing dual polarization channels and (b) the SSPA of 200 W peak transmitted power housed in (a).

9.3 Cloud and Fog Observation Radar

9.3.1 35-GHz Band Radar

General

Several 35-GHz band radars have been developed mainly for research use of cloud physics and so on (e.g., Pasqualucci 1984; Kropfli et al. 1995; Hamazu et al. 2003). A tens of kW transmitter is required to observe clouds and fogs of 20–30 km range or more. However, it is difficult in this frequency band to obtain high power by a MOPA transmitter with semiconductor devices such as transistor. Hence, the oscillating tube such as the magnetron is adopted for the transmitter. A magnetron can generate as high as 100 kW even in this frequency band. However, it is an incoherent oscillator, and the phase of transmitted rf pulse is random from pulse to pulse. Even for magnetron, if the transmitted phase is measured at an appropriate point within the transmitted pulse and the phase of received signal is referenced to it, Doppler measurement can be performed as for a coherent MOPA radar. The 35-GHz scanning Doppler radar for cloud and fog observation is the one which is developed with this method by Kyoto University and Mitsubishi Electric corporation (Hamazu et al. 2003).

Characteristics

The main characteristics of the radar are listed on Table 9.8. To observe the minute three-dimensional structure of cloud and fog, maximum observable range and range resolution is settled 30 km and 75 m (pulse width 0.5 μs). The transmitter's duty cycle of the magnetron is limited to about 0.5/1,000, thus the pulse repetition time is limited to about 1 ms at the minimum from (8.34). Furthermore, the radar wavelength is about 0.86 cm. As a result, the Nyquist velocity of the radar becomes about 2.2 m s^{-1} from (4.12) for ordinary Doppler velocity measurement, but the value is insufficient in actual use.

To avoid the above restriction of the magnetron transmission duty cycle and the limitation of the Nyquist velocity, the method of pair pulses of short and long pulse intervals, which is shown in Fig. 8.24, is adopted. Here, the interval between pairs of pulses is set at 0.22 ms and the pair of pulses are transmitted every 2.2 ms. As a result, the Nyquist velocity is expanded to 9.7 m s^{-1}. Moreover, applying the software unfolding technique, the maximum estimation range of the Doppler velocity is finally expanded more than three times, 30 m s^{-1} or more.

The simulation of echo intensity observation by the 35-GHz band radar based on Table 9.8 is shown in Fig. 9.12. Assuming that uniform fog has a depth of 600 m, a constant relative humidity (100 %), and water content (W = 0.15 g m^{-3}) as shown in Panel (a), and assuming that the vertical profile of the atmospheric temperature is given by the Standard Atmosphere discussed in Sect. 2.2.4, the measurement

Table 9.8 Basic specifications of the 35-GHz radar for cloud and fog observations [from Hamazu et al. 2003]

Radar range	30 km
Transmitted frequency	34.75 GHz
Antenna type	Cassegrain, diameter 2 m
Antenna gain	54 dB
Beam width	0.3° (half power width)
Polarization	Linear horizontal
Antenna scan speed	Azimuth 36° s^{-1} maximum Elevation 6° s^{-1} maximum
Pulse repetition frequency	900 Hz (continuous-pulse mode) 450/4,500 Hz (spaced-pair mode)
Transmitter	Magnetron
Peak transmitted power	100 kW
Average transmitted power	50 W maximum
Pulse width	0.5 μs
RF duty cycle	0.5/1,000 maximum
System temperature	450 K
Receiver dynamic range	LOG video 70 dB or more, IQ video 40 dB or more
Range data sampling	2.4 MHz (62.5 m)/1.2 MHz (125 m)
Data digitization	12 bits
Nyquist velocity	1.9 m s^{-1} (continuous-pulse mode) 9.7 m s^{-1} (spaced-pair mode)

capability of the radar reflectivity factor required to produce a SNR of 0 dB is shown in Panel (b). The equivalent radar reflectivity factor (Z_e) for SNR of 0 dB at ranges of 2.5, 5, and 10 km are approximately −38, −30, and −20 dBZ, respectively (Hamazu et al. 2003).

Configuration and Operation

The appearance and configuration of the radar is shown in Figs. 9.13 and 9.14, respectively. The transmitted pulse of magnetron output, whose initial phase is random, is radiated through the antenna, and a part of transmitted pulse (burst signal) is applied to the receiver. The burst signal is digitized in the signal processor with 12 bits, and digitally recorded in the signal processor for the phase referencing.

The received signals are passed through the low noise amplifier (LNA) in Receiver 1 and converted using the local frequency (LO) signal generated by the stabilized local oscillator (STALO) to the first IF of 176 MHz. The first IF signal is converted to the second IF signal of 32 MHz in Receiver 2, and is then applied using the coherent oscillator (COHO) signal to both the synchronous detector and the logarithmic amplifier-detector. The base band I and Q signals from the synchronous detector are applied to the signal processor to estimate the Doppler velocity.

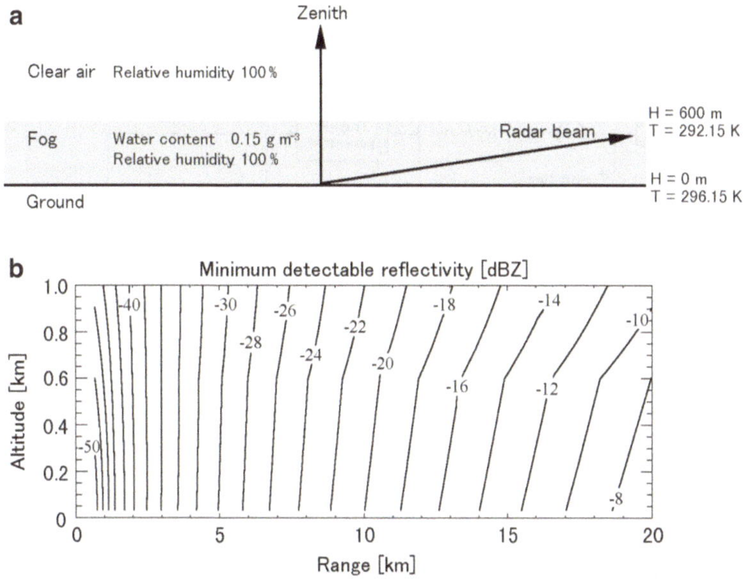

Fig. 9.12 Echo intensity observation for fog. (**a**) A fog model where the vertical profile of the atmospheric temperature given by the Standard Atmosphere, and (**b**) the measurement capability of the radar reflectivity factor for SNR of 0 dB [from Hamazu et al. 2003]

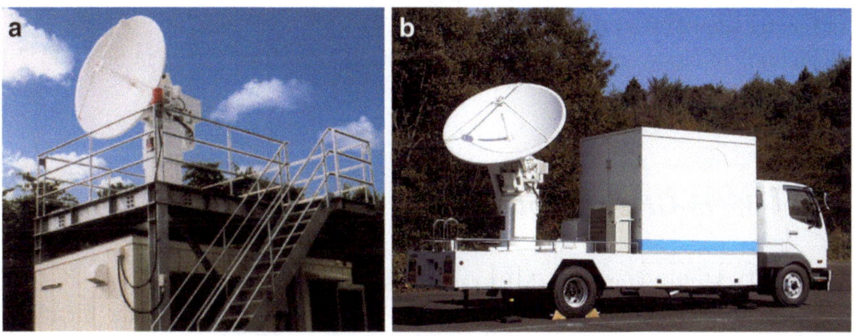

Fig. 9.13 Appearance of the 35-GHz radar for cloud and fog observations. (**a**) Ground based type and (**b**) mobile type [Courtesy of Kyoto University]

The output signals of the logarithmic detector are used to calculate the reflectivity factor. The signals are sampled and digitized with 12 bits at 0.83 μs (125 m) intervals for the simultaneous processing of echo intensity and Doppler data, or at the half interval (62.5 m) for echo intensity processing only. The phase of received signal is referenced to the phase of recorded burst signal for the Doppler processing. The Doppler data are calculated with the autocovariance processing method. The processed data, the echo intensity, the Doppler velocity, and the Doppler velocity spectrum width, are outputted to the data processing computer through the Small

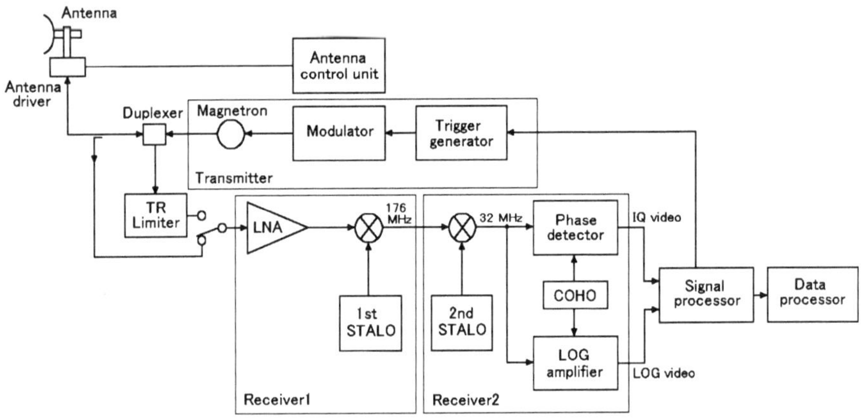

Fig. 9.14 Schematic drawing of the 35-GHz radar for cloud and fog observations

Computer System Interface (SCSI) bus. The quick-look screen display and data storage for applied analysis are available in the data processing computer. The antenna scan mode is selected with the data processing computer. This radar can scan with seven antenna scan modes as shown in Fig. 8.3, PPI, constant altitude PPI (CAPPI), range height indicator (RHI), pointing (POSITION), sector scan of limited azimuth angles and a fixed elevation angle (SECTOR), sector scan of PPI (SPPI), and sector scan of RHI (SRHI).

Pictures of cloud and fog detected with this radar will be shown in Sects. 11.7.1 and 11.7.2, respectively.

9.3.2 35/95-GHz Multiple Radar

The NIED Multi-parameter (MP) radar system is composed with the 9.4-GHz radar for precipitation observation and the 35/95-GHz multiple radar for cloud observation. The configuration of the 35/95-GHz multiple radar is shown in Fig. 9.15, and its characteristics are listed on Table 9.9. The appearance of the radar is previously shown in the left-hand side of Fig. 9.9.

General

The radar reflectivity of a precipitation particle in the Rayleigh region increases in proportion to the forth power of the radar frequency (inverse proportion to the forth power of radar wavelength λ), Thus, the radar reflectivity measurement capability of a 95-GHz radar ($\lambda = 3.2$ mm) is roughly 17 dB larger than that of a 35-GHz radar ($\lambda = 8.6$ mm) if the other radar parameters have the same values. On the other

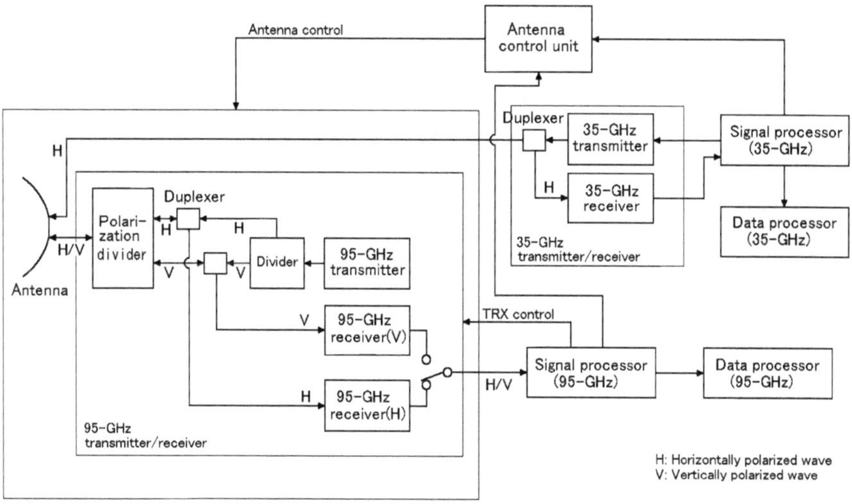

Fig. 9.15 Schematic diagram of the NIED 35/95-GHz multiple radar for cloud observations

hand, as the radar frequency increases, attenuation due to water vapor, precipitation, and cloud rapidly increase. Applying these characteristic features and observing with multiple radars of different frequencies simultaneously, the life cycle of cloud and precipitation, i.e., the generation of cloud, the generation of precipitation, its development, its weakening, and vanishing of cloud can be monitored. The NIED MP radar is one of the representative examples for the above use (Iwanami et al. 2001).

Function and Configuration

The NIED MP radar applies a shared antenna of the 35-GHz radar and the 95-GHz radar to observe the same scattering volume at the same time. The radar adopts a Cassegrain antenna and shortens the waveguide so as to suppress power attenuation minimum. The reflector size is 2 m diameter. The transmitter/receiver of the 95-GHz radar is mounted on the backside of the reflector, and the receiver of the 35-GHz radar is installed on the pedestal of the antenna. The other equipments of the 35-GHz radar are installed in the radar cubicle. The 95-GHz radar can also obtain Z_{DR}, ρ_{hv}, and K_{DP} of cloud and fog with polarimetric observation.

Table 9.9 Basic specifications of NIED 35/95-GHz multiple radar for cloud observations [from Iwanami et al. 2001]

	35-GHz radar	95-GHz radar
Radar range	30 km	30 km
Transmitted frequency	35.35 GHz	95.13 GHz
Antenna type		Cassegrain, dual frequency, diameter 2 m
Antenna gain	54 dB	59 dB
Beam width	0.3°(half power width)	0.1°(half power width)
Polarization	Linear horizontal	Linear horizontal and vertical
Beam directions		Azimuth 360° Elevation −2° to +182°
Antenna scan speed		Azimuth $\leq 24°\ s^{-1}$ Elevation $\leq 12°\ s^{-1}$
Pulse repetition frequency	400/4,000 Hz	≤ 20 kHz
Transmitter	Magnetron	Klystron (EIA)
Peak transmitted power	100 kW	2.2 kW
Pulse width	0.5 μs	0.25–2.0 μs
Minimum receive power	−108 dBm	−110 dBm
Noise figure	3.5 dB	7.5 dB
Doppler processing	Autocovariance method	FFT, autocovariance method
Radar products	$Z,\ v,\ \sigma_v$	$Z,\ v,\ \sigma_v,\ Z_{\mathrm{DR}},\ \rho_{\mathrm{hv}},\ K_{\mathrm{DP}}$
Range sample		2.4 MHz (62.5 m)/1.2 MHz (125 m)

EIA extended interaction amplifier

9.4 Satellite-Borne Radar

9.4.1 Tropical Rainfall Measuring Satellite

One of the big merits of satellite-borne radar is that much of the earth can be observed compared with the radar which is fixed on the ground. The Tropical Rainfall Measurement Mission (TRMM) satellite is the first one in the world which has on board the precipitation radar (PR), and was launched in 1997 as the Japan–U.S.A. cooperation project. The main characteristics of the TRMM PR are listed on Table 9.10. TRMM satellite takes a low earth orbit (LEO) and had been revolving the orbit at an altitude of 350 km until August 2001, and has been at 402 km since then. The orbit of TRMM satellite covers between 35° north and 35° south of the equator, and it is the non-sun-synchronous recurrent orbit, i.e., where the TRMM satellite flies over each position on the Earth's surface at a different local time each day. It takes about 90 min to complete one orbit around the Earth, and about 16 orbits are possible in a day.

Table 9.10 Basic specifications of the precipitation radar (PR) onboard the Tropical Rainfall Measuring Mission (TRMM) satellite [from Kozu et al. 2001]

Observable range	From surface to a height ≥ 15 km
Swathe width	≥ 215 km, contiguous coverage
Radar type	Active phased-array
Transmitted frequency	13.796 GHz, 13.802 GHz
	Two-channel frequency agility
Antenna type	Planar array 128-elements
	Aperture 2.1 m × 2.1 m
Beam width	0.7°
Scan angle	±17°
Pulse repetition frequency	2,776 Hz
Peak transmitted power	616 W
Pulse width	1.6 μs × 2 ch. (total 3.2 μs)
Receiver dynamic range	79 dB

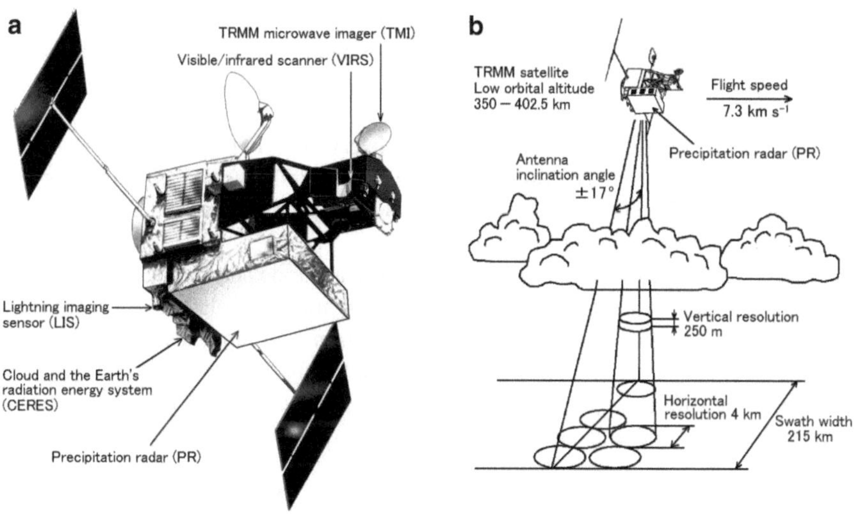

Fig. 9.16 Instruments layout of (**a**) the TRMM satellite and (**b**) concept of rainfall observation with the PR [Courtesy of Japan Aerospace Exploration Agency (JAXA)]

Precipitation Observation with Satellite-Borne Radar

Five sensors including the PR are mounted on the TRMM satellite as shown in Fig. 9.16a. In the past, the passive sensors for several microwave frequency bands were generally applied to observe precipitations from the satellite by passively receiving radiations in the several microwave bands. The horizontal distribution of precipitation is observed with these sensors. However, the quantitative estimation of precipitation intensity with this method is limited to measurements of rain over the sea, and it is not possible to perform correct observation over land.[8] On the other hand, the PR observes precipitations with the same principal as the meteorological radar on the ground. It can provide three-dimensional maps of precipitation with the combination of horizontal scanning and vertical profiling. The PR is an active phased array radar of operational frequency of 13.8 GHz. As shown in Fig. 9.16b, it has a horizontal resolution at the ground of about 3.1 miles (5 km) and a swath width of 154 miles (247 km). It can also separate precipitation echoes for vertical sample size of about 820 feet (250 m).

9.4.2 Global Precipitation Measurement Program

With the PR of TRMM satellite, it has become possible to grasp the three-dimensional structure of the typhoon in the sea where there are few observations, and the rainfall distribution over the tropical Pacific Ocean area at the time of El Niño and La Niña.

The TRMM satellite has obtained a lot of observations of precipitation in tropical and subtropical regions. Its design life has been substantially exceeded, and it is still operating. In addition, the success of the TRMM program has led to the Global Precipitation Measurement (GPM) program.

The GPM program is designed to make more accurate and frequent observation of global rainfall by expanding the area of observation to include higher latitudes. The core satellite, which is conducted in cooperation with Japan and U.S.A., carries a Dual-frequency Precipitation Radar (DPR) and a microwave radiometer, and the constellation satellites which are managed by several countries and organizations, carry microwave radiometers or microwave sounders. These satellites will enable global measurement of precipitation every 3 h.

The GPM core satellite will be launched in 2013. The DPR has two radars, the 13.6-GHz one and the 35.5-GHz one, and is expected to observe the life cycle of precipitation with a high degree of sensitivity.

---

[8]The microwave radiation from the surface of the sea is weak and moreover it is uniform. Thus if there is precipitation over the sea, it can be observed from the space. On the other hand, because the microwave radiation from the land surface is strong and non uniform, it is difficult to identify the increase of microwave radiation due to precipitation. Furthermore, it is difficult to estimate the altitude of precipitation distribution with the passive sensor.

Chapter 10
Practical Atmospheric Radars

The atmospheric radar is the general term for the mesosphere, stratosphere, and troposphere (MST) radar, the stratosphere and troposphere (ST) radar, the troposphere (T) radar, and the boundary layer (BL) radar or the BLR, etc. These radars, which originally have their own historical background and system, have been recently developed and unified to several standard systems, that individually place the observational emphasis on either the middle atmosphere (10–100 km height range), the troposphere/the lower stratosphere range, or just the lower troposphere. This development is caused due largely to striking advance of the observational techniques and to spread of networking of identical systems at multiple stations. In the present chapter, we give a general review of practical atmospheric radar, and discuss on typical operational atmospheric radars.

10.1 Characteristics of Atmospheric Radar

Due to the exponential decrease of atmospheric density (and accordingly, the exponential decrease of radio refractive index perturbations) with increasing altitude, the backscattering is too weak for useful echoes to be detected above 20–25 km in altitude. The backscattering above about 60 km, which is normally detected only during daylight hours, arises primarily from electrons that enhance the scattering from the turbulence at these altitudes (Woodman and Guillén 1974). The energy spectrum of atmospheric turbulence falls off rapidly with decreasing eddy size in the inertial subrange, and radar waves are scattered only from turbulent eddies with a scale size that satisfies the Bragg condition (i.e., one half of the transmitted wavelength). Accordingly, higher sensitivity to Bragg scatter is achieved in the very high frequency (VHF)/ultrahigh frequency (UHF) bands (30–3,000 MHz) than in the microwave frequency band generally used for meteorological radars. As shown in Fig. 7.10, the minimum size of turbulent eddies increases exponentially with

S. Fukao and K. Hamazu, *Radar for Meteorological and Atmospheric Observations*,
DOI 10.1007/978-4-431-54334-3_10, © Springer Japan 2014

increasing altitude (e.g., Hocking 1985), and so MST radars use comparatively long wavelengths, i.e., a few meters, or frequencies of approximately 50 MHz, to observe the widest possible atmospheric range.

Backscattering from the mesosphere is so weak even at 50 MHz that MST radars generally have a high peak transmitted power of the order of one megawatt (MW) and a large antenna aperture of the order of $10^4\,m^2$. Various radars with less sensitivities that operate at frequencies higher than 50 MHz are widely put to practical use for remote sensing of the lower stratosphere and the troposphere (ST radar), the troposphere (T radar), the lower troposphere (Lower Troposphere Radar or LTR), or atmospheric boundary layers (BL radar or BLR). In general, larger systems have the bigger capability to carry out varied experiments with greater degrees of complexity and flexibility, such as beam steerability, pulse coding, and data acquisition, while smaller systems tend to be designed for more dedicated purposes and often tend to specialize in particular techniques (Hocking 1997b). The small systems designed primarily for wind monitoring are called wind profilers or simply profilers and are commercially available (e.g., Ecklund et al. 1988; Vincent et al. 1998; Hashiguchi et al. 2004).

As for the operational frequency of atmospheric radars, it was decided to use the following three bands; 50-MHz band (46–68 MHz), 400-MHz band (440–450, 470–494 MHz), and 1.3-GHz band (1,270–1,295, 1,300–1,375 MHz), at the World Radiocommunication Conference (WRC) in 1997, that was sponsored by the International Telecommunication Union (ITU). In addition, 904–928 MHz-band is also applied in North America.

10.2 Large-Scale Atmospheric Radars

10.2.1 General

The heights from which one may expect to receive useful echoes at a specific radar frequency depend on the average transmitted power–aperture (PA) product. Figure 10.1 shows a schematic that illustrates average height ranges observable with different PA products for various operational radars. The LTR and the BL radar are operated with the frequency near 1.3 GHz, and other radars with near 50 MHz. A PA product exceeding $10^8\,W\,m^2$ is required to observe up to the mesosphere. The maximum observable height is significantly reduced with increasing frequency.

Atmospheric radars in the scale more than MST radar that can observe middle and upper atmosphere are currently operational in about ten places over the globe. Figure 10.2 shows representative large-scale radars that have the capability of incoherent scatter (IS) radar but also can observe the lower to upper atmosphere regions. Jicamarca radar (11.95°S, 76.87°W, Peru) shown in Panel (a) operates at the frequency of 49.9 MHz. The array antenna size is 290 m × 290 m, and the PA product is 2.0×10^{10} W m^2. Panel (b) shows Arecibo radar (18.62°N, 66.75°W,

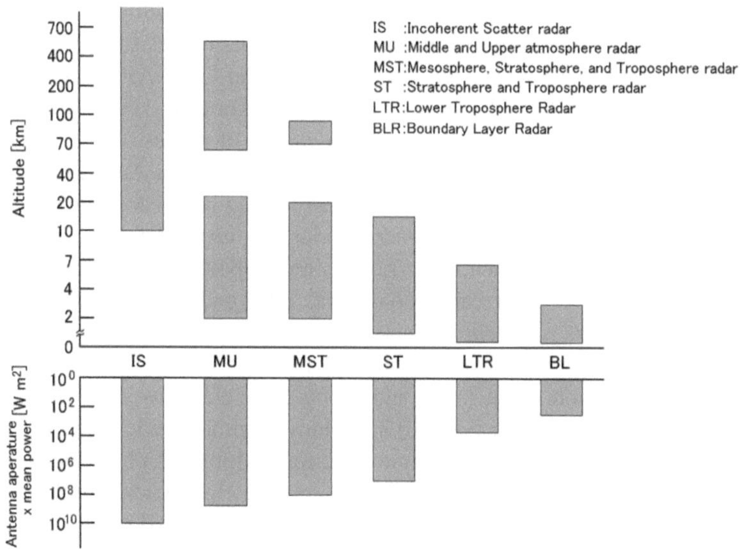

Fig. 10.1 Height ranges expected to be observed and power–aperture (PA) products for typical atmospheric radars. The LTR and the BL radar are operated with the frequency near 1.3 GHz, and other radars with near 50 MHz. Incoherent scatterer (IS) radar in the VHF band is used primarily for ionospheric observation but occasionally for MST observation purposes

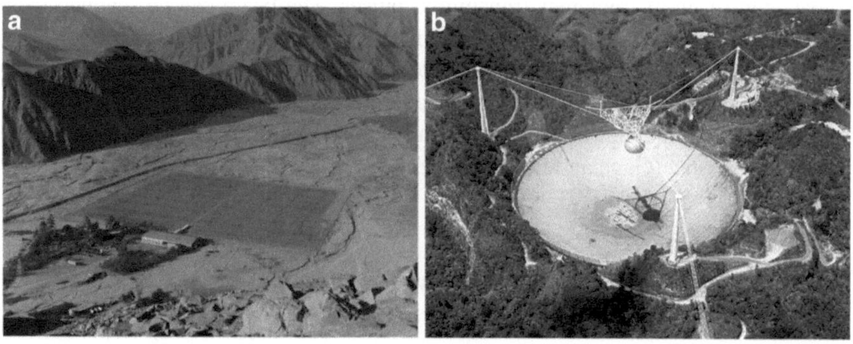

Fig. 10.2 Representative large-scale radars. (**a**) Jicamarca radar (11.95°S, 76.87°W, Peru) [http://jicamarca.ece.cornell.edu/], (**b**) Arecibo radar (18.62°N, 66.75°W, Puerto Rico) [http://www.naic.edu/]

Puerto Rico) of the operational frequency of 430 MHz. The antenna diameter is 300 m, and the PA product is 8.7×10^9 W m^2.

Some of the most sensitive MST radars are the Middle and Upper atmosphere (MU) radar in Japan (Kato et al. 1984), Gadanki radar in India (Rao et al. 1995), Aberystwyth radar in the United Kingdom (Prichard et al. 1995), SOUSY radar in Germany (now in Peru) (Röttger et al. 1978), SOUSY Mobile radar in Norway, and Poker Flat radar in Alaska (ceased operation and reassembled as four radars in the

Trans-Pacific Profiler Network in late 1980s) (Balsley et al. 1980), which, however, have finally ceased operation in early 2000s. A number of less sensitive radars are available primarily for ST research. Among them, the CLOVAR radar in Canada (Hocking 1997a) and Chung-Li radar in Taiwan (Röttger et al. 1990a) are of unique and flexible design that allows operation in a variety of modes of observation. On the other hand, some radars, e.g., the Buckland park radar in Australia, adopt the spaced antenna (SA) analysis to compensate for limited beam-steering capability.

Typical antenna types of atmospheric radar are array antenna and reflector antenna. As for antenna elements of array antenna, dipole array antenna composed of simple structural coaxial cable or coaxial tube and three or four elements Yagi Antenna of more excellent characteristics are adopted. Jicamarca radar is the representative example of the former, and adopts coaxial tubes (Ochs 1965). The MU radar is an example of the latter, and adopts three elements Yagis (Fukao et al. 1985a) as will be discussed in Sect. 10.2.3. Many of atmospheric radars with array antenna can scan the radar beam in some directions. Moreover, phased array radars, which can change the phase of each element antenna electronically in high speed, such as the MU radar are realized. In Jicamarca radar and the MU radar, the antenna can also be divided into sub-arrays, and can observe simultaneously in multi-beam.

Arecibo radar is an example which adopts large reflector antenna. Many of radars with reflector antenna rotate the reflector of parabolic surface to perform radar beam scanning, whereas the reflector surface of this radar is a sphere. The reflector is fixed on the ground, and the radar beam is scanned by mechanically moving the linear feed of multi-frequency transmitted pulses. The beam can be scanned from the zenith to within 20° in arbitrary directions (Gordon and LaLonde 1961).

One of the representative example of steerable dish radar which adopts parabolic reflector is the Chatanika radar located near Fairbanks, Alaska. The Chatanika radar operates at the frequency of 1,290 MHz and beam width of 0.6°, and the PA product is 2.6×10^7 W m^2 (Gage and Balsley 1978).

10.2.2 Radars with COCO Array Antenna

In phased array radars, there are ones which consist phased arrays of many single elements fed by a cascading network of coaxial cables, hybrids, and phase shifters, and are called the coaxial-collinear (COCO) array radars. The antenna elements in one line are fed by simply interchanging the inner and outer conductors of a coaxial cable every half wavelength. The outer conductors of the cable act as collinear dipole (Röttger and Larsen 1990). Figure 10.3a shows a schematic diagram of the complete 26-element COCO array antenna system, where A shows the messenger line and the tension line used to connect the antenna, B the antenna-feed arrangement consisting of balun and a quarter-wave matching transfer, and C the detail of connection between adjacent elements, respectively. B and C are water-shielded (Balsley and Ecklund 1972). The COCO antenna is inexpensive because coaxial cable is used for antenna elements and matching networks are simple, thus

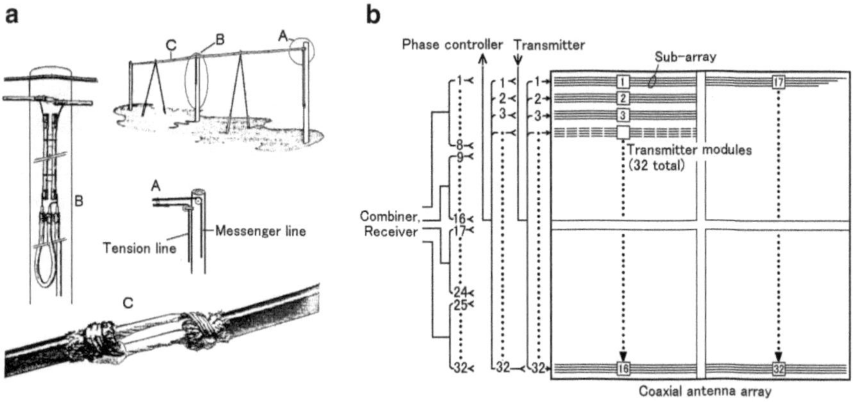

Fig. 10.3 (**a**) Schematic diagram of the complete 26-element COCO array antenna. A: the messenger line and the tension line used to connect antennas, B: antenna-feed arrangement consisting of balun and a quarter-wave matching transfer, C: connection between adjacent elements [from Balsley and Ecklund 1972]. (**b**) COCO array antenna block diagram of the Poker Flat MST radar [from Balsley et al. 1980]

this types of antenna have been adopted for many atmospheric radars of large aperture array antenna.

The Poker Flat MST radar in Alaska is a typical example which applies a COCO array antenna system, whose aperture is 200 m × 200 m and the PA product is 5.1×10^9 W m^2, operating on the transmitted frequency of 49.9 MHz with two off-vertical beam directions (Balsley et al. 1980). Figure 10.3b shows the block diagram of the COCO array antenna of the radar. The antenna consists of four quarters of subarray antenna, each constructed two pairs of 8, i.e., 16 feeding points in each subarray, totally 64 feeding points. Four coaxial cables of 48 half-wave dipoles are connected to each feeding point, thus complete array consisted of 12,288 dipoles. A pair COCO array forms the zenith and obliquely eastward radar beam, and the other pair forms the zenith and obliquely northward radar beam.

The radar is also possible to operate as four subarray radars and observe independently in different directions. After the Poker Flat project finished, the radar was divided into four subarray radars and installed in Piura (Peru) the equator neighborhood, Christmas Island (Kiribati), Pohnpei Island (Micronesia), and Biak Island (Indonesia). Balsley et al. of the National Oceanic and Atmospheric Administration (NOAA) constructed the Trans-Pacific Profiler Network (TPPN) applying those radars, and achieved the results in the research of the tropical general circulation of atmosphere according to El Nino and La Nina (e.g., Gage et al. 1991, 1996).

Fig. 10.4 A bird's eye view of the MU radar located at Shigaraki, Shiga, Japan [Courtesy of Kyoto University]

10.2.3 The MU Radar

General

The MU radar that Kyoto University established at Shigaraki, Shiga, Japan (34.85°N, 136.10°E) in 1984 is a typical sophisticated MST radar (Fukao et al. 1980a, 1985a,b, 1990). The MU radar operates at the frequency of 46.5 MHz and covers the whole middle atmosphere. In addition, the radar can observe a part of upper atmosphere as an IS radar, thus designated the Middle and Upper atmosphere (MU) radar (Kato et al. 1984). Its antenna aperture is 8,330 m^2, the diameter of 103 m, and the antenna is composed of 475 crossed three-subelement Yagis. Figure 10.4 shows a bird's eye view of the MU radar.

Feature and Specification

In conventional radar systems, a high-power transmitter feeds all array elements via an appropriate cascading feeding network. The MU radar, on the other hand, does not incorporate such a passive array but applies active array,[1] i.e., each element of the phased array is provided with a low-power amplifiers, and all the amplifiers are

[1] There were only several two-dimensional active phased array radars of 400-MHz band and 3-GHz band in the world put to practical use for military purpose by the first half of the 1980s. The MU radar may be called one of the first large-scale two-dimensional active phased array radar developed by the original technology.

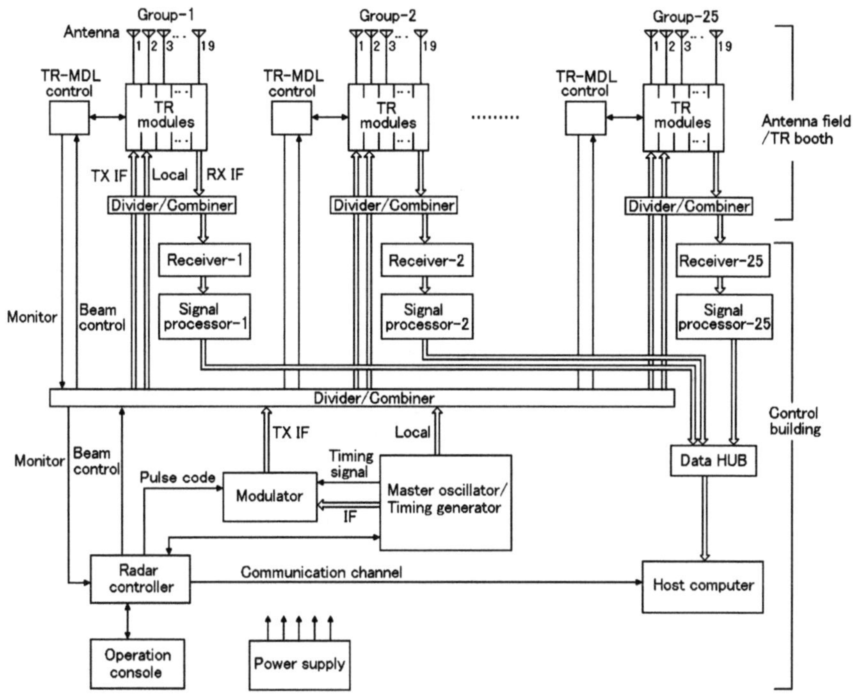

Fig. 10.5 Block diagram of the MU radar, Shigaraki, Shiga, Japan. The transmitter/receiver (TR) modules and TR module controllers (TR-MDL control) are housed in the TR booths in the antenna field. Other system equipment is installed in the control building [from Fukao et al. 1985a]

coherently driven by low-level pulses in order to produce the desires peak output power. Each Yagi is driven by the TR modules with peak output power of 2.4 kW. The maximum peak and average radiation power are 1 MW and 50 kW, respectively, and the PA product is 4.2×10^8 W m^2. The TR modules and TR module controllers (TR-MDL control) are housed in the TR booths in the antenna field. Other system equipment is installed in the control building. The block diagram and the basic specifications of the MU radar are shown in Fig. 10.5 and Table 10.1, respectively (Fukao et al. 1985a). The phase of each TR module can be controlled independently, and it is possible to steer the antenna beam in each pulse repetition time up to 200 μs virtually to any direction within 30° of the zenith under computer control. Grating lobes are formed at the zenith angle of more than 30°, thus it is unsuitable to the atmospheric observation but is used for the special observation of such as the ionosphere observation (Fukao et al. 1991; Yamamoto et al. 1991). The MU radar antenna array, the whole system consists of 25 groups of subarray, i.e., one subsystem consists of 19 array elements and 19 TR modules.

Figure 10.6a shows the MU radar antenna array. Nineteen groups are identical hexagonal subarrays composing the main part of the antenna array, while the remaining six groups are subarrays distributed around the periphery making the

Table 10.1 Basic specifications of the MU radar

Location	Shigaraki, Shiga, Japan (34.85N, 136.10°E)
Radar system	Active phased array radar
Transmitted frequency	46.5 MHz
Antenna aperture	8,330 m^2, diameter 103 m
Antenna element	475 crossed three-subelement Yagis
Beam width	3.6° (half power for full array)
Beam directions	1,657, 0–30° off zenith angle
Beam switching	200 μs (minimum)
Polarization	Linear and circular
Peak transmitted power	1 MW
Average transmitted power	50 kW (maximum)
Receiver band width	3.5 MHz
Pulse width	0.5–512 μs variable
Pulse repetition time	200 μs–65 ms variable
Receiver IF frequency	5 MHz
Data A/D conversion	14 bits × (25 + 4) channels
Pulse compression	Binary phase coding up to 32 elements Barker, complementary, and Spano codes

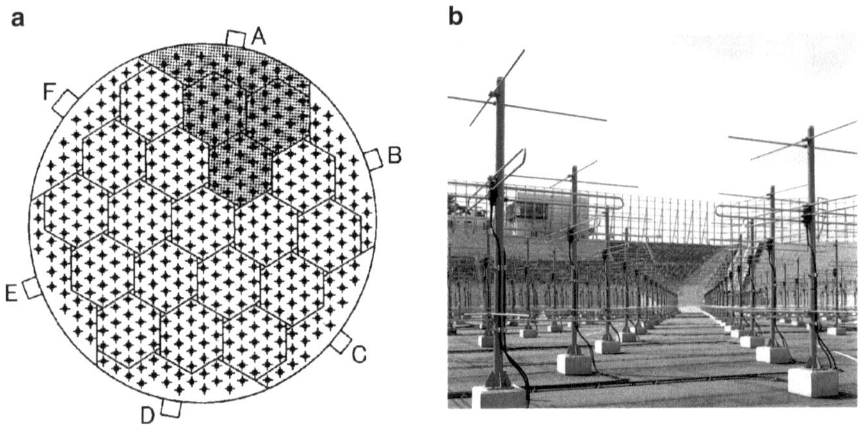

Fig. 10.6 (**a**) The MU radar antenna array. Each *cross* represents a crossed three-subelement Yagi. The whole array is composed of 19 hexagonal subarrays and 6 subarrays distributed around the periphery making the array shape nearly circular. The six boxes outside the antenna array, indicated A–F, represent the booth which accommodate the TR modules. The *shaded area* shows groups accommodated by the booth A. (**b**) Photograph taken in the antenna field, showing crossed three-subelement Yagis and rf cables [from Fukao et al. 1985a]

array nearly circular. The six boxes outside the antenna array, indicated A–F, represent the booth which accommodate the TR modules. The shaded area shows groups accommodated by booth A. Each group can be driven separately, and even individual array elements can be driven independently. At the first stage, the maximum number of separate beam is limited to four due to the number of available

Table 10.2 Basic observation parameters of the MU radar [from Fukao et al. 1990]

	Troposphere	Stratosphere	Mesosphere
Observation time zone	All day	All day	Daytime (8–16 o'clock)
Observation range (km)	2–19	5–24	60–98
Beam direction	Zenith	Zenith	Zenith
	North, 10° off zenith angle	Same as left	Same as left
	East, 10° off zenith angle	Same as left	Same as left
	South, 10° off zenith angle	Same as left	Same as left
	West, 10° off zenith angle	Same as left	Same as left
Height resolution (m)	150	150	600
Pulse repetition time (μs)	400	400	1,460
Pulse compression	Single pulse	16 bits	16 bits
	–	Complementary code	Same as left
Coherent integration	38	38	10
Incoherent integration	5	6	6
FFT points	128	128	128

detector channel. However, 25 receivers were newly added in 2004 to operate 25 groups independently. Applying these functions, e.g., to use entire subsystems for transmission and receive the scattered signal with individual subsystem at the same time, more various observations become possible. Using radar interferometry and radar imaging techniques as discussed in Sect. 7.6 and will be shown the results in Sect. 12.5, minute structures of turbulence can be known from minute phase differences of received signals in each subsystem. Table 10.2 shows the basic observation parameters for the range of troposphere to mesosphere.

Antenna Element

Figure 10.6b shows the photograph taken in the antenna field, showing antenna elements, crossed three-subelement Yagis, and rf cables. Antenna elements are allocated on the equilateral triangular grid with sides of 0.7λ ($\simeq 4.5\,\mathrm{m}$), and the faces of Yagi are pointed E–W and N–S accurately. Each element and the corresponding TR module are connected by a coaxial cable.

Master Oscillator

The MU radar aims measuring the Doppler velocity with the accuracy of about $0.1\,\mathrm{m\,s^{-1}}$ in the case of using a couple of pair pulse. Allowed frequency deviation Δf is given using the transmitted frequency f and velocity deviation Δv as

$$\Delta f = \frac{2f}{c}\Delta v, \tag{10.1}$$

where c is the speed of light. Substituting $f = 46.5 \times 10^6$ Hz and $\Delta v = 0.1\,\mathrm{m\,s^{-1}}$, from (10.1) $\Delta f = 3.1 \times 10^{-2}$ Hz is obtained. It means that the short-term frequency stability s_f is required to be $s_f \leq (3.1 \times 10^{-2})/(46.5 \times 10^6) \simeq 6.67 \times 10^{-10}$. At first, the rubidium oscillator, a high stability frequency standard, was adopted as the master oscillator (same as the local oscillator discussed in Sect. 8.3). Now it is replaced by the GPS received signal. The main reason of the replacement is to configure open system so as the other systems can easily synchronize their operation to the MU radar. The radar transmitter (TX) intermediate frequency (IF) signal and timing signals are generated by the master oscillator signal.

Modulation

The TX IF signal is modulated in the modulator with phase coding of 0° or 180° for the pulse compression. The bit length is selectable up to 32 bits, and typically the following code series are selected.

(1) Barker codes (7, 11, and 13 bits)
(2) Complementary codes (16 and 32 bits)
(3) Pseudo-Barker codes (28 bits)
(4) Spano codes (8 and 16 bits)

Analog signals, such as TX IF and timing signal, are divided to six groups and distributed to the forementioned booths A–F.

Transmitter and Receiver System

Both up-converter and down-converter are conducted in the remote booths A–F, so that the IF (5 MHz) and local (41.5 MHz) signals are transmitted from the control building to the transmitter–receivers (TRs) in the booths. The TR module mixes the IF and local signals and up-convert the IF signal to rf seed signal of 46.5 MHz and output power of 1 mW. The rf seed signal is amplified in series up to 180 W through the exciter and to the final output power of 2.4 kW through the power amplifier in the TR module. Each rf output of the TR module is sent to the corresponding array element by coaxial cable. Figure 10.7 shows (a) the TR module and (b) appearance inside the booth. Counting power attenuation through the path, the total transmitted power from the 475 antenna elements becomes 1 MW. The backscatter signal received with each antenna element is converted to the receiver (RX) IF signal through the reverse path.

Decoding

The RX IF signal is applied to the analog to digital (A/D) converter in the receiver housed in the control building and converted to 14 bits digital received signal.

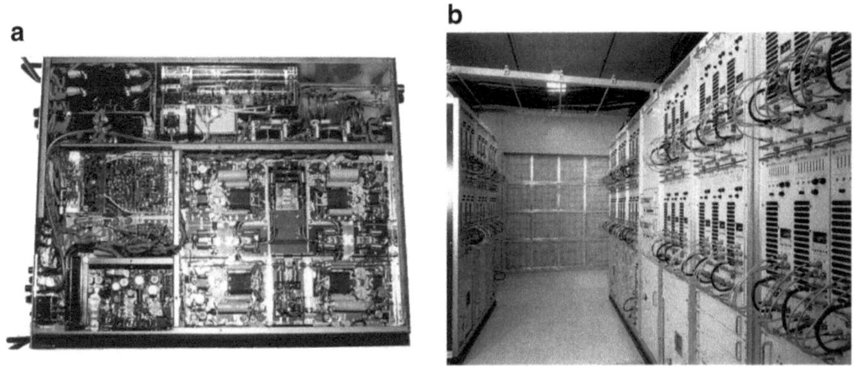

Fig. 10.7 (**a**) A single TR module of the MU radar, and (**b**) the booth that accommodates the TR modules

The digital received signal is applied phase detection, and the outputs, *I* and *Q* signals are applied pulse compression with the autocorrelators separately. Then, the *I* and *Q* signals are applied coherent integration separately. The coherent integration is performed individually according to the direction of antenna beam, the reception channel, and the measurement altitude.

Signal Processing

The signal processing such as the fast Fourier transform (FFT) and the autocorrelation are applied on the received signal after coherent integration. The power spectrum is obtained as these first processing results. In the signal processor, in addition, the detectability is improved by the incoherent integration. Various secondary processing such as to estimate echo intensity, Doppler velocity, and velocity spectrum width, are performed with the averaged spectral data.

Configuration of RASS

The MU radar equips a radio acoustic sounding system (RASS), the MU radar/RASS, that measures the atmospheric temperature. The MU radar/RASS is composed of two or more sound generator and a controller, and measures the profile of the atmospheric temperature by using the sound wave of frequency 80–108 Hz in the height range of 1.5–20 km. Figure 10.8a shows the appearance and (b) the configuration of the MU radar/RASS acoustic transmitting antenna. Two loudspeaker of 38 cm diameter are installed on the sound generator. Each maximum input power is 400 W continuously and the output power is 130 dB (9.6 W m^{-2}). The sound waves outputted from the loudspeaker in the lower right of Panel (b) are reflected by the bottom plate and radiated up in the air as spherical waves (Adachi 1996).

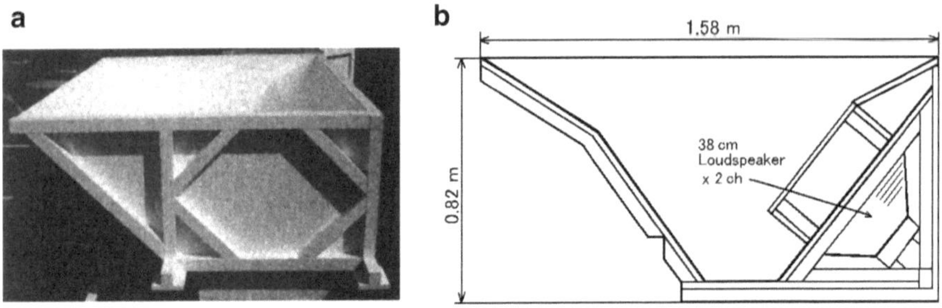

Fig. 10.8 The acoustic transmitting antenna of the MU radar/RASS. (**a**) Appearance and (**b**) configuration of loudspeaker [from Adachi 1996]

10.2.4 Equatorial Atmospheric Radar

General

The Equatorial Atmospheric Radar (EAR) is a large Doppler radar built for atmospheric observation at the equator in West Sumatra in the Republic of Indonesia (0.20°S, 100.32°E) (Fukao et al. 2003). The EAR was established by Kyoto University with the cooperation of LAPAN[2] in 2001, Fig. 10.9 shows the antenna field of the EAR. It can observe winds and turbulence in the altitude range from 1.5 to 20 km (troposphere and lower-stratosphere). It can also observe echoes from ionospheric irregularities at heights more than 90 km. The EAR observations from the vicinity of the Earth's surface to the ionosphere is applied to clarify the dynamic uniting of the upper and lower atmosphere in the equatorial region. Synthesizing this with the observation of equator atmospheric wave motion and the cumulus convection, the basic process of the Earth's atmospheric change that causes climatic variations such as water and energy transport and occurrence of El Nino is expected to be clarified.

The EAR has a circular antenna array of approximately 110 m in diameter, which consists of 560 three-subelement Yagis, and the PA product is 4.7×10^7 W m^2. The EAR is also an active phased array system with each element driven by a TR module of transmitter frequency of 47.0 MHz. The radar beam direction is switched within 200 μs in minimum. The block diagram of the EAR system is shown in Fig. 10.10, and its basic specifications are listed on Table 10.3.

[2]National Institute of Aeronautics and Space (Indonesian: Lembaga Penerbangan dan Antariksa Nasional/LAPAN).

Fig. 10.9 The Equatorial Atmospheric Radar (EAR) established by Kyoto University in West Sumatra, Indonesia

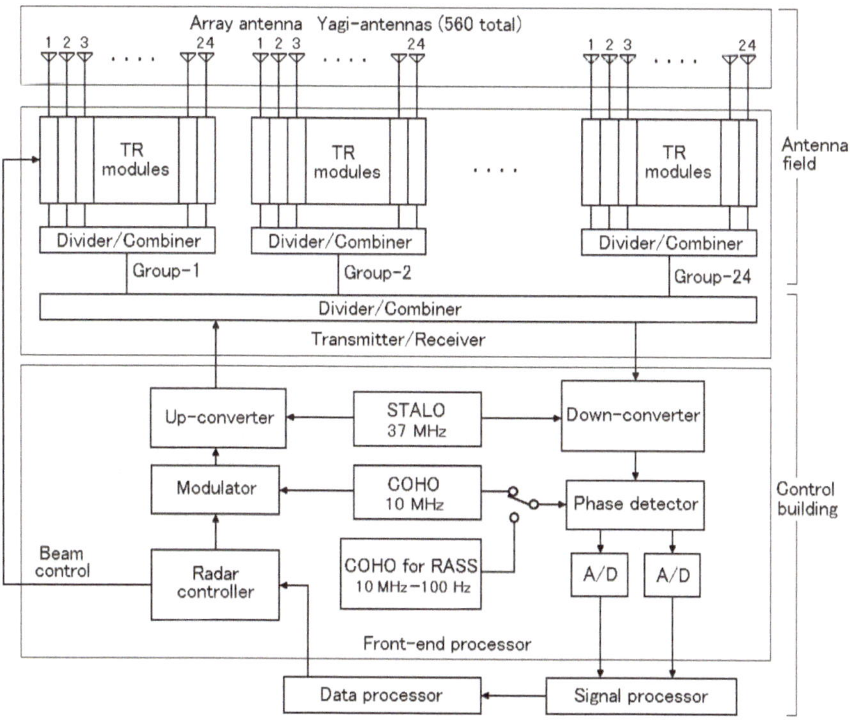

Fig. 10.10 Block diagram of the EAR system [from Fukao et al. 2003]

Table 10.3 Basic specifications of the EAR [from Fukao et al. 2003]

Location	West Sumatra in the Republic of Indonesia (0.20°S, 100.32°E)
Radar system	Active phased array radar
Transmitted frequency	47.0 MHz
Antenna aperture	Diameter 110 m
Antenna element	560 Three-subelement Yagis
Beam width	3.4° (half power of full array)
Beam direction	0°–30° off zenith arbitrary angle
Beam switching	200 μs (minimum)
Polarization	Linear
Peak transmitted power	100 kW
Average transmitted power	5 kW (maximum)
Pulse width	0.5–256 μs variable
Pulse repetition time	200 μs–10 ms variable (every 200 μs)
Data A/D conversion	14 bits × 2 channels
Pulse compression	Binary phase coding up to 16 elements Barker, complementary, and Spano codes

Antenna Element

Antenna elements, three-subelement Yagis, are allocated on the equilateral triangular grid with sides of 0.7λ ($\simeq 4.5$ m). The nominal beam width is 3.4° and the antenna gain is 33 dB or more. Each element and the corresponding TR module are connected by a coaxial cable.

Generation of Transmission Signal and Modulation

A COHO signal of 10 MHz is produced by the COHO based on the stable crystal oscillator output or the GPS signal, and a local frequency (LO) signal of 37 MHz is produced by the STALO, both in the front-end processor housed in the control building. The COHO signal is applied pulse modulation and code modulation in the modulator, and becomes TX IF. The up-converter mixes the TX IF and the LO signal and outputs it to the transmitter–receiver as the TX rf at 47.0 MHz. The TX IF is distributed to 24 feeding points by the first divider/combiner, and each TX rf is distributed again to the same number of feeding points by the second divider/combiner in the antenna field. Each output signal of the second feeding point is sent to a TR module and amplified to final transmitting power level. As a result, the original TX rf is distributed to 24 × 24 feeding points, Among these, 16 points of the edge that hardly influence on the generation of antenna pattern formation are omitted, and 560 antenna elements in total of the remainder are actually connected to the TR modules.

Fig. 10.11 TR module of the EAR. (**a**) Appearance and (**b**) configuration

TR Module

The final peak output power of the TR module is 180 W. Counting power attenuation through the path, the final maximum peak and average transmitted power are 100 and 5 kW, respectively. The TR module is attached on the base of each antenna, and is consisted of a TX amplifier, an RX amplifier, a semiconductor TX/RX switch, and a phase shifter for the phase control of the TX and RX signals. The appearance and configuration of the TR module are shown in Fig. 10.11a, b, respectively.

Signal Processing

The received signal, RX rf, is amplified by the TR module, down-converted and phase detected and converted to digital *I* and *Q* signals by the front-end processor. The digital *I* and *Q* signals are applied pulse compression, decoding, coherent integration, FFT, and incoherent integration, and finally becomes power spectrum by the signal processor.

Data Processing

The power spectrum data of the signal processor output is applied to the data processor. The echo intensity, the Doppler velocity, and the velocity spectrum width are derived from the power spectrum data. The wind vector is calculated based on the results of several radar beam scanning and recorded. The results of radar observation and radar system monitoring record are sent to Japan through the satellite communication channel.

10.2.5 The Antarctic Syowa MST/IS Radar: PANSY

The program to establish a large scale atmospheric radar at the Syowa Station in the Antarctic (69°00′S, 39°35′E) has been ongoing since 2009 and is supported by the National Institute of Polar Research, Japan. The radar, named Program of the Antarctic Syowa MST/IS Radar (PANSY) (Sato et al. 2006), has a bit larger PA product than that of the MU radar. The PANSY radar has wide observation ranges from the troposphere to the mesosphere. In addition, it can observe a part of upper atmosphere as an IS radar. The PANSY radar is also an active phased array radar that operates at the frequency of 47.0 MHz. Its antenna aperture is 18,000 m^2, and the antenna is composed of 1,045 crossed three-subelement Yagis. The maximum peak and average radiation power are 500 and 25 kW, respectively, and the PA product is 4.5×10^8 W m^2 (the MU radar: 4.2×10^8 W m^2. See Sect. 10.2.3). The receiver is composed of 55 channel multi-receiving system and each sub-receiver can operate independently.

The PANSY radar is going to start operation by middle of the 2010s and is expected to elucidate the dynamics and effects of polar atmospheric waves on the Earth's climate through the observations of global-scale atmospheric circulation and interactions of the neutral atmosphere with the ionosphere and magnetosphere, so on.

10.3 Wind Profiler

The technical results of the MST/ST radar developed by the National Oceanic and Atmospheric Administration (NOAA) and the Max-Planck-Institut für Radioastronomie (MPIfR; German) are marketed as "Wind Profiler" by the manufacturers who have jointly developed with them. Moreover, the group of Adelaide University in Australia also established a company and dispenses atmospheric radars such as the ST radar all over the world. In these radars, cost down for manufacturing and installation is tried by suppressing output power and resolution, adopting COCO array antenna, and so on.

10.3.1 The NOAA Profiler Network

General

The NOAA has established a wind profiler network, the NOAA Profiler Network (NPN), consisting of 35 Doppler radars of around 404 MHz located in 18 central

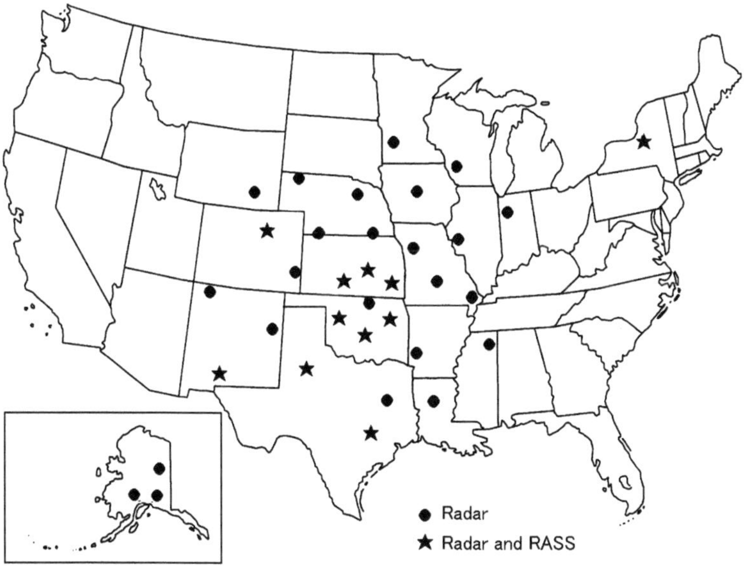

Fig. 10.12 Site location of the NPN radars

US and Alaska[3] and has been operating since 1992. Figure 10.12 shows the site location of the NPN radars and RASSs, and Fig. 10.13 shows an example of the appearance of the NPN radar and RASS system at Platteville, Colorado. Wind profilers measure vertical profiles of horizontal wind speed and direction to above the tropopause. The data produced by the NPN is distributed in real time to the National Weather Service (NWS), environmental research groups, and is utilized to improve the accuracy of short-term weather prediction. It is also distributed every hour to the Global Telecommunication System (GTS) and is open to the public on the Internet.

System Configuration

The block diagram of the NPN radar and RASS system is shown in Fig. 10.14. The radar wavelength is 74 cm, and the radar adopts a COCO array antenna. The transmitter, the receiver, the signal processor, the data processor, and peripherals are housed in the white cabinet shown in Fig. 10.13 middle right. The transmitter provides 6 kW of peak power, and the transmission pulse width is selected 3.3 μs (low mode) or 20 μs (high mode). The low mode samples the lower atmosphere

[3]Thirty-two radars of 404 MHz are located in central US, and three radars of 449 MHz are in Alaska. The frequency of 404 MHz is used for the salvage signal of the space satellite, thus it is discussed to shift the frequency to 409 MHz so as to avoid mutual interferences.

Fig. 10.13 NPN radar and RASS system at Platteville, Colorado [http://profiler.noaa.gov/npn/]

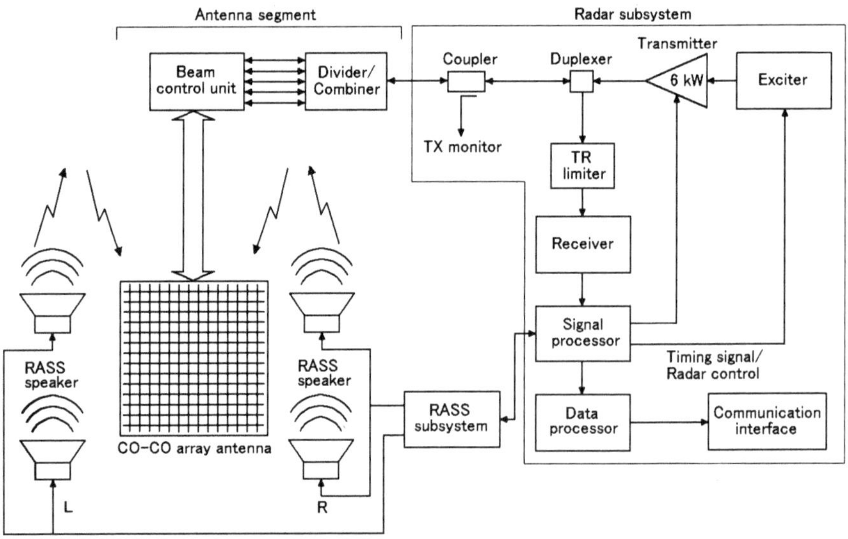

Fig. 10.14 Block diagram of the NPN radar and RASS system

between the height of 500 m and 9.25 km above ground level (AGL), and the high mode between 7.5 and 16.25 km AGL, respectively, with 36 range gates spaced every 250 m in the vertical. The radar uses a 128-point FFT to perform spectral analysis and extrapolate velocity estimates.

Antenna

The aperture of NPN radar antenna is 40 ft×40 ft and the antenna is a passive array one consists of two crossed COCO arrays. One array forms the zenith and obliquely eastward radar beam, and the other array the zenith and obliquely northward one. Beam directions are about 16° off zenith angles, respectively. Each COCO array consists of 20 coaxial cables. The transmitted power of 6 kW is first distributed five and again distributed four, and supplied to those cables.

NPN RASS

The sound waves outputted from the loudspeaker are reflected by the dish reflector and radiated to the space. The frequency of the sound is about 850–900 Hz which corresponds to the half-wavelength of the NTP radar (about 37 cm). The processor in the RASS subsystem processes the RASS received signal, cooperates with the NPN radar in low mode, and produces the profile of vertical wind speed and temperature in real time.

10.3.2 Wind Profiler Network in Europe

In Europe, experimental works to configure the wind profiler network for actual use have been developed by European Cooperation in Science and Technology (COST). The observation network, named EUPROF, is composed of various wind profilers from 13 countries, and the collected data are delivered by British Weather Bureau. Moreover, the data are open on the Internet. Several meteorological organizations utilize single wind profiler data or multiple ones for their actual weather forecasting use. Figure 10.15 shows the wind profiler network (EUPROF) in Europe. In recent years, the coordination of efforts to the frequency allocation has been promoted internationally. Thus, it is expected that more wind profiler networks will be established.

10.4 Lower Troposphere Radar

Although ST and T radars are miniaturized compared with MST radars, they still radiate large power. In these radars, it is in fact impossible to observe near the Earth's surface, about 1 km height from the ground and the major region of the atmospheric boundary layer, due to the leakage of transmitted power to receiver. In addition, it has been commonly recognized that the atmospheric boundary layer is critical to human activities. Thus, the research and development have been advanced to miniaturize ST radar further since the end of the 1980s.

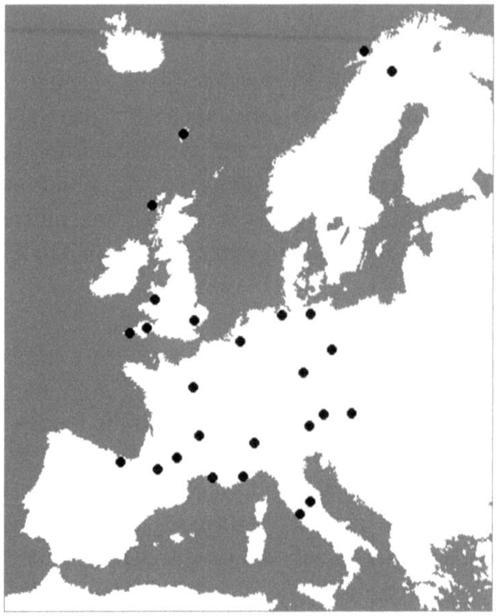

Fig. 10.15 Wind profiler network (EUPROF) in Europe [http://www.met-office.gov.uk/research/interproj/cwinde/profiler/index.html]

10.4.1 Boundary Layer Radar (BLR)

The design and development of the boundary layer radar (BL radar or BLR) were advanced to give priority for the ease of transportation and the small size. Hence, the transmitted frequency of the BLR is selected as 900-MHz band to microwave band, and the subsystems, the antenna, the transmitter, and the receiver, are composed of small and inexpensive components. Furthermore, the RASS is integrated into the system since the early stage of development to obtain the temperature profile. Ecklund et al. (1988) of the NOAA, USA completed the BLR first. It adopts the UHF band, 915 MHz, for the transmitted frequency. In Japan, both groups of Kyoto University and the National Institute of Information and Communications Technology (NICT) developed BLRs of L-band, 1,357 MHz, independently in around 1990 (Hashiguchi et al. 1992; Nakamura and Masuda 1992).

Figure 10.16 shows the BLR developed by Kyoto University at the first time. It has one transmitter of 1 kW peak power and three parabolic antennas of 2 m diameter, and switch the radiation of transmission pulse in sequence to obtain the profile of wind vector continuously.

Figure 10.17 shows the antenna of the BLR which operates at higher frequency than the former, 3.05 GHz (Yamamoto et al. 2002). It adopts a planar patch array antenna, and is highly transportable and strong in the impact such as the vibrations due to transportation.

In the radar frequencies of 900-MHz, 1.3-GHz, and 3.05-GHz bands, precipitation echo becomes predominant, thus the observation of vertical wind velocity in

Fig. 10.16 The first BLR developed by Kyoto University [from Hashiguchi et al. 1995a]

Fig. 10.17 Appearance of the 3.05 GHz BLR. (**a**) Antenna module and (**b**) the closeup of the module [from Yamamoto et al. 2002]

precipitation becomes difficult. However, precipitation particles can be used as the tracer of wind if particle fall speeds are negligible.

About 100 BLRs have already been operated by the state governments and local fire stations in the U.S.A., and are utilized to observe inversion layers and to forecast forest fires, etc. These radars are called Cooperative Agency Profilers (CAPs).

10.4.2 Turbulent Eddy Profiler (TEP)

With the increasing need of detailed measurements of the boundary layer, the 915-MHz turbulent eddy profiler (TEP) was developed by the University of Massachusetts, applying the beamforming procedure and the radar imaging technique.

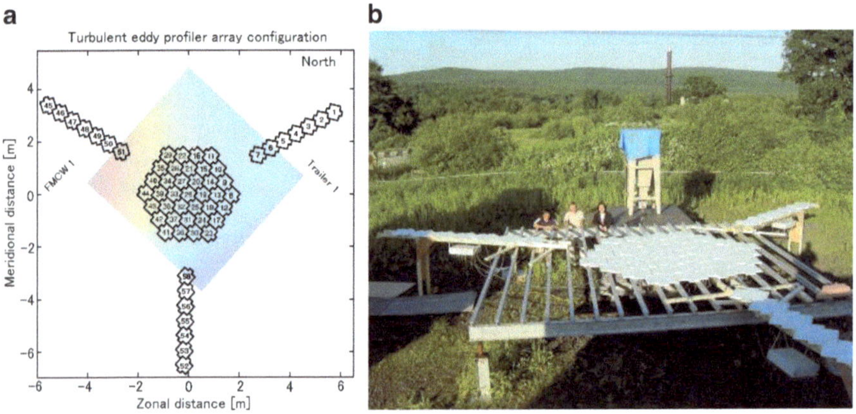

Fig. 10.18 (**a**) The receiver array configuration of TEP. Shading on the array configuration provides the area where precise measurements of the element heights were made for doing precise phase calibration. (**b**) The appearance of the TEP with the transmitting horn antenna in the background [from Palmer et al. 2005]

The TEP consists of a transmitter with wide radar beam of 25°, and a receiver array of up to 64 microstrip patch elements separated by approximately 0.57 m (Mead et al. 1998; Lopez-Dekker and Frasier 2004). The peak transmitted power is 4 kW. Using the receiver array that operates as the independent receivers, coherent radar imaging (CRI) methods can be implemented to produce high resolution images of the boundary layer with temporal resolution of approximately 2.5 s.

Another attempt to achieve higher angular resolution, the so called Big-Y configuration is adopted as the schematic and appearance is shown in Fig. 10.18 (Palmer et al. 2005). The TEP radar with Big-Y is configured by building three outrigger arms and placing seven array elements on each. In Panel (a), the center portion, the array numbers 8–44, consists the original TEP radar with 37 elements which has an aperture of approximately 3.4 m diameter and beam width of approximately 5.4°. With the sparse array of the three outrigger arms, the TEP radar can achieve the angular resolution expected from a filled array with an aperture of 11.3 m and beam width of approximately 2.7°, although the sidelobe level increase and the gain become small. Shading on the array configuration in Panel (a) provides the area where precise measurements of the element heights were made for doing precise phase calibration. Phase calibration can become problematic for sparse arrays when segments of the array are separated by several wavelengths, as is the case with the Big-Y configuration. At present, successful calibration has been accomplished for only the center portion (Palmer et al. 2005). Palmar et al. used spectral sorting with the TEP radar to study the interaction of precipitation and clear-air turbulence. Exploiting the Capon based imaging, the authors were able to produce separate three-dimensional images of precipitation and clear-air echoes with unprecedented spatial and temporal resolution. The TEP radar is also used in a detailed study of the DBS method by comparing the radar imaging technique results to various DBS beam configurations (Cheong et al. 2008). Such studies would not be possible without the unique design of the TEP radar.

10.4.3 Lower Troposphere Radar (LTR)

After the research and developments of several atmospheric radars, Kyoto University developed the lower troposphere radar (LTR) to observe the whole troposphere by enlarging the 1.3-GHz BLR, but preserving its transportability (Hashiguchi et al. 2004). The same radar system is adopted in the wind profiler network, the Wind Profiler Data Acquisition System (WINDAS), of the Japan Meteorological Agency (JMA).

System Configuration

The LTR is an active phased array radar, and consists of the antenna unit, the transmitter–receiver module unit, the transmitter unit, the receiver unit, and the signal processor unit. The basic characteristics of the LTR is listed on Table 10.4. A phased array antenna of $4\,\mathrm{m}\times 4\,\mathrm{m}$ aperture is adopted to the antenna unit as shown in Fig. 10.19a. The radar beam can be electrically steered sequentially to the vertical and at zenith angles of $0-45°$ in the north, south, east, and west vertical planes. The structure of the antenna elements are shown in Panel (b). The element is an electromagnetically coupled coaxial dipole (ECCD) antenna which was originally developed for the base station of the Personal Handy-phone System (PHS). The ECCD array antenna is a modification of the COCO array antenna which was discussed in Sect. 10.2.2. It has the ECCD as the radiating element which is composed of a half wavelength metallic circular pile fed electromagnetically by an annular ring slot on the outer conductor of the feeding coaxial cable. The metallic circular piles act as radiating dipoles and their collinear arrangement in-phase excitation gives an array performance (Miyashita et al. 1999). The physical arrangement of the antenna is 24 rows (x direction) and 24 columns (y direction) of elements, each of which is sub-divided into 2 sub-elements, thus a total of 96 sub-elements are used. An active module is connected to four sub-elements in such a way that a module controls two sub-elements each for one row and one column, thus a total of 24 modules of each peak output power of 100 W are used. These modules are housed in the transmitter–receiver unit in the bottom of Panel (a).

A total of peak output power of all modules accounting combination loss is 2 kW, Furthermore, by applying a pulse compression technique which uses binary phase coding up to eight elements, the peak transmitted power is equivalently 16 times high in maximum compared with the first BLR shown in Fig. 10.16. Moreover, the transmitter–receiver unit is installed just beside the antenna, and the power attenuation in the transmission line is suppressed to the minimum. To reduce the ground clutter echoes, the height of radiation point is empirically chosen about 20 cm. As for the phase coding in the pulse compression, Spano codes discussed

Table 10.4 Basic specifications of the LTR [from Hashiguchi et al. 2004]

Transmitted frequency	1357.5 MHz
Antenna type	Phased array antenna
Antenna aperture	About 16 m^2 (4 m × 4 m)
Beam width	4.0° (half power)
Beam directions	5 directions (zenith, north, south, east, and west) 0°–45° off zenith angle
Antenna gain	33 dB
Polarization	Linear
Peak transmitted power	2 kW
Average transmitted power	428 W (maximum)
Receiver band width	10 MHz
Pulse width	0.67, 1.0, 1.33 μs (switching)
Pulse repetition time	25, 50, 100 μs (switching)
Height resolution	100 m (minimum)
Time resolution	30 s–1 min
Observation range	100 m to several km

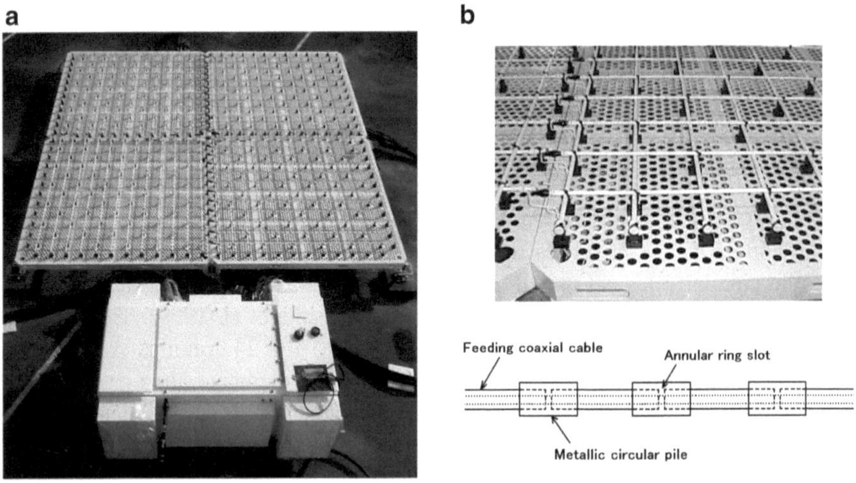

Fig. 10.19 (**a**) Antenna unit and transmitter–receiver module unit of the LTR. Antenna aperture is 4 m×4 m. (**b**) Structure of the antenna elements [from Miyashita et al. 1999]

in Sect. 8.3.4 are applied, thus the truncated range is eliminated and the data of the vicinity of the radar also can be obtained. In addition, by the offset of COHO frequency, observations of atmospheric temperature are also performed by using the radar acoustic sounding system (RASS).

Fig. 10.20 Photographs of the WINDAS LTRs. (**a**) One of the WINDAS LTRs, (**b**) same but covered by a semi-globe radome made of FRP to prevent from heavy snow accumulation over the antenna [from Ishihara et al. 2006]

10.4.4 WINDAS of Japan

The Japan Meteorological Agency (JMA) has established the Wind Profiler Data Acquisition System (WINDAS). The WINDAS consists of thirty-three 1.3-GHz LTRs which are located with dense spatial resolution of 130 km on the average over the main islands of Japan, and the Control Center is located in Tokyo. Twenty-five LTRs started to be in operation in 2001, and eight LTRs have been added by 2011. The photographs of the WINDAS are shown in Fig. 10.20. Panel (a) shows one of the WINDAS LTRs, (b) same but a semi-globe radome made of fiber reinforced plastics (FRP) is installed, which is located in the heavy snowfall area to prevent heavy snow accumulation over the antenna (Ishihara et al. 2006). In the case to house near the building, the electromagnetic shielding fence is installed to avoid electromagnetic wave interference and to suppress ground clutters (Rao et al. 2003). The site location of the WINDAS LTRs is shown in Fig. 10.21.

The data processing height of the WINDAS is between 400 m and 9 km. Height averages of wind measurement are 6–7 km in summer, 3–4 km in winter, and 5.3 km on the average through a year, and wind vector profile of the height resolution of 100–600 m is obtained every 10 min. The observed data by 33 WINDAS LTRs are sent to the Control Center every 1 h. The data are used to provide upper-air wind data to the numerical weather prediction (NWP) of the JMA, particularly to the hydrostatic (till 2004) and non-hydrostatic (from 2004) mesoscale numerical model (MSM) (Ishihara et al. 2006).

Fig. 10.21 Site location of the WINDAS LTRs [Courtesy of the Japan Meteorological Agency]

Chapter 11
Observations by Meteorological Radar

Most of the meteorological disasters are caused by such disturbance phenomena as typhoon, local severe rain, thunderstorm, tornado, wind shear, and so on. Meteorological radar observations play an important role in understanding the details of these phenomena when observational data are assimilated into mesoscale numerical models (e.g., Houze 1993, Chap. 4). In this chapter, representative examples of these phenomena which were observed by meteorological radars with various antenna scan modes are presented.

11.1 Precipitation Observation by Meteorological Radar

A conceptual diagram for observation of precipitation by a meteorological radar is shown in Fig. 11.1. The electromagnetic wave radiated from the radar antenna, installed in places having a good field of view, is scattered by raindrops or cloud particles. Waterdrops in cloud at high altitudes freeze and become ice particles such as graupel, hail, and snowflakes. Large snowflakes melt when falling through the melting layer[1] of several hundreds meter thickness, which exists immediately beneath the 0°C height. Partially melted snowflakes have ice cores and appear as large drops that are unstable and break apart into numerous small raindrops. After a snowflake begins to melt, its shape is reduced to a smaller raindrop or even broken up into a number of small raindrops, but in the first stage of melting, only the surface of the particle melts, and it looks as if a raindrop of larger diameter exists. Because the refractive index of water is larger than that of ice, radar echoes from the melting layer are observed stronger than those from above this layer. Because the drop's backscatter cross section is proportional to the 6th power of its diameter, and because large raindrops break apart as they fall below the melting layer, reflectivity

[1]The altitude of melting layer at mid latitudes is about 2,000–4,000 m. However, it changes in height with the seasons, and with passages of cold fronts.

S. Fukao and K. Hamazu, *Radar for Meteorological and Atmospheric Observations*,
DOI 10.1007/978-4-431-54334-3_11,

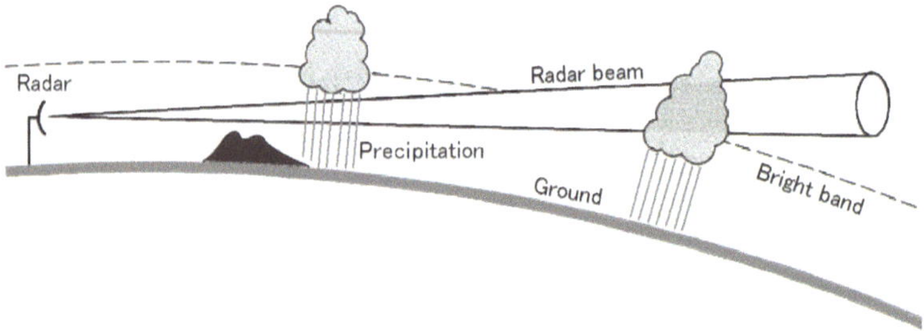

Fig. 11.1 Conceptual diagram for observation of precipitation by a meteorological radar

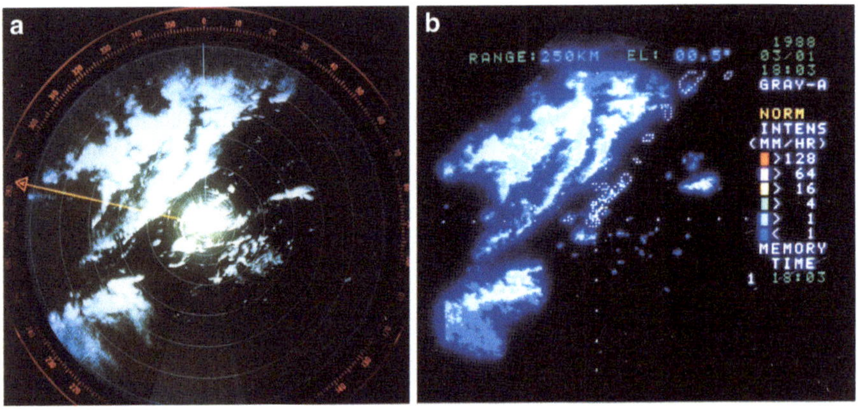

Fig. 11.2 Extratropical cyclone observed by a meteorological radar. (**a**) Raw PPI display and (**b**) digitally processed display after ground clutter suppression [Provided by the Japan Meteorological Agency]

decreases below this layer. Although there are more drops below the melting layer, the increase in the number of drops is insufficient to overcome the decrease of reflectivity due to the decrease in drop cross sections. Thus the reflectivity is weaker above and below the melting layer. For this reason, the layer is called the bright band.

Radar Echo Indication

The advantage of meteorological radar is to be able to observe precipitation in a wide area continuously in time. Figure 11.2 shows the plane distribution of rain area caused by an extratropical cyclone which is observed by a meteorological radar. Panel (a) shows the plan position indicator (PPI) plot of the 300 km radius by an analog data display which was widely used in the past. The radar is located at the center of the screen, and the white color in the screen indicates the area of intense radar echoes. The analog data is composed of weather echoes and ground clutter which cannot be visually separated. Panel (b) shows the digitally processed data. Ground clutter echoes which are included in Panel (a) are suppressed, whereas only precipitation echoes are displayed.

Intensification and Dissipation of Precipitation

A series of schematic of the structure of a mesoscale precipitation feature as viewed by meteorological radar is shown in Fig. 11.3 (Leary and Houze 1979). The right panels are vertical cross sections, and the left ones are horizontal cross sections during (a) formative, (b) intensifying, (c) matured, and (d) dissipating stages of its life cycle. The outside contour for radar reflectivity is the weakest detectable echo, and the inner contours are for successively higher reflectivity values. The highest intensity in the left and right panels are the same level, whereas lower intensity levels in the left panels are rounded up to the second highest level.

11.2 Mesoscale Rain

11.2.1 Structure of Extratropical Cyclone and Front

Rain Bands

Line echo regions in connection with cyclones which are commonly observed by meteorological radars may have comparatively strong intensity, and sometimes bring localized heavy rains. Extratropical cyclones are lows that progress along the westerlies belt at mid-latitude. Major rain bands generated near lows or fronts are as follows (Houze et al. 1976; Hobbs 1978):

1. Type 1. Warm frontal. Rain bands of approximately 50 km in width and oriented parallel to warm front. It is found toward the leading edge of a frontal cloud shield.
2. Type 2. Warm sector. Rain bands of typically 50 km in width, found in the intersection of the surface warm and cold fronts. It tends to be parallel to cold fronts.
3. Type 3. Cold frontal-wide. Rain bands of approximately 50 km in width, oriented parallel to cold front. It is found toward the trailing edge of a frontal cold shield.
4. Type 4. Cold frontal-narrow. Extremely narrow rain band (up to 5 km in width) coinciding with surface cold front.
5. Type 5. Post frontal. Rain bands located in the convective cloud field behind a frontal cloud shield.

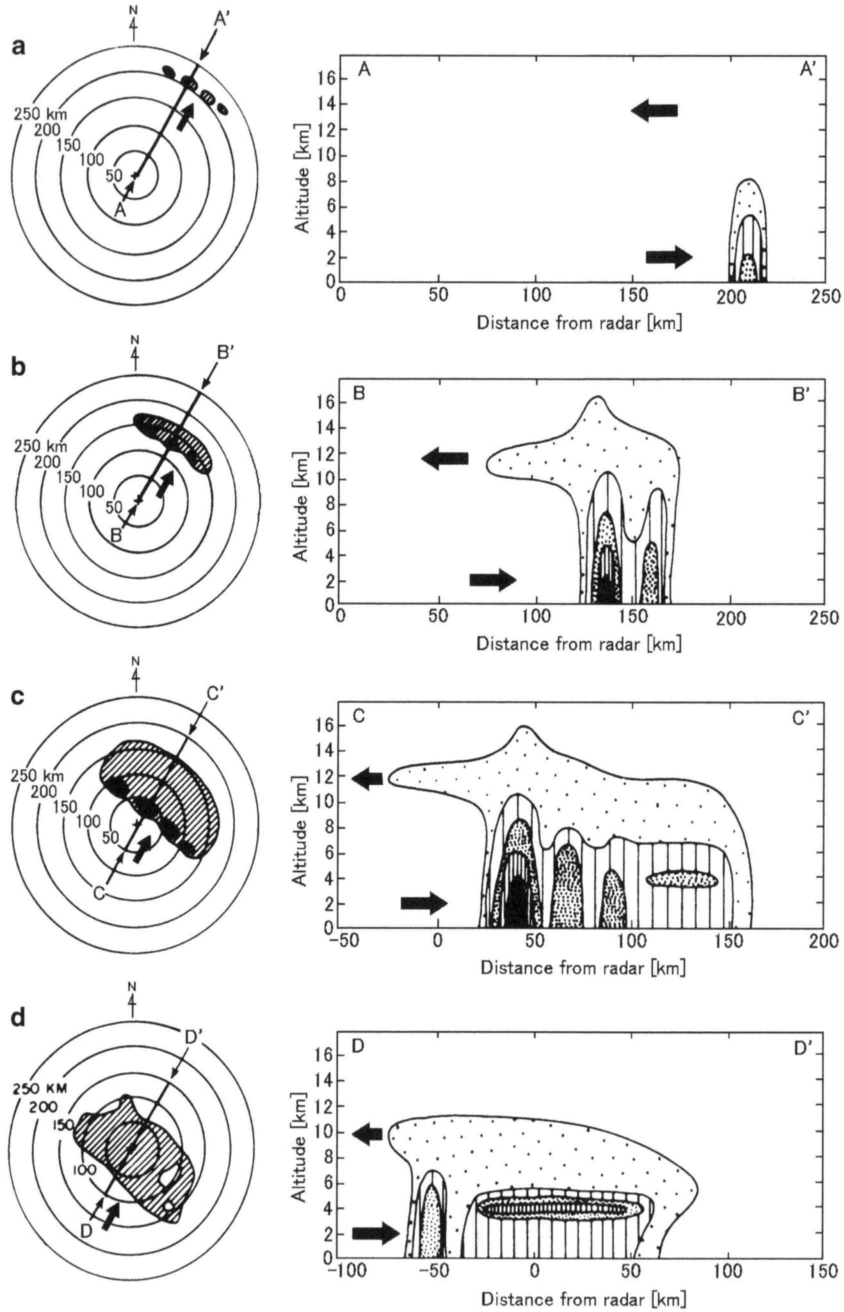

Fig. 11.3 Schematic of the structure of a mesoscale precipitation feature as viewed by meteorological radar. The *vertical cross sections* in the *right panels* along A–A' to D–D' correspond to the baselines with the same marks in the *horizontal plots* on the *left* during (**a**) formative, (**b**) intensifying, (**c**) matured, and (**d**) dissipating stages of its life cycle. *Heavy arrows* in the *horizontal plots* indicate direction of low-level winds. *Arrows* on *vertical cross sections* indicate directions of the low- and upper-level winds [from Leary and Houze 1979]

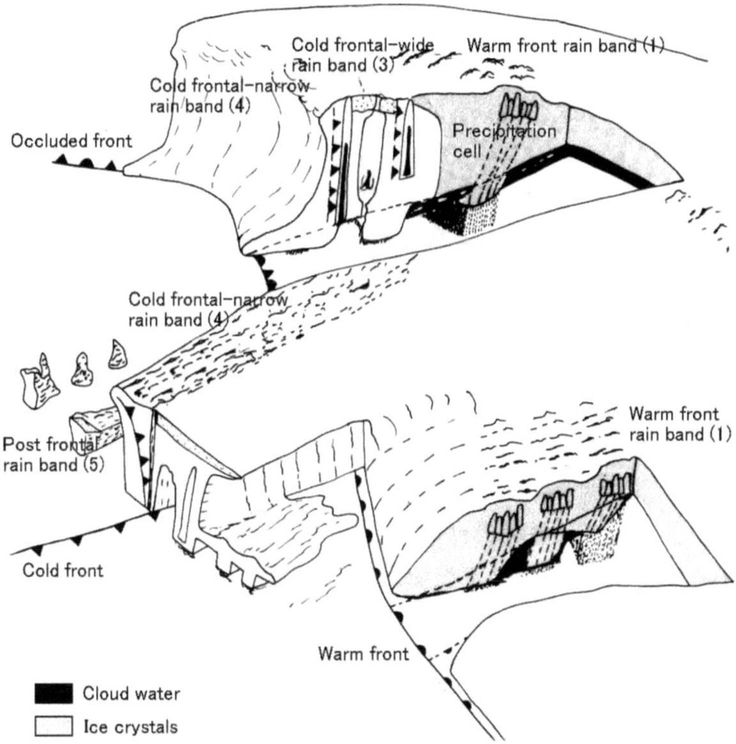

Fig. 11.4 Schematic three-dimensional structure of an extratropical cyclone [from Hobbs 1978]

Figure 11.4 shows the schematic three-dimensional structure of an extratropical cyclone by Hobbs (1978). Numbers (1)–(5) in the figure correspond to above mentioned rain band Type 1–5, respectively. The length of each rain band is several-hundreds kilometer. It is known that convective precipitation cells, which bring stronger rain, exist inside of these mesoscale rain bands. The horizontal distribution of the precipitation area is observed by meteorological radars with PPI scan, and its vertical structure is observed with RHI scan.

Warm Conveyor Belts

Extratropical cyclones are occasionally generated around warm conveyor belts (WCB), and the associated rainfalls have been commonly observed by meteorological radar, where rainfall intensity is relatively strong, and regional heavy rain may sometimes follow. It develops along a belt of mid-latitude westerlies. Figure 11.5 shows a conceptual diagram of typical distribution precipitations generated around

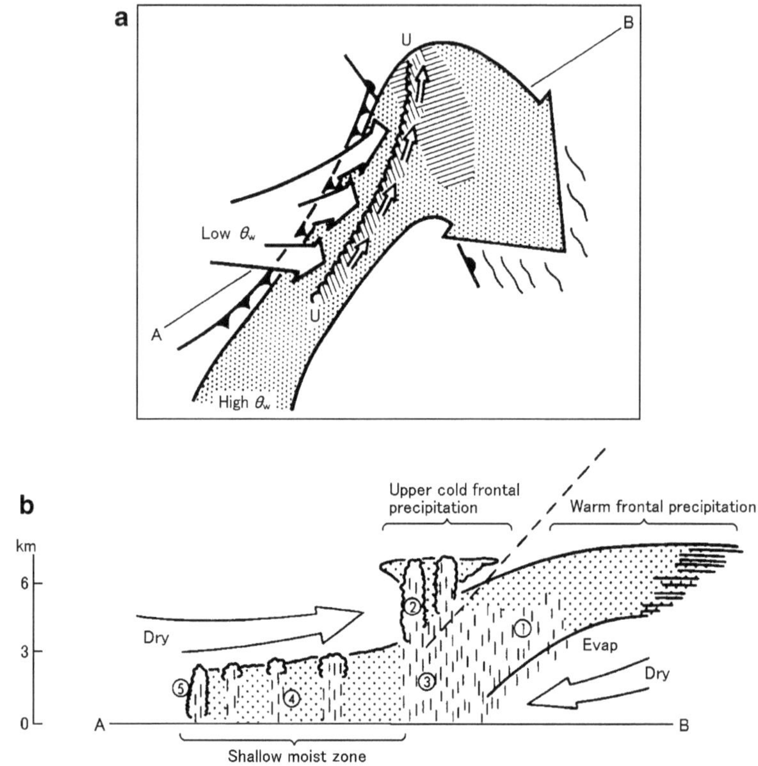

Fig. 11.5 Conceptual diagram of typical precipitations generated in association with a split front; (**a**) plan view; (**b**) vertical section along AB in (**a**). The hatched shading along UU and ahead of the warm front represents precipitation associated with the upper cold front and warm front, respectively. In (**b**), *numbers* represent precipitation types: (1) warm frontal precipitation, (2) convective precipitation-generating cells associated with the upper cold front, (3) precipitation from the upper cold frontal convection descending through an area of warm advection, (4) shallow moist zone between the upper and surface cold fronts, and (5) a shallow layer of precipitation at the surface cold front itself [from Browning and Monk 1982; Browning 1986]

such extratropical cyclones–frontal zone systems. (a) Plan view and (b) vertical section along the baseline of AB in (a) are shown. In (a), UU represents the upper cold front. The upper cold front is ahead of the surface warm front, and thus is referred to a split front. Numbers in (b) represent precipitation types as follows: (1) warm frontal precipitation, (2) convective precipitation-generating cells associated with the upper cold front, (3) precipitation from the upper cold frontal convection descending through an area of warm advection, (4) shallow moist zone between the upper and surface cold fronts, and (5) shallow precipitation at the surface cold front itself (Browning and Monk 1982; Browning 1986).

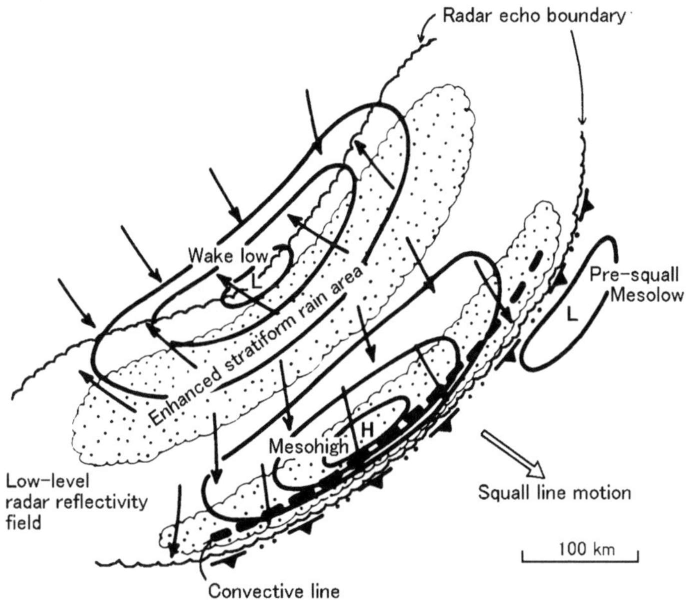

Fig. 11.6 Schematic diagram of plan view of surface winds and precipitations, which produce the wake low at the rear of a squall line with stratiform precipitation. The *arrows* represent wind relative to the squall line [from Johnson and Hamilton 1988]

The horizontal distribution of convective cells inside these mesoscale[2] precipitation areas can be observed by meteorological radar with PPI scan, and the details of the vertical structure can be seen with RHI scan.

11.2.2 Horizontal Structure of Precipitation

Figure 11.6 schematically indicates an actual mesoscale weather and the corresponding observational results by meteorological radar (Johnson and Hamilton 1988). Surface winds relative to movement of the squall line are indicated by arrows in the figure together with radar echo region at the lower height, meso-high, meso-low, and so on. There is a region of strong convective rain at the tip of the squall line. By watching the temporal variation, precipitation cells reached a maturity while proceeding and were eventually weakened. Behind the convective region,

[2]The horizontal range of mesoscale spreads from around 2 to 200 km, and the nominal designation of mesoscale is divided into three ranges: meso-α, meso-β, and meso-γ. Meso-α weather is the phenomena in scales around 200 km and more. Meso-β weather and meso-γ weather are the phenomena in scales of around 20–200 km and 2–20 km, respectively (Orlansky 1975).

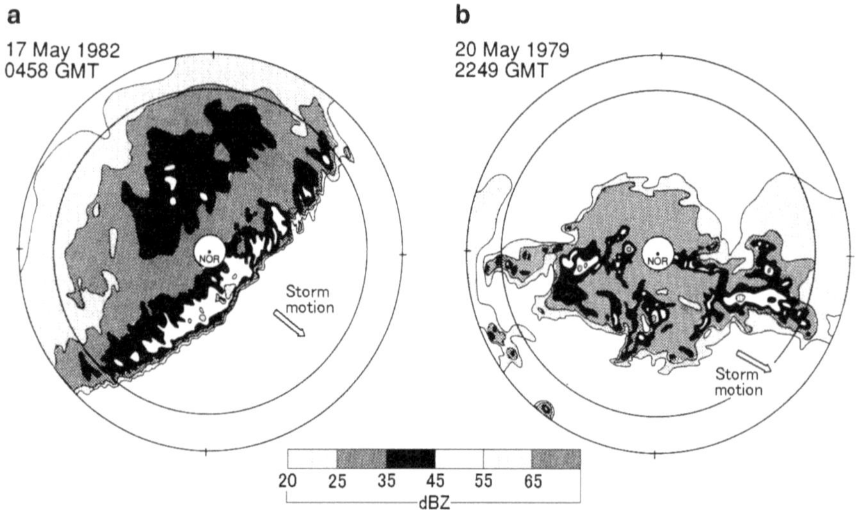

Fig. 11.7 PPI observational examples of the mesoscale convective system at mid-latitudes, which were observed on different days by the NSSL in Norman, Oklahoma. Range rings are at 20, 200, and 240 km, respectively. (**a**) Mesoscale convective system whose echo patterns closely resemble to the schematic diagram in Fig. 11.5, and (**b**) more chaotic example where a number of convective cells are distributed in the precipitation area [from Houze et al. 1990]

the stratiform clouds widely developed, and weak precipitation was observed. The region from the squall line to the edge of the following stratiform rain region was identified by radar echoes.

The horizontal distribution of radar echo can be obtained by the observation with the PPI, Sector PPI (SPPI), and constant altitude PPI (CAPPI) scan modes. Typical examples of mesoscale convective systems at mid-latitudes, which were observed with the PPI scan mode by the National Severe Storms Laboratory (NSSL) in Norman, Oklahoma, are shown in Fig. 11.7, where two panels of radar reflectivity at low elevation angle on different days are shown. Panel (a) shows the result almost identical to the schematic diagram in Fig. 11.6. The high radar reflectivity factor of more than 45 dBZ is observed along the leading edge of the meso-high convective line, whereas that of following wide stratiform rain area is comparatively low, 20–45 dBZ. In such cases, usually a heavy rain shower is generated at the passage of squall line, and weak precipitation continues for about 1 or 2 h afterward. Panel (b) shows the more chaotic example where a number of convective cells are distributed in the precipitation area.

11.2.3 Vertical Structure of Precipitation

Figure 11.8 shows a schematic diagram of the vertical section along the proceeding direction of a squall line (Houze et al. 1989). The squall line proceeds from left to

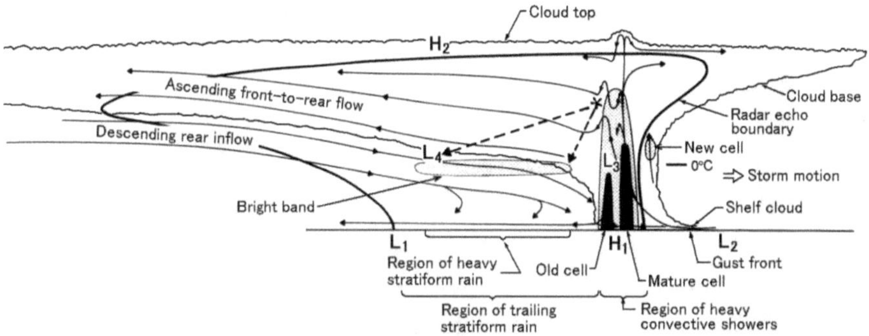

Fig. 11.8 Vertical section along the proceeding direction of a squall line. The *wavy line* of the outermost edge and the *thick line* indicate the areas of cloud and radar echo, respectively. *Broken line* shows the trajectory of falling precipitation particles. H and L are the areas of relatively high and low atmospheric pressure, respectively [from Houze et al. 1989]

right. The thin solid lines show winds relative to the squall line. The wavy line of the outermost edge shows the contour of visual clouds, and the thick line indicates the area of radar echo which appears in SRHI or RHI observation. Black and pale shadows are strong and slightly strong echo areas, respectively. The broken line in the figure shows the trajectory of falling precipitation particles. H and L are the areas of relatively high and low atmospheric pressure, respectively. The pale shadow area in the stratiform cloud following a strong convective rain area is the bright band which is formed just under the 0 °C height, and has the thickness of a few hundred meters. As mentioned in Sect. 11.1, a snow particle, which has grown by merging and/or aggregating cloud particles while falling down from the upper layer, melts from its surface as it passes through the 0 °C height. Hence, the melting snow is observed as the intenser radar reflectivity layer than the regions above and below. By observing such vertical cross sections, it is possible to discriminate stratiform rain from convective rain, and to obtain various information on convective phenomena.

Figure 11.9 shows the vertical section of a squall line and the following stratiform rain area observed by meteorological radar (Biggerstaff and Houze 1993). This figure shows the radar reflectivity factors and vertical velocities, which are composed and averaged by the results of multiple Doppler radars installed within the distance of 60 km in the direction of the squall line motion. Vertical velocities are indicated with the contours of −0.9, −0.45, −0.15, 0.15, 0.45, 0.9, 1.5, 2.4, 3.6 m s^{-1}, respectively. The negative speed (toward radar) is shown by the broken line, while the positive speed (away from the radar) is shown by the solid line. The squall line proceeded from left to right in the figure. The precipitation cell in the area along the squall line is dominated by the upward flow, and the maximum upward flow region corresponds to the echo top.

Figure 11.10 shows a cold front observed by the 9.8-GHz transportable Doppler radar of the Meteorological Research Institute (MRI) in Japan (Sakakibara et al. 1991). In the figure, the vertical cross section along the baseline of nearly southeast

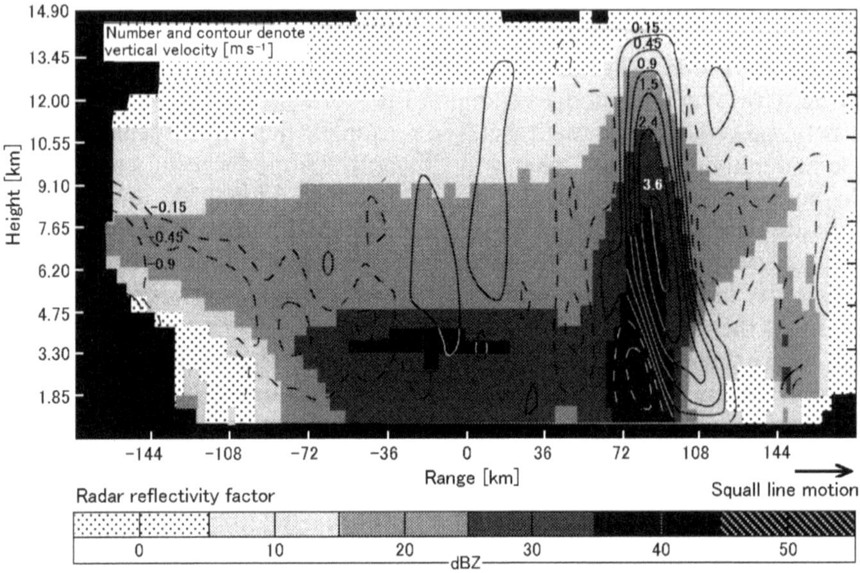

Fig. 11.9 Vertical cross section of squall line and the following stratiform rain area observed by multiple Doppler radars installed within the distance of 60 km in the direction of motion of the squall line. The results are composed and averaged [from Biggerstaff and Houze 1993]

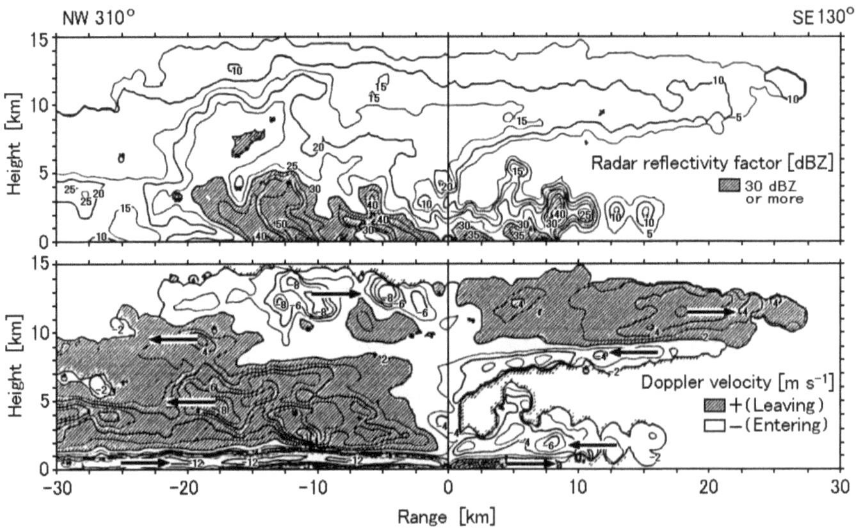

Fig. 11.10 Cold front observed by the 9.8-GHz transportable Doppler radar of the Meteorological Research Institute in Japan, where the *vertical cross section* along the baseline from nearly southeast to northwest, perpendicular to the cold front. *Upper panel*; radar reflectivity factor, and *lower panel*; Doppler velocity [from Sakakibara et al. 1991]

to northwest which is perpendicular to the cold front is shown, where radar reflectivity factor and Doppler velocity are indicated in the upper and lower panels, respectively. The gray area in the upper panel denotes radar reflectivity factors equal to or more than 30 dBZ, and that in the lower panel denotes wind velocity away from the radar. The front moved to the southeast as a whole. Anvil cloud echo can be seen in the upper layer, and small scale convective cloud echoes can be seen over the lower layer. Furthermore, the most active convection echo is generated around 10 km northwest from the radar. In the distribution of Doppler velocity in the lower panel, cold air that enters from the tip of the undermost layer in the northwest, and above the layer warm air moving from southeast to northwest, can be seen.

11.3 Typhoon

Atmospheric lows generated over the ocean in the tropics and subtropics are called tropical cyclones (TCs). Among the TCs generated in the North Pacific region between 100° and 180°E, whose wind velocity increases to more than 17 m s^{-1}(34 knots), are called typhoons.[3] A typhoon is a vortex which extends to all layers in the troposphere, and characterized with its circular arc echo pattern.

11.3.1 Horizontal Structure

Figure 11.11 shows PPI echo patterns of a typhoon and a hurricane seen in the northern hemisphere observed by 2.8-GHz band radars. Panel (a) indicates Typhoon 6717[4] observed by Mt. Fuji radar which had been operated at the top of Mt. Fuji, 3,776 m above sea level, for 35 years between 1964 and 1999 in Japan (Kodaira 1990). Panel (a) was obtained with the PPI scan at the elevation angle of −1.5° so that the position of the radar is off-center in the picture and indicated by a cross. The typhoon was located at 100–600 km apart from the radar in the south-southwest. Ground clutters are observed in the vicinity of the radar, and sea clutters are also seen within the range up to about 150 km in the southeast. In the central area of the typhoon or the typhoon eye, no echo was received in the tens of kilometers

[3]TC is called by different names in different areas; typhoon in the North Pacific Ocean, hurricane in the North Atlantic Ocean, or willy-willy in Australia.

[4]In Japan, a typhoon is numbered by composing the last 2 digits of the Christian era and the serial number of typhoon occurrence every year. Thus Typhoon 6717 means the 17th typhoon generated in 1967.

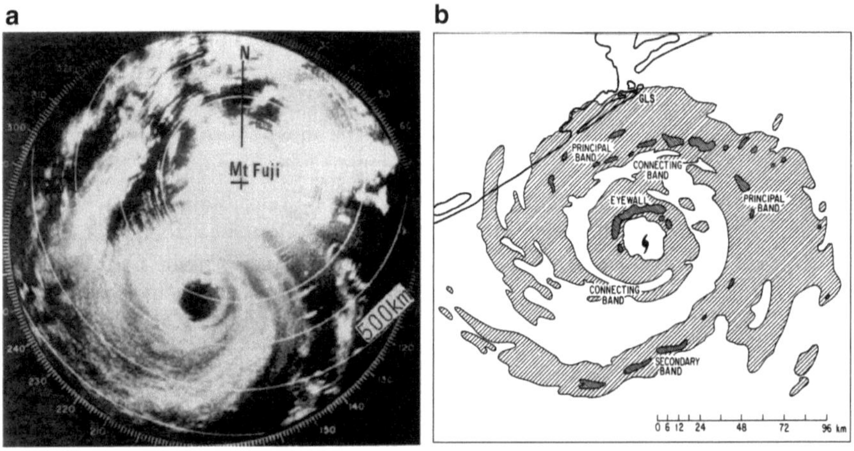

Fig. 11.11 PPI echo patterns observed by 2.8-GHz band radars. (**a**) Typhoon 6717 observed by Mt. Fuji radar in Japan (Kodaira 1990), (**b**) Hurricane "Alicia" observed by the WSR-57 in the United States on 18 August 1983 [from Marks and Houze 1987]

diameter, It is surrounded by the cumulonimbus and stratiform precipitation belt of counterclockwise vortex swirling from the left corner to the upper part in the picture. As indicated by the intensity of sea clutter, the wind velocity on the right side of the eye with respect to the direction of typhoon motion is larger than that on the left side. This is because the moving velocity of the typhoon is added to the vortex speed of the typhoon on the right side.

Nowadays, applications of geostationary meteorological satellite (GMS) and weather observation satellite to research are making steady progress, and it has become possible to observe a typhoon in a wide area. Moreover, long range observing radars such as Mt. Fuji radar at the top of the highest mountain in Japan, whose observation area was 800 km in radius, have been superseded by multiple smaller meteorological radars installed at lower heights. Figure 11.12a shows the picture of Typhoon 0205 observed by the microwave radiometer AMSR-E which was developed in Japan and on board the Aqua satellite of the National Aeronautics and Space Administration (NASA). In this panel, the brightness temperatures at 89.0-GHz (vertically and horizontally polarized waves) and that at 23.8-GHz (vertically polarized wave) were synthesized, where the bright part is precipitation area, and the change of the hue corresponds to an increase of water vapor and clouds in the atmosphere. The satellite observations were made every 12 h when passing over the vicinity of Japan twice a day in the middle of night and afternoon, and this picture was obtained at 0226 Japan standard time (JST) on 4 July 2002. Panel (b) is the cloud image of the same typhoon as (a) taken at 1800 JST on 3 July by the infrared radiation (IR) sensor on board the GMS-5 "Himawari". Panel (c) shows the echo intensity of the precipitation area of the same typhoon obtained by Ishigakijima

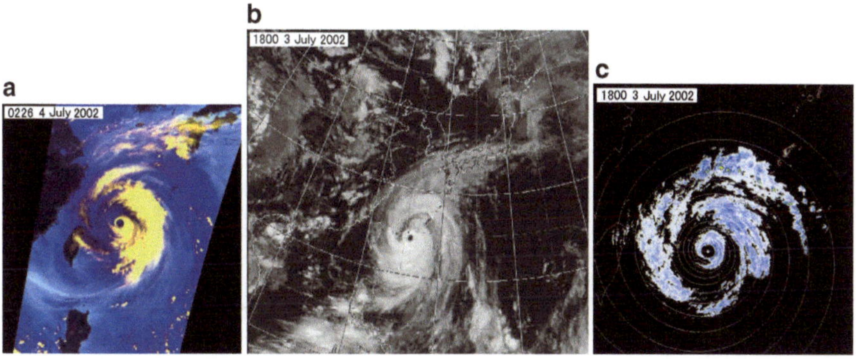

Fig. 11.12 Typhoon 0205 observed by (**a**) the microwave radiometer AMSR-E on board the Aqua satellite of NASA, where the brightness temperatures at 89.0-GHz (vertically and horizontally polarized waves) and that at 23.8-GHz (vertically polarized wave) are synthesized [courtesy of the Japan Aerospace Exploration Agency], (**b**) by the infrared radiation (IR) sensor on board the GMS-5 "Himawari" [courtesy of the Japan Meteorological Agency], and (**c**) by Ishigakijima meteorological radar [courtesy of the Japan Meteorological Agency], respectively

meteorological radar of the Japan Meteorological agency (JMA) at the same time as (b). Panel (a) was observed from the synchronous quasi-recurrent orbit satellite at the distance of 705 km, and (b) from the geostationary meteorological satellite at the distance of 36,000 km from the earth surface, respectively, whereas (c) was observed by the ground based radar with almost horizontal radar beam scanning.

Center of the Typhoon Observed by Two Doppler Radars

Figure 11.13 shows the horizontal structure near the eye of the typhoon, which passed between two Doppler radars. These pictures were obtained by the independent scanning method mentioned in Sect. 4.3.1 when Typhoon 8514 passed between the 5.3-GHz ground base meteorological radar of Meteorological Research Institute of Japan in Tsukuba city and the 9.8-GHz transportable Doppler radar placed at about 28 km southwest from the institute. CAPPI pictures at the 1.5, 2.5, and 3.5 km height are shown from left to right panels, respectively. Wind direction and speed in the horizontal plane are given by arrows, and the contour of radar reflectivity factor is overlaid on the panels. The contour is drawn every 2.5 dB in the area of more than 15 dBZ, and the areas of more than 35 dBZ are shaded. The cloud wall of the eye of the typhoon and the wind that turns around the typhoon center appear clearly from the center of the typhoon to the northeast.

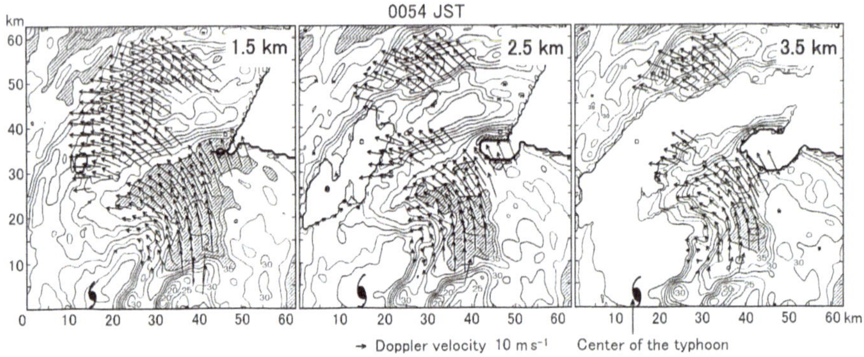

Fig. 11.13 Horizontal structure near the eye of Typhoon 8514 which passed between two Doppler radars. These pictures were obtained by independent scanning mode. Wind direction and speed in the horizontal plane are given by *arrows*, and the contour of radar reflectivity factor is overlaid. The contour is drawn every 2.5 dB in the area of more than 15 dBZ, and the areas of more than 35 dBZ are shaded [from Tabata et al. 1992]

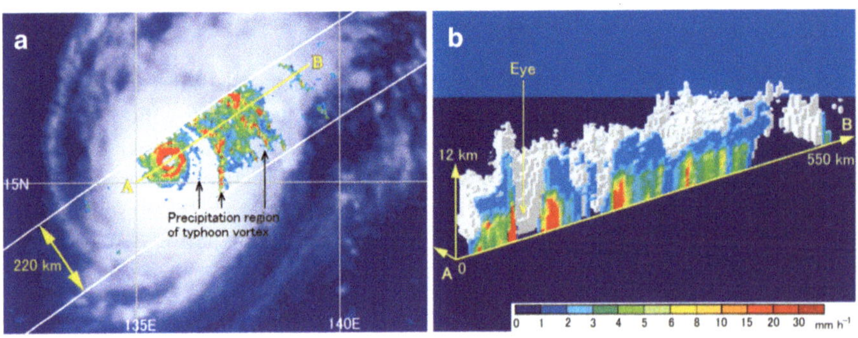

Fig. 11.14 Typhoon 9728 observed by the PR on board the TRMM satellite. (**a**) Radar echo intensity (converted to rainfall rate) overlaid on the visible image obtained by the geostationary meteorological satellite, and (**b**) spatial structure of the same typhoon [Courtesy of the Japan Aerospace Exploration Agency]

11.3.2 Spatial Structure

Figure 11.14 shows a typhoon observed by the precipitation radar (PR) on board the Tropical Rainfall Measurement Mission (TRMM) satellite mentioned in Sect. 9.4.1. Panel (a) indicates the radar echo intensity (converted to rainfall rate) overlaid on the visible image of Typhoon 9728, which was obtained by GMS, and (b) indicates the spatial structure of the same typhoon. There was no precipitation in the eye, whereas it was observed in the east side of the eye. Moreover, a heavy precipitation area of high echo top surrounds the eye, and the typhoon vortex can be seen in the east side of the eye. If a typhoon passes immediately over the atmospheric radar, its vertical structure can be clarified in more detail as will be discussed in Sect. 12.2.2.

11.4 Cumulus Convection

11.4.1 Multicell Thunderstorms

Thunderstorms are characterized with extremely high cloud top and shapes of cumulonimbus developed by intense convective activities. The vertical profile of radar echoes reflects these features clearly. Figure 11.15 schematically shows the vertical profile of a large-scale thunderstorm with multicells of precipitation based on the radar observations in Colorado in the United States (Browning et al. 1976). The storm is generated and developed in order of $n-2$, $n-1$, n, and $n+1$, while moving from the left to the right (north to south).

The solid and broken lines in the figure show the wind flow relative to the storm motion. The actual wind flow is three dimensional, and consequently, its direction perpendicular to the vertical section is discriminated by different lines; wind blowing from the outside toward the inside of the vertical section is shown by a broken line, and that blowing oppositely is shown by a solid line. Consecutive white circles denote the trajectory of a hailstone, which is generated at the cloud bottom in the right, while it grows, matures, and falls. The light gray area, dark gray area, and the inside white area represent radar reflectivity factors of 35, 45,

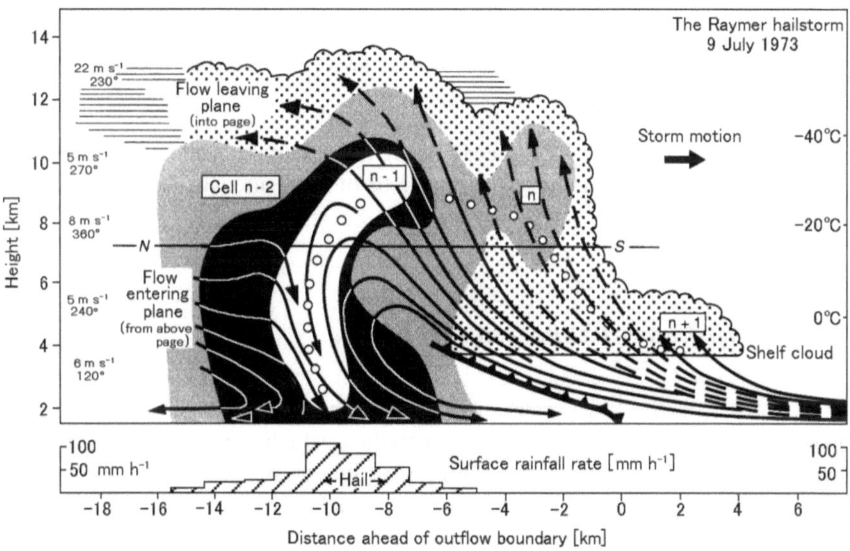

Fig. 11.15 Vertical profile of a multicell thunderstorm schematically shown based on the radar observations in Colorado in the United States. The storm is generated and developed in order of $n-2$, $n-1$, n, and $n+1$, while moving from left to right (from the north to the south). Consecutive *white circles* denote the trajectory of a hailstone. *Light gray*; 35 dBZ, *dark gray*; 45 dBZ, and *white* area inside the *dark gray* area; 50 dBZ, respectively [from Browning et al. 1976]

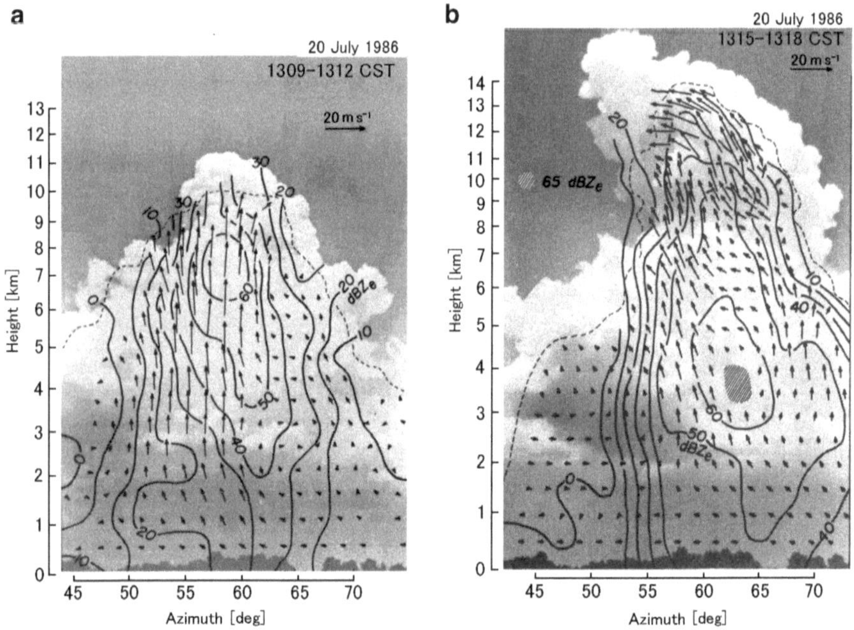

Fig. 11.16 Wind vectors and radar reflectivity factors in a thundercloud obtained with radar superimposed on the photograph of the thundercloud, where the wind vectors indicate the component in the cross section passing through the center of the storm and perpendicular to the line of sight during (**a**) growing and (**b**) matured stages from Doviak and Zrnić (2006, pp. 302–303)

and 50 dBZ, respectively. Wind speed relative to storm motion and direction in each height are indicated along the left vertical axis of the figure (Browning et al. 1976; Houze 1993, p. 276).

11.4.2 Ordinary Thunderstorms

In contrast with the large-scale multicell thunderstorms, ordinary thunderstorms have comparatively simple convective activities, which are composed of upward and downward flows. In general, the lifetime of upward flow is up to about 1 h. The activity of upward flow becomes weak gradually along with the activation of precipitation in the thundercloud, and finally the precipitation reaches ground with downdraft. Figure 11.16 shows the wind vectors and radar reflectivity factors in a thundercloud obtained with radar superimposing in the photograph of the thundercloud (Kingsmill and Wakimoto 1991; Doviak and Zrnić 2006, pp. 302–303). The wind vector, in cross section passing through the center of the storm

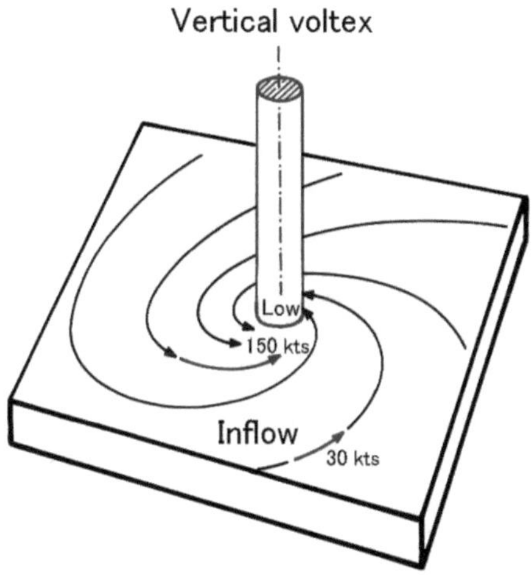

Fig. 11.17 Simplified model of a vertical vortex in a tornado [from Fujita 1985, p. 14]

and perpendicular to the line of sight, was obtained by the independent scanning mode (see Sect. 4.3.1) by three Doppler radars. The radar reflectivity factor was obtained by the CP-2 radar[5] observation in a cross section passing through the center of the storm and orthogonal to the line of sight from the photographer. The contour of equal radar reflectivity factor is drawn every 10 dB. Panel (a) shows the growing stage of a storm which contains updraft speed up to $18\,\mathrm{m\,s^{-1}}$, where the radar reflectivity factor in the center of the cell at the height of about 7 km becomes 60 dBZ, and (b) shows the mature stage after about 6 min from (a), where the radar reflectivity factor becomes the maximum value of 65 dBZ and the highest reflectivity area has descended to around 3.5 km. In the following stage, the downdraft becomes superior.

11.4.3 *Tornado*

The tornado is a high-speed whirlwind that rotates rapidly around an almost vertical axis. It is known that many tornadoes are generated from a large-scale cumulonimbus, i.e., developing multicell thunderstorm as discussed in Sect. 11.4.1. It usually has diameters ranging from ten meters to over several hundred meters. Figure 11.17 shows a simplified model of a vertical vortex in a tornado (Fujita 1985, p. 14). Many tornadoes are generated in the United States, especially in the central part. In some cases, the maximum wind speed exceeds $140\,\mathrm{m\,s^{-1}}$, and causes

[5] The dual-frequency and polarimetric radars of 2.8 and 9.4 GHz of the National Center for Atmospheric Research (NCAR) in the United States. This observational result was derived by the 2.8-GHz radar.

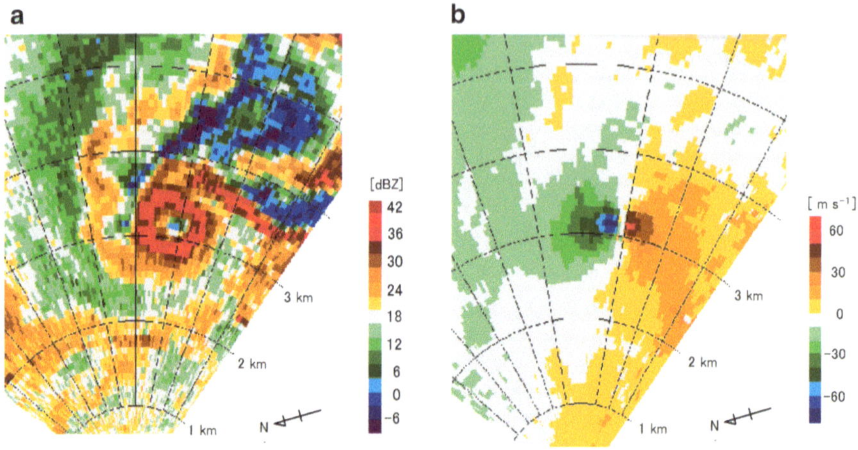

Fig. 11.18 Horizontal cross section of a tornado observed in Dimmitti, Texas at 0103 UTC on 3 June 1995, by the mobile Doppler radar with magnetron transmitter at the frequency of 9.37–9.38 GHz and the radar beam width of 0.93°. (**a**) The radar reflectivity field and (**b**) the Doppler velocity field in radial direction [from Wurman et al. 1996]

significant disasters. In other countries, tornadoes of smaller scale are generated from lower convective clouds developing in the cold front.

Figure 11.18 shows the horizontal appearance of a tornado observed in Dimmitti, Texas, at 0103 UTC[6] on 3 June 1995 by the mobile Doppler radar with magnetron transmitter of the frequency of 9.37–9.38 GHz and the radar beam width of 0.93° (Wurman et al. 1996). The radar reflectivity field and the Doppler velocity field in radial direction are shown in panels (a) and (b), respectively. The center of the tornado was located about 3 km from the radar. The radar reflectivity factor of near 40 dBZ and the radial velocity of around 60 m s^{-1} were observed in the wall region around the center, whereas the center region was calm and the radar reflectivity in the center was about 30 dB lower than that of the maximum intensity region.

Figure 11.19 shows the vertical cross section, whose orientation is nearly perpendicular to the radar beam in order to capture maximum Doppler velocity, of the same tornado as that of Fig. 11.18 (a) The radar reflectivity field and (b) the Doppler velocity field (Wurman et al. 1996). The center of the tornado is clearly indicated in both pictures.

[6]Coordinated Universal Time, the time standard by which the world regulates clocks and time. It is almost the same as Greenwich Mean Time (GMT) based on astronomical observation, but the leap second to adjust the gap with Temps Atomique International (TAI) that counts the elapsed time in GMT from 0000:00, 1 January 1958.

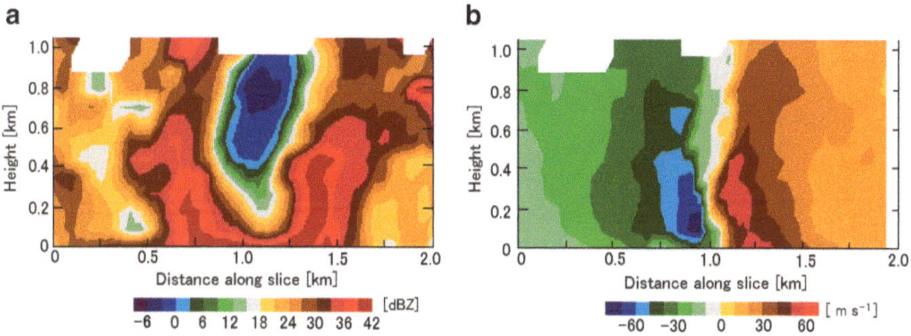

Fig. 11.19 Vertical cross section, whose orientation is nearly perpendicular to the radar beam, of the same tornado as Fig. 11.18, when it approached nearer the radar. (**a**) The radar reflectivity field and (**b**) the Doppler velocity field [from Wurman et al. 1996]

11.4.4 Downburst

Abrupt and furious downdrafts are often generated in the latter half of the growth of cumulonimbus. This is induced by the cooled heavy air mass which is caused by the falling of precipitation particles that draw in dry ambient air in which ice water and/or liquid water sublimate and evaporate, cooling the ambient air. The sizes of downbursts vary from less than 1 km to tens of kilometers. Fujita (1985, p. 8) subdivides them into macrobursts and microbursts according to the horizontal scale of damaging winds as follows. (As for "microburst", same as defined in Sect. 9.2.3. Two classes of microburst, "dry microburst" and "wet microburst" are also discussed in Sect. 9.2.3.)

- Macroburst: A large downburst with its outburst winds extending in excess of 4 km (2.5 miles) in horizontal dimension. An intense macroburst often causes widespread tornado-like damage. Damaging winds, lasting 5–30 min, could be as high as the wind speed of 60 m s^{-1} (134 mph).
- Microburst: A small downburst with its outburst, damaging winds extending only 4 km (2.5 miles) or less. In spite of its small horizontal scale, an intense microburst could induce damaging winds as high as 75 m s^{-1} (168 mph).

Microburst

In particular, microbursts wreak a lot of damage to the aircraft taking off and landing; thus, specialized radar such as the Terminal Doppler Weather Radar (TDWR) are developed to detect them as discussed in Sect. 9.2.3. A simplified model of a horizontal vortex in a microburst is shown in Fig. 11.20. A descending microburst is schematically shown in Fig. 11.21 in three stages according to the time series: midair microburst, touchdown, and surface microburst.

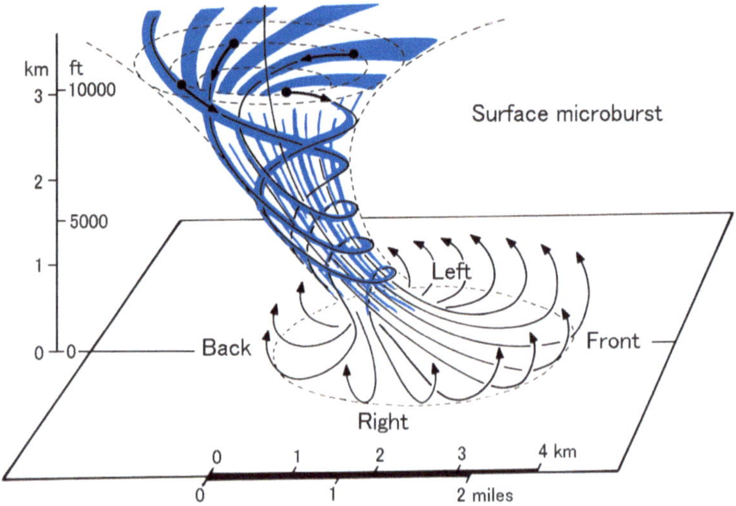

Fig. 11.20 Schematic view of a three-dimensional microburst. Convergence aloft, rotating downdraft, and surface divergence are included [from Fujita 1985, p. 75]

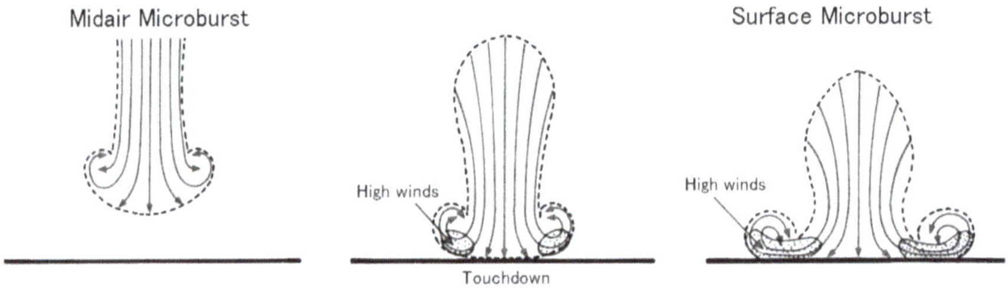

Fig. 11.21 Schematics of descending microburst at three stages according to the time series: midair microburst, touchdown, and surface microburst [Fujita 1985, p. 73]

Figure 11.22 shows the vertical cross sections (RHI pictures) of a microburst observed by the CP-3 radar, a 5.6-GHz Doppler radar of National Center for Atmospheric Research (NCAR), Denver, Colorado. In panels (a) and (b), radar reflectivity factor and Doppler velocity are indicated, respectively (Fujita 1985).

In panel (a), it appears clearly that the downdraft with precipitation descends and spread in the vicinity of surface of the earth. In the Doppler velocity field in panel (b), the winds at higher altitude converge on the center of the downdraft, i.e., Doppler velocity (green patch) on the right side at the top of the downburst shows an approaching radial wind whereas on the other side (yellow patch) shows receding radial velocities. However, the direction of the winds near the earth's surface diverges from the center of the downdraft.

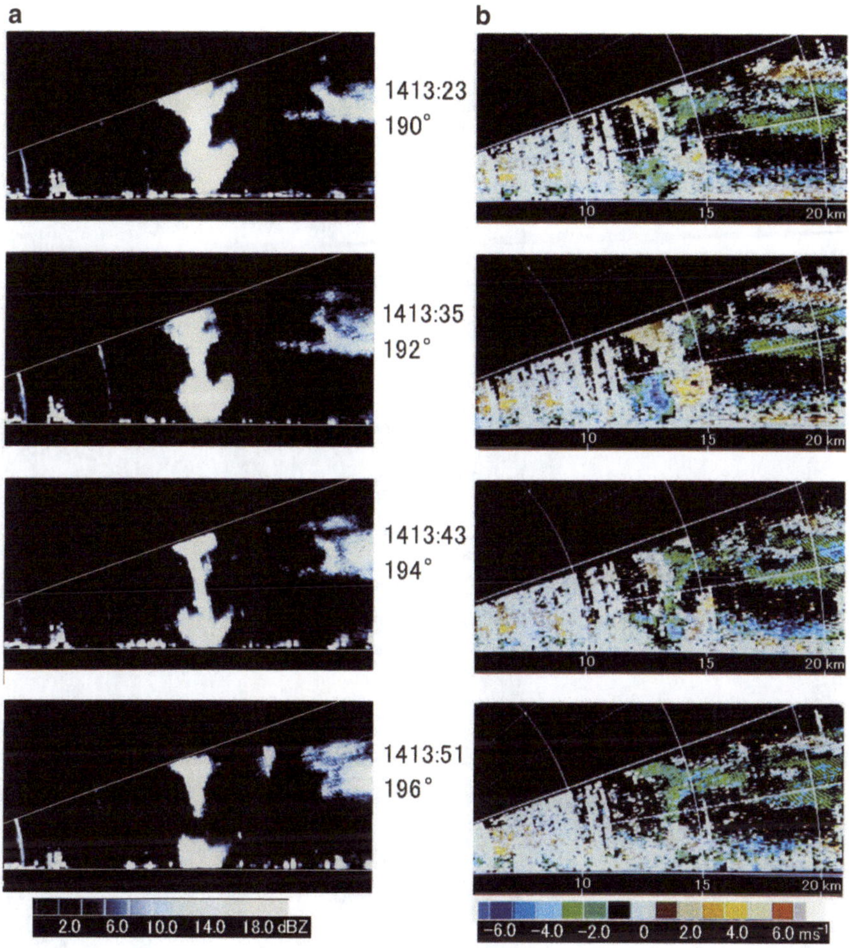

Fig. 11.22 Vertical cross sections of a microburst observed by the CP-3 radar. (**a**) Radar reflectivity factor and (**b**) Doppler velocity [from Fujita 1985, p. 17]

Figure 11.23 shows (a) horizontal distributions and (b) vertical distributions of two individual microbursts (Fujita and McCarthy 1990). In panel (a), the Hickory Ridge microburst was observed by the FL-2 radar, a 2.8-GHz radar developed and operated by Lincoln Laboratory for the Federal Aviation Administration (FAA) on 26 June 1985 near Memphis, Tennessee, where (top) spectrum width, (middle) radar reflectivity factor, and (bottom) Doppler velocity, respectively. In panel (b), the Monrovia microburst was observed by the CP-4 radar, a 5.5-GHz Doppler radar of the NCAR, on 20 July 1988 near Huntsville, Alabama, where (top) the cloud photo, (middle) the radar reflectivity factor, and (bottom) the Doppler velocity. The top panel of (b) gives an impression that lower half of the cloud began sinking fast is due to downdraft. In the bottom of (b), updraft is seen in the vicinity of the cloud top

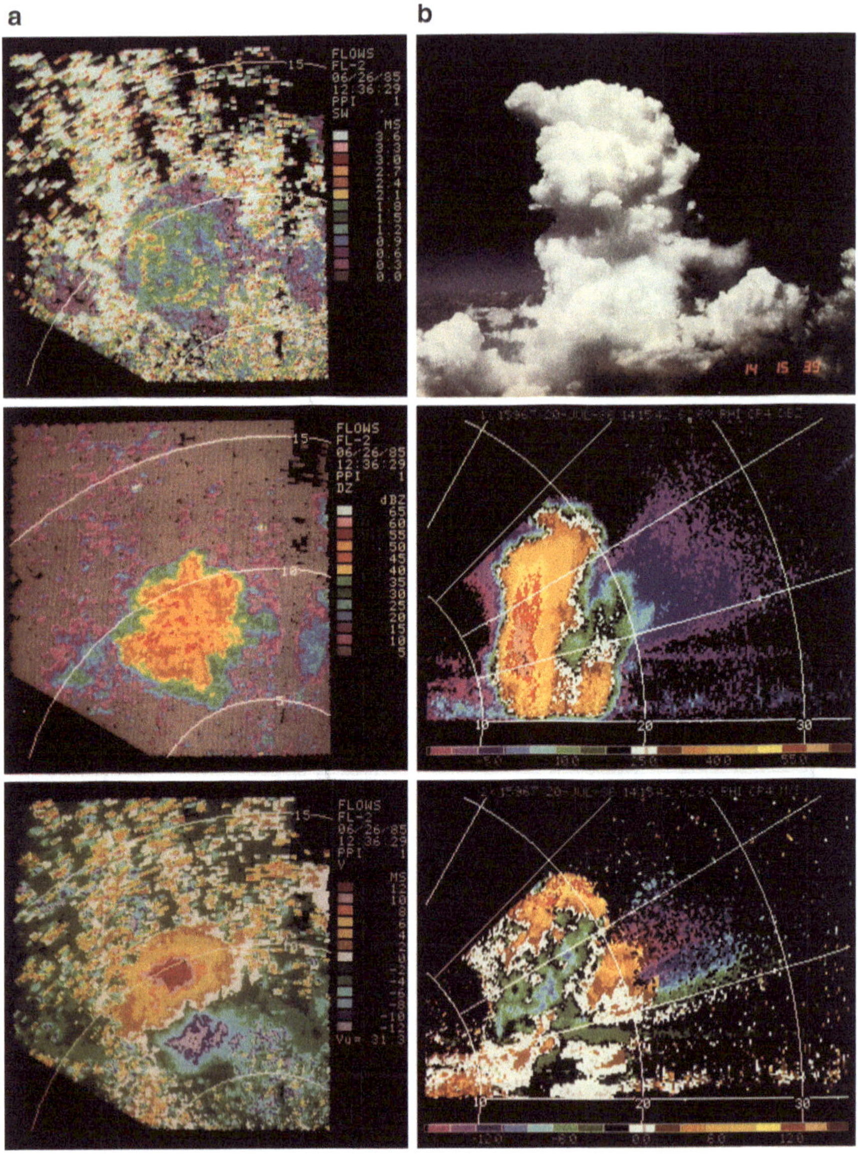

Fig. 11.23 (**a**) Horizontal distribution of the Hickory Ridge microburst observed by the FL-2 radar near Memphis, Tennessee on 26 June 1985, showing (*top*) the spectrum width, (*middle*) the radar reflectivity factor, and (*bottom*) the Doppler velocity. (**b**) Vertical distribution of the Monrovia microburst observed by the CP-4 radar near Huntsville, Alabama on 20 July 1988, showing (*top*) the cloud photo, (*middle*) the radar reflectivity factor, and (*bottom*) the Doppler velocity [from Fujita and McCarthy 1990]

and the upper right area, whereas the central area and the lower area are occupied with downdraft, and it reaches the vicinity of the earth's surface. Moreover, a part of the downdraft which is rolled up after horizontal extension is seen in the left area of the cloud in the vicinity of ground.

11.5 Polarimetric Radar Observations

11.5.1 Polarimetric Parameters

Ice Water and Melting Zone

As discussed in Sect. 6.4, various polarimetric parameters can be observed by polarimetric radar. The vertical cross sections of ice cloud are shown in Fig. 11.24a and b which are obtained by the 2.8-GHz band Chilbolton radar in U.K. on 6 July 1983 at different azimuth angles and different time. In each panel, radar reflectivity factor Z (top) and differential reflectivity Z_{DR} (bottom) are shown. These quantities are utilized for judging if precipitating particles are frozen or melted. In panel (a), a formation of melting zone is at a height of about 2.5 km. The Z_{DR} value is about 1 dB or less in regions above the melting layer, whereas it is around 2–4 dB in the region below this height. It is understood that ice particles change to water droplets, i.e., a melting zone is formed at the height of around 2.5 km. In panel (b), an intrusion of substantial Z_{DR} (liquid or very wet icy hydrometeors) above the melting level in a convective storm is shown. A part of the melting zone reaches up to around 5 km due to a strong updraft inside the cloud (Jameson and Johnson 1990).

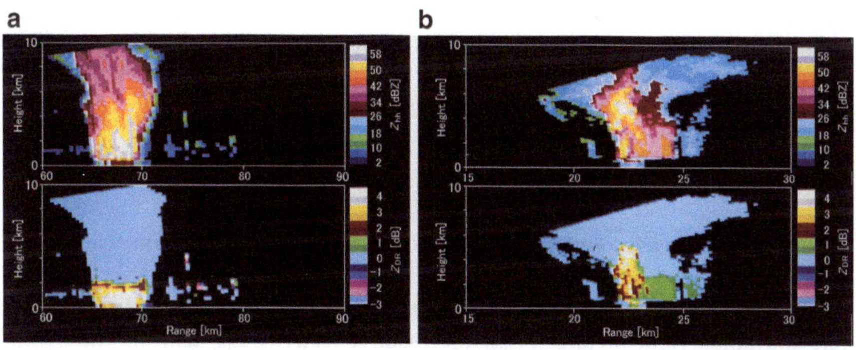

Fig. 11.24 Z (*top*) and Z_{DR} (*bottom*) obtained by the 2.8-GHz band Chilbolton radar in U.K. on 6 July 1983 at different azimuth angles and different time. (**a**) Ice water discrimination, i.e., formation of melting zone and (**b**) an intrusion of substantial Z_{DR} (liquid or very wet icy hydrometeors) above the melting level in a convective storm are shown [from Jameson and Johnson 1990]

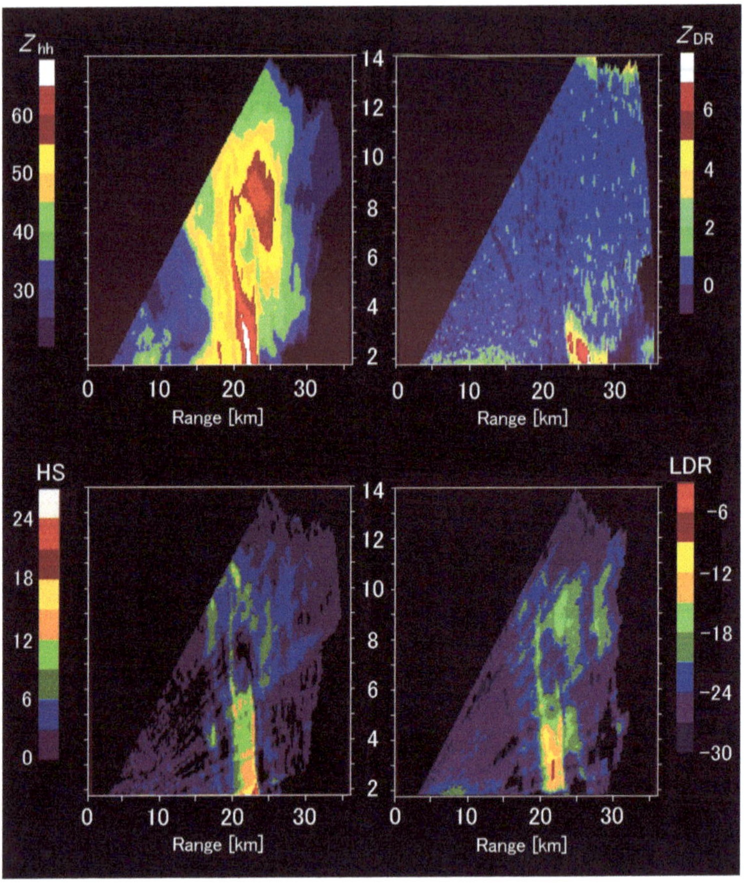

Fig. 11.25 Z_{hh} and Z_{DR} observed by the CP-2 10-cm wavelength radar (*upper pictures*). The hail signal (HS) derived from the two equivalent radar reflectivity factors of the CP-2 3-cm and 10-cm wavelength radars, and the LDR observed at 3-cm wavelength (*lower pictures*) [from Jameson and Johnson 1990]

Hail

Figure 11.25 shows the vertical cross section of a significant hailstorm near Denver, Colorado observed by the CP-2 radar, the formerly mentioned dual wavelength (of 10-cm and 3-cm wavelengths) dual polarization radar of the NCAR. The upper left and upper right pictures show the radar reflectivity Z_{hh} and the differential reflectivity Z_{DR}, respectively, both obtained by the 10-cm wavelength radar. If the target particles are hail of around 10 mm diameter, $\pi D/\lambda \simeq 0.3$ at 10-cm wavelength, where D is the diameter of the particle, and λ the radar wavelength; thus the backscattering for this particle is in the Rayleigh region as discussed in Sect. 3.2.2, whereas $\pi D/\lambda \simeq 1.0$ at 3-cm wavelength and is in the Mie region. As a result, two equivalent radar reflectivities Z_{e3} and Z_{e10} differ significantly. The lower left

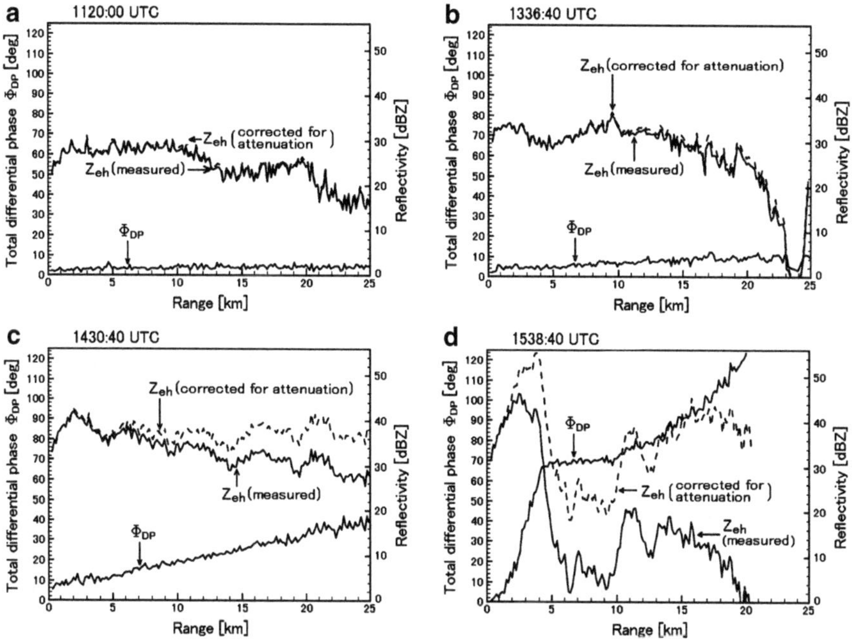

Fig. 11.26 Radar reflectivity factor Z and differential propagation phase Φ_{DP} in rain of various intensities, (**a**) very light, (**b**) light, (**c**) moderate, and (**d**) heavy rains, observed by the NOAA ETL 3-cm wavelength radar deployed at the NASA Wallops Island facility in one long event when the rain increased from very light to very heavy at the direction of (azimuth, elevation) = (134 °, 1.8°) on 21 March 2001. *Solid* and *broken Z lines* indicate measured and corrected values, respectively, using K_{DP} [from Matrosov et al. 2002]

picture shows the hail signal (HS), where the HS is defined as Z_{e3}/Z_{e10}. The lower right picture shows the linear depolarization ratio (LDR) observed at 3-cm wavelength. The LDR value is adjusted to remove the effects of attenuation at the 3-cm wavelength. The region of HS ≥ 5 dB is estimated to show the existence of hailstones. As is evident from the upper right and lower right pictures, while Z_{DR} is very uniform in locations of icy hydrometeors, LDR exhibits extensive structure that is quite similar but not identical to that for the HS in the lower left picture. The substantial Z_{DR} at low level is likely an artifact caused by different beam patterns for horizontal and vertical polarizations [from Jameson and Johnson 1990].

11.5.2 Attenuation Correction

As discussed in Sect. 6.4.4, the attenuations of Z and Z_{DR} due to precipitation along the propagation path can be corrected using the specific differential phase K_{DP}. Figure 11.26 shows radar polarimetric measurements, Z and the differential phase

Φ_{DP}, in rain of various intensities, (a) very light, (b) light, (c) moderate, and (d) heavy rains, observed by the National Oceanic and Atmospheric Administration (NOAA) ETL 3-cm wavelength radar deployed at the NASA Wallops Island facility at the direction of (azimuth, elevation) = (134°, 1.8°) on 21 March 2001 (Matrosov et al. 2002). All these diagrams were obtained in one long lasting event when the rain increased from very light to very heavy. In (a), according to the ground based raingauge measurement at 3.3 km range, a rainfall rate near the time of the radar measurement was about 2.4 mm h^{-1}. In this case, the Φ_{DP} value is very small and the mean K_{DP} values is about 0.1° km^{-1}. As a result, the attenuation correction for the 20 km range is only about 1 dB. In (b), the ground based raingauge measurement indicated about 4 mm h^{-1}. The mean K_{DP} values is about 0.25° km^{-1}, and the maximum value of Z correction is about 2 dB. As a result, the attenuation correction for 20 km range is only about 2 dB. In (c), the ground based raingauge measurement indicated about 10 mm h^{-1}. The K_{DP} value is about 0.8° km^{-1}, and consequently, the attenuation correction for Z at the longer ranges reaches about 10 dB. In (d), a very heavy rain was observed near the radar up to about 4 km range, and the measurement of the high-resolution raingauge at the radar measurement time corresponded to about 110–130 mm h^{-1}. The Φ_{DP} value increased significantly to about 4.5 km range and K_{DP} reached up to 12–14° km^{-1}. As a result, Z correction becomes around 35–40 dB at the range of about 20 km where measured Z ranges about 0 dBZ.

Figure 11.27 shows PPI images of polarimetric variables observed by the 3-cm wavelength MP-X radar of the National Research Institute of Earth Science and Disaster Prevision (NIED), Japan when a significant attenuation occurred (Park et al. 2005b). In (a), a reflectivity factor over 50 dBZ was observed in region "A" adjacent to the radar site. Because attenuation increases when electromagnetic waves propagate through intense precipitation, the measured radar reflectivity factor decreases as a function of distance from the radar. In (b), Z_{DR} values also decrease behind the high reflectivity region in the same direction as (a), the values become negative, whereas these values are positive in the direction of weak echo such as seen to the north. In (c) and (d), K_{DP} and ρ_{hv} are shown, respectively. In (c), high K_{DP} values above 7° km^{-1} are observed in the intense reflectivity region at "A". In region "B" about 20 km from the radar, high K_{DP} values are about the same as in region "A", on the other hand, Z values in region "B" are not so strong (i.e., around 40 dBZ in maximum) as shown in (a), and Z_{DR} values in the same region are also small, mostly negative, as shown in (b). In a similar way, high K_{DP} values are observed in the direction around AZ 294°; however, Z and Z_{DR} values on the same direction are rather small because of attenuation. In (e) and (f), corrected Z and Z_{DR} images are shown, respectively, where the correction is performed using K_{DP} as mentioned in Sect. 6.4.4. As a result, Z (e) and Z_{DR} (f) after correction are consistent with K_{DP} which is not affected by attenuation (Park et al. 2005b).

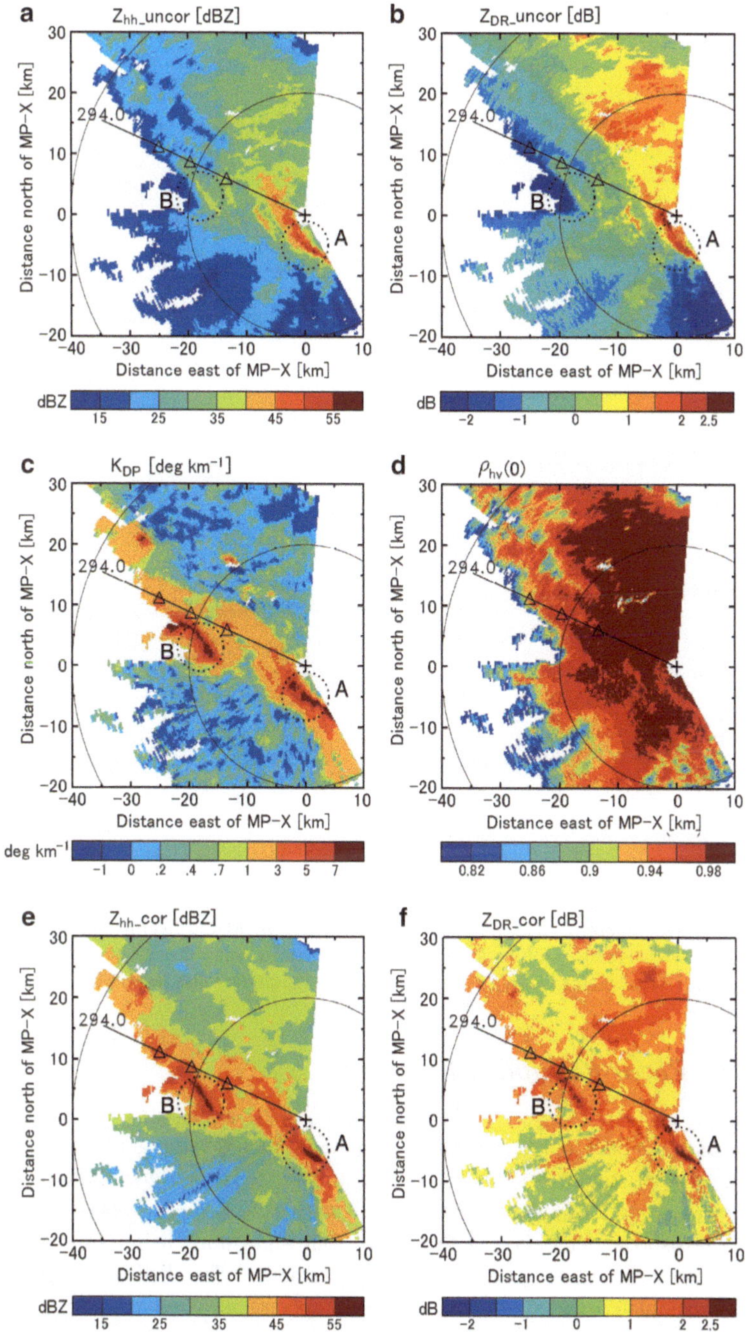

Fig. 11.27 PPI images observed by the 3-cm wavelength MP-X radar of NIED, Japan. Measured values (**a**) Z, (**b**) Z_{DR}, (**c**) K_{DP}, (**d**) $\rho_{hv}(0)$, (**e**) Z, and (**f**) Z_{DR}. The values of (**e**) and (**f**) are corrected by using K_{DP}. The azimuthal angle is given by rotating clockwise from the north [from Park et al. 2005b]

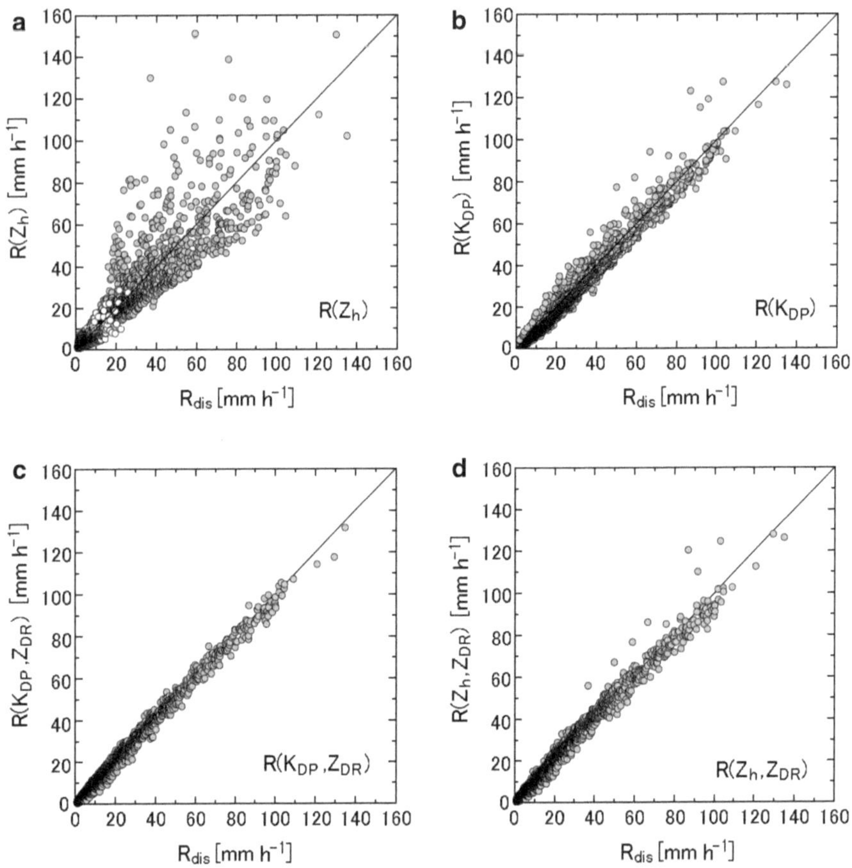

Fig. 11.28 Comparisons of disdrometer rain rates R_{dis} calculated based on the measured drop size distribution and simulated radar measurements (**a**) $R(Z)$, (**b**) $R(K_{\mathrm{DP}})$, (**c**) $R(K_{\mathrm{DP}}, Z_{\mathrm{DR}})$, and (**d**) $R(Z_{\mathrm{h}}, Z_{\mathrm{DR}})$, where polarization parameters are calculated assuming that the precipitations of the measured DSD are obtained with a 3-cm wavelength polarimetric radar [from Maki et al. 2005]

11.5.3 Radar Rainfall Estimation

As discussed in Sect. 6.4.6, rainfall rate can be estimated using algorithms based on polarization parameters besides $R(Z)$ based only on Z. These algorithms are $R(K_{\mathrm{DP}})$, $R(K_{\mathrm{DP}}, Z_{\mathrm{DR}})$, and $R(Z_{\mathrm{h}}, Z_{\mathrm{DR}})$. Figure 11.28 shows the comparisons of rainfall rate based on disdrometer data and the above four algorithms using simulated radar measurements based on disdrometer measurements. The disdrometer rain rates R_{dis} are calculated based on the drop size spectra of 7664 raindrops obtained from 1 min of disdrometer data, whereas the polarization parameters are calculated assuming that the calculated drop size distribution (DSD) are obtained with a 3-cm wavelength polarimetric radar (Maki et al. 2005). The scatter plots of $R(Z)$, $R(K_{\mathrm{DP}})$,

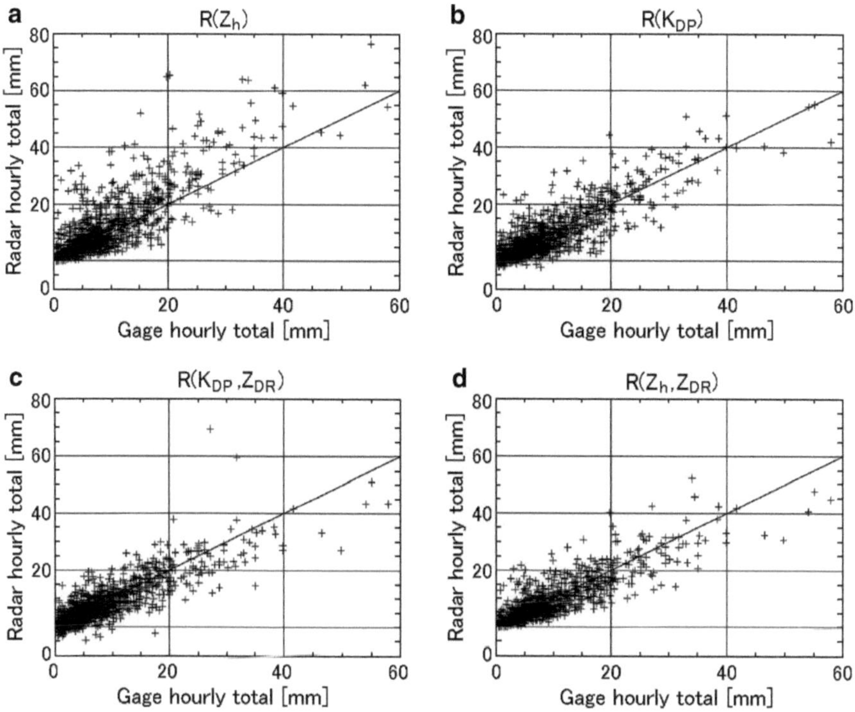

Fig. 11.29 Comparisons between 1-h gauge rain accumulation and the actual observational data of the KOUN radar, the WSR-88D of NSSL, Oklahoma. (**a**) $R(Z)$, (**b**) $R(K_{\mathrm{DP}})$, (**c**) $R(K_{\mathrm{DP}}, Z_{\mathrm{DR}})$, and (**d**) $R(Z_{\mathrm{h}}, Z_{\mathrm{DR}})$ [from Ryzhkov et al. 2005]

$R(K_{\mathrm{DP}}, Z_{\mathrm{DR}})$, and $R(Z_{\mathrm{h}}, Z_{\mathrm{DR}})$ compared with R_{dis} are shown in Panel (a), (b), (c), and (d), respectively. The plots suggest that $R(K_{\mathrm{DP}}, Z_{\mathrm{DR}})$ is the most accurate of the four estimator from the view point that does not have the sensitivity to the variations in the DSD.

Maki et al. (2005) quantitatively examined the uncertainty of the four estimators due to the variations in the DSD of above samples and calculated that the normalized errors of $R(Z)$, $R(K_{\mathrm{DP}})$, $R(K_{\mathrm{DP}}, Z_{\mathrm{DR}})$, and $R(Z_{\mathrm{h}}, Z_{\mathrm{DR}})$ for R_{dis} of all DSD samples are 25%, 14%, 9%, and 10%, respectively. For rain rates larger than $10\,\mathrm{mm\,h^{-1}}$, the average normalized errors of $R(Z)$, $R(K_{\mathrm{DP}})$, $R(K_{\mathrm{DP}}, Z_{\mathrm{DR}})$, and $R(Z_{\mathrm{h}}, Z_{\mathrm{DR}})$ are 25%, 9%, 5%, and 7%, respectively, and the error of $R(K_{\mathrm{DP}}, Z_{\mathrm{DR}})$ becomes minimum.

Figure 11.29 shows the comparison of the four algorithms, but applied to actual radar measurements made with the KOUN radar, the WSR-88D, of the National Severe Storms Laboratory (NSSL), Oklahoma and 1-h gauge rain accumulation. The polarimetric rain rate algorithms (a) $R(Z)$, (b) $R(K_{\mathrm{DP}})$, (c) $R(K_{\mathrm{DP}}, Z_{\mathrm{DR}})$, and (d) $R(Z_{\mathrm{h}}, Z_{\mathrm{DR}})$, are plotted, respectively (Ryzhkov et al. 2005). These results are obtained from actual observations with the 2.7-GHz radar, and the tendency of the uncertainty for the estimators looks similar to that of the simulation for 3-cm

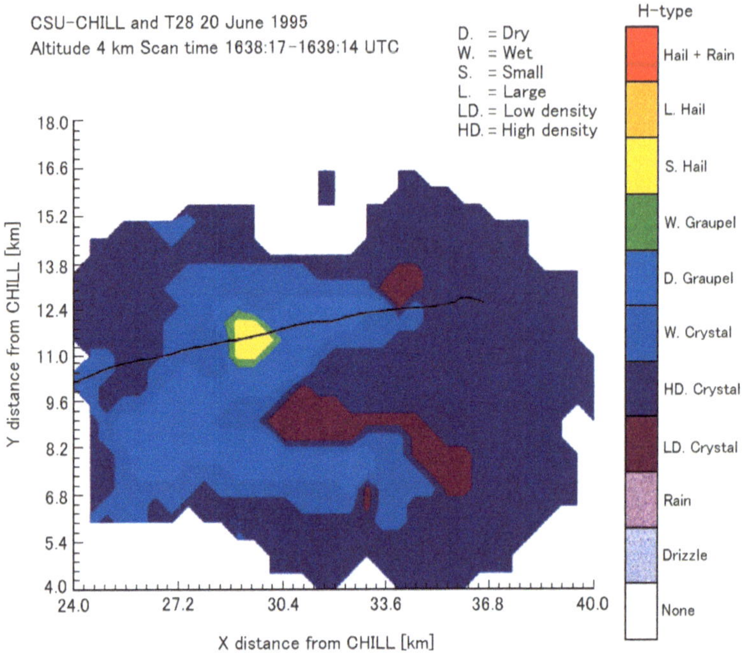

Fig. 11.30 CAPPI image of the classification of a storm cell at the altitude of 4 km collected by the CSU-CHILL radar. The fuzzy logic with membership functions are applied using five polarimetric parameters, Z_{h}, Z_{DR}, K_{DP}, LDR, and $\rho_{\mathrm{hv}}(0)$. The *solid line* is the T-28 aircraft track, along which data from the High-Volume Particle Spectrometer (HVPS) probe was collected for in situ verification of hydrometeor classification [from Liu and Chandrasekar 2000]

wavelength radar in Fig. 11.28. Ryzhkov et al. (2005) indicated that the synthetic rainfall algorithm $R(Z, K_{\mathrm{DP}}, Z_{\mathrm{DR}})$, which capitalizes on the relative merits of the $R(Z_{\mathrm{h}})$, $R(K_{\mathrm{DP}}, Z_{\mathrm{DR}})$, and $R(K_{\mathrm{DP}})$ relations for different categories of rain rate, is most robust with respect to radar calibration errors, DSD variations, uncertainty of raindrop shapes and possible presence of hail, and shows the best performance.

11.5.4 Hydrometeor Classification

Various hydrometeor classification of applying polarimetric parameters has been tried. Figure 11.30 shows an example of CAPPI image of the classification of a storm cell at an altitude of 4 km. The radar data were collected between 1638:17 and 1639:14 UTC on 20 June 1995 by the Colorado State University (CSU)-CHILL radar, an S-band transportable polarimetric radar in Greeley, Colorado, The fuzzy logic with membership functions as discussed in Sect. 6.4.8 are applied using five polarimetric parameters, Z_{h}, Z_{DR}, K_{DP}, LDR, and $\rho_{\mathrm{hv}}(0)$. The solid line in the figure is the T-28 aircraft track, along which data from the High-Volume Particle

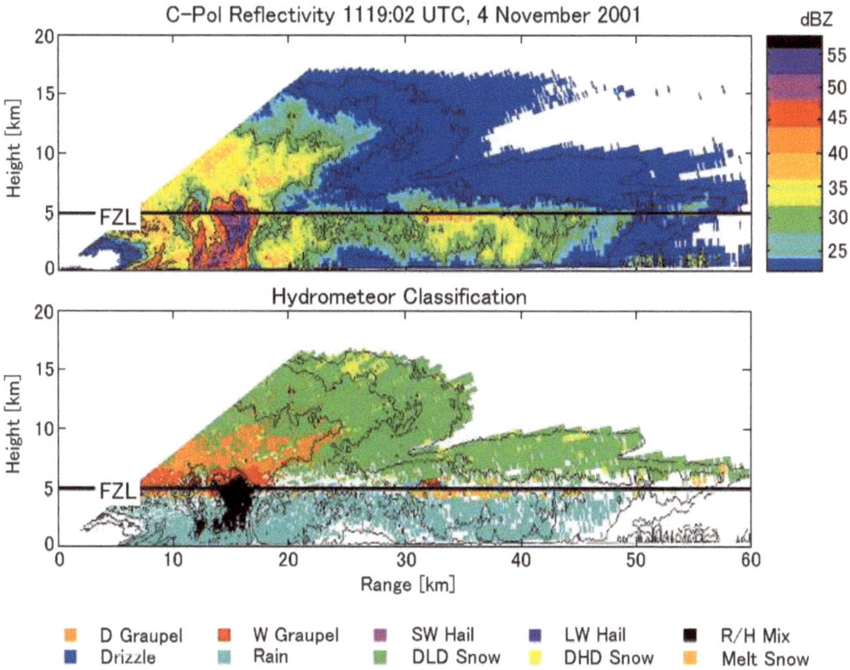

Fig. 11.31 Hydrometeor classification made by applying the fuzzy logic method. The data was obtained by the C-Pol radar in Darwin, Australia. The radar reflectivity factor (*top panel*) and hydrometeors classified (*bottom panel*), where contours of the radar reflectivity factor are overlaid in each panel. The *horizontal line* labeled FZL is the freezing level [from May and Keenan 2005]

Spectrometer (HVPS)[7] probe was collected for in situ verification of hydrometeor classification. In the picture, the aircraft tracks are in predominantly graupel region and a region of small hail, and this inference agrees very well with the T-28 in situ aircraft observations (Liu and Chandrasekar 2000).

Figure 11.31 shows an RHI image of the classification which is also applied the fuzzy logic as Fig. 11.30. The data was obtained by the C-band polarimetric (C-Pol) radar in Darwin, Australia, with the RHI scan mode at 1119 UTC 4 November 2001. The radar reflectivity factor and the result of classification are shown in the top and bottom panels, respectively. These panels show active convective elements with a rain-hail mixture and a large area of wet graupel around the cell core. In addition, there is a narrow melting zone of snow, i.e., the bright band, that is resolved in the stratiform region farther from the radar. Contours of radar reflectivity factor are overlaid on the two panels. As for the validation of the classifications, a combination of 50- and 920-MHz wind profiler estimates of rain and hail over the profilers are adopted (May and Keenan 2005).

[7]The HVPS can measure particles size up to 4.5 cm, by taking two-dimensional picture of hydrometeors that pass through a plane normal to the direction of aircraft flight.

11.6 Clear Air Observations

11.6.1 High Power Large Radar Observation

There are two types of turbulence that are detected in the atmosphere; one is clear air turbulence (CAT) that is not accompanied with cloud, and the other is generated near convective cloud such as thunderclouds. Both types are always accompanied by significant change of wind velocity, i.e., wind shear. As CAT has no water particles, in general, it is difficult to observe CAT with meteorological radar at centimeter or millimeter wavelengths, whereas it is possible to estimate the intensity and motion of CAT by using atmospheric radar at the very high frequency (VHF) band. In the case of turbulence in a convective cloud, it is accompanied with precipitation, and thus can be observed with not only atmospheric radar but also with meteorological radar.

The main objectives of meteorological radar are various precipitation associated phenomena, however, radars with large antenna and high power operating at 2.8-GHz band sometimes receive echoes from CAT. In this frequency band, the radar wavelength is around 10 cm, and thus the half-wavelength scale belongs to the inertial subrange in the troposphere as shown in Fig. 7.10. Therefore, eddies with the half-wavelength scale in the troposphere generate perturbation of the refractive index with the same scale, which are observed by atmospheric radar.

Figure 11.32a shows a horizontal cross section of a CAT in the convective boundary layer observed with the 2.8-GHz band radar of the Joint Air Force and NASA (JAFNA), with the antenna diameter of 18.4 m, and the transmitted power of 3 MW, at the 10.7-cm wavelength, and with transmitted pulse width of 2 μs, in Wallops Island, Virginia (Hardy and Ottersten 1969). It was observed with the SPPI scan mode as previously shown in Fig. 8.3 at 3° elevation angle at 1052 Eastern Standard Time (EST) on 15 August 1967. Horizontal cross sections of convection cells are seen like doughnuts because echo strength in the central area of the cells is comparatively weak. Panel (b) shows the conceptual diagram to explain the mechanism that doughnuts-like cells are observed with radar. As air near the ground is warmed by the heated surface and moistened by evaporation, it becomes buoyant; thus it rises and cools adiabatically. But the interior of the warmed cell has relatively uniform properties (i.e., very little turbulence to mix different air masses, a process that produces large fluctuations in refractive index). Thus the reflectivity in the interior is relatively low. The warm air continues its upward acceleration until it passes the level of zero buoyancy (the altitude where the temperature of the cell air equals the ambient temperature). Air parcels carried above the level of zero buoyancy will be cool and moist relative to the air surrounding the updraft. Because turbulence is mixing these two air masses of very different refractive index, strong perturbations of refractive index with exist along the boundaries of the updraft (Doviak and Zrnić 2006, p. 481). Thus convective cells will produce doughnut-like reflectivity structure seen in the SPPI displays. The size of each small cell is approximately 200 m in diameter at the early stage, and it gradually changes with time and reach approximately several thousand meters (Konrad 1970).

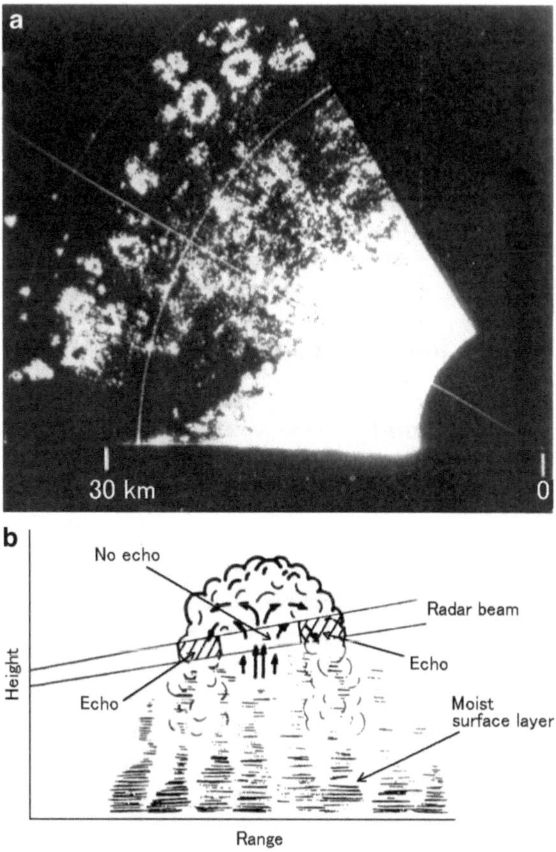

Fig. 11.32 A CAT in the convective boundary layer obtained by the 2.8-GHz band radar of the Joint Air Force and NASA (JAFNA) in Wallops Island. (**a**) *Horizontal cross section* observed with the SPPI scan at 3° elevation angle at 1052 EST on 15 August 1967, and (**b**) conceptual diagram to explain the mechanism that doughnuts-like cells are observed with radar [from Hardy and Ottersten 1969]

11.6.2 FMCW Radar Observation

As discussed in Sect. 8.1.1, the FMCW radar continuously transmits waves which are frequency modulated; thus two antennas are required, one that continuously transmits the continuous wave, and a second one that continuously receives echoes. Because there is no transmit/receive switch used by pulsed radar with one antenna for both transmission and reception, the FMCW radar can observe even within the range of several tens meters from the radar antenna which is blinded by transmitted pulse for usual pulse radar. Because observations can be made at close ranges, minute structures of the lower atmosphere can be observed. Figure 11.33 shows an example of Kelvin-Helmholtz (KH) instability caused by atmospheric gravity wave that will be discussed in Sect. 12.3.1 in detail (Gossard et al. 1970; Gossard and Strauch 1983, Chap. 1). It is a time-height cross section of echo intensity where the radar beam is fixed to the zenith direction, and the height resolution is about 1 m. It is clearly shown in the picture that turbulence due to breaking wave is formed in thin atmospheric layer of the thickness of about 100 m.

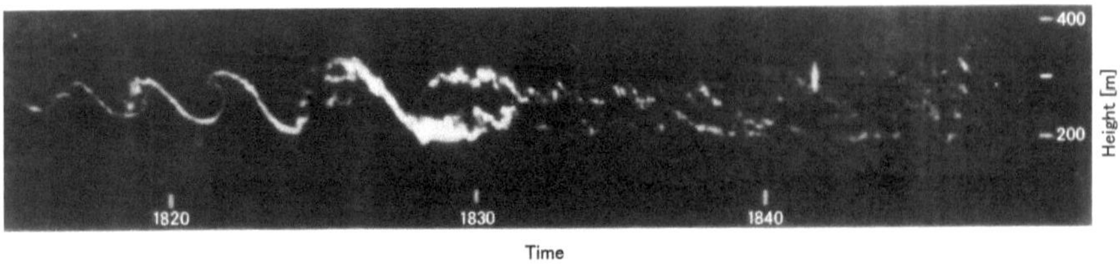

Fig. 11.33 Kelvin-Helmholtz (KH) instability observed with the FMCW radar [from Gossard and Strauch 1983, p. 49]

11.7 Cloud and Fog Observations

11.7.1 Cloud

Ice Cloud

Tiny water drops in cloud and fog have sizes of approximately several tens of μm, i.e., approximately two orders of magnitude smaller than that of raindrops. Millimeter wavelength radars operating in the 35- and 95-GHz bands are optimal for the observations of these non-precipitating particles due to the characteristics of scatters as discussed in Sect. 3.2.2. Figure 11.34 shows a dual layer ice cloud observed with the 35-GHz radar, mentioned in Sect. 9.3.1, when it was located at the MU Observatory, Shigaraki, Japan, on 17 July 1997. The radar reflectivity field (a) and the radial Doppler velocity field (b) are shown. The upper layer extends at the height of around 8 km with about 1,500 m thickness, and the lower layer at the height of around 6 km with about 500 m thickness. The radial Doppler velocity right above the radar was almost 0 m s^{-1} thus it was estimated that the vertical wind was small, whereas the horizontal wind from the southwest of approximately 25 m s^{-1} was observed (Hamazu 2002, p. 168).

Snow Band

Figure 11.35 shows the snow band observed with two Doppler radars on the shores of Ishikari Bay, Hokkaido, Japan, on 19 January 1992. It is a typical example of meso-β and meso-γ scale eddy echoes which are often observed in the Sea of Japan coast in winter. Panel (a) shows the CAPPI picture at 300 m height which is overlaid with the distribution of horizontal winds obtained by two Doppler radars, and Panel (b) shows the three dimensional image of the radar echo of 18 dBZ isosurface in the square frame of Panel (a). This eddy was generated in a large horizontal shear which was formed between the two wind systems in the opposite direction each other, the southwester in the Ishikari Bay and the northeaster that appeared offshore of the bay, and the developing process was observed for many hours. It is understood from

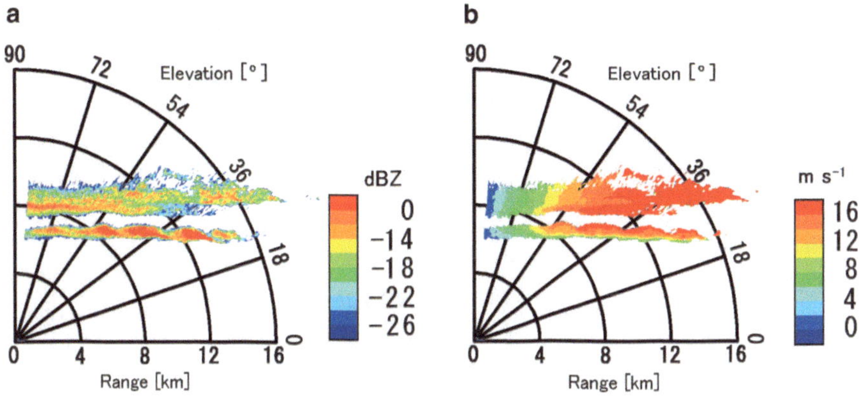

Fig. 11.34 RHI picture of a dual-layer ice cloud observed with the 35-GHz radar mentioned in Sect. 9.3.1 at the MU Observatory, Shigaraki, Japan, on 17 July 1997. (**a**) Radar reflectivity field and (**b**) radial Doppler velocity field [from Hamazu 2002, p. 168]

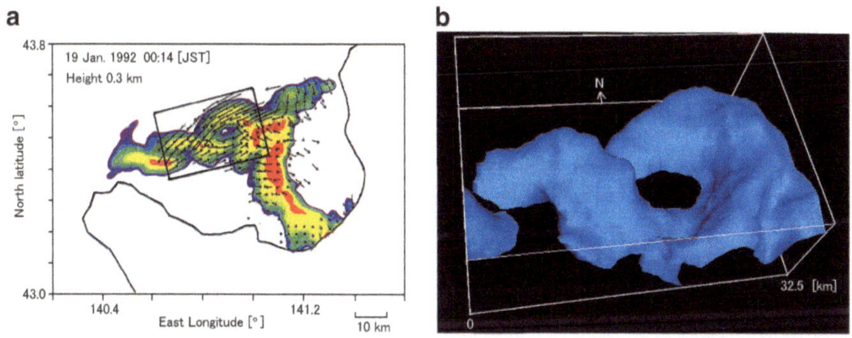

Fig. 11.35 Snow band observed with two Doppler radars in Ishikari Bay, Hokkaido, Japan, in winter. (**a**) CAPPI picture of 300 m height overlaid with the distribution of horizontal winds obtained by two Doppler radars, and (**b**) three dimensional image of the radar echo of 18 dBZ isosurface in the square frame of (**a**) [from Fujiyoshi 2001]

the three dimensional analysis of wind velocity that the disappearance of the radar echo in the central area is due to downward flow. The upward flow existed along the eddy echo region, and became weak while going from the south (this side of the paper) to the north (back of the paper) (Fujiyoshi et al. 1998; Fujiyoshi 2001).

11.7.2 Fog

Fog is, in simple terms, a cloud near the earth's surface. Annually, especially in summer, there are many foggy days along the eastern Pacific coast of Hokkaido

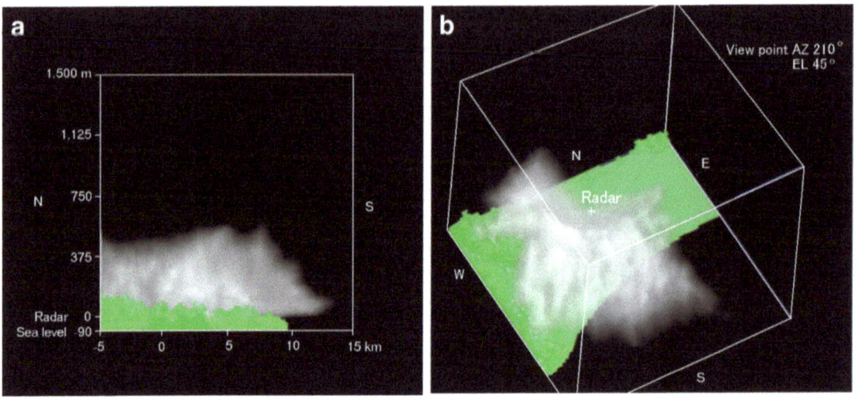

Fig. 11.36 Sea fog observed in Kusiro, the east coast of Hokkaido, by the 35-GHz band radar. Radar reflectivity factor of more than −40 dBZ are shown in (**a**) side view and (**b**) view from an oblique direction. The lack of data in the east of the radar is due to blockage of radar beam by the radar vehicle and trees [from Hamazu et al. 2003]

in Japan. Sometimes, certain areas are covered with fog all day long. Figure 11.36 shows the three dimensional appearance of sea fog observed in Kusiro, on the east coast of Hokkaido, by the 35-GHz band radar of Fig. 9.11b. The echo volumes of radar reflectivity factor of more than −40 dBZ are shown in the display volume of 20 km (N–S), 20 km (E–W), and 1.5 km (height). Two different views from the radar are shown, (a) side view and (b) view from an oblique direction. The brighter the echo, the larger the reflectivity. The lack of data to the east of the radar is due to blockage of radar beam by the radar vehicle and trees (Hamazu et al. 2003). Continuous observations show clearly the feature of the advection fog that moves from the coast to the inland.

Figure 11.37 shows the north–south cross section of sea fog echo observed at the same location as Fig. 11.36. A periodic structure of about 1 km in the horizontal wave length was observed inside of the fog as it was advecting inland. Although the periodic structure became more obvious inland at the distance of several km from the radar, it was understood that the initial stages of this developing structure formed over the ocean.

11.8 Retrieval of Heating Distribution in a Cloud

The wind velocity field observed with Doppler radar relates to the atmospheric pressure field and the buoyancy field, and decides the distribution of precipitation particles. Meanwhile, the distribution of the observed precipitation particles is related to the atmospheric pressure field and wind velocity field through the buoyancy field. The thermodynamic quantity and the quantities of cloud physics

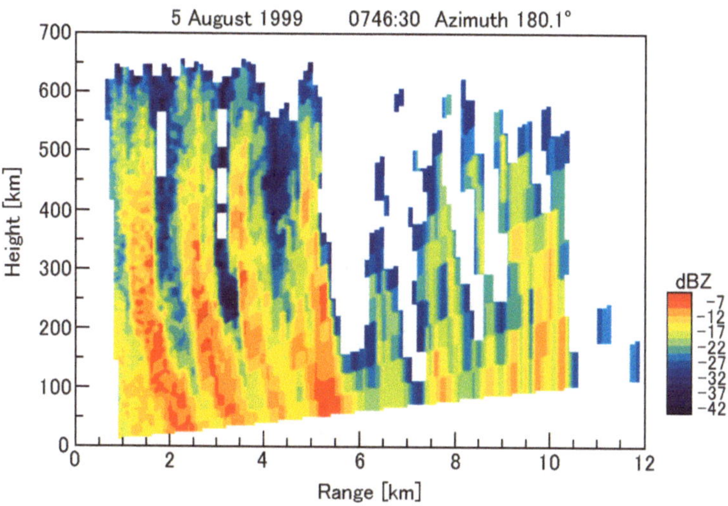

Fig. 11.37 North–south cross section of sea fog echo observed at the same location as Fig. 11.36

can be evaluated diagnostically by analyzing the relation between the above fields using the equations of motion, thermodynamics, and continuity for the atmosphere and water substances. This analysis technique is called the retrieval method, which is divided roughly into two methods; thermodynamic retrieval method and microphysical retrieval method. The former evaluates the temperature and the atmospheric pressure, and the later evaluates cloud water contents and the atmospheric pressure. Recently, both methods are combined (e.g., Hauser et al. 1988; Geerts and Hobbs 1991) to estimate both the thermodynamic quantity and the quantities of the cloud physics at the same time. Various types of thermodynamic and cloud physics structure of precipitation system have been clarified from the mesoscale wind velocity fluctuation observed with Doppler radar. For example, the effectiveness of retrieval of thermodynamic quantity and quantities of cloud physics of precipitation system that changes in time is examined using the results of numerical experiment of multicell storm (e.g., Yoshizaki and Seko 1994), and the retrievals concerning the thermochemical structure of rainband according to the seasonal rain front observed with Doppler radar are performed (e.g., Kawashima et al. 1995).

The latent heat release according to the dynamic convective activity in the tropical West Pacific Ocean has played an important role for the global atmospheric general circulation. The influence of the convection system on a large scale field can be estimated by clarifying the internal structure of the precipitation system and the distribution of latent heat release in this region. For instance, Ushiyama et al. (2003) analyzed the data, observed with two Doppler radars between November 1992 and February 1993 in Manus Island (2°S, 147°E) in Papua New Guinea as a part of the

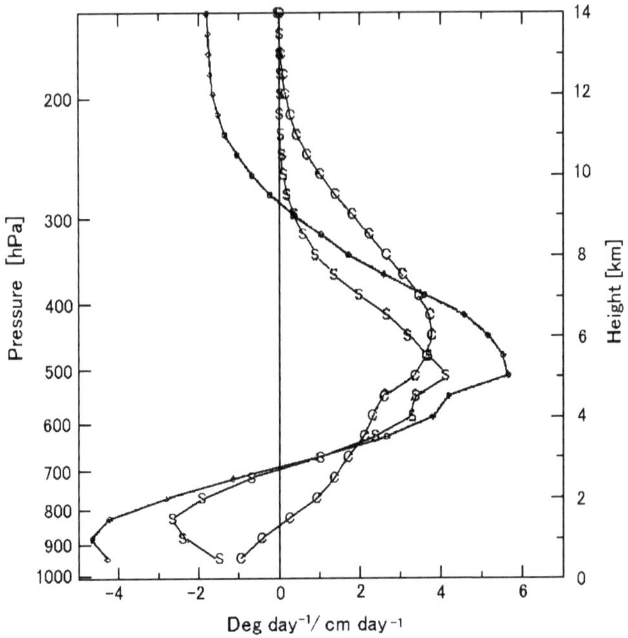

Fig. 11.38 Latent heating profile in the precipitation system, observed with two Doppler radars between November 1992 and February 1993 in Manus Island in Papua New Guinea as part of the TOGA-COARE project, using the thermodynamic retrieval method. The *solid line* with *black circles* indicates latent heating rate averaged over the analytical domain with all sides of around 50 km including a radiative cooling effect. Lines labeled C and S indicate the reference profiles in convective and stratiform parts, respectively [from Ushiyama et al. 2003]

TOGA-COARE[8] project, using the thermodynamic retrieval method. Based on the above results, the internal structure and the heating distribution were clarified for the precipitation of 16 cases that were generated or passed in the analytical domain with all sides of around 50 km. Figure 11.38 shows the latent heating profiles in the precipitation system in the observation area, where the resultant distribution is averaged for all observation area through the observation period and normalized by the ground precipitation amount.

In the case of the stratified precipitation system, the layer below 3 km (cloud base) was cooled, and the layer between 3 and 9 km was heated by the cloud system. The heating rate between 4 and 7 km was especially large. The convective part of the domain was heated in the entire volume up to the height of 14 km except for a shallow layer near the earth's surface, and the heating rate between 2 and 10 km

[8]Tropical Ocean Global Atmosphere—Coupled Ocean Atmosphere Response Experiment; one of the international joint observation project of World Climate Research Programs (WCRP) that had been executed for 10 years between 1985 and 1994. Many countries including U.S.A. and Japan participated.

was especially large. Furthermore, the thermal dose and its fluctuation in the entire observation area were estimated by multiplying the average heating rate shown in Fig. 11.38, where the ratio of stratified and convective precipitation systems in the TOGA-COARE observation area were estimated using the infrared radiation (IR) sensor on board the geostationary meteorological satellite of Japan, the GMS-5 "Himawari" (see Fig. 11.2b). As a result, it was clarified that the change of the entire heating rate was corresponding well to the change of the heating rate estimated from the actual measurements of atmospheric temperature change and the radiosonde observation with the time scale of about 30 days (Ushiyama et al. 2003).

The time resolution of the Doppler radar observation is finite and the phenomena change even in the sub-time domain of the observation. Therefore, unsolved problems have been left as the estimated error grows when the retrieval method is applied to the phenomenon with significant time fluctuation such as convection precipitation system.

Chapter 12
Observations by Atmospheric Radar

In general, atmospheric radars are capable of continuously monitoring echo power spectrum in the vicinity of the zenith. From the spectrum, the echo intensity, the Doppler velocity, and the velocity spectrum width are derivable over the wide altitude range 1–100 km in the Earth's atmosphere.

The antenna of atmospheric radar is commonly fixed on the ground and can point several directions; thus two dimensional data vs height and time in each direction can be obtained. Assuming that the atmosphere structure does not change but simply advects above the radar, the observed time-altitude cross section can be interpreted as a distance-altitude cross section. With this method the tangential, radial, and vertical winds are obtained with temporal and altitude resolutions of about 1–3 min and 150 m, respectively.

12.1 Wind Measurements

Wind vector profiles are calculated from the radial Doppler shifts in at least three non-coplanar directions, assuming spatial uniformity of the wind field within the beam's steered area at each height. Atmospheric radars can be used to observe wind velocities continuously during all weather conditions, and are thus well suited to the investigation of wind systems and atmospheric waves on various scales in the lower and middle atmospheres. In addition, MST radars are sensitive enough to measure precipitation as well (Fukao et al. 1985c). The most important capability of atmospheric radars is routine continuous monitoring of three-dimensional atmospheric winds over an altitude range including most of the troposphere, lower stratosphere and mesosphere. In particular, there has been no other direct means to continuously measure profiles of vertical wind velocities. Large efforts have been put forth to investigate the reliability and accuracy of the atmospheric radar technique. Comparisons with in-situ measurements have been conducted mostly with balloons, but also with satellite, rockets, airplanes, towers, etc. The radar gives an Eulerian estimate of the wind vector in time simultaneously over all altitudes at

S. Fukao and K. Hamazu, *Radar for Meteorological and Atmospheric Observations*, DOI 10.1007/978-4-431-54334-3_12, © Springer Japan 2014

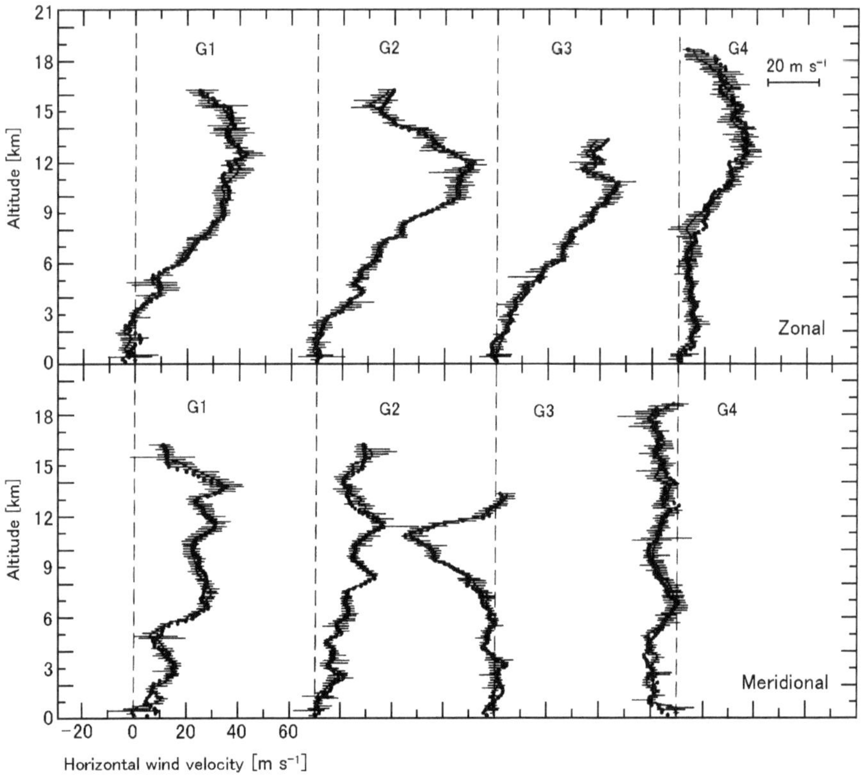

Fig. 12.1 Comparisons between the MU radar (*solid lines*) and GPS (chain of *circles*) zonal (*top*) and meridional (*bottom*) wind components for four balloon flights, labeled G1, G2, G3, and G4 conducted, respectively, on 10, 12, 14, and 25 May 2000. Note that *solid lines* and chains of *circles* are too close to discriminate. The *horizontal bars* indicate the variability (and to a lesser extent the estimation errors) of the wind data used for averaging [from Luce et al. 2001a]

a given location, while the balloon gives a "Lagrangian" value at various altitudes at different times and at different positions. Nevertheless, the comparisons are, in general, fairly good irrespective of the type of balloon, indicating that the radar wind measurements are meaningful and can be used for meteorological application. Figure 12.1 demonstrates one of such comparisons between horizontal velocities measured by the MU radar and by instrumented balloons using global positioning system (GPS) rawin-sondes (GPS radiosondes) (Luce et al. 2001a). Comparisons between the MU radar (solid thin lines) and GPS (chain of circles and thick lines) zonal (top) and meridional (bottom) wind components for four balloon flights are shown in the panel, where groups G1, G2, G3, and G4 are conducted, respectively, on 10, 12, 14, and 25 May 2000. The MU radar velocities were obtained with the Doppler beam swinging (DBS) technique discussed in Sect. 7.2.1. The horizontal bars indicate the variability (and to a lesser extent the estimation errors) of the wind data used for averaging. The maximum and minimum horizontal distances between

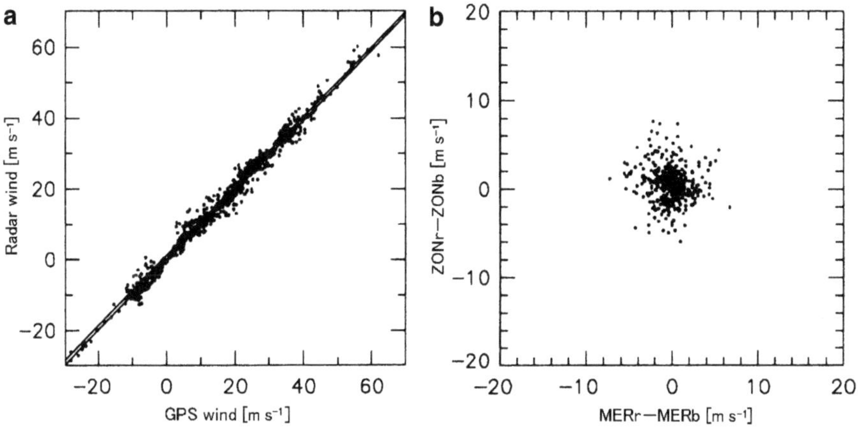

Fig. 12.2 (**a**) Zonal and meridional wind components measured by GPS radiosondes vs. that deduced from MU radar measurements. (**b**) Differences between MU radar (ZON_r, MER_r) and GPS (ZON_b, MER_b) wind component measurements in (**a**), where the differences between the zonal and meridional components are given by the vertical and horizontal coordinates, respectively [from Luce et al. 2001a]

a baloon and the radar site during the ascent were less than to 120 km and 6.8 km, respectively. In general, the balloon drifts with wind, thus the distance between the radar and the balloon becomes farther apart as the measuring altitude increases. Although the radar data and the GPS data are almost corresponding, differences tend to grow at comparatively higher altitude due to above reason.

In Fig. 12.2a, the MU radar-deduced zonal and meridional wind components are plotted against those measured by GPS radiosondes. The linear regression curve has a slope very close to 1, and the two components are well corresponding. In (b), differences between MU radar (ZON_r, MER_r) and GPS (ZON_b, MER_b) wind component measurements in (a) are plotted, where the differences between the zonal and meridional components are given by the vertical and horizontal coordinates, respectively. The standard deviation of the differences of the two winds is about $2\,\mathrm{m\,s^{-1}}$ (from Luce et al. 2001a).

It is not easy usually to verify the radar observational results of the mesosphere or higher region where balloons do not reach. However, a lot of comparative verifications for velocities observed by atmospheric radars have been performed although intermittently through simultaneous observations with the meteor radar or rockets (e.g., Fukao et al. 1980b) and with space satellites (e.g., Takeda et al. 2001).

12.2 Mesoscale Convective System

With ST/MST radars, wind velocities and echo intensities up to the tropopause or higher region can be obtained continuously timewise and spatially. A variety of mesoscale phenomena are researched with these data not obtained with past

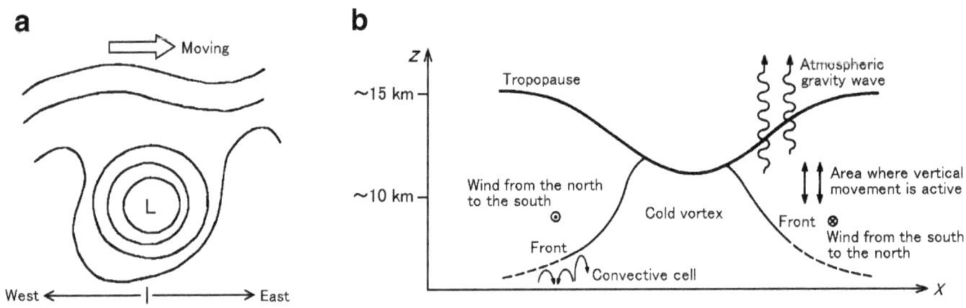

Fig. 12.3 Cold vortex shown in schematic. (**a**) Top view, and (**b**) an east-west vertical cross section of the vortex through its center

meteorological instruments. For instance, most atmospheric phenomena advect in the east west direction because the prevailing winds are from the west. Thus, if paying attention to the movement of these phenomena, their vertical and east-west horizontal structures can be obtained by converting time axis to horizontal distance. In general, atmospheric radars are sited in fixed locations and their beams are aimed to the vicinity of zenith, thus an atmospheric radar can observe only above the radar. However, if a multitude of the phenomena of the same kind but different north–south locations are observed over long periods of time, the three-dimensional structure of mesoscale convective system can be estimated by synthesizing these observation results.

12.2.1 Cold Vortex

A cold vortex and a funnel-shaped tropopause observed with the MU radar is discussed bellow (Fukao et al. 1989). As shown schematically in Fig. 12.3a, a cold vortex is a cold low which is separated from a jet stream that meanders greatly when the trough of the upper layer develops, and is an extratropical cyclone whose atmospheric temperature in the central is lower than surroundings. In the northern hemisphere, the winds rotates counterclockwise around the center of the low. Figure 12.3b shows the vertical profile of the vortex including its center. The tropopause tips down at the center, and the temperatures of the frontal surface in the constricted area of the funnel-shape are higher than surroundings. The density of air in this area is comparatively small, thus the atmospheric pressure of the upper and middle layer becomes lower than surroundings although the atmospheric temperature of the upper and middle layer (e.g., the tropopause) is lower than surroundings.

On the front, it is known that echoes from the partial reflection mechanism prevails compared with echoes from the Bragg scatter mechanism. The relative strength of these echoes prevailing can be known by taking the ratio of echo

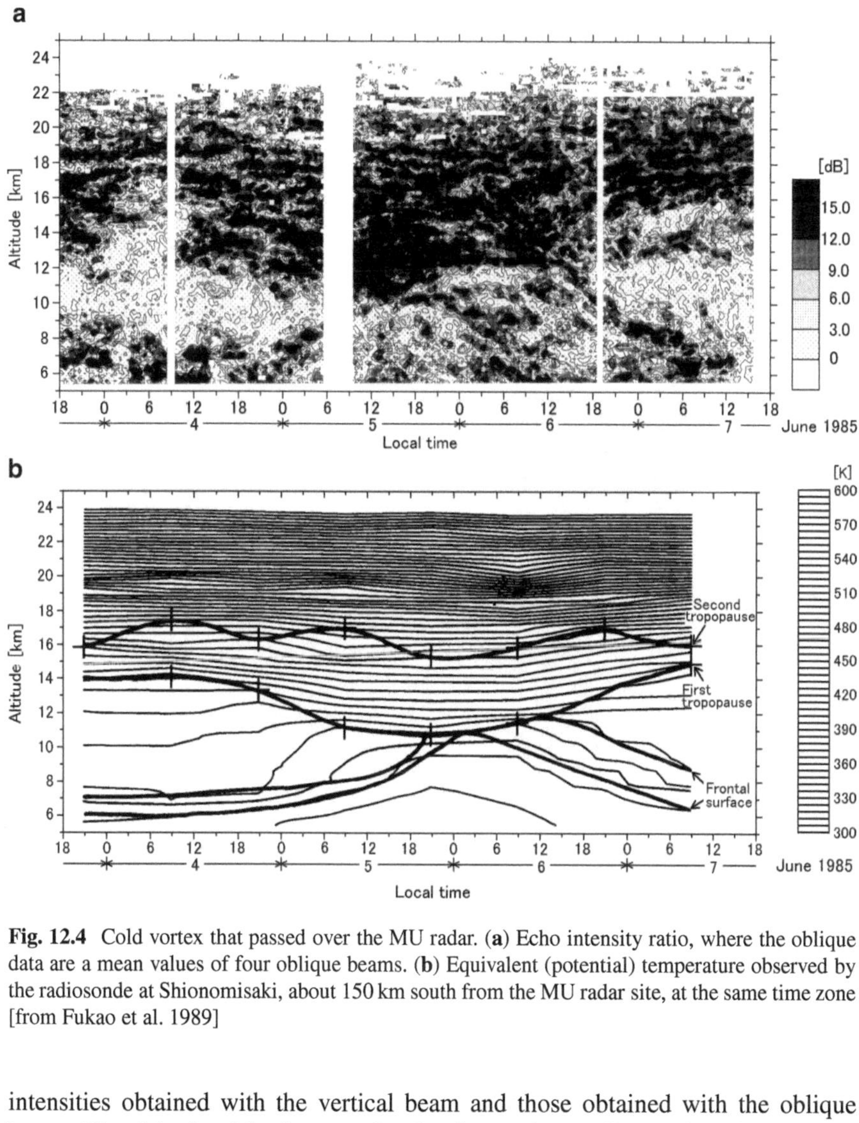

Fig. 12.4 Cold vortex that passed over the MU radar. (**a**) Echo intensity ratio, where the oblique data are a mean values of four oblique beams. (**b**) Equivalent (potential) temperature observed by the radiosonde at Shionomisaki, about 150 km south from the MU radar site, at the same time zone [from Fukao et al. 1989]

intensities obtained with the vertical beam and those obtained with the oblique beams. The altitude of the front can be also detected according to the echo intensity ratio (e.g., Röttger 1979).

Figure 12.4 shows a cold vortex that passed over the MU radar from west to east on 3–7 June, 1985 (Fukao et al. 1989). If the observed time axis of the figure is reversed, the observed time-altitude cross section can be assumed to express the east and west cross section. Panel (a) shows the echo intensity ratio, where the oblique data are a mean values of four oblique beams. Panel (b) shows the equivalent

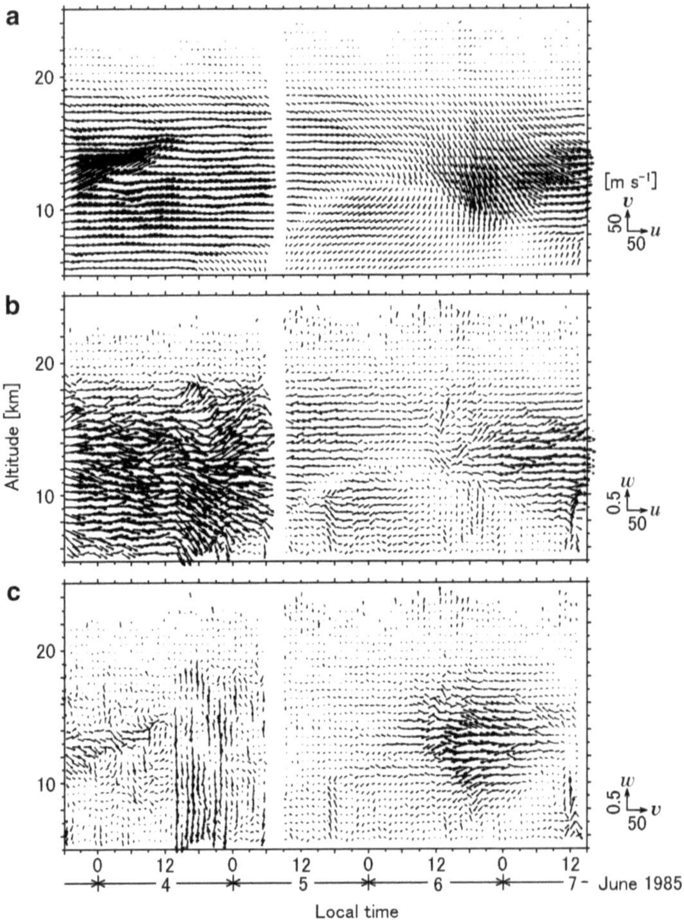

Fig. 12.5 Wind vectors observed by the MU radar during the same time period as that shown in Fig. 12.4, (**a**) zonal-meridional winds, (**b**) zonal-vertical winds, and (**c**) meridional-vertical winds [from Fukao et al. 1989]

(potential) temperature[1] observed by the radiosonde at Shionomisaki, about 150 km south from the MU radar site, at the same time zone. In Panel (b), upper two lines are the tropopauses, where the second tropopause is the coexisting tropical tropopause, and lower two lines show the frontal surfaces.

Figure 12.5 shows the wind vectors observed by the MU radar at the same time period as in Fig. 12.4, (a) zonal-meridional winds and (b) zonal-vertical winds, and (c) meridional-vertical winds. The following features are clearly shown in this figure. That is, as shown in Fig. 12.3, the velocity fluctuations are especially intense

[1] Refer to Sect. 7.3.3, footnote 7.

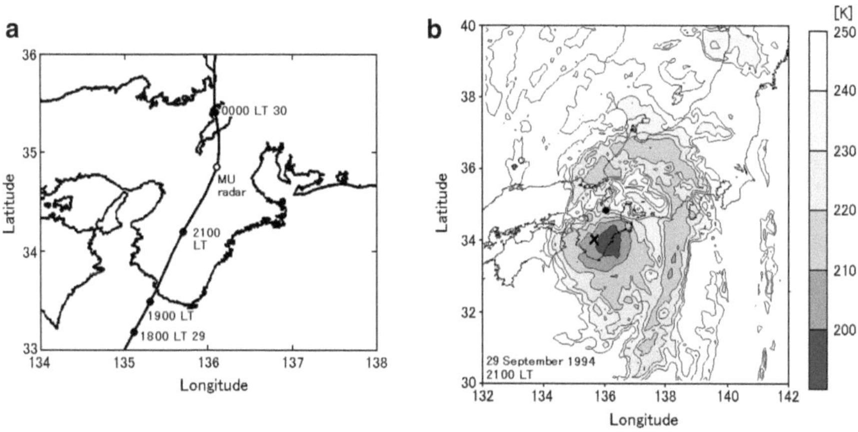

Fig. 12.6 (**a**) Trajectory of the TC 9426. (**b**) Horizontal distribution of taller clouds based on T_{BB} of GMS IR at 2100 LT 29, where *crosses* shows the center of the TC [from Shibagaki et al. 2003]

along the frontal boundary of the cold vortex (left side of Fig. 12.5). This is where the jet stream meanders first to the south on the west side of the low and then to the north on the low's east side. Moreover, many small-scale convection cells are generated under the front in rear (right side of the same figure) part of this chart. Furthermore, evidences are obtained in another observation that an atmospheric gravity wave is excited (Fukao et al. 1989).

12.2.2 Tropical Cyclone

Tropical cyclones (TC or typhoons) have been observed extensively with the MST radars in Japan and Taiwan (e.g., Fukao et al. 1985d; Sato 1993; May et al. 1994; Teshiba et al. 2001). A mature-stage TC forms an 'eye' where no clouds and precipitation exist, which generally makes it difficult to observe echoes from the eye with microwave meteorological radars. Assuming that the typhoon structure does not change during passage, the observed time-altitude cross section can be interpreted as a distance-altitude cross section of the TC. With this method the tangential, radial, and vertical winds relative to the typhoon center can be obtained up to the lower stratosphere with high temporal and altitude resolution, e.g., 3 min and 150 m. Shibagaki et al. (2003) observed the passage of the eye of mature-stage typhoon T9426 (Orchid) right above the MU radar during 29–30 September 1994. Figure 12.6a shows the trajectory of the TC. Panel (b) of this figure shows the horizontal distribution of taller clouds based on T_{BB} (equivalent black body temperature) of the GMS (geostationary meteorological satellite) IR (infrared) at 2100 LT 29, where "crosses" shows the center of the TC. It accompanied a wall cloud of the eye in the vicinity of the center and cloud band to the north side.

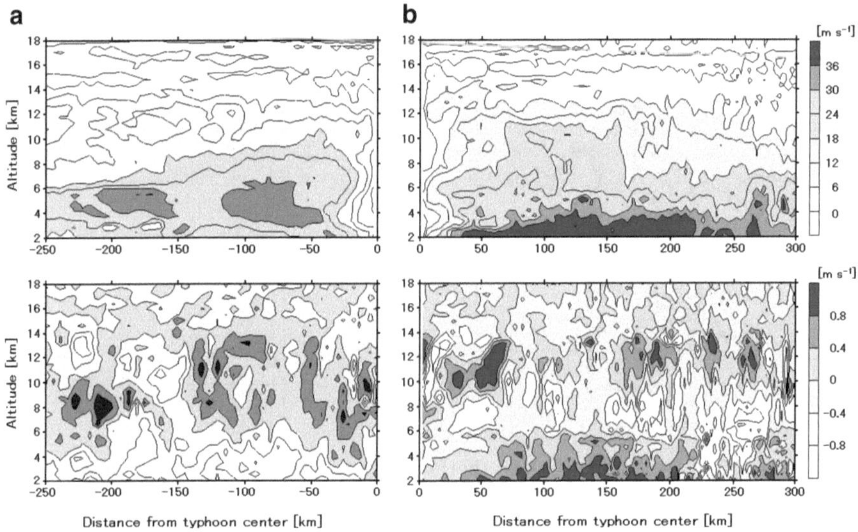

Fig. 12.7 Vertical cross sections of (**a**) rear and (**b**) front of the typhoon observed with the MU radar. In both (**a**) and (**b**), *upper panels* show the tangential winds, and *lower panels* show the vertical winds. The typhoon center is the 0 km point in each panel, and the horizontal axis shows the distance from the typhoon center converted from the transit time [from Shibagaki et al. 2003]

Figure 12.7 shows vertical cross sections of (a) the rear and (b) the front of the typhoon observed with the MU radar. In both (a) and (b), upper panels show the tangential winds[2] and lower panels show the vertical winds. The typhoon center is the 0 km point in each panel, and the horizontal axis shows the distance from the typhoon center converted from the transit time. The kinematic structures in front and rear of the typhoon were significantly different because of its transition from mature to decaying stages and an asymmetric distribution of clouds and precipitation. In the front of the typhoon, cyclonic circulations which become maximum at the bottoms were observed, and updrafts were predominant within precipitating clouds below 6 km altitude. In altitude range of 10–14 km, strong updraft regions were observed. They corresponded to inner cloud and band-shaped cloud regions. Meanwhile, the maximum speed of the cyclonic wind was less and the magnitude of the updraft was smaller than that in the front.

Figure 12.8 shows the enlarged view of the vertical structure of the typhoon center in Fig. 12.7, where (a) the tangential winds and (b) the vertical winds (Shibagaki et al. 2003). The tangential winds around the typhoon center show the spiral structure in an altitude direction. This is thought to be the result that the

[2]The mean radial wind within a diameter of ∼150 km (corresponding to time period of 3 h), calculated at each distance, is used as the environmental wind. Positive (negative) values in tangential and radial winds relative to the environmental wind are cyclonic (anti-cyclonic) rotation and outflow (inflow), respectively.

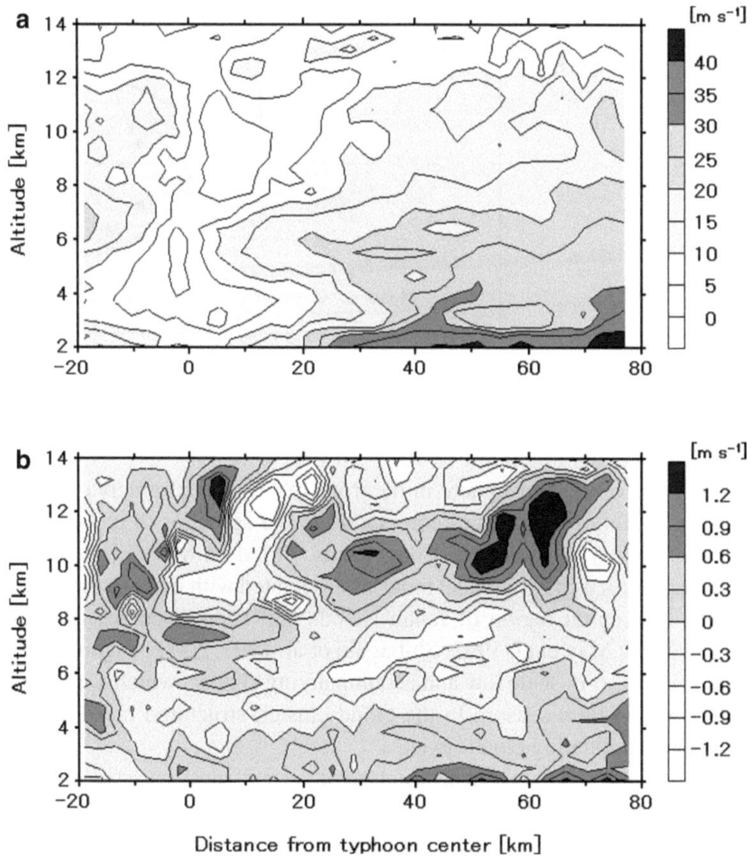

Fig. 12.8 Enlarged view of the vertical structure of the typhoon center in Fig. 12.7. (**a**) Tangential winds and (**b**) vertical winds [from Shibagaki et al. 2003]

weakened typhoon eye was changed. As shown in Panel (b), downdrafts were caused in the eye region of the typhoon along with this spiral structure. This was confirmed for the first time by the direct observation by the MU radar.

Teshiba et al. (2005) observed passage of the eye of T0310 (Etau) over the lower troposphere radar (LTR) of 1.3 GHz at Naze, Okinawa in Fig. 10.21, Japan during 7–8 August 2003. In the classical models of the typical typhoon eye, weak downdrafts are considered to be distributed uniformly. However, this observation has revealed that both weak updrafts and downdrafts exist in the middle troposphere, and downdrafts are dominant in the lower troposphere. The updraft did not form clouds and precipitation because stable and relatively dry air existed in the middle troposphere. Furthermore, the movement of a typhoon was observed by multiple LTRs at the same time, and the time change of the spatial structure was observed directly. As a result, it is becoming possible to discuss the asymmetries of the front–rear and the left–right structures of the typhoon.

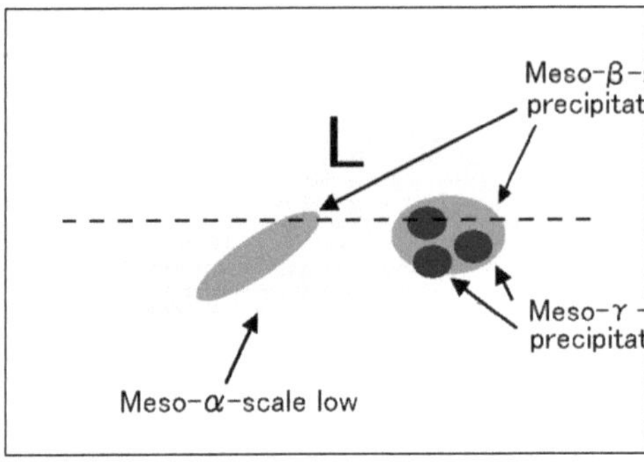

Fig. 12.9 Schematic of meso-α-scale low and accompanying cloud clusters

12.2.3 Convection

Vertical motions of the atmosphere in the troposphere are extremely complex. It is estimated that a lot of turbulences correspond to the activities of cloud convection except for mountain wave during strong winds. Based on MU radar observations, the multi-scale structures of vertical motions associated with meso-α-scale cyclones that passed the central region of Japan islands had been discussed (Fukao et al. 1989; Sumi 1989; May et al. 1992; Shibagaki et al. 1997, 2000). Figure 12.9 shows a schematic of meso-α-scale low and accompanying clusters which appear in Japan islands during the Baiu season. In the cloud cluster, structured meso-β-scale and meso-γ-scale precipitating clouds exist.

Figure 12.10a shows the vertical structure of vertical motions associated with the passage of a cyclone during 4–5 July 1991 (Shibagaki et al. 1997, 2000). Two strong updraft regions, lasting to 6 h, are observed within the cyclone. One is associated with a meso-β-scale precipitating cloud in front of the center of cyclone. The other is associated with a westerly trough tilting westward behind the center, which is not accompanied by precipitating systems. It is also found that the meso-β-scale updraft consists of several meso-γ-scale strong updrafts (lasting for 30–60 min) within the precipitating cloud. In panel (b), the wind circulation (relative to the environmental flow) within a meso-γ-scale precipitating cloud is presented in detail. A significant updraft occurs with the convergence of zonal wind in the lower troposphere, which leads to outflow in the upper troposphere.

Rotor

A rotor circulation with the zonal, meridional, and vertical scales to 40 km, to 150–200 km, and to 2 km, respectively, was directly detected by the MU radar near the Baiu front in the lower troposphere (Watanabe et al. 1994). It was identified as a meso-β-scale depression generated right under and between banded precipitation clouds of a meso-α-scale cyclone. Figure 12.11 shows the rotor detected by the MU radar, (a) zonal wind vs vertical wind and (b) meridional wind vs vertical wind.

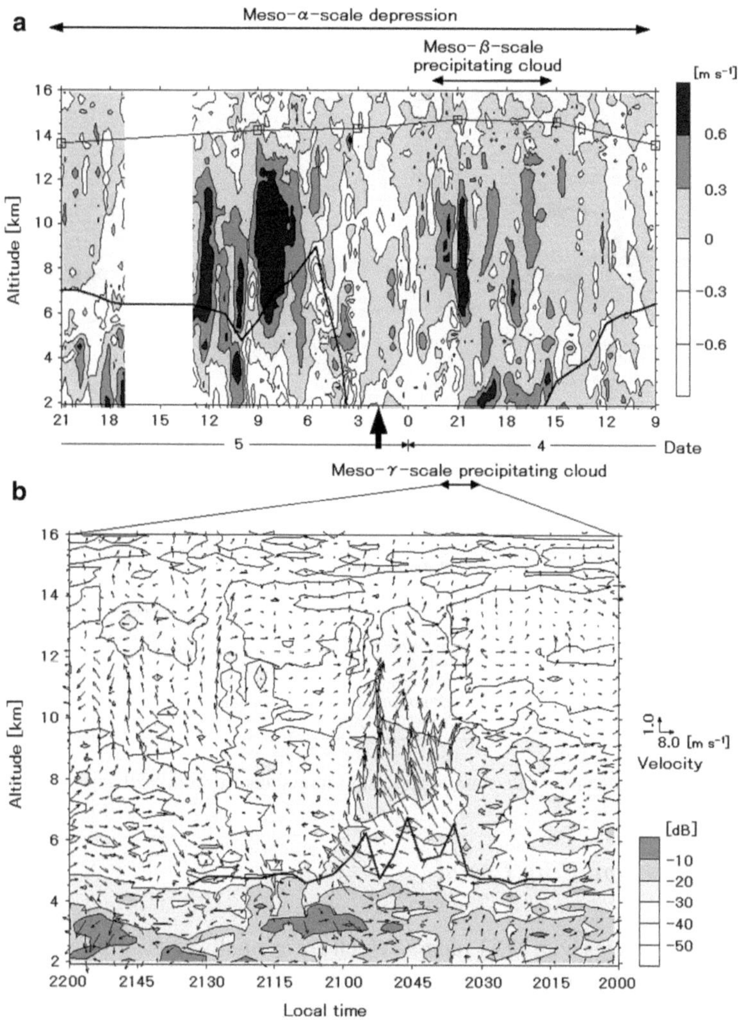

Fig. 12.10 (**a**) Time-altitude section of vertical velocity within a meso-α-scale cyclone observed by the MU radar during 4–5 July 1991. The center of a cyclone passed over the MU radar site around 0100 LT on 5 July. *Thick solid lines* show a frontal surface identified from operational rawinsonde data, the global objective analysis (GANAL), and vertical wind shear measured with the MU radar. *Squares* show the tropopause level. (**b**) Zonal-vertical wind vectors (*arrows*) relative to the environmental wind. The contours show vertical echo intensity, which depends on the high humidity associated with meso-γ-scale precipitating clouds. The *thick solid line* is the top level of precipitation echoes identified from the MU radar observations [from Shibagaki et al. 1997, 2000]

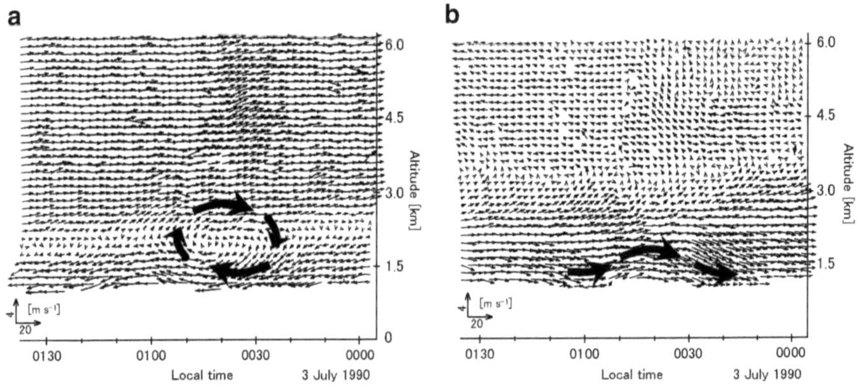

Fig. 12.11 Rotor detected by the MU radar. (**a**) Zonal wind vs. vertical wind and (**b**) meridional wind vs vertical wind [from Watanabe et al. 1994]

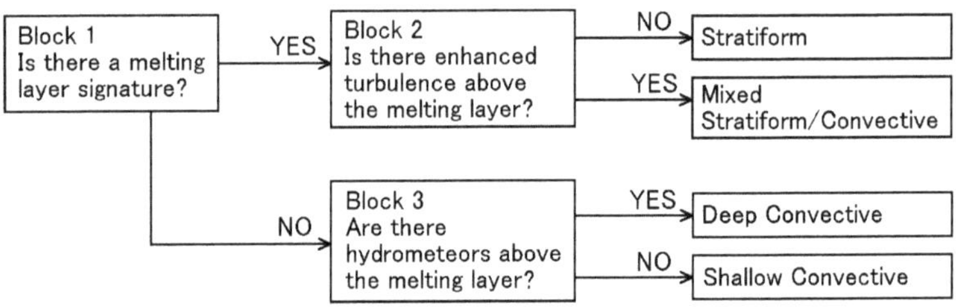

Fig. 12.12 Flow diagram of rain classification algorithm [adapted from Williams et al. 1995]

12.2.4 *Precipitating Cloud*

Precipitation studies with MST radars, which are normally considered to be clear-air instruments, have proved also to be quite interesting (e.g., Hocking 1997b). Fukao et al. (1985c) presented the first observations of rain echoes at very high frequency (VHF), and this spawned a large number of follow-up papers in this area.

The release of latent heat in convective systems in the tropics drives global circulations. Consequently, the proper specification of realistic vertical distributions of diabatic heating is crucially important to models of the atmosphere. An algorithm has been developed that classifies precipitating clouds into either stratiform, mixed stratiform/convective, deep convective, or shallow convective clouds by analyzing the vertical structure of reflectivity, velocity, and spectrum width derived from measurements made with the vertically pointing beam of UHF wind profilers (Williams et al. 1995; Renggono et al. 2001). The criteria used to classify the precipitating clouds into the four regimes are shown in the flow diagram in Fig. 12.12. This algorithm is based on the following criteria: presence of melting layer (bright band, see Sect. 11.1), presence of enhanced turbulence above the melting layer, and the presence of hydrometeors above the melting layer.

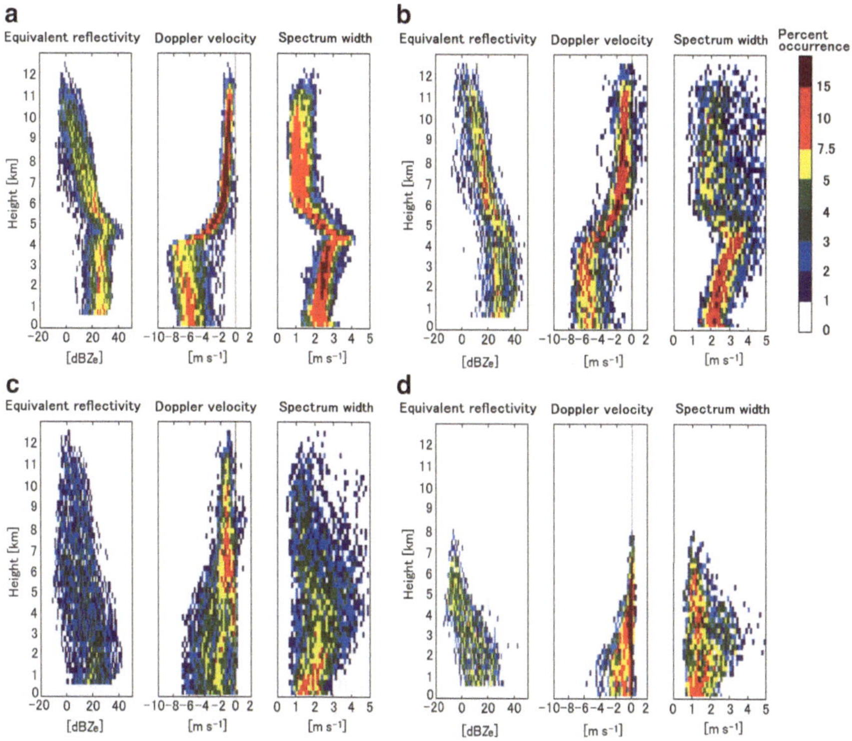

Fig. 12.13 Frequency distribution of Doppler spectral moments for precipitating clouds classified as (**a**) stratiform clouds, (**b**) mixed stratiform/convective clouds, (**c**) deep convective clouds, and (**d**) shallow convective clouds, measurements with a 915-MHz Doppler wind profiler at Manus Island from May 1992 through February 1993. In each panel, the *left*, *center*, and *right* picture shows the equivalent radar reflectivity factor (Z_e in dB), the Doppler velocity, and the velocity spectrum width, respectively. The *color bar* shows the frequency distribution (occurrence) in percent [adapted from Williams et al. 1995]

Figure 12.13 depicts the frequency distribution of Doppler spectral moments for precipitating clouds classified as (a) stratiform clouds, (b) mixed stratiform/convective clouds, (c) deep convective clouds, and (d) shallow convective clouds. In each panel, the left, center, and right picture shows the equivalent radar reflectivity factor (Z_e in dB, see Sect. 3.3.5), the Doppler velocity, and the velocity spectrum width, respectively. The data are derived from measurements with a 915-MHz Doppler wind profiler at Manus Island from May 1992 through February 1993. The color bar shows the frequency distribution (occurrence) in percent as a function of height (Williams et al. 1995).

In Panel (a), an increase in reflectivity is visible near 4.5 km showing the melting layer is easily identifiable. The Doppler velocity frequency distribution is very narrow above the melting layer, and its largest concentration occurs near $-2\,\mathrm{m\,s^{-1}}$ at a height of 6 km, whereas the largest concentration is about $-7\,\mathrm{m\,s^{-1}}$ at a height of 4 km. The increase in fall speed (i.e., larger negative Doppler velocities) through the melting layer is expected due to the formation of large drops. The spectral width of above 6 km is centered around 1.25 $m\,s^{-1}$, indicating that the profiles are not turbulent above the melting layer. These features well agree with the classification to the statiform cloud. In Panel (b), the most striking feature is the broad distribution of spectral width above the melting layer. The next prominent feature is the upward Doppler velocities above 6 km. Above the melting layer, there are profiles with upward motions exceeding 1 $m\,s^{-1}$. There are also downdrafts producing Doppler velocities of $-3\,m\,s^{-1}$ above the melting layer although the downdrafts are not as frequent as the updrafts. These features suggest the existence of mixed stratiform/convective clouds. In Panel (c), the reflectivity has a very broad distribution with values ranging from 0 to 40 dBZ_e, whereas the Doppler velocity distribution does not have the knee in velocities that is predominant in the stratiform and mixed stratiform/convective clouds. In Panel (d), the warm rain processes that do not involve frozen hydrometeors are presented (Williams et al. 1995). Each panel is well in conformity with the classification of the four regimes in Fig. 12.12.

12.2.5 Orographic Rainfall

There are a lot of intense rainfalls, especially in rainy seasons like Meiyu/Baiu, in mountainous regions in Japan. They are affected and strengthened by the moist air being uplifted over mountains, and have already been classified into several types of rainfalls, which are based on the Froude Number F_r ($= U/Nh$, where U, N, and h represent horizontal wind speed, Brunt Väisälä frequency, and mountain height, respectively) (Houze 1993, pp. 535, 536). Additionally, orographically-affected rainfalls around Japan sometimes make a line of rainfall along the direction of horizontal wind. The line of rainfall results in continuous and localized precipitation, causing occasionally heavy disasters. Umemoto et al. (2004) conducted the case study of orographic rainfall on the west side of Kyushu Island of Japan. Figure 12.14a shows radar rain intensities at the height of 2 km observed by the Japan Meteorological Agency (JMA) operational meteorological radars of 5.3-GHz band at 0200 JST on 29 June 1999. The orographic rainfall started from a small island and moved to the north-eastward. The trigger of this rainfall is the updraft caused by the small islands (Point crosses in Panel (a)) which are several hundred meters in height, because the Froude number is over 1 which means that the air mass passes beyond the islands. Moreover, this line-shaped rainfall is caused in the warm sector of the low pressure, because there is vertical shear of horizontal wind. The development of this orographic rainfall corresponded to the strengthening of vertical shear as shown in Fig. 12.14b. In the figure, The hodograph is starting from the lowest height (200 m; triangle) up to 2.5 km in every 1 km (cross). Each profile is obtained by averaging the profiles observed by two 1.3-GHz band wind profilers while the orographic rainfall was observed by the operational meteorological radar. A, B and C mean the wind profiles while line-shaped rainfalls were seen. N means

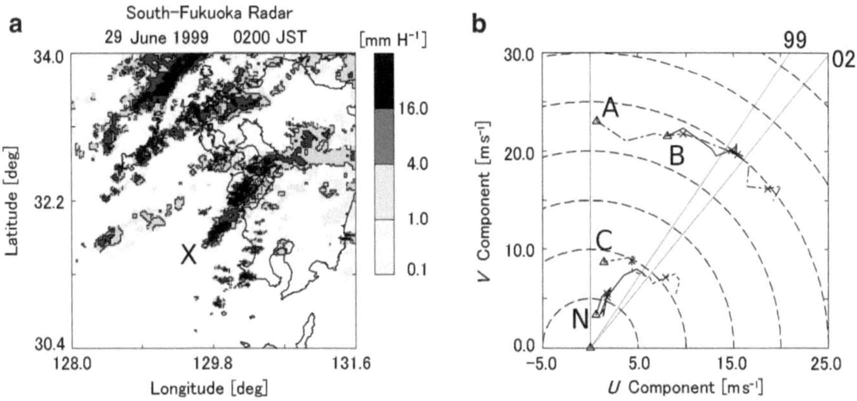

Fig. 12.14 (**a**) Radar rain intensity at the height of 2 km observed by operational meteorological radars. The orographic rainfall started from a small island (Point *crosses*) and moved to the north-eastward. (**b**) Hodograph starting from the lowest height (200 m; *triangle*) up to 2.5 km in every 1 km (*cross*). Each profile is obtained by wind profilers. A, B and C mean the wind profiles while line-shaped rainfalls were seen. N means the profile without rainfall. As a reference, the directions of line-shaped rainfalls in 1999 (as shown "99") and 2002 ("02") are also shown in *solid lines* [from Umemoto et al. 2004]

the profile without rainfall. As a reference, the directions of line-shaped rainfalls in 1999 (as shown "99") and 2002 ("02") are also shown in solid lines. Both the existence and strength of the vertical shear result in the continuous supply of humidity at the bottom of clouds and the movement of clouds toward leeward of the mountains. As a result, this strong rainfall continued for a couple of hours at the same position while a lot of clouds passing over.

12.2.6 Echoes from Precipitation

The rainfall rate changes significantly with the drop size distribution (DSD), which is closely concerned with the rain formation process involving breakup and coalescence. Since echoes from both precipitation particles and ambient air have comparable intensity in the 50-MHz band, and are discernible even at 400 MHz, they are discriminated according to their Doppler velocities (Fukao et al. 1985c). This ability is used to measure the terminal velocity of raindrops accurately, which enables estimation of the DSD continuously in real rain (Wakasugi et al. 1986). The technique has been used to determine absolute precipitation rate (e.g., Sato et al. 1990; Maguire and Avery 1995). Figure 12.15 shows the power spectrum of precipitation and atmospheric echoes obtained from the MU radar observations (Fukao et al. 1985e). The spectral data are 6 min mean values. In a period of no precipitation (Panel (a)), only the power spectrum of atmospheric echoes from refractive index perturbations (i.e., Bragg scatter) is obtained. On the other

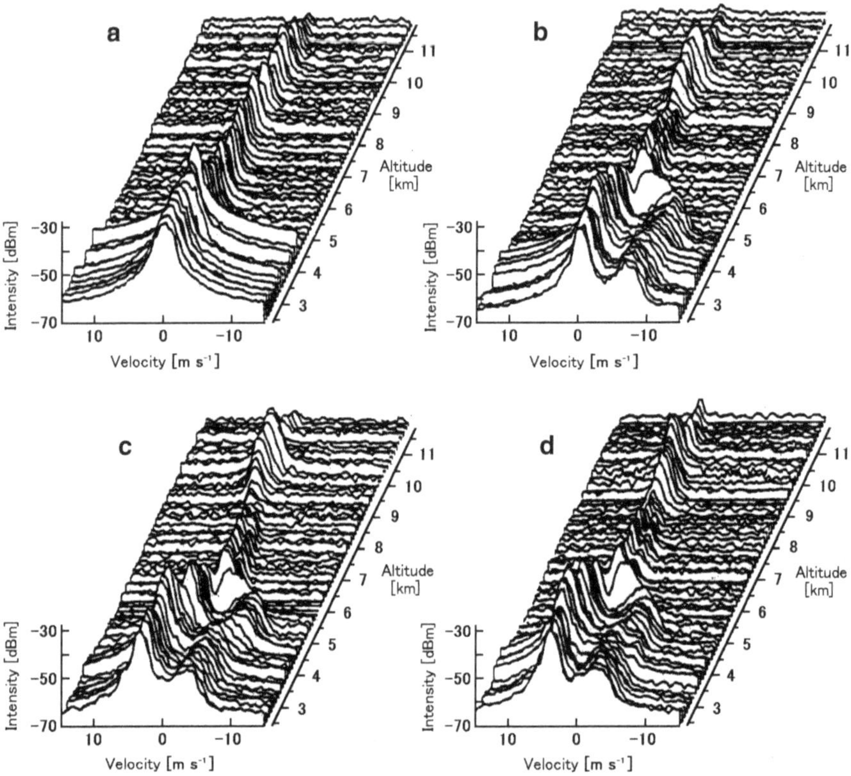

Fig. 12.15 Height profiles of the power spectrum of precipitation and atmospheric echoes obtained from the MU radar observations. Radar beams of (**a**) and (**b**) are to the zenith. Those of (**c**) and (**d**) are oblique ones to the north and to the east, respectively. Positive Doppler velocity is to the upper direction. (**a**) No precipitation, and (**b**)–(**d**) 1 mm h^{-1} rainfall were observed on the ground [from Fukao et al. 1985e]

hand, during periods of precipitation (Panels (b)–(d)) when 1 mm h^{-1} rainfall were observed on the ground, in addition to the power spectrum of atmospheric echoes whose peaks appeared left (positive), the power spectrum of precipitation echoes appeared right (negative) at the altitude below 6 km. Furthermore, a melting layer was observed at the altitude of around 6 km, and echoes that seem to be ice crystals appeared at the altitude between 6 and 8 km.

Figure 12.16 shows the power spectrum of convective rain obtained from the 404-MHz wind profiler observation (Kobayashi and Adachi 2001). The radar beam was pointed to the zenith, and the maximum precipitation on the ground was 50 mm h^{-1}. The power spectrum is the 1 min mean value. A small spectral peak around zero Doppler velocity is the one associated with Bragg scatter, and the large broad spectral peak is that associated with precipitation. It is the one that the spectrum shape has changed remarkably in a short period of time and the change in the drop

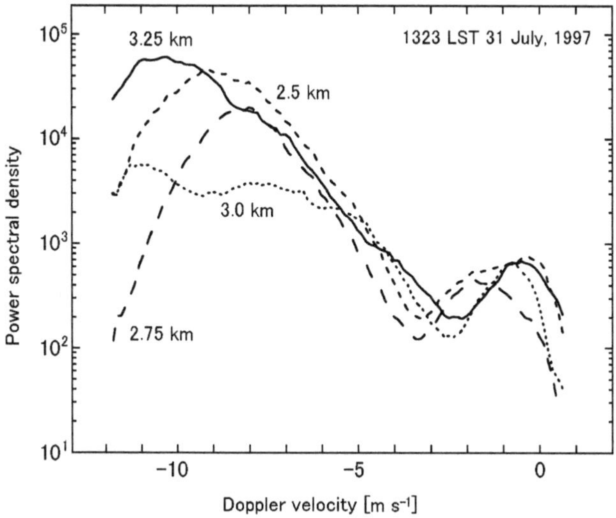

Fig. 12.16 Power spectrum of convective rain obtained from the 404-MHz wind profiler observation. The radar beam was pointed to the zenith. The power spectrum is the 1 min mean value. Positive Doppler velocity is upward motion [from Kobayashi and Adachi 2001]

size distribution is apparent. The remarkable change of spectrum shape in a narrow range of altitude between 2.5 and 3.25 km reflect the change of the DSD. Kobayashi and Adachi (2001) reported a significant altitude variation of the DSD observed aloft with a 404-MHz wind profiler in convective rain at Tsukuba, Japan. Figure 12.17 shows (left) the altitude variations of the DSD from the 404-MHz wind profiler and (right) simulated raindrop images corresponding to the DSD, where the circle size corresponds to the raindrop diameter. Giant raindrops larger than 6 mm in diameter existed at the altitude of 3.25 km but almost disappeared below 2.75 km, presumably due to raindrop breakup.

12.3 Atmospheric Gravity Waves

The rising surface of wavelets and undulations of the sea is pulled downward by the gravity of the Earth, and the neighbor surface of the water is lifted due to the reaction. The height changes of the water's surface are transmitted along the sea surface to long distances. These waves, generated by the gravity which works as the restoring force, are called gravity waves. Similar gravity waves also exist in the stratified atmosphere and below the sea surface. For instance, a row of clouds might be seen to run along the row of mountains. These phenomena can be formed by the gravity wave generated on lee side when flows pass the mountain row. Atmospheric gravity waves (hereafter gravity waves) are generated by cumulus

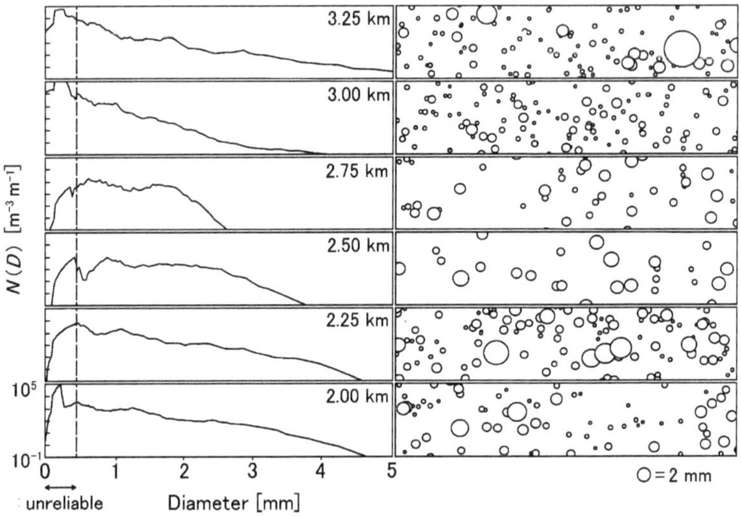

Fig. 12.17 (*Left*) Altitude variations of the DSD from the 404-MHz wind profiler and (*right*) simulated raindrop images corresponding to DSD shown in the *left*. The *circle* size corresponds to the raindrop diameter [from Kobayashi and Adachi 2001]

convections, the jet streams, fronts, typhoons, and so on. MST radars are the most suitable instruments for studying gravity waves due to their capability of continuous measurement of three-dimensional winds with fine time resolution simultaneously at many altitudes. Each stage of the scenario has been demonstrated by MST radar observations.

12.3.1 Wave Propagation

Unlike the surface of the sea, the boundary is not clear in the atmosphere. Thus, gravity waves generated at lower altitude is transmitted to the upper layer as shown in Fig. 12.18 (Geller 1983). The right panel shows the exponential growth of horizontal velocity U and temperature T with height, accompanying the exponential decrease of the atmospheric density ρ_a. Horizontal shears also increase while the height increases. The waves become locally unstable which is called shear instability or Kelvin-Helmholtz (KH) instability at the height where the horizontal shear becomes maximum[3] and break. That action is called wave breaking. The turbulence generated by the wave breaking causes the vertical transport of heat and the mixture

[3]Strictly, the height where the Richardson number R_i becomes equal to or smaller than 1/4. It may not necessarily be the same height where the horizontal shear becomes maximum or the atmospheric stability becomes minimum.

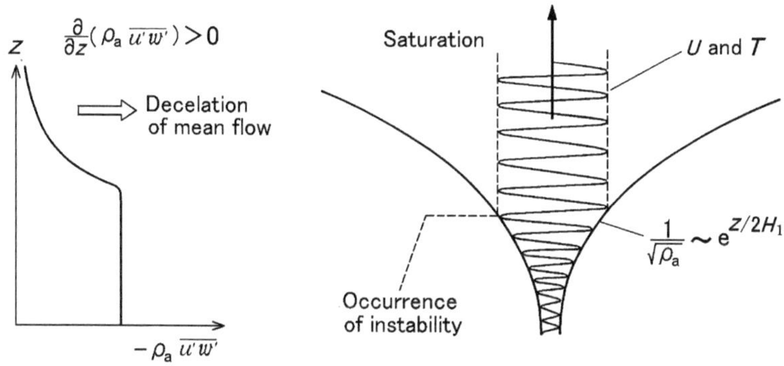

Fig. 12.18 Exponential growth of a gravity wave, its saturation, and wave breaking (*right panel*), and the resulting vertical flux of zonal momentum (*left panel*). U and T are the horizontal velocity and the temperature, respectively. Values u' and w' are the horizontal and vertical fluctuations from turbulence, respectively. $\rho_a \overline{u'w'}$ is the momentum flux (Reynolds stress) that the wave carries upward [adapted from Geller 1983]

of atmospheric minor components. Moreover, when the temperature gradient of the atmosphere locally becomes larger than the adiabatic lapse rate, convective instability is generated. The instability of gravity waves are generally assumed to be caused due to convective instability at the higher frequency region, and due to shear instability at the lower frequency region. The growth of the amplitude of a gravity wave stops and saturates above the height where the instability occurs. In Fig. 12.18, u' and w' are the horizontal and vertical fluctuations from turbulence, respectively. The product of the wave amplitude and the mass is the momentum that the gravity wave carries. Moreover, the product of the momentum and the variance of the vertical flow of gravity wave, $\rho_a \overline{u'w'}$, is the momentum flux (Reynolds stress) that the wave carries upward. As shown in the left panel of this figure, the momentum flux decays exponentially to the height where the instability occurs.

12.3.2 Dispersion Equation

The propagation speed of gravity waves is different depending on wavelength. The expression that provides for the relation among the horizontal wavenumber k, vertical wavenumber m, and the angular frequency ω is the dispersion equation[4] and is (e.g., Kato 1980)

---

[4]Equation of motion, mass conservation law, and the 1st law of thermodynamics are solved at time and space for the 1st perturbation of the harmonic function in an adiabatic and frictionless atmosphere. Here, only the gravity wave is accounted and the sound wave is excluded.

$$m^2 = \frac{k^2(N^2 - \omega^2)}{\omega^2 - f_\mathrm{i}^2}, \tag{12.1}$$

where N is the Brunt Väisälä frequency and its period is about 10 min and 5 min in the troposphere and the stratosphere, respectively. The value f_i is the inertial frequency, and is given by $f_\mathrm{i} = 2\Omega \sin \vartheta$ for the latitude ϑ, where Ω is the angular velocity of the Earth's rotation ($7.292 \times 10^{-5}\,\mathrm{s}^{-1}$). The value f_i in period is about 21 h for $\vartheta = 35\,^\circ$N. The waves whose period is longer than the Brunt Väisälä period and shorter than the inertial period can be propagated as gravity waves. The gravity wave whose period is long and strongly influenced by the Coriolis force is distinguished and called the inertia-gravity wave (IGW).

A significant feature of gravity wave is that the group velocity and the phase velocity become the same direction in horizontal direction whereas they become opposite in vertical direction. It means that the group velocity (direction where energy is transmitted) whose phase propagates downward (iso-phase height descends) is upward. In accordance with the custom, ω is assumed to be positive, and the value m becomes negative ((positive) when the energy propagates upward (downward). The polarization relations between each element of the inertia gravity is given by

$$v' = -\frac{jf_\mathrm{i}}{\omega}u', \tag{12.2}$$

$$w' = -\frac{k}{m}u', \tag{12.3}$$

where $j = \sqrt{-1}$ is the imaginary unit, and v' is the fluctuation components of wind perpendicular to the direction of wave travel. From (12.2), it is understood that the horizontal component of gravity wave becomes an elliptically polarized wave with the long axis in the direction of propagation. In the northern hemisphere, if $u \propto \exp(jmz)$, then $v \propto \exp(jmz - \pi/2)$. In the northern hemisphere, from above relation, if m is negative then the polarization direction of the horizontal component becomes clockwise with the increase of height, and if m is positive then counter-clockwise. This direction reverses in the southern hemisphere. The frequency of the wave is decided from the ratio of the long axis and the short axis of the ellipse. Meanwhile, (12.3) shows that the wave number vector is orthogonal to the direction of the oscillation, i.e., the gravity wave is a transverse wave. Moreover, using (12.3), the direction of the horizontal propagation of a gravity wave can be uniquely decided from the direction of the vertical component. This technique is named the hodograph analysis, and is frequently applied on the analysis of inertia-gravity wave (e.g., Maekawa et al. 1984; Hirota and Niki 1986; Gavrilov et al. 1996).

The gravity wave that is excited in the lower atmosphere and is propagated upward increases its amplitude with the increase of height. The amplitude of the prevailing component in the mesosphere increases to about ten times that of the troposphere. Figure 12.19 shows the north–south wind component in the mesosphere (the height of 60–90 km) from the MU radar observations between 13 and 31 October 1986 (Tsuda et al. 1990a). At these altitudes, atmospheric radars can

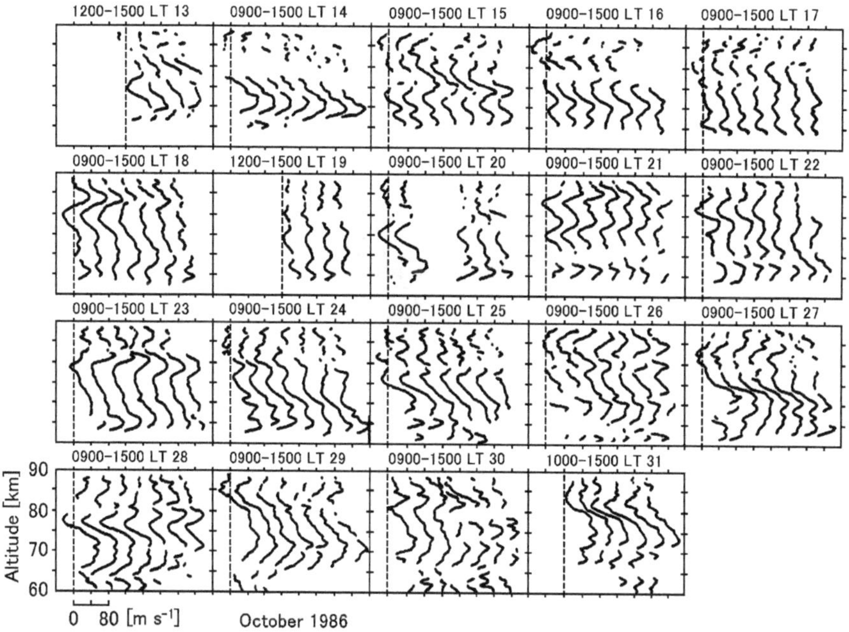

Fig. 12.19 North–south wind components in the mesosphere (the height of 60–90 km) from the MU radar observations between 13 and 31 October 1986. Only eastward components are shown [from Tsuda et al. 1990a]

observe gravity waves only in the daytime (0900 to 1500 LT) when free electrons exist, thus 1 h mean profiles in daytime are shown. The sine wave oscillations in each profile are due to the prevailing gravity waves.

If the background winds does not change horizontally and timewise, the angular frequency ω and the horizontal wavenumber k of the gravity wave observed by a ground based atmospheric radar are preserved. In the case of the gravity wave in a uniform flow, ω in the dispersion equations (12.1) and the polarization relations (12.2)–(12.3) can be replaced with ω_{i} in the following equation

$$\omega_{\mathrm{i}} = k(c - \bar{u}), \tag{12.4}$$

where c is the observed phase velocity, $c = \omega/k$, $\bar{u}$ is the mean zonal wind, and ω_{i} is the angular frequency on the coordinate system in the uniform flow, and is called the intrinsic angular frequency.

12.3.3 Critical Layer

When a gravity wave that is transmitted upward in the atmosphere where the background wind changes with height (i.e., wind shear exists) and reaches the height

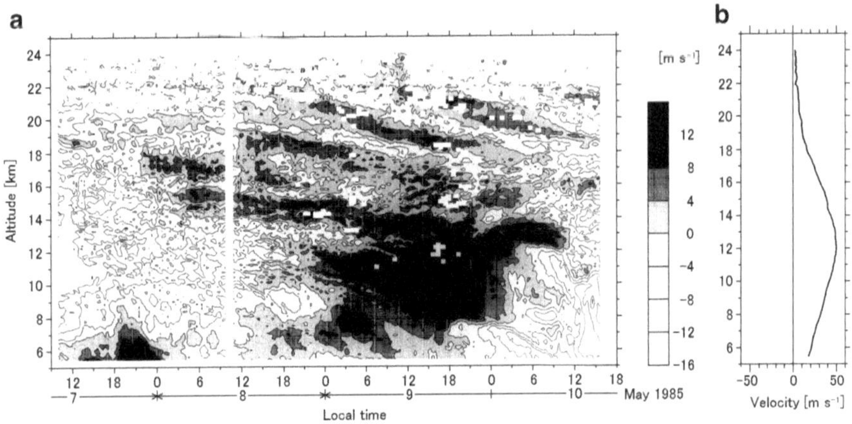

Fig. 12.20 Height profiles of (**a**) velocity fluctuations and (**b**) mean zonal winds obtained from the MU radar observations [from Yamanaka et al. 1989]

where c becomes equal to $\bar{u}$, its vertical group velocity becomes zero and the gravity wave stops upward transmission. The height range is called the critical layer.

The appearance of a gravity wave caught in a critical layer was obtained from the MU radar observations (Yamanaka et al. 1989). Figure 12.20 shows the observed height profiles of (a) velocity fluctuations and (b) mean zonal winds. It is clear from the figure that waves of comparatively long period propagating downward are dominant in the lower stratosphere (height of between about 14 and 24 km). The periods of the waves are about 22 h and their vertical wavelengths are 2–3 km. The parameters of gravity waves can be estimated by averaging the data 30 min and applying hodograph analysis in sequence. In this observation, the mean value of the intrinsic period of the gravity waves is about 7 h.

Figure 12.21a shows the histogram of eastward phase velocity c_x, and (b) shows the circumstances schematically that the ICW is caught by the critical layer. The thick line in Panel (a) shows the mean zonal winds $\bar{u}$ in Fig. 12.20b. It is clear from the panel that $c_x \simeq \bar{u}$ in the lower region of a jet stream. It is the evidence that the wave is caught with the mean zonal winds. On the other hand, $c_x \leq \bar{u}$ in the upper region of jet stream. It means that the gravity waves generated by a jet stream are caught with the critical layer on the way to propagate upward.

12.3.4 Gravity Wave Spectra

It is rare that a gravity wave is detected as a quasi-monochromatic wave as shown in Fig. 12.18, but is usually observed as the composition of many waves. Hence, the frequency spectrum or the vertical wavenumber spectrum is often analyzed based on the time change and the height profile of the wind velocity. Shears generally grow

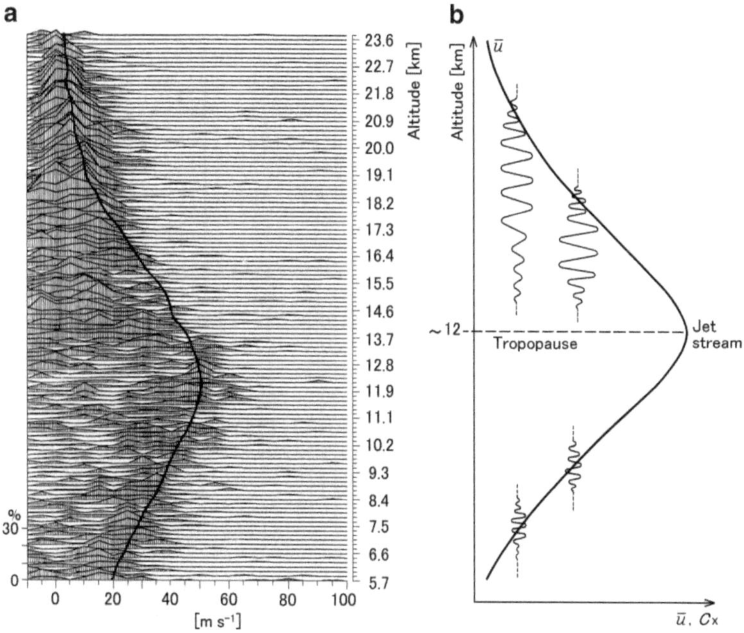

Fig. 12.21 (**a**) Histogram of eastward phase velocity c_x obtained from the MU radar observations at the same time with Fig. 12.20, and (**b**) schematic that gravity waves are caught by the critical layer [from Yamanaka et al. 1989]

as the vertical wavenumber becomes larger (i.e., the wavelength becomes shorter), and gravity waves become unstable and saturated even at small amplitudes. Hence, the vertical wavenumber spectra are saturated at higher frequencies. The spectral density of vertical wavenumber is proportional to the -3th power of the vertical wavenumber m in the saturated region, and its gradient does not depend on the height. On the other hand, gravity waves of lower wavenumber are not saturated and can be propagated to upper altitudes. Thus, in an unsaturated region, the gradient of the spectral density is zero, and its power spectrum depends on the square of the Brunt Väisälä frequency N, and increases with height.

Figure 12.22a shows the theoretical model of the vertical wavenumber spectra (Smith et al. 1987). It is estimated based on the value N^2 obtained with the temperature profile measured by the rawinsonde in the radar observation period. In the model, the ratio of power spectrum of gravity waves in the mesosphere and that in the troposphere becomes about 10^4 times in the lower frequency region. It is due to the decrease of atmospheric density in higher altitude. The reason that the observed ratio is smaller than the theoretical one is due to the wave breaking and saturation on the upward propagation of gravity waves. Panel (b) shows the vertical wavenumber spectra of velocity fluctuation obtained from the MU radar observations on October 1986, where M, S, and T denote the mesosphere, the

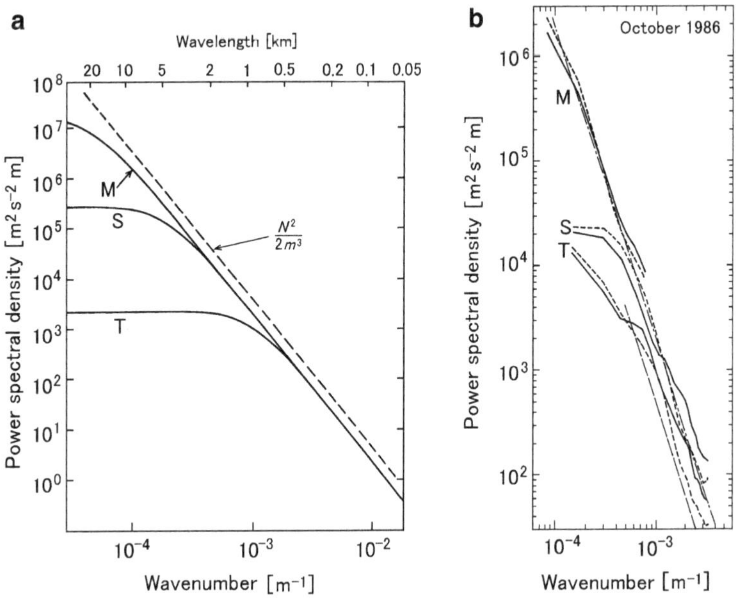

Fig. 12.22 Vertical wavenumber spectra of gravity waves. (**a**) Theoretical model and (**b**) obtained from the MU radar observations on October 1986, where M, S, and T denote the mesosphere, the lower stratosphere, and the troposphere, respectively. The zonal and meridional spectral curves are indicated by *solid* and *dotted lines*, respectively. The model saturated spectrum (Smith et al. 1987) is indicated by *straight lines dashed* for the troposphere and *dot-dashed* for the mesosphere and lower stratosphere, respectively [(from Tsuda et al. 1989b)]

lower stratosphere, and the troposphere, respectively. In the mesosphere, the shapes of the zonal and meridional spectra agree very well with the model, whereas both are slightly different in the lower stratosphere and in the troposphere. It is considered that gravity waves were not completely saturated in the observation period, because there is an annual modulation in the activity of gravity waves in the lower stratosphere, and the observations were performed in autumn when the activity level was low. Moreover, in the troposphere, due to the contribution of fluctuations other than gravity waves, the actual observed curve exceeded the theoretical model (Tsuda et al. 1989b).

12.3.5 Momentum Flux

The momentum flux of gravity waves are measured with the method of using two pairs of radar beams as shown in Sect. 7.3.1 (e.g., Reid and Vincent 1987). The MU radar has continuously observed wind systems at the altitudes between 60 and 85 km for several days a month. The momentum flux of gravity waves obtained with the above observations of 3 years between 1985 and 1988 are shown in Fig. 12.23, where (a) shows seasonal change and (b) height profiles. The upper panel

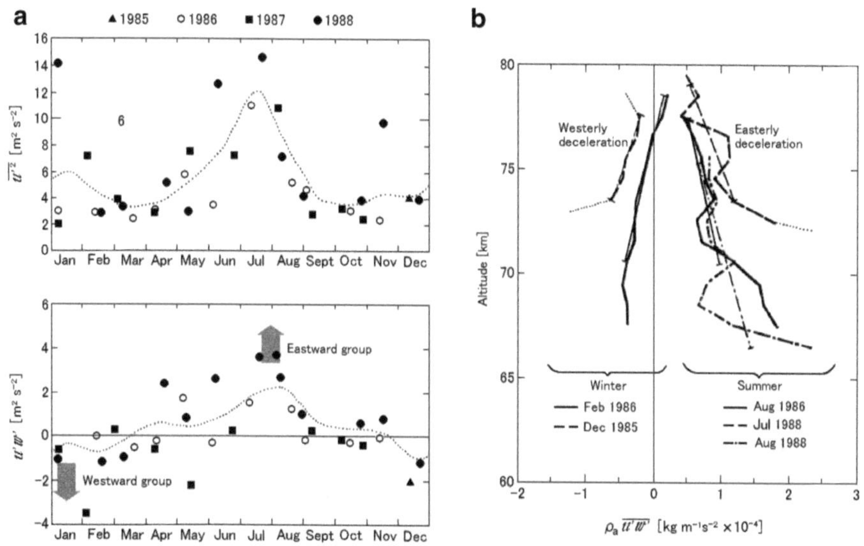

Fig. 12.23 Momentum flux of gravity waves obtained from the MU radar observations of 3 years between 1985 and 1988. (**a**) Monthly mean values of the variance of eastward line of sight velocity (*upper panel*) and the momentum flux (*lower panel*) with cycle of 5 min to 2 h for zenith angle of 10°. (**b**) Height profile of the momentum flux for each unit volume obtained by multiplying the density of standard atmosphere (CIRA-86). *Thin lines* show the best fit for vertical gradient of momentum flux [from Tsuda et al. 1990b]

of (a) shows monthly mean values of the variance of eastward line of sight velocity with cycle of 5 min to 2 h for zenith angle of 10°. In each year, the maximum values of 10–14 $m^2 s^{-2}$ are obtained in summer, and comparatively large values are also obtained in winter. On the other hand, the values become minimum in vernal and autumnal seasons. As a whole, the momentum flux changes at the cycle of half a year. The lower panel of (a) shows the momentum flux estimated with (7.37), where eastward winds are positive. It is shown in this panel that the momentum flux changes annually, eastward $(2\,m^2 s^{-2})$ in summer and westerly $(-1.5\,m^2 s^{-2})$ in winter. This means that distinguishing selections are performed with seasons when the gravity waves propagate from the lower layer to the upper layer.

Figure 12.23b shows the height profile of the momentum flux for each unit volume obtained by multiplying the density of standard atmosphere air (CIRA[5]-86). Thin lines show the best fit for vertical gradient of momentum flux. Its amplitude decreases with the increase of altitude, and it is suggested that the momentum flux transferred to background winds. Power to accelerate or decelerate mean wind velocities is calculated from the height gradient of the momentum flux. The values are eastward 7–13 $m\,s^{-1}\,day^{-1}$ in summer and westerly 8–11 $m\,s^{-1}\,day^{-1}$ in winter.

[5]The Committee on Space Research (COSPA) International Reference Atmosphere.

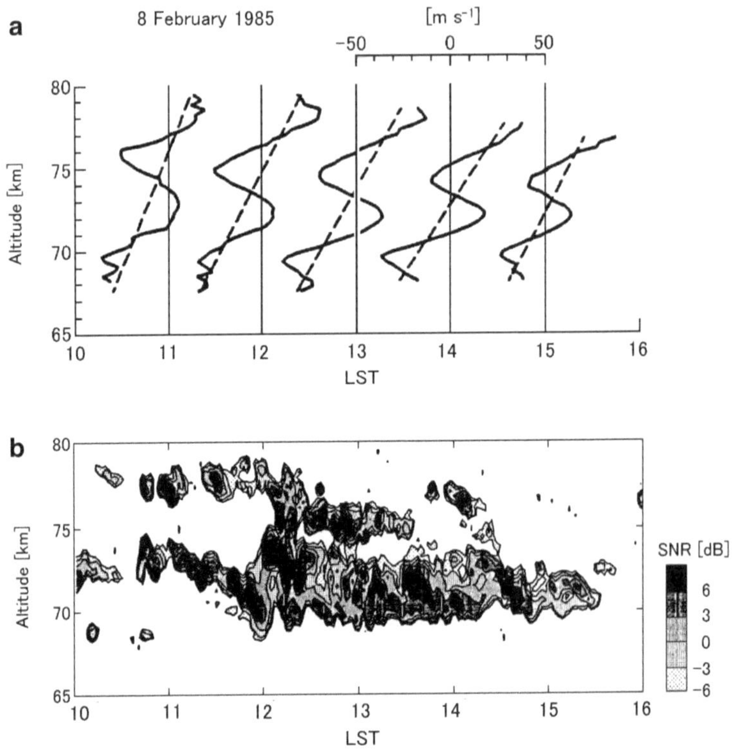

Fig. 12.24 **(a)** Northward wind profile in the mesosphere region observed by the MU radar. Gravity waves of vertical wavelength of several km to 10km are superimposed on the background wind radar. **(b)** Time-height profile of echo intensity observed by the MU radar at the same time zone as **(a)** [from Yamamoto et al. 1987]

Both works to decelerate mean wind, and the process is assumed to be the evidence that gravity waves give significant influence on the large-scale circulation.

12.3.6 Turbulence

The process with which atmospheric instability that originates in gravity waves generates turbulence can be caught in detail by observing echo intensity simultaneously with the wind system [e.g., Yamamoto et al. 1987, 1988; Muraoka et al. 1987]. Figure 12.24a shows northward wind profile in the mesosphere region observed by the MU radar. Gravity waves of vertical wavelength of several km to 10 km are superimposed on the background wind shear. Vertical shears between the height of 71–73 km are strengthened by this effect, and the height is descending with time. The same panel (b) shows the time-height profile of echo intensity at the same time zone. A strong scattering layer exists almost at the same height of the strong shear, and it is descending along with the phase propagation of gravity waves. Yamamoto et al. (1987) found that the Richardson number becomes zero and vertical shears of the frequency close to Brunt Väisälä frequency are generated in this region. It is estimated that a convective instability was generated and it caused small-scale velocity turbulence to bring out above phenomena.

Larger-scale waves such as atmospheric tides and planetary waves also actually change background winds. It is thought that atmospheric instability is caused when gravity waves are superimposed on these large-scale fluctuations and shear is strengthened locally, and that turbulence can then be generated . It is necessary to pay attention to the fact that the generation of turbulence is not uniform in time and space and thus the transport of heat, momentum, and material[6] are intermittent.

Long-term fluctuations of turbulence can be examined by analyzing long-term data with the method discussed in Sect. 7.3.2 (Fukao et al. 1994; Kurosaki et al. 1996). Figure 12.25a shows the seasonal change of vertical eddy diffusivity in the mesosphere region observed by the MU radar. The vertical eddy diffusivity is especially large in summer, and is also somewhat large at the altitude higher than 70 km in winter. These tendencies correspond to the seasonal change of gravity waves. It suggests that the generation of turbulence is due to the wave breaking of gravity waves. Moreover, though the height profile of the vertical eddy diffusivity shows wide dispersion as shown in Fig. 12.25b, the median value (solid line) of the turbulence is one digit smaller than the model value by Ogawa and Shimazaki (1975). There is some possibility that the dominant scale of the turbulence is larger than the scattering volume of the radar, and is not fully caught by atmospheric radars.

12.3.7 Wave Sources

Various observations that examine the wave sources of gravity waves are tried with atmospheric radar. The primary source is a variety of dynamic processes in the lower atmosphere. First, mountain waves that are generated on lee side when flows pass over mountains are a typical source (e.g., Nastrom and Fritts 1992; Lott and Teitelbaum 1993). Sato (1990) showed from the MU radar observations that vertical wind variations were intensified by topographically forced gravity waves caused by a strong westerly. A second source is shear instability in jet streams (e.g., Hines 1968; Lindzen 1984; Fritts 1984; Fritts and Nastrom 1992). It has been confirmed

---

[6]Strictly speaking, vertical eddy diffusivity K discussed in Sect. 7.3.3 is an isotropic eddy diffusivity of energy (temperature). This value is not necessarily corresponding to the eddy diffusivity of the material or the momentum in the same scattering volume. However, it has not come to distinguish the diffusivities of different physical values in the accuracy of current research, and they are thought to be almost equal order as the first approximation.

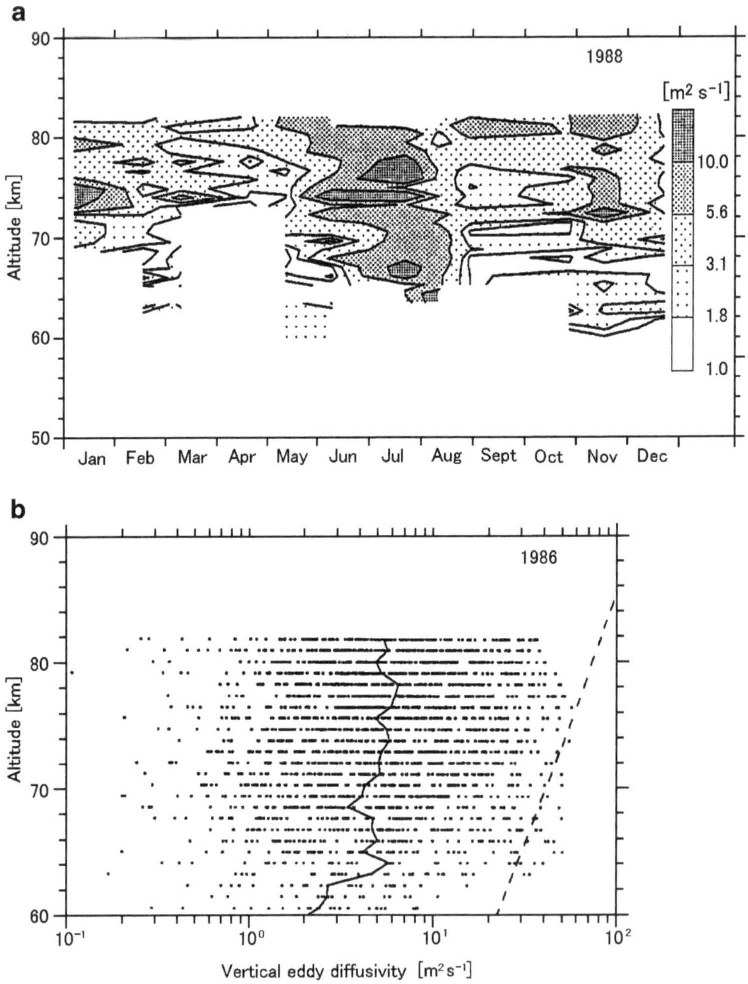

Fig. 12.25 (**a**) Seasonal change and (**b**) height profile of the vertical eddy diffusivity in the mesosphere region observed by the MU radar. In (**b**) the *solid line* shows the median value, and the broken line the model value by Ogawa and Shimazaki (1975) [from Fukao et al. 1994]

by Murayama et al. (1994) that gravity wave activity observed with the MU radar in the troposphere and lower stratosphere is well correlated with the intensity of the jet stream, strongest in winter, weakest in summer. Third, a good correlation has been noticed between gravity wave intensity and the passage of cold/warm fronts (e.g., Einaudi et al. 1978). A few observations have been presented from MST radars, in particular, with the SOUSY radar in Germany by Röttger (1979) and with the MU radar by Fukao et al. (1989) and Sato (1989). Among other gravity wave sources are convection [Townsend 1966; Röttger 1980b; Manzini and Hamilton 1993; Sato et al. 1995] and typhoons (Pfister et al. 1993), which have also been demonstrated by MST radar observations (Sato et al. 1991; Sato 1993; Dhaka et al. 2003). The generation of long-period inertia-gravity waves by synoptic-scale motions and jet streams is described by the theory of spontaneous adjustment that the wave motions appear as the result of permanent competition between the violation of equilibrium of pressure and velocity and the tendency of the atmosphere to establish the quasi-geostrophic balance of these fields [e.g., Gill 1982, Chap. 8; O'Sullivan and Dunkerton 1995]. The emission of meso-scale gravity waves is provided by meso-scale turbulent motions (e.g., Townsend 1965; Gavrilov 1992). The initial observations with MST radars were provided by Fritts and Luo (1992) and Gavrilov et al. (1999).

In the equatorial area where the wind is weak, mountain waves are not easily generated. Instead, it is thought that solar radiation energies, given to the atmosphere in the vicinity of the ground, are carried upward by means of active cumulus convections and latent heat release in the form of water vapors and generate gravity waves. However, observations are not yet enough. Despite the important role of gravity waves in formation of the general circulation, thermal regime, and composition of the middle atmosphere, their origin is still an issue in need of further investigation.

12.4 Boundary Layer and Equatorial Atmosphere

12.4.1 Boundary Layer

Recently, observations of the planetary boundary layer (PBL) of the Earth with BLRs (boundary layer radar) and T radars have been extensively analyzed, and significantly new information has been obtained. For instance, the low-level jet (LLJ) around the 400 m height was closely investigated (Zhong et al. 1996).

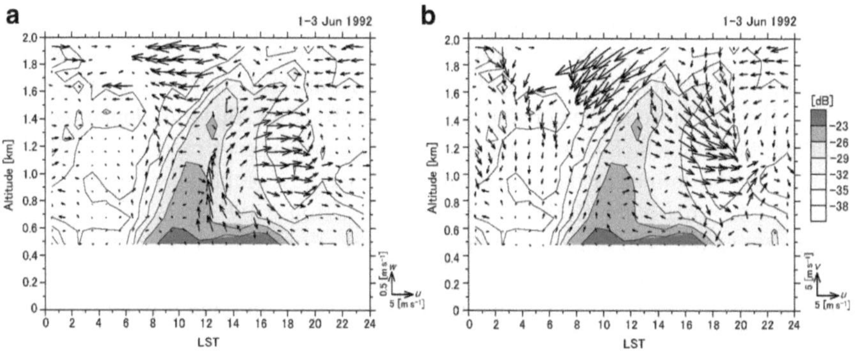

Fig. 12.26 The typical diurnal variation of the PBL. (**a**) Zonal-vertical and (**b**) zonal-meridional winds, averaged for 3 days during 1–3 June 1992 with upward and southerly winds positive. The darker the shading, the stronger is the echo intensity [from Hashiguchi et al. 1995a]

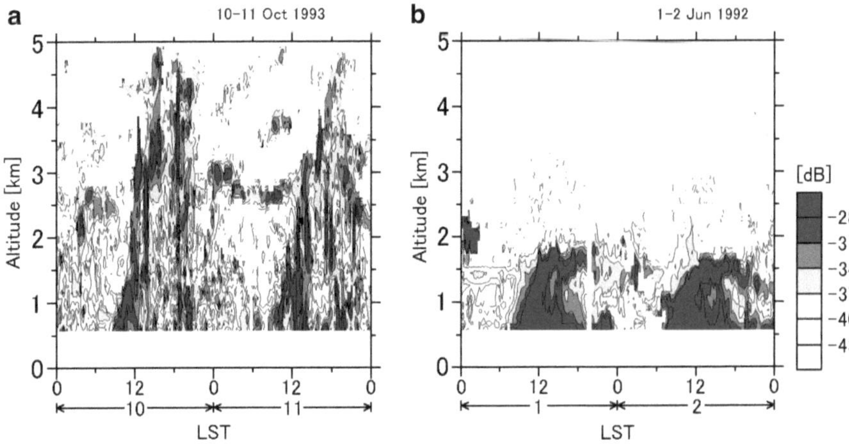

Fig. 12.27 Comparison of the echo intensities observed by the identical BLR at (**a**) Serpong Indonesia and (**b**) Shigaraki Japan. The darker the shading, the stronger is the echo intensity [from Hashiguchi et al. 1995a,b]

Figure 12.26 shows diurnal variations of the three components of the wind velocity which is overlaid on a time-height cross section of echo power multiplied by the square of the range (Hashiguchi et al. 1995a). Zonal-vertical and zonal-meridional winds, averaged over 3 days during 1–3 June 1992, are shown in panel (a) and (b), respectively. The typical diurnal variation of the PBL at the radar site on clear days is remarkably demonstrated in these panels. The enhanced echo region rapidly rose upward with time from around 07 LT, indicating that the PBL started to develop with height of the sun. Upward velocities much larger than the ascent speed of the echo layer (approximately 0.07 $\mathrm{m\,s^{-1}}$) were observed in the height range of 0.8–1.5 km near 08 LT when the PBL was most rapidly rising. Also, it is noted that northerly wind was dominated at heights above 1 km. Turbulence intensity estimated from the width of the Doppler spectra (Hocking 1985) was strong accompanying the strong updraft. This suggests that the PBL is well mixed by convection and turbulence.

Figure 12.27 compares the diurnal variations on clear days between a tropical and mid-latitude sites (Hashiguchi et al. 1995a,b). Panel (a) is for Serpong Indonesia (6.4S, 106.7E) close to the equator, while Panel (b) is for Shigaraki Japan, at mid-latitude, both of which were obtained with the same type of radar. If the PBL is idealized to be defined by the Ekman layer, the thickness of the layer is proportional to $f^{-1/2}$, where f is the Coriolis parameter (e.g., Houze 1993, p.67). The Coriolis parameter varies as the sine of the latitude. Thus, f becomes smaller and the PBL becomes thicker at lower latitude. The observed result clearly demonstrates that the thickness of the PBL in the tropical region is 3–5 km, although it is limited up to 1–2 km at mid-latitude.

12.4.2 Equatorial Atmosphere

The high temporal and altitude resolution data obtained by the Equatorial Atmospheric Radar (EAR) in Indonesia have contributed to improve our understanding of tropical convective systems. These systems significantly affect the circulation of energy, momentum and constituents (Fukao 2006). For instance, gravity waves generated by convection (Alexander et al. 2006; Dhaka et al. 2006), variations and fluctuations of vertical drafts (Mori et al. 2006; Nishi et al. 2007), and rain drop size distribution of precipitating clouds (Kozu et al. 2005; Renggono et al. 2006) have been extensively investigated. Moreover, detailed structure of the atmospheric temperature and humidity can be monitored with the EAR/RASS (Furumoto et al. 2006).

The tropical tropopause is considered a place where airs are exchanged between the troposphere and the stratosphere. Figure 12.28a shows the time-height section of the turbulence component of the spectrum width observed by the EAR during November 10–30, 2001. The spectrum width is proportional to the kinetic energy of turbulence, as described in Sect. 7.3, where the contribution to spectrum width from other spectrum width broadening mechanisms has been removed to obtain the turbulent component. The spectrum width due to turbulence in the height range between 15.5 and 17.0 km significantly increased during November 19–23, when the tropopause height (crosses in Fig. 12.28a) rapidly jumped from around 16 km to 17–18 km. Turbulence intensity became 6 times larger there than that in quiet periods. Zonal winds in the upper troposphere and lower stratosphere observed by the EAR are indicated in Fig. 12.28b. In the right panel, the average zonal wind profile measured by the EAR (eastward positive) during the same period as in (a) is indicated. In the left panel, the zonal wind anomaly from the average zonal wind of the right panel is indicated. In (a) and (b), the height of the tropopause indicated by crosses is defined by the temperature minimum measured by the on-site radiosondes (Fujiwara et al. 2003).

A significant zonal wind oscillation of $20\,\mathrm{m\,s^{-1}}$ was observed and identified as Kelvin waves. Jump of the tropopause was caused by the downward progression of the maximum eastward phase of equatorial Kelvin wave. The Kelvin wave was saturated and broken, and the convective turbulence was generated there. The wave breaking turbulence in the tropopause region merged into the turbulence below, and caused effective and irreversible transport of lower stratospheric air mass deeply into the troposphere (Fujiwara et al. 2003).

It has been found from the EAR observations that the intense vertical shear of zonal wind near the tropical tropopause consistently excites KH instabilities, which are considered quite effective in exchanging air mass between the troposphere and the stratosphere (Yamamoto et al. 2003a).

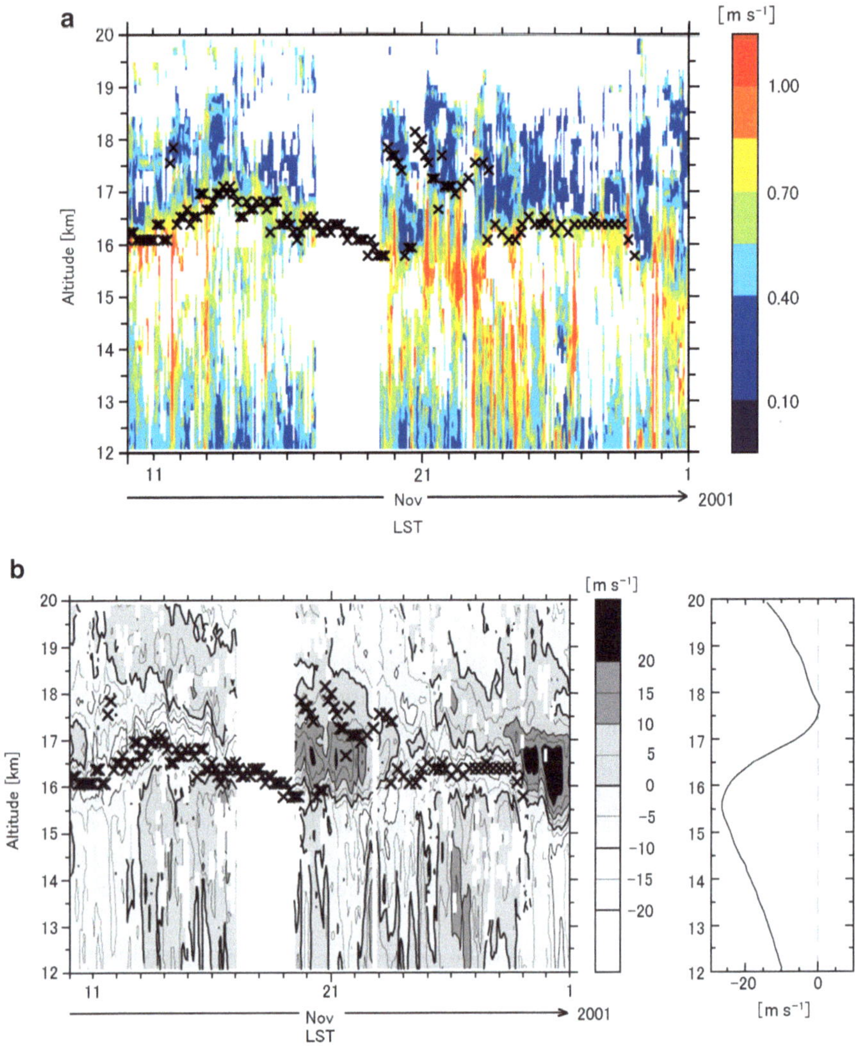

Fig. 12.28 (**a**) Time–height section of the turbulence component of the spectrum width observed in the northward beam of the EAR during 10–30 November 2001. No observation was conducted on 17, 18, and the first half of 19 November. Other *white* regions indicate no estimation is available due to low signal-to-noise ratio of the measurement. (**b**) *Right*: the average zonal wind profile measured by the EAR (eastward positive) during the same period as in (**a**). *Left*: the zonal wind anomaly from the average zonal wind of the *right panel*. In (**a**) and (**b**), the height of the tropopause indicated by *crosses* is defined by the temperature minimum measured by the on-site radiosondes [from Fujiwara et al. 2003]

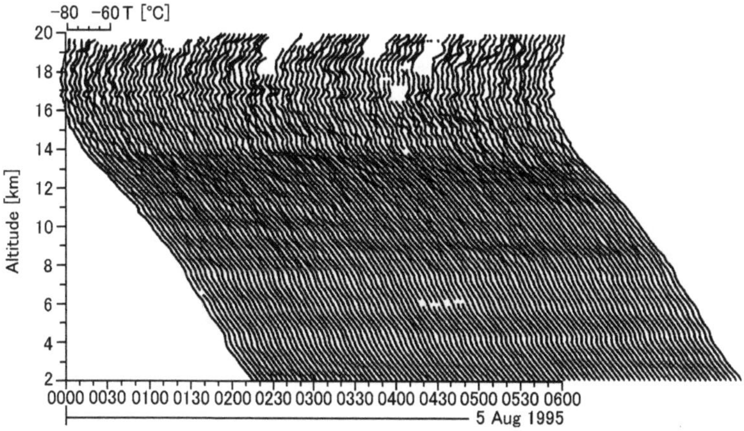

Fig. 12.29 Virtual atmospheric temperature profiles determined by the MU radar/RASS on 5 August 1995 [from Tsuda 2001]

12.4.3 Atmospheric Temperature and Water Vapor Content

Atmospheric radar has been already highly regarded as a quasi-standard tool for measuring wind velocity aloft. Efforts have been continued for establishing operational networks of atmospheric radars. In addition, the radio acoustic sounding system (RASS) technique has been widely incorporated in many atmospheric radars, enabling to monitor the atmospheric temperature and moisture from the surface.

Atmospheric Temperature Measurement

The operational measurement of atmospheric temperature by radiosonde is conducted by meteorological bureaus, in general, twice or four times a day. However, the recent incorporation of RASS in atmospheric radars has made the temperature measurements feasible with the temporal resolution extremely higher than radiosonde measurements. Figure 12.29 shows the (virtual) temperature profiles determined by the MU radar/RASS on 5 August 1995 (Tsuda 2001). The height resolution is 150 m. They are successively determined and shown in the diagram every 3.69 min. The temperature scale corresponding to the first profile is given in the top left of the diagram. The temperature lapse rate up to the tropopause is, on the average, 6.67 $^{\circ}\mathrm{C\,km^{-1}}$. There are fluctuating components of temperature. It has been revealed that superposition of many gravity waves with different periods, which include the dominant long-period components with period of 6–10 h and short-period components in small amplitudes, produced the complicated structure. The observed temperature profiles obtained with RASS agree quite well with the

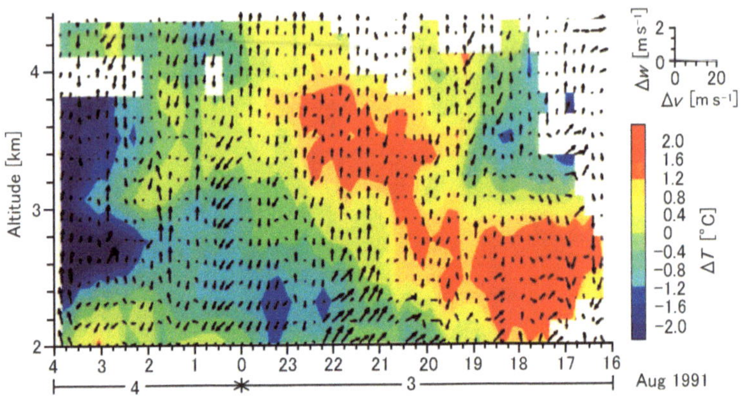

Fig. 12.30 A baseline-height structure of cold front measured with the MU radar/RASS during 3–4 August 1991. The fluctuating component (ΔT) of the temperature from the mean value during the observation period is shown by color contour, while the fluctuating components of the northward and vertical wind components (Δv, and Δw, respectively) from the mean values are indicated by vectors [from Tsuda 2001]

virtual temperature derived from radiosonde measurements in the height range below 7 km where the mean difference between the temperatures was approximately 0.3 °C (Tsuda et al. 1994).

The atmospheric wind velocity and temperature are supplemental, and their data that are simultaneously obtained are especially important to understand any meteorological phenomena. The vertical profiles of fluctuations of temperature and wind from the mean values during the observation period associated with the cold front which passed over the MU radar are shown in Fig. 12.30 (Tsuda 2001). The time axis in the diagram points to the right following convention. Hence, the time-height cross section approximately resembles a baseline-height cross section of the phenomenon which passed over the MU radar from the west to the east. The overall wind and temperature structures which represent the cold front are clearly demonstrated, and it is also obviously shown that cold air crawled under warm air, generating the intense updraft at its tip.

Estimates of Humidity

Humidity is related to absorption and release of latent heat energy, and is one of the most important parameters for meteorological studies. It is known that almost all water vapor exists in the lower part of the troposphere below 5 km in altitude, and varies significantly from time to time and from place to place. Figure 12.31 compares (b) the relative humidity measured by the MU radar with time and height resolution of 10 min and 150 m in the period from 29 to 30 July 1999, and (a) that observed with radiosondes launched approximately every 3 h during the same

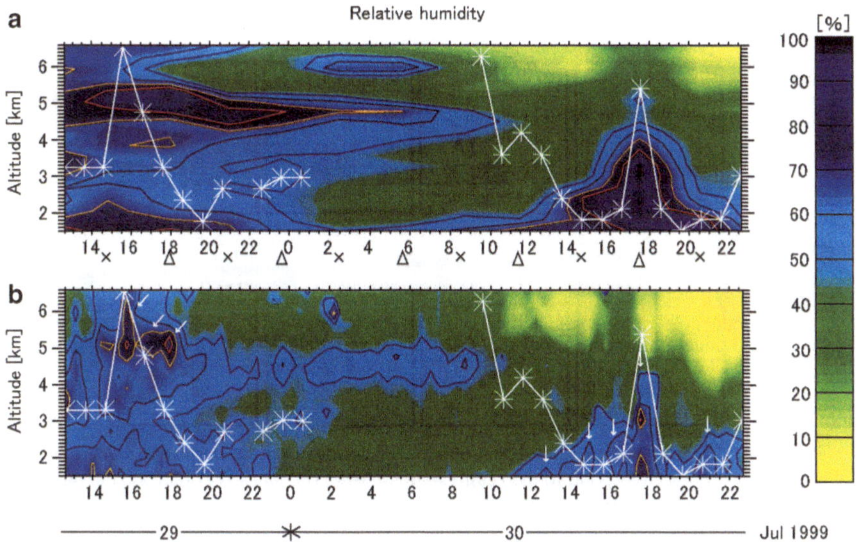

Fig. 12.31 Time-height sections of the relative humidity observed during the period from 1240 on 29 July to 2300 LT on 30 July 1999. (**a**) Results with radiosondes, and (**b**) that by the MU radar. Launching times of the radiosondes are indicated by *crosses* and *triangle* in (**a**) [from Furumoto et al. 2003]

observation period. The marks (crosses and triangle) in the bottom of (a) indicate the time of radiosonde launches (Furumoto et al. 2003).

The data of radiosondes launched at triangle were used as the initial guess necessary for the MU radar estimate. Both results shows the overall consistency. Owing to the high time resolution, the MU radar results contain a short-period (1–1.5 h) fluctuation that was not detected by radiosonde observations and considered to be of gravity wave origin. Moreover, the symbols *asterisks* and white lines in the diagram show changes of the cloud top height observed by the meteorological satellite, which show a good correlation with the distribution of humidity obtained by the MU radar.

12.5 Beam Swinging and Radar Imaging Techniques

A variety of radar imaging techniques have been developed in rapid succession. Some of the observational results obtained by applying these advanced techniques are reviewed in the following sections.

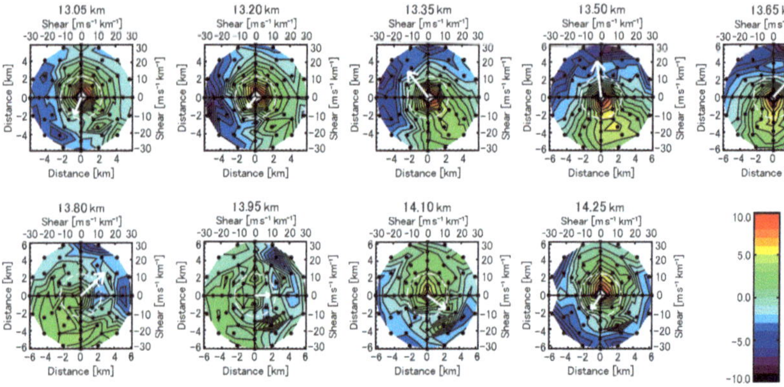

Fig. 12.32 Horizontal maps of echo power from 13.05 to 14.25 km averaged over the period from 2239 to 0046 LT. *White arrows* indicate the maximum wind shear direction and magnitude. The *white dashed circles* correspond to the zenith angle of 10°. The *black dots* indicate the directions of the radar beams [from Hirono et al. 2004]

12.5.1 Scattering Layer Observations

Observations with Beam Swinging

Studies on the detailed structure and dynamics of the atmosphere can be made by observations of the scattering layers using the radar beam swinging technique. For example, the bean swinging technique can be used for estimating the aspect sensitivity of the radar echo power (defined in Sect. 3.4.3). By measuring the aspect ratio (i.e., the ratio of the echo power between the vertical and oblique beams), it was demonstrated that ground based radars are the excellent tools for measuring the altitude of the thermal tropopause (e.g., Gage and Green 1979; Yamamoto et al. 2003b).

Moreover, several recent works focused on the azimuthal dependency of the radar echo power (e.g., Tsuda et al. 1997; Worthington et al. 1999; Hirono et al. 2004). Figure 12.32 shows an example of the horizontal distribution of the radar echo power at different altitudes (above the tropopause) using the fast-beam steering capability of the MU radar (Hirono et al. 2004). The observations were performed with 65 beams successively steered from the zenith down to 24° off zenith at different azimuths. The total acquisition time was about 10 min. Six consecutive maps were averaged to get a horizontal map in approximately 1 h of observations. The figure clearly indicates a zenithal and azimuthal dependency of the echo power. On each panel, the arrows indicate the direction and modulus of the vertical shear of horizontal wind. It can be seen that the shear is rotating clockwise with increasing altitude. As already mentioned, the rotation of the wind vector is supposed to result from the variations of the horizontal wind due to gravity waves, which transport energy and momentum upward. From this figure, it is also interesting to note that larger echo power area appears in the opposite direction to the maximum wind shear direction. This property is observed not only within the lower stratosphere but also in the troposphere (e.g. Worthington et al. 1999). It was suggested that the azimuthal dependency of the echo power results from atmospheric shearing instabilities generated by low frequency gravity waves tilting the scatterers in the opposite direction to the vertical shear of horizontal wind.

Interferometry and Radar Imaging Observations

Numerous studies on the detailed radar echo structure have been made from the application of the spatial domain interferometry (SDI), frequency domain interferometry (FDI), coherent radar imaging (CRI), and range imaging (RIM) techniques (see Sect. 7.6).

1. Spatial domain interferometry (SDI)

Palmer et al. (1991) studied the advantage of the SDI technique over the DBS technique in measuring vertical wind velocity. The SDI technique based upon the expression (7.113) and the DBS method using the vertical beam were compared. They showed that tilted scattering layers prevent the estimation of the correct vertical wind velocity with the DBS method when the scattering layers are tilted even slightly.

2. Frequency domain interferometry (FDI)

The dual FDI technique provides its effectiveness in improving the range resolution. Figure 12.33 shows observations with the MU radar using the dual FDI and DBS techniques alternately (Kilburn et al. 1995). The left panel shows a time-height cross section of echo power from the middle troposphere to the lower stratosphere in the DBS mode at a range resolution of 300 m. The right panel shows the corresponding time-height cross section of echo power after dual FDI processing with a 500-kHz frequency separation. For example, it can be noted that two thin scattering layers are detected with the FDI technique around the altitude of 6.45 km, whilst they were not distinguishable by the standard range resolution of 300 m. Their vertical apparent motions are also more clearly visible. The FDI technique also sheds light on sporadic echoing structures e.g., around 8.0 km in Fig. 12.33 and on the thin multi-layering of the atmosphere that can persist for several consecutive hours without significant changes in appearance. These observations are important for understanding the mechanisms of constituent transport through turbulent diffusion. However, due to the hypotheses used (a single Gaussian-distributed scattering layer) for applying the dual FDI technique, this technique just provides apparent structures consistent with the model used. Consequently, we must be cautious when interpreting the results of dual FDI observations. Range imaging techniques using multiple frequencies are now more frequently used.

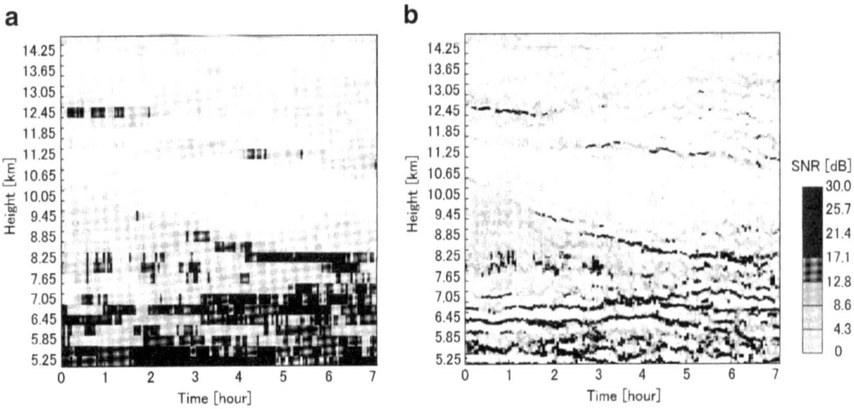

Fig. 12.33 Signal-to-noise ratio (SNR) observed by the vertical beam of the MU radar. The abscissa shows hours after the start of observation. (**a**) The time-height cross section obtained in the DBS mode at a range resolution of 300 m. (**b**) Corresponding time-height cross section after a dual FDI processing with a 500-kHz frequency separation was applied [from Kilburn et al. 1995]

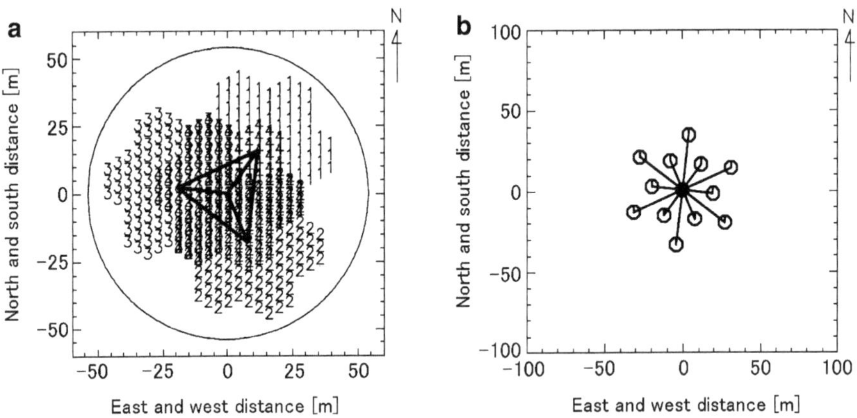

Fig. 12.34 (**a**) The locations of the individual antennas, where the numbers represent the receiver to which the antenna signal is fed. The three outer receivers form an equilateral triangle, while the fourth was placed at the center of the array. (**b**) The spatial sampling resulting from the receiver locations is depicted. The length scales are different between (**a**) and (**b**) [from Palmer et al. 1998]

3. Coherent radar imaging (CRI)

CRI using multiple receivers is a technique providing a detailed view of the horizontal distribution of the echo power within the radar receiving beam. Figure 12.34a shows an example of arrangement of the MU radar receiving antenna subarrays for applying the radar imaging technique with Fourier-based and Capon's methods. The subarrays numbered from 1 to 4 refer to the antenna subarrays in the same size used for the reception.

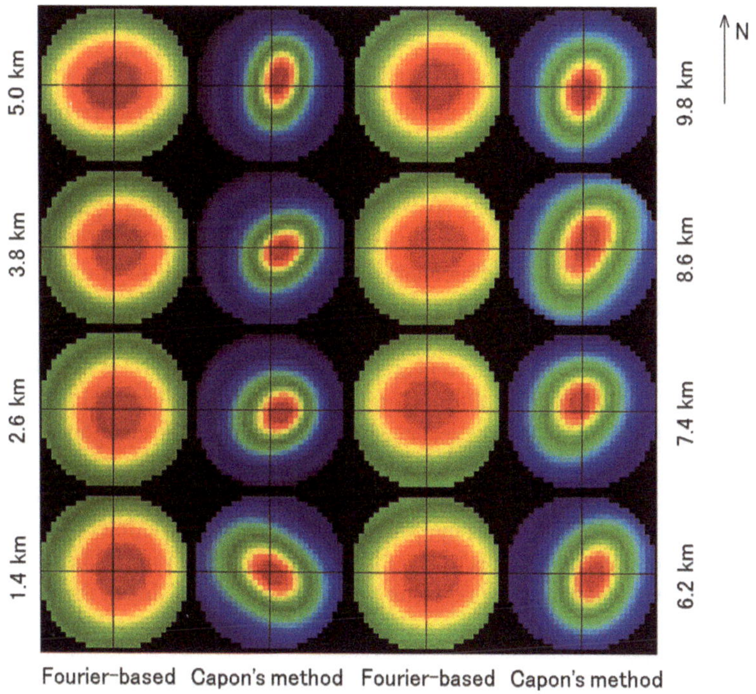

Fig. 12.35 Brightness distribution averaged over the velocity range $\pm 1\,\mathrm{m\,s^{-1}}$ within $\pm 4°$ above the MU radar at eight arbitrarily selected altitudes obtained using (*left*) Fourier-based images and (*right*) Capon imaging techniques, respectively [from Palmer et al. 1998]

The three outer subarrays form an equilateral triangle with separation of 33.97 m. The fourth subarray at the center is separated by 19.62 m from the other three. These baseline lengths correspond to angular aliasing of $\pm 5.44°$ and $\pm 9.44°$, respectively. However, because the full antenna is used for transmission giving rise to a beamwidth of 3.6°, substantial angular aliasing can be neglected. According to the disposition of the receiving antennas, Fig. 12.34b shows the spatial sampling according to the receiver locations. The vectors represent lengths and azimuth angles of each baseline, providing a view of how well the ground refraction pattern is sampled.

Figure 12.35 shows examples of horizontal maps of brightness distribution averaged over the velocity range $\pm 1\,\mathrm{m\,s^{-1}}$ within $\pm 4°$ above the MU radar, with the CRI technique using Fourier-based and Capon iamging techniques. Red (purple) colors correspond to the largest (smallest) brightness values. It should be noted that the midpoint between the peaks of the images corresponding to positive and negative velocities should be equivalent to the location of peak observed over the entire frequency range (Palmer et al. 1998). Even if Fourier and Capon brightness distributions are both maximum near the zenith, strong differences must be recognized. It is noticed that Capon's method exhibits better resolution than the

Fourier-based method. In particular, the brightness maps obtained from Capon's method show narrower maxima, generally located slightly off zenith and with a clear azimuthal anisotropy, suggesting quasi partial reflection from slightly tilted layers. The Fourier method is limited in angular resolution by its frequency response of the filter (a sinc function).

CRI with Capon's method (hereafter called the CRI Capon's method) was also applied at ultrahigh frequency (UHF) for three-dimensional imaging. Palmer et al. (2005) showed some impressive results with the 915-MHz turbulent eddy profiler (TEP, see Sect. 10.4.2) radar which consists of a transmitting horn antenna and a receiving array up to 64 microstrips hexagonal patch elements separated by about 0.57 m. The results shown in Fig. 12.36 with the CRI Capon's method were obtained from the combination of 37 receivers at a range resolution of 33.3 m (a pulse width of 0.222 μs). Using a Doppler filtering procedure, clear air and precipitation echoes could be separated from the Doppler spectra calculated from the complex time series reconstructed with Capon filter in an angular domain of $\pm 12.4^\circ$. Figure 12.36 shows a series of images within the cone of observations at a time resolution of about 10 s over about 2.5 min. Yellow and blue colors show the 3D distribution of the clear air and precipitation echoes, respectively, inside the radar beam up to the altitude of 1.8 km. These observations should be useful for the understanding of the interaction between clear air turbulence and the formation or inhibition of some precipitations due to turbulence.

A composite figure from several time-height cross section of echo power after the CRI Capon processing with the 915-MHz TEP radar is shown in Fig. 12.37 (Palmer et al. 2005). Panel (a) shows 1-h observations of vertical echo power revealing both clear-air and precipitation echoes (total echo power). Precipitation echoes are so strong that clear air echoes could not be distinguished (e.g., around 2035 UTC near the ground). After high-pass and low-pass filtering procedure, Panels (b) shows the time-height cross section of precipitation echo power before the imaging process (precipitation echo power), and (c) the echo power from refractive index perturbations after CRI Capon processing (Bragg scatter echo power). The refractive index echoes reveal some structures which resemble KH billows, especially after 2045 UTC.

Section 7.6.2 explained the capability of the CRI Capon's method to reject ground and moving clutter. Cheong et al. (2006) experimentally demonstrated this capability with the 915-MHz TEP radar using 57 receiving patch elements. Figure 12.38 shows time-height cross section of radar echo power and radial velocity associated with refractive index perturbations. Panels (a) and (b) show the results after the Fourier and Capon processing, respectively. Panel (a) shows the presence of sporadic echoes which can mainly be attributed to birds (or other biological targets). A possible KH instability can also be noted around the altitude of 500 m before 0200 LT. Because Capon filter is adaptive, sidelobes are reduced in the direction of the unwanted outliers. Panel (b) shows the result after Capon processing and it is clearly seen that almost all the biological interferences have been removed. Similar experiments which demonstrate remarkable clutter suppression were conducted with the MU radar using 25 receiving sub-arrays combined to range image with

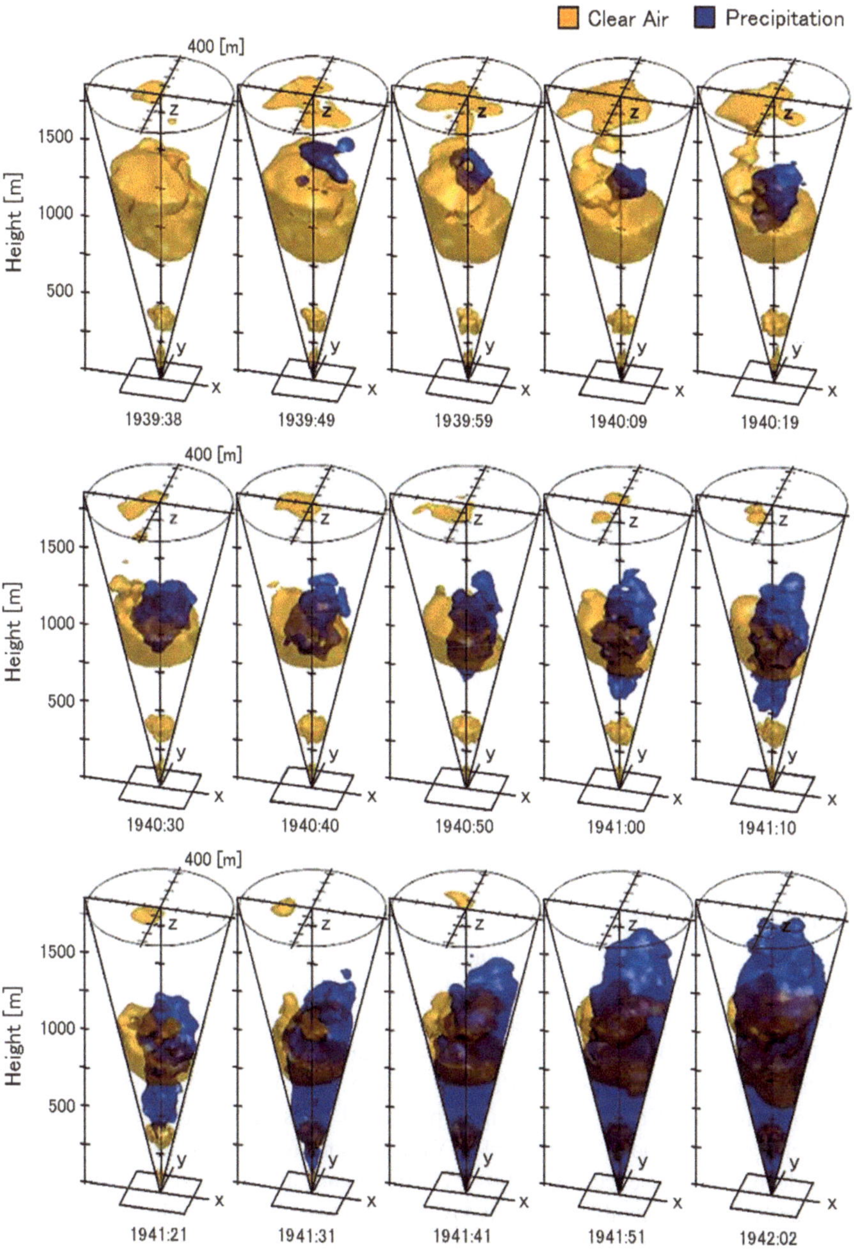

Fig. 12.36 Conical images obtained every 2.5 min (21 June 2003, 1939:38–1942:02 UTC) with the CRI Capon's method. The clear air and isolated precipitation echoes are shown in *blue* and *yellow colours*, respectively [from Palmer et al. 2005]

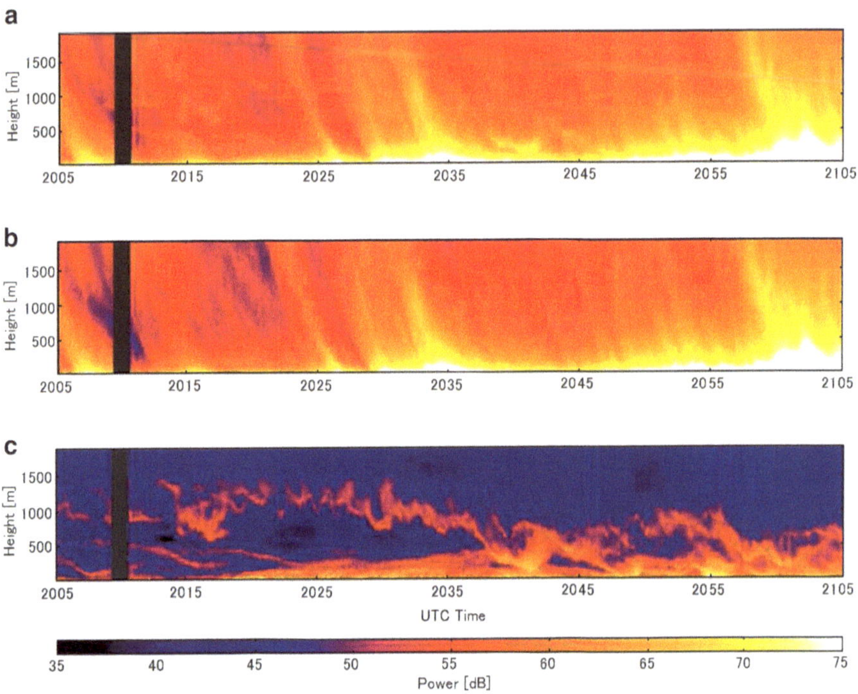

Fig. 12.37 Time-height cross section of total echo power during 2005–2105 UTC on 23 June 2003. (**a**) Total echo power, (**b**) precipitation echo power, and (**c**) Bragg scatter echo power [from Palmer et al. 2005]

5 frequencies by Yu et al. (2010). Figure 12.39 (a) shows the time-height cross section of echo power obtained in the standard mode at a range resolution of 150 m with the MU radar. Panel (b) shows the corresponding plot after applying CRI with 25 receivers and RIM with 5 frequencies and Capon processing. The contamination by moving clutter is apparent in (a) at heights above approximately 6 km, whereas (b) clearly shows that moving clutterer echoes are strongly attenuated.

Hélal et al. (2001) also applied angular radar imaging by using the sequential postset beam steering mode (SPBS) technique and various processing methods as Fourier-based, Capon, and the multiple signal classification (MUSIC) with the French Provence radar. They used a Yagi antenna for transmission resulting in a wide beamwidth exceeding 100°. The receiving antennas were composed of 16 bands of 26 coaxial-collinear (COCO) antennas. With a single receiver, they collected the signals consecutively from 16 receiving arrays and estimated the brightness distribution in a wide angular domain in a given vertical plane. The results with the Fourier method are shown in Fig. 12.40. The figure reveals not only the vertical distribution of multiple scattering layers but also clear information about the nature of the atmospheric layers. For example, the echoing layer around the

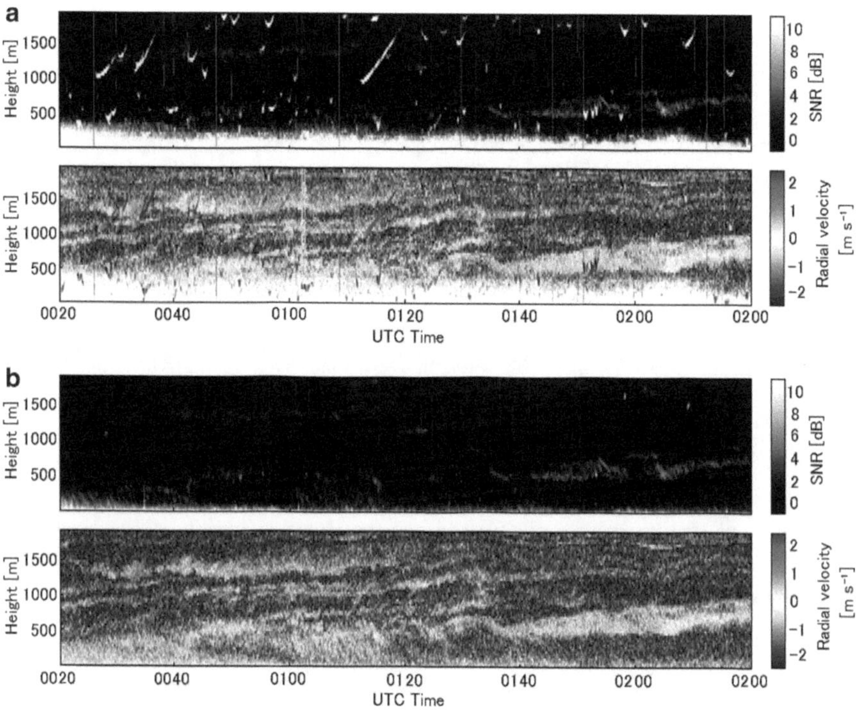

Fig. 12.38 Time-height cross sections of vertical echo power and radial velocity. (**a**) Generated with Fourier processing, while (**b**) with Capon processing [from Cheong et al. 2006]

altitude of 3,600 m exhibits a large horizontal extent likely due to the presence of a layer of refractive index perturbations at the Bragg scale.

Another interesting echoing structure can also be observed around the altitude of 4,800 m. Contrary to the previous case, the horizontal extent of the echoing layer does not exceed the horizontal extent corresponding to the beamwidth of the full antenna array (about $\pm 5°$). At this altitude, the mechanism of partial reflection from a horizontal layer is very likely the cause of this echo. The SPBS technique also provides the radial velocity. An example of vertical cross section of radial velocity along the North–South axis obtained with the Provence radar is shown in Fig. 12.41 (Hélal et al. 2001). As expected, the pattern is nearly symmetric with respect to the vertical line of sight (not exactly because of vertical wind). It also shows some altitude variations due to the vertical variations of the horizontal wind. Such a technique would be useful for studying the small-scale inhomogeneities of the wind field.

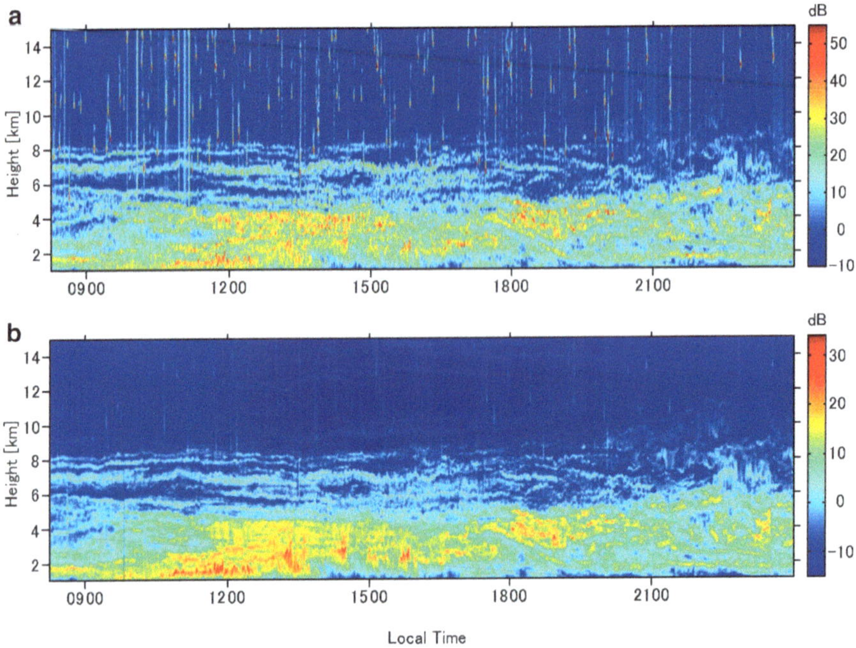

Fig. 12.39 (**a**) Time-height cross section of echo power obtained in the standard mode at a range resolution of 150 m with the MU radar. (**b**) The corresponding plot after applying CRI with 25 receivers and RIM with 5 frequencies and Capon processing [from Yu et al. 2010]

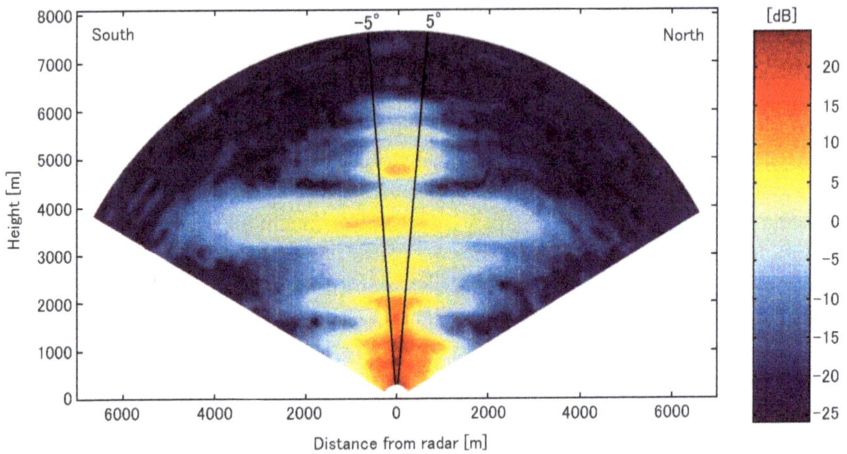

Fig. 12.40 SNR (in decibels) obtained with the Province radar in France on 4 May 1999. The observed zenith angles are between −60 and 60°, while the main lobe half-power beamwidth is indicated by the radial lines at −5 and 5° [from Hélal et al. 2001]

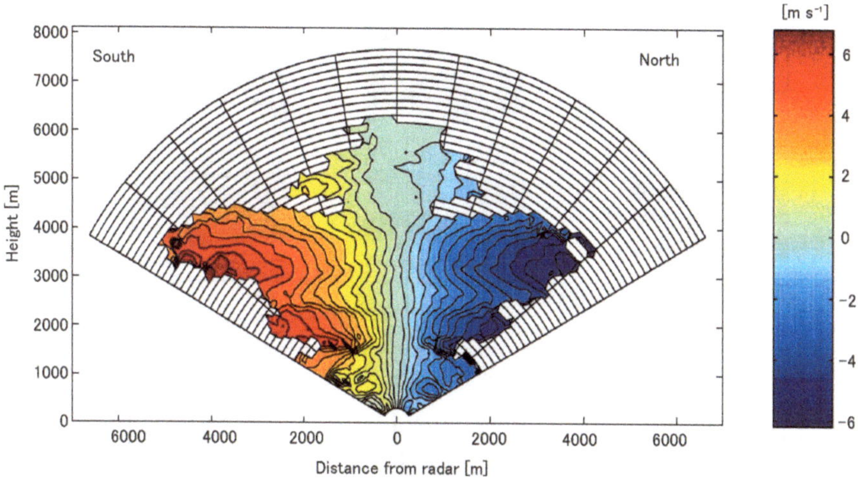

Fig. 12.41 Radial wind velocities in the same configuration as Fig. 12.40. The background grid indicates the size of the cells for which the velocities are estimated [(from Hélal et al. 2001)]

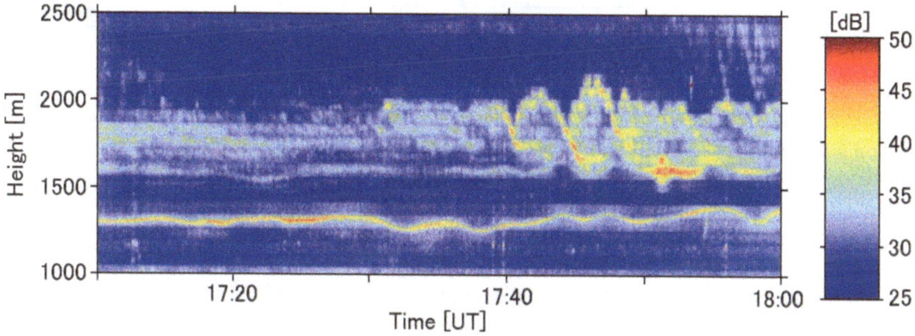

Fig. 12.42 Brightness (in decibels) distribution after applying Capon's method with the 915-MHz radar at Platteville, the U.S.A. [from Chilson et al. 2003]

4. Range imaging (RIM)/Frequency domain interferometric imaging (FII)

Applications of range imaging for improving the range resolution were conducted as shown below [RIM; Palmer et al. 1999, FII; Luce et al. 2001b]. Figure 12.42 shows the vertical height-time cross section of echo power after doing the range imaging technique with the 915-MHz Platteville radar (the U.S.A.) for about 1 h around 1730 UTC in the height range 1.0–2.5 km (Chilson et al. 2003). From 1740 to 1800 UTC and below the altitude of 2.0 km, echoing structures similar to those detected by FMCW radars (see Sect. 11.6.2) are clearly seen. It is very likely that these structures, also reported by FMCW radars, result from KH instabilities. Also, a long lived thin scattering layer can be noted around 1.35 km. Its thickness was about 12 m and can be related to a thin and strong gradient of temperature (and/or humidity). Such a thin layer and its small apparent vertical oscillations cannot be detected at the initial range resolution of 75 m.

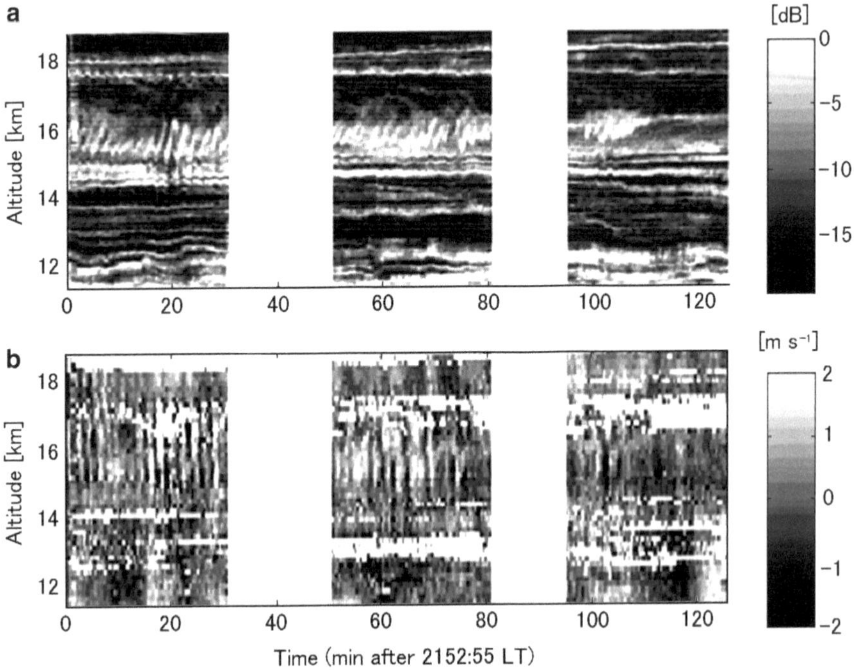

Fig. 12.43 (**a**) SNR (in decibels) obtained with the vertically oriented beam and processed by Capon's method between 11.4 and 18.9 km during 2152 and 2358 LT on 13 November 2005. (**b**) The corresponding plots of radial velocity measured with the vertical beam at a range resolution of 150 m [from Luce et al. 2008]

Another example of KH billow observations from the MU radar in the FII mode can be seen in Fig. 12.43. Panel (a) shows vertical time height contours of Capon brightness (SNR) for about 2 h and in the height range 11.4–18.9 km (Luce et al. 2008). It is one of the first direct observations of KH billows with a VHF (46.5 MHz) radar. For these observations, the subpulse width was 1 μs (i.e., the initial range resolution was 150 m), five equally spaced frequencies were transmitted pulse to pulse from 46 to 47 MHz and the time resolution was about 33 s with sampling every 16.5 s. The corresponding time-height cross section of vertical wind velocities are shown in Panel (b). It reveals vertical oscillations which are fully consistent with the presence of a KH instability. The image of radar echo power shows slanted "S" structures or braids characteristics of vortices or billows. The braids result from the stretching and compression of the isentropes at the edges of the billows. As in Fig. 12.43, the image reveals multiple thin scattering layers in the tropopause region. Some of them (in particular those located near the billows) exhibit vertical oscillations and temporal variations of echo power likely due to tilted facets of laminar structures (e.g., just above 14 km below 18 km). The distortions of these layers are very likely due to gravity waves propagating vertically and generated

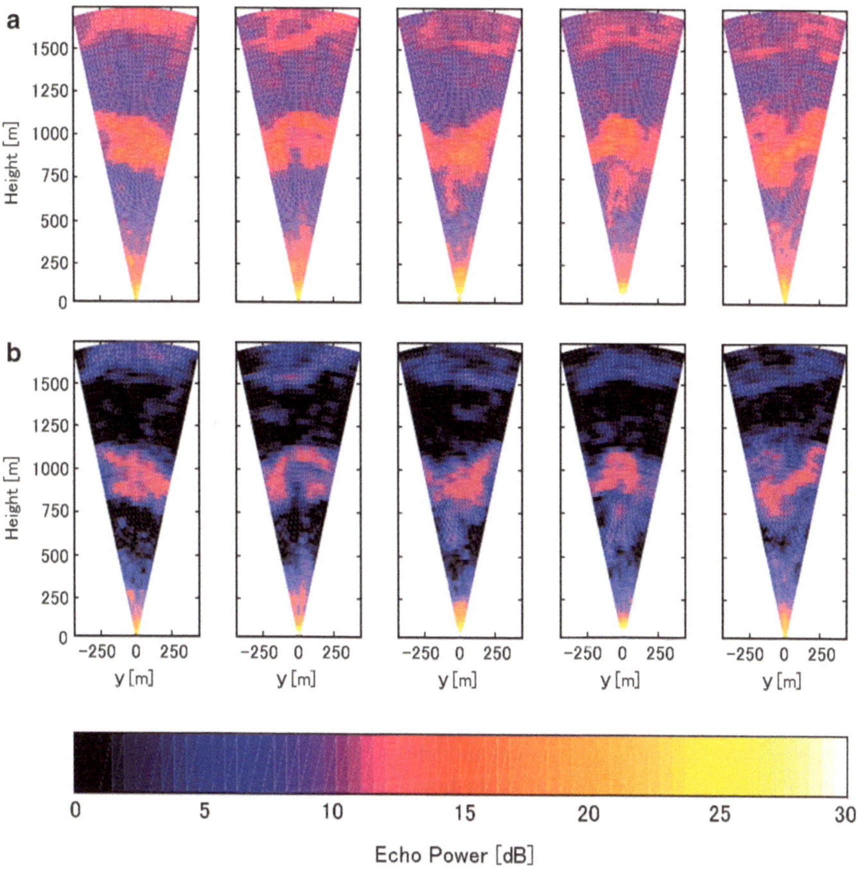

Fig. 12.44 Comparison between measured powers processed with (**a**) Capon's method and (**b**) the ML method using the 915-MHz TEP radar [from Le et al. 2010]

by the KH billows. These structures are associated with a quite strong shear of horizontal wind of about 35 $\mathrm{m\,s^{-1}\,km^{-1}}$ due to the association of the decrease of the jet-stream and the presence of an inertia-gravity wave (Luce et al. 2008).

5. Multi-lag method

As shown in Sect. 7.6.2, the results of Capon's method depend on the SNR. Thus, improvements are not expected with respect to the Fourier method for very low SNR. Recently, Le et al. (2010) proposed the multilag (ML) method based on the 1- and 2-lag covariance matrix estimate. Figure 12.44 shows a comparison of (a) Capon's method and (b) the ML method with the 915-MHz TEP radar. A better dynamics is obtained with this new method, which is promising for detecting atmospheric structures at low SNR.

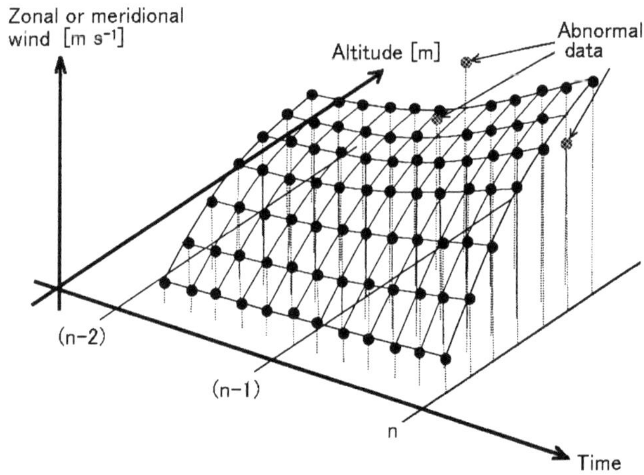

Fig. 12.45 Conceptual diagram of the quality control algorithm for the WINDAS [courtesy of the Japan Meteorological Agency]

12.6 Wind Profiler Network

Recognizing the unique capability of wind profilers to monitor wind aloft continuously in time and height, attempts have been made to integrate wind profiler observational data into numerical models of atmospheric general circulation to improve the accuracy of large scale field forecasting and simulation. The Wind Profiler Data Acquisition System (WINDAS) of the Japan Meteorological Agency (JMA), described in Sect. 10.4.4, is operated with the goal of improving short-term forecast of meso-scale weather. With the introduction of WINDAS, the data of upper wind taken into the Meso-Scale Model (MSM) has increased rapidly. The WINDAS data are used as an initial value in the MSM. The data assimilation in the MSM is made using a full forecast-analysis system with the four dimensional variational (4D-VAR) method. This is the technique to assimilate observations having high time and spatial resolution into numerical models with the aim of improving the accuracy of model initialization.

12.6.1 Quality Control and Actual Operation

When wind profiler data are employed for actual use, the quality control that excludes abnormal data is significantly important. As discussed in Sect. 8.4.2, it is reported that various data errors are caused by ground clutter, aircrafts, migrating birds, and swarms of insects (e.g., Wilczak et al. 1995) In the WINDAS, the quality control algorithm shown in Fig. 12.45 is applied in consideration that wind variance in time and height is within expected values based upon past experience. That is,

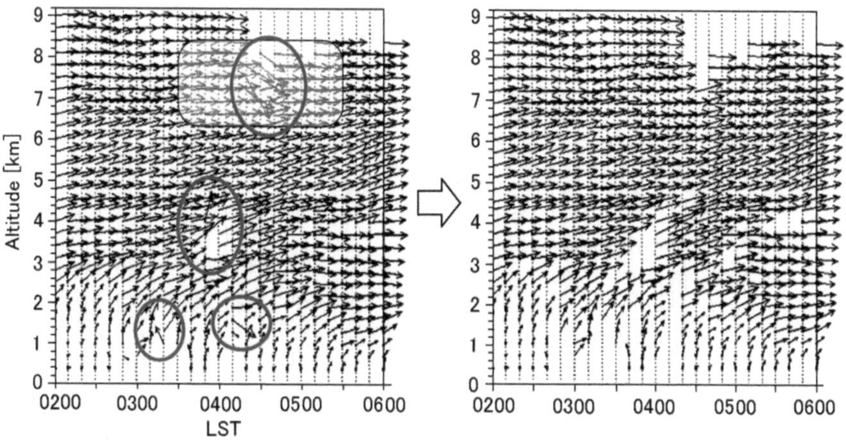

Fig. 12.46 Height-time cross section of horizontal winds before and after the application of the Q-surface quality control program. Data in enclosures in the *left panel* is judged to be abnormal, and excluded and replaced as shown in the *right panel* [courtesy of the Japan Meteorological Agency]

the quadric surface (Q-surface) as the mean value is first decided by using the data of 12 time-points (2 h by every 10 min) in seven adjoined altitudes (1.8 km height zone by every 300 m) according to the least squares method. Next, the continuity of horizontal velocity of the wind is assumed, and if the difference between the actual data value and the Q-surface exceeds a threshold, that data is deleted and replaced with one that fits on the Q-surface. Figure 12.46 shows an example of the height-time cross section of horizontal winds before and after the application of above quality control program. Data in enclosures in the left panel is judged to be abnormal, and excluded and replaced as shown in the right panel.

One of the performance measure of wind profilers is the attained height in observation. Figure 12.47 shows the result of the WINDAS LTR observation during April 2002 to March 2003 in the standard observation mode (height resolution of 300 m) (Ishihara et al. 2006). Data of 25 points are averaged, and the results are shown in three groups, precipitation, no rain, and all of them. As for the attained height in no rain, the maximum value of monthly averages is 6.7 km in summer, and minimum 3.8 km in winter. In precipitation, the values are 1.5–2.0 km higher than those in no rain in each month. This is a reflection that the intensity of echoes from Bragg scatterers in the troposphere depends on the water vapor content.

12.6.2 Application for Short-Term Forecasting

An impact experiment, to examine the effect that the WINDAS data brings to numerical forecasting, was conducted with the MSM using the WINDAS data and the 4D-VAR method for a severe rainstorm that occurred in western Japan

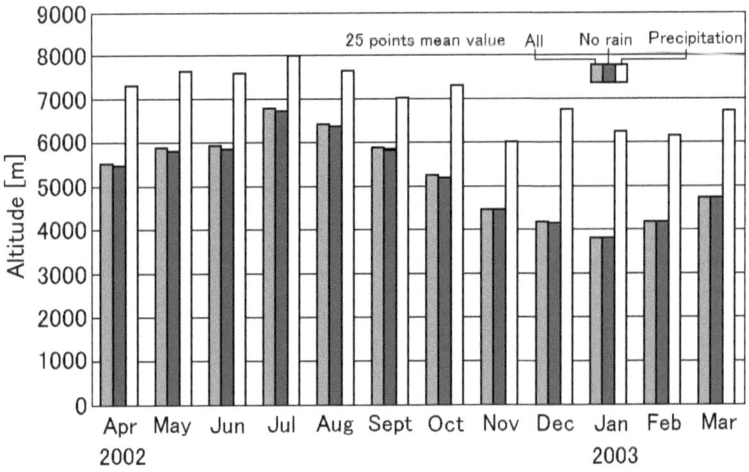

Fig. 12.47 WINDAS LTR observed during April 2002 to March 2003 in the standard observation mode (height resolution of 300 m). Data of 25 points are averaged [from Ishihara et al. 2006]

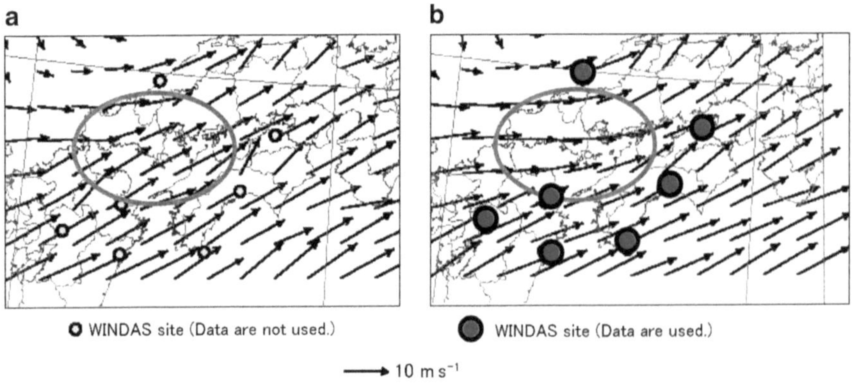

Fig. 12.48 Winds of 3 h forecast of MSM at 850 hPa at 1200 UTC on 19 June 2001, (**a**) forecasted with radiosondes and (**b**) with radiosondes and the WINDAS. The ellipses in *both panels* surround the storm [from Ishihara et al. 2006]

in June 2001 (Ishihara et al. 2006). Figure 12.48 shows the winds of 3 h forecast of the MSM at 850 hPa at 12 UTC (Coordinated Universal Time) on 19 June, (a) forecasted with radiosondes and (b) with radiosondes and the WINDAS. The northerly component of winds around the storm (inside of the ellipse in (a)), forecasted from only radiosondes, was larger than that in the case of adding the WINDAS data (inside of the ellipse in (b)). Figure 12.49 shows (a) rainfall amount for 3 h following the start of a MSM forecast using the 4D-VAR method at 12 UTC on the same day with radiosonde data, but without the WINDAS data, (b) using the 4D-VAR scheme with both radiosonde data and the WINDAS data, and (c) actual surface data during 12–15 UTC.

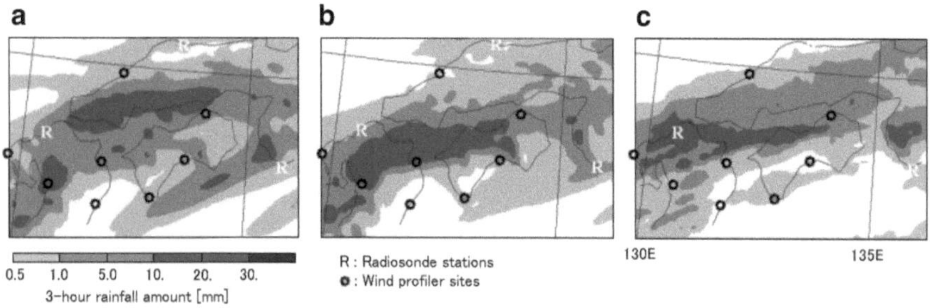

Fig. 12.49 Rainfall amount for 3 h following the start of a MSM forecast at 1200 UTC on the same day as Fig. 12.48 using the 4D-VAR assumption scheme. (**a**) With radiosonde data but without the WINDAS data, (**b**) with both radiosonde and the WINDAS data, and (**c**) actual surface data during 1200 to 1500 UTC [from Ishihara et al. 2006]

As shown in Figs. 12.48 and 12.49, when only conventional upper-air observational data were used, the rainstorm was forecast to occur in a region −60 km to the north of the true rainstorm region. On the other hand, the region forecast with the WINDAS data agrees well with the true rainstorm region. This is because the southerly component of winds observed with the WINDAS and assimilated by the MSM was actually smaller than that forecast by using only conventional data. From similar case studies it has been confirmed that the WINDAS data improve the accuracy of the MSM for mesoscale weather systems particularly for heavy rainfall events.

In general, the data from these wind profiler networks are distributed around the world via Global Telecommunication System (GTS) in real time.

Errata

Radar for Meteorological and Atmospheric Observations

Shoichiro Fukao and Kyosuke Hamazu

S. Fukao and K. Hamazu, Radar for Meteorological and Atmospheric Observations,
DOI 10.1007/978-4-431-54334-3, © Springer Japan 2014

DOI 10.1007/978-4-431-54334-3

FM

Front matter, page xxvi, line 3, EIK should read EIA.

Chapter 3

Radar Measurements and Scatterer Parameters

Page 49, Line 1, "... can be calculated from (3.38)" should read "... can be calculated from (3.40)".

Page 50, Line 1, "... when multiplied by the incident power, gives the total power taken from the incident wave (Battan 1973, p. 67)" should be read "... when multiplied by the incident power, the total power taken from the incident wave (Battan 1973, p. 67)".

Page 65, Footnote 19, "... in Sect. 7.3.2" should read "... in Sect. 7.3.3".

The online version of the original chapter can be found at
http://dx.doi.org/10.1007/978-4-431-54334-3_3

Chapter 5
Reception and Processing of Signals

Page 108, Equation 5.10, "$\Delta P_{\mathrm{no}} = 7gkT_{\mathrm{e}}B_{\mathrm{n}}$" should read "$\Delta P_{\mathrm{no}} = gkT_{\mathrm{e}}B_{\mathrm{n}}$".

Page 163, Line 19, "If samples are independent, var$([\hat{v}_{\mathrm{t}}])$ is expressed ..." should read "If samples are independent, var$[\hat{v}_{\mathrm{t}}]$ is expressed...".

The online version of the original chapter can be found at
http://dx.doi.org/10.1007/978-4-431-54334-3_5

Chapter 6
Radar Observations of Precipitation

Page 188, Equation 6.69, "$A_{\mathrm{prp}} = a_0 R_0^b$" should read "$A_{\mathrm{prp}} = a_0 R^{b_0}$".

Page 217, Line 17, "... can be expressed as $a_0 R_0^b$ =..." should read "... can be expressed as $a_0 R^{b_0}$ =...".

The online version of the original chapter can be found at
http://dx.doi.org/10.1007/978-0-387-22677-4_6

Chapter 8

Overview of Radar

Page 279, Line 11, Equation number 8.8 "$g_D \simeq \frac{4100}{\vartheta_B \varphi_B}$" should read "$g_D \simeq \frac{41000}{\vartheta_B \varphi_B}$".

Page 296, Line 12, "Figure 8.18a shows the received noise power of Cassiopeia measured with ..." should read "Figure 8.18a shows the received noise power of Cassiopeia A measured with ...".

Ppage 333, Line 10, "Appendix B" should read "Appendix C".

Page 337, Line 17, "(8.61), where ..." should read "(8.64), where ...".

The online version of the original chapter can be found at
http://dx.doi.org/10.1007/978-0-387-22677-4_8

Appendix A
Mie Coefficients

As discussed in Sect. 3.3.3, the Mie coefficients a_n and b_n are precisely given by Bessel functions.[1] On the other hand, Deirmendjian (1969) showed that they were also derivable by means of the recursion formula. The derivation has been reviewed in varied ways by predecessors (e.g., Ulaby et al. 1981, pp. 290–291, 1986, pp. 2129–2132). Present appendix will be another collection which briefly presents the derivation by recursion formula after Deirmendjian (1969, Chap. 2).

These coefficients can be expressed by the Riccati–Bessel functions,[2] $\psi_n(x)$ and $\zeta_n(x)$, as (Van de Hulst 1957, p. 123)

---

[1]Bessel functions are defined as the canonical solutions $y(x)$ of Bessel's differential equation;

$$x^2\frac{\mathrm{d}^2y}{\mathrm{d}x^2}+x\frac{\mathrm{d}y}{\mathrm{d}x}+(x^2-\alpha^2)y=0.$$

If α is integer n, the solutions are given by Bessel function of the first kind $J_n(x)$ and that of the second kind $Y_n(x)$. Moreover, for the spherical Bessel differential equation

$$x^2\frac{\mathrm{d}^2y}{\mathrm{d}x^2}+2x\frac{\mathrm{d}y}{\mathrm{d}x}+x^2-n(n+1)y=0,$$

two linear independent solutions are expressed by the spherical Bessel functions $j_n(x)$ and $y_n(x)$, which are in the following relationships with $J_n(x)$ and $Y_n(x)$, respectively;

$$j_n(x)=\sqrt{\frac{\pi}{2x}}J_{n+1/2}(x),\ \ y_n(x)=\sqrt{\frac{\pi}{2x}}Y_{n+1/2}(x)=(-1)^{n+1}\sqrt{\frac{\pi}{2x}}(-1)^{n+1}J_{-n-1/2}(x).$$

[2]Riccati–Bessel functions are expressed using Bessel function and the spherical Bessel functions as

$$\psi_n(x)=xj_n(x)=\sqrt{\frac{\pi x}{2}}J_{n+1/2}(x),\ \ \xi_n(x)=-xy_n(x)=\sqrt{\frac{\pi x}{2}}Y_{n+1/2}(x),\ \ \zeta_n(x)=\psi_n(x)+j\xi_n(x).$$

S. Fukao and K. Hamazu, *Radar for Meteorological and Atmospheric Observations*,
DOI 10.1007/978-4-431-54334-3,

$$a_n - \frac{A_n(y)\psi_n(x) - m\psi_n'(x)}{A_n(y)\zeta_n(x) - m\zeta_n'(x)}, \tag{A.1}$$

$$b_n = \frac{mA_n(y)\psi_n(x) - \psi_n'(x)}{mA_n(y)\zeta_n(x) - m\zeta_n'(x)}, \tag{A.2}$$

where $'$ denotes derivative. Functions $\psi_n(x)$ and $\zeta_n(x)$ can be expressed using $(n+1/2)$th Bessel function of the first kind[3] $J_{n+1/2}$, and that of the second kind $Y_{n+1/2}$ as

$$\psi_n(x) = xj_n(x) = \sqrt{\frac{\pi x}{2}}J_{n+1/2}(x), \tag{A.3}$$

$$\begin{aligned}\zeta_n(x) &= \sqrt{\frac{\pi x}{2}}J_{n+1/2}(x) - jY_{n+1/2}(x) \\ &= \sqrt{\frac{\pi x}{2}}[J_{n+1/2}(x) + (-1)^n jJ_{-n-1/2}(x)].\end{aligned} \tag{A.4}$$

In (A.1) and (A.2), $A_n(y)$ is adopted to separate the functions with argument y from those with argument x, and defined as

$$A_n(y) \equiv \frac{\psi_n'(y)}{\psi_n(y)}, \quad y \equiv mx = m\frac{\pi D}{\lambda}, \tag{A.5}$$

where m is the complex refractive index, D the diameter of the sphere, and λ the wavelength in free space. For absorbing spheres such as raindrops, m is complex. Thus, $A_n(y)$ depends on Bessel functions of complex argument.

Using the characteristics of Bessel function of the first kind for half integer, derivatives of (A.3) and (A.4) become as follows;

$$\psi_n'(x) = \frac{\mathrm{d}}{\mathrm{d}x}\left[\sqrt{\frac{\pi x}{2}}J_{n+1/2}(x)\right] = \sqrt{\frac{\pi x}{2}}\left[J_{n-1/2}(x) - \frac{n}{x}J_{n+1/2}(x)\right], \tag{A.6}$$

---

[3] Bessel function of the first kind $J_{n+1/2}(x)$ for half integer $n+1/2$ can be expressed as

$$J_{1/2}(x) = \sqrt{\frac{2}{\pi x}}\sin x, \quad J_{-1/2}(x) = \sqrt{\frac{2}{\pi x}}\cos x,$$

$$J_{3/2}(x) = \frac{1}{x}J_{1/2}(x) - J_{-1/2}(x) = \sqrt{\frac{2}{\pi x}}\left(\frac{\sin x}{x} - \cos x\right),$$

$$J_{-3/2}(x) = -\frac{1}{x}J_{-1/2}(x) - J_{1/2}(x) = \sqrt{\frac{2}{\pi x}}\left(\frac{\cos x}{x} + \sin x\right),$$

$$J_{n+1/2}(x) = \frac{2(n-\frac{1}{2})}{x}J_{n-1/2}(x) - J_{n-3/2}(x), \quad J_{n+1/2}'(x) = \frac{1}{2}J_{n-1/2}(x) - \frac{1}{2}J_{n+3/2}(x),$$

$$\frac{\mathrm{d}}{\mathrm{d}x}\left[\sqrt{\frac{\pi x}{2}}J_{-n-1/2}(x)\right] = -\sqrt{\frac{\pi x}{2}}\left[J_{-n+1/2}(x) + \frac{n}{x}J_{-n-1/2}(x)\right], \tag{A.7}$$

$$\zeta_n'(x) = \sqrt{\frac{\pi x}{2}}\left\{J_{n-1/2}(x) - \frac{n}{x}J_{n+1/2}(x) - (-1)^n j\left[J_{-n+1/2}(x) + \frac{n}{x}J_{-n-1/2}(x)\right]\right\}. \tag{A.8}$$

Substituting (A.3)–(A.8) into (A.1) and (A.2), the Mie coefficients become

$$\begin{aligned} a_n(m,\,x) = &\left\{\left[\frac{A_n(y)}{m} + \frac{n}{x}\right]J_{n+1/2}(x) - J_{n-1/2}(x)\right\} \\ &\times\left\{\left[\frac{A_n(y)}{m} + \frac{n}{x}\right]\left[J_{n+1/2}(x) + (-1)^n jJ_{-n-1/2}(x)\right]\right. \\ &\left. - \left[J_{n-1/2}(x) - (-1)^n jJ_{-n+1/2}(x)\right]\right\}^{-1} \end{aligned} \tag{A.9}$$

and

$$\begin{aligned} b_n(m,\,x) = &\left\{\left[mA_n(y) + \frac{n}{x}\right]J_{n+1/2}(x) - J_{n-1/2}(x)\right\} \\ &\times\left\{\left[mA_n(y) + \frac{n}{x}\right]\left[J_{n+1/2}(x) + (-1)^n jJ_{-n-1/2}(x)\right]\right. \\ &\left. - \left[J_{n-1/2}(x) - (-1)^n jJ_{-n+1/2}(x)\right]\right\}^{-1}. \end{aligned} \tag{A.10}$$

The Bessel functions of real argument which appear in (A.9) and (A.10) can be expressed with circular functions by defining following recursion formula;

$$w_n(x) = \frac{2n-1}{x}w_{n-1}(x) - w_{n-2}(x), \tag{A.11}$$

with

$$w_0(x) = \sin x - j\cos x = \sqrt{\frac{\pi x}{2}}[J_{1/2}(x) + jJ_{-1/2}(x)], \tag{A.12}$$

$$w_{-1}(x) = \cos x - j\sin x = \sqrt{\frac{\pi x}{2}}[J_{1/2}(x) - jJ_{1/2}(x)]. \tag{A.13}$$

For example,

$$w_1(x) = \frac{1}{x}w_0(x) - w_{-1}(x) = \sqrt{\frac{\pi x}{2}}[J_{3/2}(x) - jJ_{3/2}(x)], \tag{A.14}$$

$$w_2(x) = \frac{3}{x}w_1(x) - w_0(x) = \sqrt{\frac{\pi x}{2}}[J_{5/2}(x) + jJ_{5/2}(x)], \tag{A.15}$$

and $w_n(x)$ is generally expressed using finite term functions $J_{n+1/2}(x)$ and $J_{-n-1/2}(x)$ as

$$w_n(x) = \sqrt{\frac{\pi x}{2}}[J_{n+1/2}(x) + (-1)^n jJ_{-n-1/2}(x)]. \tag{A.16}$$

Comparing (A.16) with (A.9) and (A.10), and eliminating $\sqrt{\frac{\pi x}{2}}$ which is common to the numerators and denominators, the Mie coefficients can be rewritten as

$$a_n(m, x) = \frac{\left[\frac{A_n(y)}{m} + \frac{n}{x}\right]\mathrm{Re}[w_n(x)] - \mathrm{Re}[w_{n-1}(x)]}{\left[\frac{A_n(y)}{m} + \frac{n}{x}\right]w_n(x) - w_{n-1}(x)}, \tag{A.17}$$

$$b_n(m, x) = \frac{\left[mA_n(y) + \frac{n}{x}\right]\mathrm{Re}[w_n(x)] - \mathrm{Re}[w_{n-1}(x)]}{\left[mA_n(y) + \frac{n}{x}\right]w_n(x) - w_{n-1}(x)}, \tag{A.18}$$

where $A_n(y)$ is arranged using the Riccati–Bessel functions in (A.3), and their derivatives (A.6) and (A.7) to lead the following equation

$$A_n(y) = \frac{J_{n-1/2}(y) - \frac{n}{y}J_{n+1/2}(y)}{J_{n+1/2}(y)} = -\frac{n}{y} + \frac{J_{n-1/2}(y)}{J_{n+1/2}(y)}. \tag{A.19}$$

If the above function is transformed into the following recursion formula:

$$\begin{aligned} A_n(y) &= -\frac{n-1}{y} + \frac{J_{n-3/2}(y)}{J_{n-1/2}(y)} \\ &= -\frac{n-1}{y} + \frac{\frac{2n-1}{y}J_{n-1/2}(y) - J_{n+1/2}(y)}{J_{n-1/2}(y)}, \end{aligned} \tag{A.20}$$

or

$$\frac{J_{n+1/2}(y)}{J_{n-1/2}(y)} = \frac{n}{y} - A_{n-1}(y). \tag{A.21}$$

Substituting (A.21) into (A.19), the final recursion formula of $A_n(y)$ can be expressed as

$$A_n(y) = -\frac{n}{y} + \left[\frac{n}{y} - A_{n-1}(y)\right]^{-1}, \tag{A.22}$$

and $A_0(y)$ becomes

$$A_0(y) = \frac{J_{-1/2}(y)}{J_{1/2}(y)} = \cot y. \tag{A.23}$$

Putting the real and imaginary parts of the complex refractive index be n_r and k_a, respectively, and $y \equiv mx = n_\mathrm{r}x - jk_\mathrm{a}x = p - jq$, where $p = n_\mathrm{r}x$ and $q = k_\mathrm{a}x$. Substituting these values into (A.23) and arranging the expression, $A_0(y)$ becomes

$$A_0(y) = \frac{\sin p \cos p + j \sinh q \cosh q}{\sin^2 p + \sinh^2 q}. \tag{A.24}$$

Using (A.22) and (A.24), actual $A_n(y)$ can be calculated. In addition, if m is real (nonabsorbing sphere), the imaginary parts of $A_0(y)$ and all $A_n(y)$ become 0.

Appendix B
Autocovariance Analysis

B.1 Mean Doppler Frequency

The autocorrelation function of a complex time series signal $R(\tau)$ is a complex number obtained by the inner product computation of complex signal and is expressed as

$$R(\tau) = A(\tau)\mathrm{e}^{j\theta(\tau)}, \tag{B.1}$$

where $A(\tau)$ denotes amplitude. As $R(\tau) = R(-\tau)^*$,

$$A(\tau) : \text{Real even } \; A(\tau) = A(-\tau), \tag{B.2}$$

$$\theta(\tau) : \text{Real odd } \; \theta(\tau) = -\theta(-\tau), \; \theta(0) = 0, \tag{B.3}$$

and the supersubscript * denotes complex conjugate. From Wiener–Khinchin's theorem, following relations are formed as shown in (5.86) and (5.87),

$$S(f) = \int_{-\infty}^{\infty} R(\tau)\mathrm{e}^{-j2\pi f\tau}\mathrm{d}\tau, \tag{B.4}$$

$$R(\tau) = \int_{-\infty}^{\infty} S(f)\mathrm{e}^{j2\pi f\tau}\mathrm{d}f. \tag{B.5}$$

Differentiating (B.5) with τ, and following relation is obtained,

$$\frac{\mathrm{d}R(\tau)}{\mathrm{d}\tau} = R'(\tau) = j2\pi\int_{-\infty}^{\infty} fS(f)\mathrm{e}^{j2\pi f\tau}\mathrm{d}f. \tag{B.6}$$

If $\tau = 0$, (B.5) and (B.6) become

$$R(0) = \int_{-\infty}^{\infty} S(f)\mathrm{d}f, \tag{B.7}$$

S. Fukao and K. Hamazu, *Radar for Meteorological and Atmospheric Observations*,
DOI 10.1007/978-4-431-54334-3,

$$R'(0) = j2\pi \int_{-\infty}^{\infty} fS(f)\mathrm{d}f, \tag{B.8}$$

respectively. $R(0)$ is the autocorrelation function (real) of time lag zero. From (5,107), Equations (B.7) and (B.8) are related with the first-order moment of power spectrum, $\mu_1 = E[f]$, as

$$\frac{R'(0)}{R(0)} = j2\pi E[f], \tag{B.9}$$

where $E[\,]$ denotes ensemble mean value. Differentiating (B.1) with τ,

$$R'(\tau) = A'(\tau)\mathrm{e}^{j\theta(\tau)} + jA(\tau)\mathrm{e}^{j\theta(\tau)}\theta'(\tau). \tag{B.10}$$

As $A(\tau)$ is an even function, $A'(0) = 0$ and $R'(0) = jA(0)\theta'(0)$. Hence, using $R(0) = A(0)$, (B.9) becomes

$$E[f] = \frac{\theta'(0)}{2\pi}. \tag{B.11}$$

If $\theta'(0)$ is approximated by the difference for the sample time interval T_s as

$$\theta'(0) \simeq \frac{\theta(T_\mathrm{s}) - \theta(0)}{T_\mathrm{s}} \tag{B.12}$$

and $\theta(0) = 0$ is applied, $E(f)$ becomes

$$E[f] \simeq \frac{\theta(T_\mathrm{s})}{2\pi T_\mathrm{s}}. \tag{B.13}$$

Now, $\theta(T_\mathrm{s})$ is the argument of the autocorrelation function $R(T_\mathrm{s})$, the mean Doppler frequency $\overline{f}$ is obtained by

$$\overline{f} = E[f] \simeq \frac{1}{2\pi T_\mathrm{s}} \arctan\left[\frac{\mathrm{I}_m[R(T_\mathrm{s})]}{\mathrm{R}_e[R(T_\mathrm{s})]}\right]. \tag{B.14}$$

B.2 Doppler Frequency Spectrum Width

From (5.108), second order derivative of $R(\tau)$, $R''(\tau)$, is related with the second order moment of power spectrum at $\tau = 0$ as

$$\frac{R''(0)}{R(0)} = -(2\pi)^2 E[f^2]. \tag{B.15}$$

The variance σ_{fd}^2 of the mean Doppler frequency $E[f]$ is expressed as

$$\sigma_{fd}^2 = E[f^2] - E[f]^2. \tag{B.16}$$

Substituting (B.9) and (B.15) into (B.16),

$$\sigma_{fd}^2 = -\frac{1}{(2\pi)^2}\left[\frac{R''(0)}{R(0)} - \left(\frac{R'(0)}{R(0)}\right)^2\right] \tag{B.17}$$

is obtained. On the other hand, from (B.10),

$$\begin{aligned} R''(\tau) =& A''(\tau)\mathrm{e}^{j\theta(\tau)} + 2jA'(\tau)\mathrm{e}^{j\theta(\tau)}\theta'(\tau) \\ &- A(\tau)\mathrm{e}^{j\theta(\tau)}\theta'(\tau)^2 + jA(\tau)\mathrm{e}^{j\theta(\tau)}\theta''(\tau) \end{aligned} \tag{B.18}$$

is derived. As discussed previously, $A'(0) = 0$ and $R(0) = A(0)$, thus (B.18) becomes

$$\frac{R''(0)}{R(0)} = \frac{A''(0)}{A(0)} - \theta'(0)^2 + j\theta''(0). \tag{B.19}$$

Furthermore, from (B.9), (B.11), and (B.19),

$$\left(\frac{R'(0)}{R(0)}\right)^2 = -\theta'(0)^2 \tag{B.20}$$

is obtained. Substituting (B.19) and (B.20) into (B.17),

$$\sigma_{fd}^2 = -\frac{1}{(2\pi)^2}\left[\frac{A''(0)}{A(0)} + j\theta''(0)\right] \tag{B.21}$$

is obtained.

The term $\theta''(0)$ in (B.21) can be assumed as the acceleration component of the Doppler velocity, thus if $\theta''(0)$ is approximated to $\theta''(0) \simeq 0$, (B.21) becomes

$$\sigma_{fd}^2 \simeq -\frac{1}{(2\pi)^2}\frac{A''(0)}{A(0)}. \tag{B.22}$$

Using the Taylor series with center $= 0$ (the Maclaurin series of $A(T_\mathrm{s})$ for the sample time interval T_s around 0), we obtain

$$A(T_\mathrm{s}) = A(0) + T_\mathrm{s}\frac{\mathrm{d}A(0)}{\mathrm{d}\tau} + \frac{T_\mathrm{s}^2}{2}\frac{\mathrm{d}^2A(0)}{\mathrm{d}\tau^2} + \frac{T_\mathrm{s}^3}{6}\frac{\mathrm{d}^3A(0)}{\mathrm{d}\tau^3} + \cdots. \tag{B.23}$$

Neglecting the third and higher terms of (B.23) and accounting $A'(0) = 0$,

$$A(T_s) = A(0) + \frac{T_s^2}{2} A''(0), \tag{B.24}$$

and

$$\frac{A''(0)}{A(0)} = \frac{2}{T_s^2} \left[\frac{A(T_s)}{A(0)} - 1 \right]. \tag{B.25}$$

Substituting (B.25) into (B.22), we obtain

$$\sigma_{fd}^2 \simeq \frac{1}{2\pi^2 T_s^2} \left[1 - \frac{A(T_s)}{A(0)} \right]. \tag{B.26}$$

As $A(T_s)$ is the absolute value of $R(T_s)$, $R(0) = A(0)$, thus the squared Doppler frequency spectrum width (variance) is given by

$$\sigma_{fd}^2 \simeq \frac{1}{2\pi^2 T_s^2} \left[1 - \frac{|R(T_s)|}{R(0)} \right]. \tag{B.27}$$

Appendix C
The Fast Fourier Transform (FFT) Algorithm

C.1 Decimation-in-Time (DIT) FFT Algorithm

There are already a lot of documents about the high-speed calculation algorithm of the discrete Fourier transform (DFT). Here, we summarize the fast Fourier transform (FFT) based on Ziemer et al. (1998, Chap. 10) as follows. In the expression of discrete signals discussed in Sect. 5.5.3, the DFT is expressed from (5.121) as

$$V(k) = \sum_{m=0}^{M-1} v(m) W_M^{mk} \quad k = 0, 1, \cdots, M-1, \tag{C.1}$$

where

$$W_M = \mathrm{e}^{-j2\pi/M}. \tag{C.2}$$

For a discrete time signal series $v(m)$ of length M, the sum (C.1) gives a discrete Frequency signal series $V(k)$ of length M. If M is large such as several tens or several hundreds, number of operations according to (C.1) becomes huge as discussed in Sect. 8.4.4. Therefore, a great reduction of number of operations is desirable from the point of practical use.

The FFT is an algorithm to decrease the number of operations by limiting when length M becomes $M = 2^n$ (n: integer), and using the periodicity of (C.2). Consider the right-hand side (C.1) with the summation carried out separately over the even- and odd-index terms in the sum. If the indexes are $m = 2s$ for the even-index terms and $m = 2s + 1$ for the odd-index terms, (C.1) is written as

$$\begin{aligned} V(k) &= \sum_{s=0}^{M/2-1} v(2s) W_M^{2sk} + \sum_{s=0}^{M/2-1} v(2s+1) W_M^{(2s+1)k} \\ &= \sum_{s=0}^{M/2-1} v(2s) W_M^{2sk} + W_M^k \sum_{s=0}^{M/2-1} v(2s+1) W_M^{2sk}. \end{aligned} \tag{C.3}$$

S. Fukao and K. Hamazu, *Radar for Meteorological and Atmospheric Observations*, DOI 10.1007/978-4-431-54334-3, © Springer Japan 2014

Applying the relation

$$W_M^{2sk} = \mathrm{e}^{-j(2\pi/M)(2sk)} = \mathrm{e}^{-j[(2\pi)/(M/2)](sk)} = W_{M/2}^{sk}, \tag{C.4}$$

(C.3) becomes

$$\begin{aligned} V(k) &= \sum_{s=0}^{M/2-1} v(2s) W_{M/2}^{sk} + W_M^k \sum_{s=0}^{M/2-1} v(2s+1) W_{M/2}^{sk} \\ &= G(k) + W_M^k H(k) \quad k = 0, 1, \cdots, M-1, \end{aligned} \tag{C.5}$$

where $G(k)$ and $H(k)$ are

$$G(k) = \sum_{s=0}^{M/2-1} v(2s) W_{M/2}^{sk}, \tag{C.6}$$

$$H(k) = \sum_{s=0}^{M/2-1} v(2s+1) W_{M/2}^{sk}, \tag{C.7}$$

and are $(M/2)$-point DFT of the even- and odd-index points of the discrete time signal series $v(m)$, respectively. Since $G(k)$ and $H(k)$ are periodic in k with period $M/2$, only $M/2$ values of $G(k)$ and $H(k)$ are needed to compute (C.5). Assuming that an algorithm is available for computing an $(M/2)$-point DFT, the N-point DFT of $v(m)$ may be provided by combining the $(M/2)$-point DFTs, $G(k)$ and $H(k)$, in accordance with (C.5).

Figure C.1a shows the flow graph of the DIT decomposition for the DFT of $M = 8$, based on the combination of two $(M/2)$-point DFTs of the even- and odd-index points. In the figure, the lines with arrows indicate the quantities to be added, with the power of W_M along the arrow indicating the multiplication of $H(k)$ in (C.5) by W_M^k. Arrows with nothing beside them are multiplications by unity, i.e., the $G(k)$s are transmitted as they appear at the output of the DFT boxes, e.g., $V(0) = G(0) + W_M^0 H(0)$. Furthermore, we can break up each of the $M/2$-point DFTs in Panel (a) into two $M/4$-point DFTs, as shown in Fig. C.1(b), where the relation $W_{M/2}^s = W_M^{2s}$ is used in combining the sums $G(k)$ and $H(k)$. The same procedure can be continued until a series of $M/2$-point DFTs results for the first stage of the M-point DFT computation. Figure C.2 shows the procedure to the final stage for $M = 8$, which is a precise flow graph of the FFT. Since this algorithm separates the input time signal samples into successively smaller sets, the resulting algorithm is referred as a decimation-in-time (DIT) FFT. The schematic of W_M^k for $M = 8$ is shown in Fig. C.3, and 2-point DFT flow graph is shown in Fig. C.4.

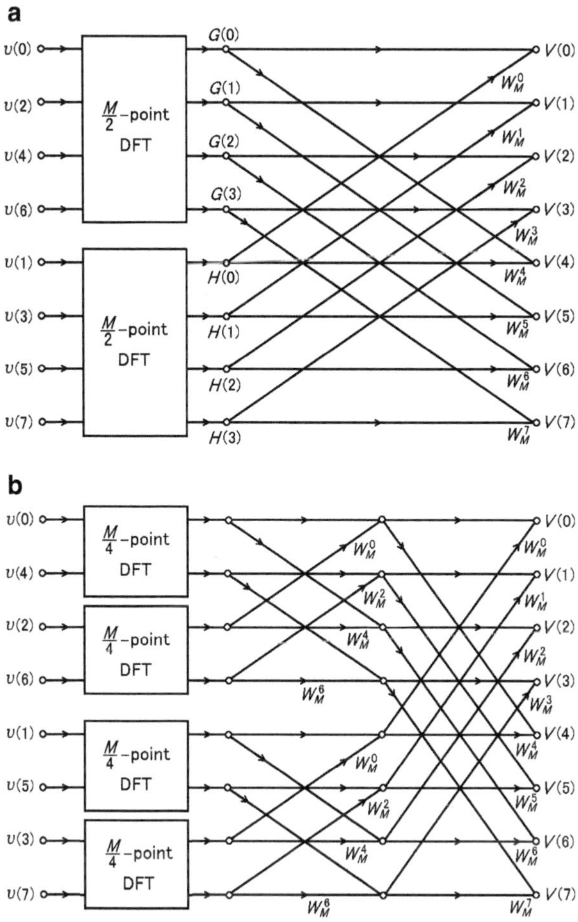

Fig. C.1 Flow graph of the DIT decomposition for the DFT of $M = 8$, based on the combination of (**a**) two $(M/2)$-point DFTs of the even- and odd-index points and (**b**) that of four $(M/4)$-point DFTs

C.2 Decimation-in-Frequency (DIF) FFT Algorithm

Here we discuss another type of algorithm for finding the DFT. Consider the DFT sum (C.1) with the sum carried out over the first half and the last half of the input ample separately, the result becomes

$$\begin{aligned} V(k) &= \sum_{m=0}^{M/2-1} v(m)W_M^{mk} + \sum_{m=M/2}^{M-1} v(m)W_M^{mk} \\ &= \sum_{m=0}^{M/2-1} v(m)W_M^{mk} + W_M^{(M/2)k} \sum_{m=0}^{M/2-1} v\left(m+\frac{M}{2}\right) W_M^{mk}. \end{aligned} \tag{C.8}$$

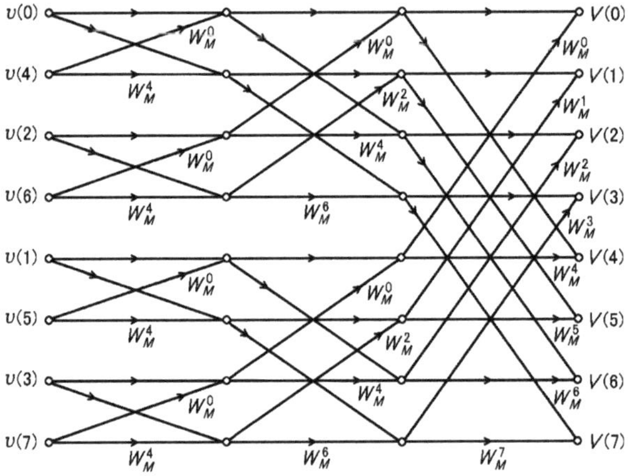

Fig. C.2 Procedure to the final stage for $M = 8$, which is a precise flow graph of the FFT

Fig. C.3 Schematic of W_M^k for $M = 8$

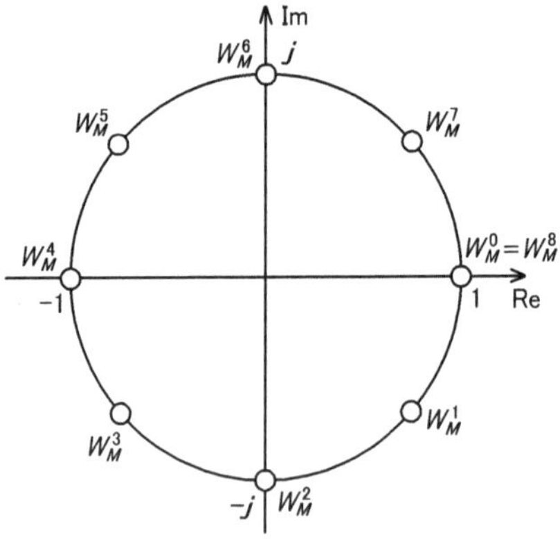

Fig. C.4 Flow graph of 2-point DFT, $M = 2$

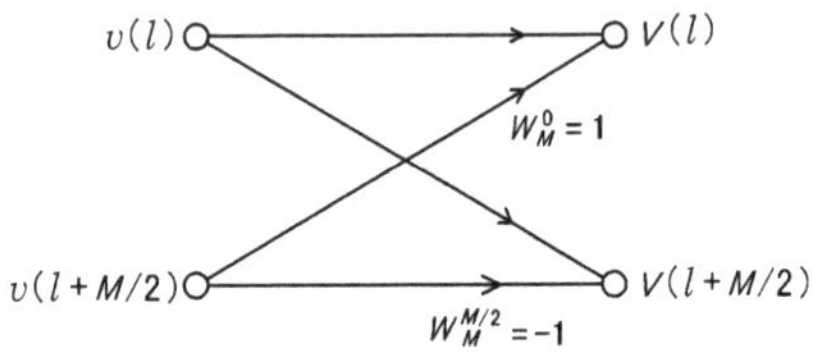

Using the fact that $W_M^{(M/2)k} = (-1)^k$ to combine the two sums in (C.8), and we obtain

$$V(k) = \sum_{m=0}^{M/2-1} \left[v(m) + (-1)^k v\left(m + \frac{M}{2}\right) W_M^{mk} \right]. \tag{C.9}$$

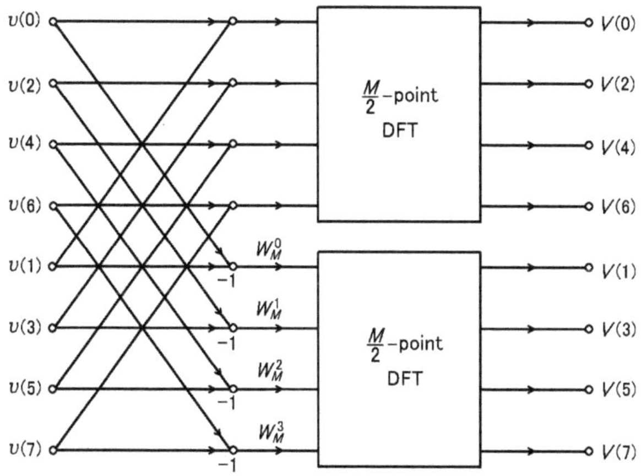

Fig. C.5 Flow graph for the first separation to two $(M/2)$-point DFTs of 8-point DFT in the DIF FFT algorithm

If k is treated even and odd separately as $k = 2s$ and $k = 2s+1$, respectively, the even- and odd-index output points of (C.9) are indicated for $s = 0, 1, \cdots, M/2-1$ as

$$V(2s) = \sum_{m=0}^{M/2-1} \left[v(m) + v\left(m + \frac{M}{2}\right) \right] W_M^{2sm}, \tag{C.10}$$

$$V(2s+1) = \sum_{m=0}^{M/2-1} \left[v(m) - v\left(m + \frac{M}{2}\right) \right] W_M^{m} W_M^{2sm}. \tag{C.11}$$

Since $W_M^{2m} = W_{M/2}^{m}$ as noted previously, (C.10) and (C.11) are both $M/2$-point DFTs. The flow graph for obtaining $V(k)$ in this fashion is shown in Fig. C.5. This process can be expanded to replace each $(M/2)$-point DFT by two $(M/4)$-point DFTs, and so on until only two points are left in each DFT. The final flow graph for 8-point DFT becomes precisely the reverse of Fig. C.2, with all arrows reversed and input and output interchanged. Since the frequency points were subdivided to obtain this algorithm, it is called as a decimation-in-frequency (DIF) FFT algorithm.

Appendix D
Radar Equation for RASS Echo

The radar equation for the radio acoustic sounding system (RASS) was first given by Marshall et al. (1972). Clifford and Wang (1977) discuss the equation based on them. The following discussion are based on Clifford and Wang.

Assuming that both radar antenna and acoustic wave source are at the origin of coordinates, and that the radar range resolution and the beam width are ΔR and $2\theta_{\mathrm{r}}$, respectively as shown in Fig. D.1, a volume element dv at (r, ϕ, θ) in the sampling volume scatters the radar waves and generates a RASS echo at $(r_0, \phi_0, \pi/2)$. The electric field at the point $\boldsymbol{r_0}$ in the xy-plane generated by the RASS echo is derived as follows.

The electric field generated at $\boldsymbol{r_0}$ due to the refractive index perturbations $\Delta n(\boldsymbol{r})$ is expressed by the volume integration of the electric field $E(\boldsymbol{r})$ and $\Delta n(\boldsymbol{r})$ in the scattering volume as

$$E_{\mathrm{r}}(\boldsymbol{r_0}) = \frac{k^2}{2\pi} \int_V \frac{\exp(jk|\boldsymbol{r_0} - \boldsymbol{r}|)}{|\boldsymbol{r_0} - \boldsymbol{r}|} \Delta n(\boldsymbol{r}) E(\boldsymbol{r}) \mathrm{d}v, \tag{D.1}$$

where k is the radar wavelength. Using the transmission electric field E_0, $E(r)$ can be expressed as

$$E(r) = \frac{E_0}{r} \exp(jkr), \tag{D.2}$$

whereas the refractive index perturbation $\Delta n(r)$ of a sine acoustic wave becomes

$$\Delta n(r) = \begin{cases} \dfrac{A_{\mathrm{a}}}{r} \exp(jk_{\mathrm{a}} r) & R - \Delta R/2 < r < R + \Delta R/2, \\ 0 & \text{otherwise,} \end{cases} \tag{D.3}$$

where A_{a} is a constant given by (7.89) and is determined by characteristics of sound source and condition of the atmosphere, and k_{a} is the wavenumber of the acoustic wave.

S. Fukao and K. Hamazu, *Radar for Meteorological and Atmospheric Observations*, DOI 10.1007/978-4-431-54334-3,

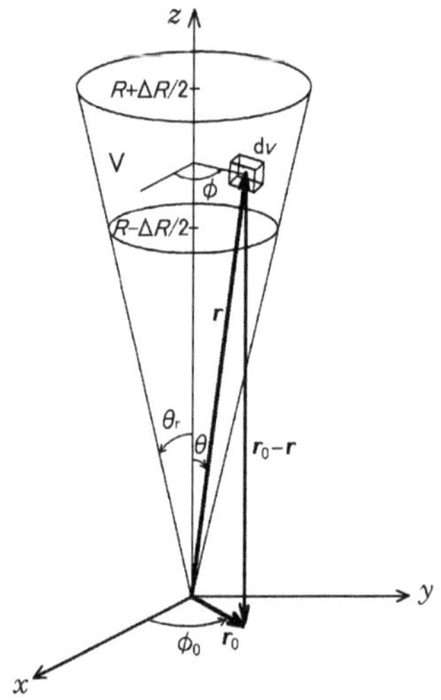

Fig. D.1 Geometric configuration for a radar equation of RASS echo. Both radar antenna and acoustic wave sources are located at the origin and transmit the waves to the zenith. The radar range resolution and the beam width are ΔR and $2\theta_r$, respectively. A volume element dv at (r, ϕ, θ) in the sampling volume scatters the radar waves and generate a RASS echo at $(r_0, \phi_0, \pi/2)$ [from Marshall et al. 1972]

For simplicity, put the denominator $|\boldsymbol{r_0} - \boldsymbol{r}|$ of (D.1) as $|\boldsymbol{r_0} - \boldsymbol{r}| = R$ and approximate the numerator $|\boldsymbol{r_0} - \boldsymbol{r}|$ as

$$|\boldsymbol{r_0} - \boldsymbol{r}| = r - r_0 \sin\theta \cos(\phi - \phi_0). \tag{D.4}$$

As the RASS echo extends isotropically in the azimuthal direction, the value $|\boldsymbol{r_0} - \boldsymbol{r}|$ is represented by the value for $\phi_0 = 0$, thus (D.4) is replaced with

$$|\boldsymbol{r_0} - \boldsymbol{r}| = r - r_0 \sin\theta \cos\phi. \tag{D.5}$$

From Fig. D.1, dv is expressed in the polar coordinate as

$$\mathrm{d}v = r^2 \sin\theta \mathrm{d}\theta \mathrm{d}\phi \mathrm{d}r. \tag{D.6}$$

Substituting these equations into (D.1), and

$$E_r(\boldsymbol{r_0}) = \frac{k^2 E_0 A_a}{2\pi R} \int_0^{\theta_r} \sin\theta \mathrm{d}\theta \int_0^{2\pi} \exp[jk(-r_0 \sin\theta\cos\phi)]\mathrm{d}\phi$$
$$\times \int_{R-\Delta R/2}^{R+\Delta R/2} \exp[j(2k + k_a)r]\mathrm{d}r. \tag{D.7}$$

Applying that $\sin\theta \simeq \theta$ if $\theta \ll 1$, (D.7) is expressed as

$$E_r(\boldsymbol{r_0}) = \frac{k^2 E_0 A_a}{2\pi R} 2\pi \int_0^{\theta_r} \theta J_0(kr_0\theta)\mathrm{d}\theta \int_{R-\Delta R/2}^{R+\Delta R/2} \exp[j(2k + k_a)r]\mathrm{d}r$$
$$= \frac{k^2 E_0 A_a}{2\pi R}(\pi\theta_r^2)\frac{2J_1(kr_0\theta_r)}{kr_0\theta_r} \int_{R-\Delta R/2}^{R+\Delta R/2} \exp[j(2k + k_a)r]\mathrm{d}r, \tag{D.8}$$

where $J_0(kr_0\theta)$ and $J_1(kr_0\theta_r)$ are 0th and 1th Bessel function of the first kind, respectively. $2J_1(kr_0\theta_r)/kr_0\theta_r$ in the right-hand side of (D.8) shows the radiation pattern of the radar antenna that generates the electromagnetic wave of beam width of $2\theta_r$. In the RASS method, the same antenna is used for transmission and reception of radar waves, thus the received power of RASS echo P_{ar} is given by

$$P_{ar} = \int_0^{r_e} \int_0^{2\pi} \frac{|E_r(r_0)|^2}{Z_0} r_0 \mathrm{d}r_0 \mathrm{d}\phi_0, \tag{D.9}$$

where r_e and Z_0 are the radius of radar antenna and its characteristic impedance, respectively. Substituting (D.8) into (D.9), P_{ar} becomes

$$P_{ar} = \frac{1}{Z_0}\left(\frac{k^2 E_0 A_a}{2R}\theta_r^2\right)^2 \int_0^{2\pi} \mathrm{d}\phi_0 \int_0^{r_e} \left[\frac{2J_1(kr_0\theta_r)}{kr_0\theta_r}\right]^2 r_0 \mathrm{d}r_0 \times \left\{\int_{R-\Delta R/2}^{R+\Delta R/2} \exp\left[j(2k+k_a)r\right]\mathrm{d}r\right\}^2. \tag{D.10}$$

The integration for ϕ_0 and r_0 in the right-hand side of (D.10) expresses the maximum value of the radar cross section, and can be calculated by assuming that the effective antenna aperture is infinite as

$$\int_0^{2\pi} \mathrm{d}\phi_0 \int_0^{\infty} \left[\frac{2J_1(kr_0\theta_r)}{kr_0\theta_r}\right]^2 r_0 \mathrm{d}r_0 = \frac{4\pi}{(k\theta_r)^2} \tag{D.11}$$

$$= g\frac{\lambda^2}{4\pi}, \tag{D.12}$$

where g and λ are the antenna gain and the wavelength of the radar, respectively. The transmission electric field E_0 and the transmitted peak power P_t are related as

$$\frac{E_0^2}{Z_0 R^2} = \frac{P_t g}{4\pi R^2}. \tag{D.13}$$

Substituting (D.11)–(D.13) into (D.10), and P_{ar} becomes

$$P_{ar} = A_a^2 \frac{4\pi^2 P_t}{\lambda^2 R^2} \left\{\int_{R-\Delta R/2}^{R+\Delta R/2} \exp\left[j(2k+k_a)r\right]\mathrm{d}r\right\}^2. \tag{D.14}$$

The integration of the right-hand side of (D.14) is

$$\int_{R-\Delta R/2}^{R+\Delta R/2} \exp\left[j(2k+k_a)r\right]\mathrm{d}r = \frac{\Delta R}{2}\left\{\cos[(2k+k_a)R]\,\mathrm{sinc}\left[(2k+k_a)\frac{\Delta R}{2}\right] + \cos[(2k-k_a)R]\,\mathrm{sinc}\left[(2k-k_a)\frac{\Delta R}{2}\right]\right\}, \tag{D.15}$$

where $\mathrm{sinc}x = \sin x/x$. If $2k \simeq k_\mathrm{a}$, $\mathrm{sinc}[(2k+k_\mathrm{a})\Delta R/2]$ becomes small, thus the term that include $2k+k_\mathrm{a}$ in the right-hand side of (D.15) becomes negligible, and (D.15) is simplified as

$$\int_{R-\Delta R/2}^{R+\Delta R/2} \exp\left[j(2k+k_\mathrm{a})r\right]\mathrm{d}r = \frac{\Delta R}{2}\mathrm{sinc}\left[(2k-k_\mathrm{a})\frac{\Delta R}{2}\right]. \tag{D.16}$$

Substituting (D.16) into (D.14), the radar equation for the RASS echo is expressed as

$$P_\mathrm{ar} = A_\mathrm{a}^2\frac{4\pi^2 P_\mathrm{t}}{\lambda^2 R^2}\left(\frac{\Delta R}{2}\right)^2 \mathrm{sinc}^2\left[(2k-k_\mathrm{a})\frac{\Delta R}{2}\right]. \tag{D.17}$$

It is obvious from (D.17) that the received power of the RASS echo is proportional to the square of the range resolution of the radar and inverse proportional to the square of the distance from the radar. Replacing R with r and ΔR with Δr, respectively in (D.17), (7.90) is obtained.

References

Adachi, T. 1996. *Detailed temperature structure of meteorological disturbances observed with RASS (Radio Acoustic Sounding System)*, 173 pp. Ph.D. dissertation. Kyoto University.

Alexander, S., T. Tsuda, J. Furumoto, T. Shimomai, T. Kozu, and M. Kawashima. 2006. A statistical overview of convection during the CPEA-I campaign. *Journal of the Meteorological Society of Japan* 84: 57–93.

Alvarez, H., J. Aparici, J. May, and F. Olmos. 1997. A 45-MHz continuum survey of the southern hemisphere. *Astronomy and Astrophysics Supplement Series* 124: 315–328.

Armijo, L. 1969. A theory for the determination of wind and precipitation velocities with Doppler radars. *Journal of the Atmospheric Sciences* 26: 570–573.

Atlas, D. and C.W. Ulbrich. 1977. Path- and area-integrated rainfall measurement by microwave attenuation in 1–3 cm band. *Journal of Applied Meteorology* 16: 1322–1331.

Atlas, D., R.S. Srivastava, and R.S. Sekhon. 1973. Doppler radar characteristics of precipitation at vertical incidence. *Reviews of Geophysics and Space Physics* 11: 1–35.

Balsley, B.B., and W.L. Ecklund. 1972. A portable coaxial collinear antenna. *IEEE Transactions on Antennas and Propagation* AP-20: 512–516.

Balsley, B.B., and K.S. Gage. 1980. The MST radar technique: Potential for middle atmospheric studies. *Pure and Applied Geophysics* 118: 452–493.

Balsley, B.B., W.L. Ecklund, D.A. Carter, and P.E. Johnston. 1980. The MST radar at Poker Flat, Alaska. *Radio Science* 15: 213–223.

Barratt, P. and I. C. Browne. 1953. A new method of measuring vertical air currents. *Quart. J. Roy. Meteor. Soc.* 79: 550.

Battan, L.J. 1973. *Radar observation of the atmosphere*, 324 pp. Illinois: The University of Chicago Press.

Bean, B.R., and E.J. Dutton. 1966. *Radio meteorology*. National Bureau of Standards, Monograph 92, Supt. Doc. U.S. Govt., Printing Office, Washington.

Beard, K.V., and C. Chuang. 1987. A new model for the equilibrium shape of raindrops. *Journal of the Atmospheric Sciences* 44: 1509–1524.

Benoit, A. 1968. Signal attenuation due to neutral oxygen and water vapor, rain and clouds. *Microwave Journal* 11: 73–80.

Berger, T., and H.L. Groginsky. 1973. Estimation of the spectral moments of pulse trains. In *International conference on information theory (preprints)*, Tel Aviv, Israel.

Bienvenu, G., L. Kopp. 1983. Optimality of high resolution array processing using the eigensystem approach. *IEEE Transactions on Acoustics, Speech, & Signal Processing* 31: 1235–1248.

Biggerstaff, M.I., and R.A. Houze, Jr. 1993. Kinematics and microphysics of the transition zone of the 10–11 June 1985 squall line. *Journal of the Atmospheric Sciences* 50: 3091–3110.

Booker, H.G., and W.E. Gordon. 1950. A theory of radio scattering in the troposphere. *Proceedings of the IRE* 38: 401–412.

S. Fukao and K. Hamazu, *Radar for Meteorological and Atmospheric Observations*,
DOI 10.1007/978-4-431-54334-3,

Borgeaud, M., R.T. Shin, and J.A. Kong. 1987. Theoretical models for polarimetric radar clutter. *Journal of Electromagnetic Waves and Applications* 1: 73–89

Brandes, E.A. 1977. Flow in severe thunderstorms observed by dual-Doppler radar. *Monthly Weather Review* 105: 113–120.

Briggs, B.H. 1984. The analysis of spaced sensor records by correlation technique. In *Handbook for MAP*, vol. 13, 166–186. Urbana: ICSU Scientific Committee on Solar-Terrestrial Physics (SCOSTEP).

Briggs, B.H., and R.A. Vincent. 1992. Spaced-antenna analysis in the frequency domain. *Radio Science* 27: 117–129.

Bringi, V.N., and V. Chandrasekar. 2001. *Polarimetric Doppler weather radar*, 636 pp. New York: Cambridge University Press.

Bringi, V.N., R. Hoferer, D.A. Brunkow, R. Schwerdtfeger, V. Chandrasekar, S.A. Rutledge, J. George, and P.C. Kennedy. 2011. Design and performance characteristics of the new 8.5-m dual-offset Gregorian antenna for the CSU-CHILL radar. *Journal of Atmospheric and Oceanic Technology* 28: 907–920.

Browning, K.A. 1986. Conceptual models of precipitation systems. *Weather and Forecasting* 1: 23–41.

Browning, K.A., and G.A. Monk. 1982. A simple model for the synoptic analysis of cold fronts. *Quarterly Journal of the Royal Meteorological Society* 108: 435–452.

Browning, K.A., and R. Wexler. 1968. A determination of kinematic properties of a wind field using Doppler radar. *Journal of Applied Meteorology* 7: 105–113.

Browning, K.A., J.C. Fankhauser, J.P. Chalon, P.J. Eccles, R.G. Strauch, F.H. Merrem, D.J. Musil, E.L. May, and W.R. Sand. 1976. Structure of an evolving hailstorm, Part V: Synthesis and implications for hail growth and hail suppression. *Monthly Weather Review* 104: 603–610.

Capon, J. 1969. High-resolution frequency-wavenumber spectrum analysis. *Proceedings of the IEEE* 57: 1408–1418.

Caughey, S.J., B.A. Crease, D.N. Asimakapoulos, and R.S. Cole. 1978. Quantitative bistatic acoustic sounding of the atmospheric boundary layer. *Quarterly Journal of the Royal Meteorological Society* 104: 146–161.

CCIR. 1991. CCIR, Propagation data and prediction methods required for terrestrial line-of-site systems, CCIR Reports 338-6, V, ITU, Geneva, 1991.

Cheong, B.L., M.W. Hoffman, R.D. Palmer, S.J. Fraiser, and F.J. López-Dekker. 2004. Pulse pair beamforming and the effects of reflectivity field variations on imaging radars. *Radio Science* 39:RS3014. doi:10.1029/2002RS002843.

Cheong, B.L., M.W. Hoffman, R.D. Palmer, S.J. Frasier, and F.J. López-Dekker. 2006. Phased-array design for biological clutter rejection : Simulation and experimental validation. *Journal of Atmospheric and Oceanic Technology* 23: 585–598.

Cheong, B.L., T.-Y. Yu, R.D. Palmer, K.-F. Yang, M.W. Hoffman, S.J. Frasier, and F.J. Lopez-Dekker. 2008. Effects of wind wield inhomogeneities on Doppler beam swinging received by an imaging radar. *Journal of Atmospheric and Oceanic Technology* 25: 1414–1422.

Chilson, P.B., T.Y. Yu, R.G. Strauch, A. Muscinski, and R.D. Palmer. 2003. Implementation and validation of range imaging on a UHF radar wind profiler. *Journal of Atmospheric and Oceanic Technology* 104: 987–996.

Cho, J.Y.N. 2009. Moving clutter spectral filter for terminal Doppler weather radar. In *34th International conference on radar meteorology (preprints)*. Williamsburg: American Meteorological Society, P5.2.

Clifford, S.F., and T.I. Wang. 1977. The range limitation on radar-acoustic sounding system (RASS) due to atmospheric refractive turbulence. *IEEE Transactions on Antennas and Propagation* 25: 319–326.

Cohn, S. 1995. Radar measurements of turbulent eddy dissipation rate in the troposphere: A comparison of techniques. *Journal of Atmospheric and Oceanic Technology* 12: 85–95.

Cole, A.E., A. Court, and A.J. Kantor. 1965. Model atmospheres, In *Handbook of geophysics and space environment*, ed. S.L. Valley. Bedford: Office of Aerospace Research, USAF, Cambridge Research Laboratories, Chapter 2.

Cooley, J.W., and J.W. Tukey. 1965. An algorithm for the machine calculation of complex Fourier series. *Mathematics of Computation* 19: 297–301.
Costa, E., and F. Fougere. 1988. Cross-spectral analysis of spaced-receiver measurements. *Radio Science* 23: 129–139.
Crane, R.K. 1980. A review of radar observations of turbulence in the lower stratosphere. *Radio Science* 15: 177–193.
Crocker, S.C. 1988. *TDWR PRF selection criteria*, 57 pp. Project Rep. ATC-147, DOT/FAA/PM-87-25, MIT Lincoln Laboratory.
Czechowsky, P., G. Schmidt, and R. Rüster. 1984. The mobile SOUSY Doppler radar: Technical design and first results. *Radio Science* 19: 441–450.
Dalaudier, F., C. Sidi, M. Crochet, and J. Vernin. 1994. Direct evidence of "sheet" in the atmospheric temperature field. *Journal of the Atmospheric Sciences* 51: 237–248.
Davies-Jones, R.P. 1979. Dual-Doppler coverage area as a function of measurement accuracy and spatial resolution. *Journal of Applied Meteorology* 18: 1229–1233.
de Elía, R., and I. Zawadzki. 2000. Sidelobe contamination in bistatic radars. *Journal of Atmospheric and Oceanic Technology* 17: 1313–1329.
de Elía, R., and I. Zawadzki. 2001. Optimal layout of a bistatic radar network. *Journal of Atmospheric and Oceanic Technology* 18: 1184–1194.
Deirmendjian, D. 1969. *Electromagnetic scattering on spherical polydispersions*, 290 pp. New York: Elsevier.
Dhaka, S.K., M. Takahashi, Y. Kawatani, S. Malik, Y. Shibagaki, and S. Fukao. 2003. Observations of deep convective updrafts in tropical convection and their role in the generation of gravity waves. *Journal of the Meteorological Society of Japan* 81: 1185–1199.
Dhaka, S.K., M.K. Yamamoto, Y. Shibagaki, H. Hashiguchi, S. Fukao, and H.-Y. Chun. 2006. Equatorial Atmosphere Radar observations of short vertical wavelength gravity waves in the upper troposphere and lower stratosphere region induced by localized convection. *Geophysical Research Letters* 33: L19805. doi:10.1029/2006GL027026.
Dicke, R.H. 1946. The measurement of thermal radiation at microwave frequencies. *Review of Scientific Instruments* 17: 268–275.
Doviak, R.J. 1972. Comparison of bistatic and monostatic radar detection of clear air atmospheric targets. *AIAA paper*, 8 pp. No. 72–175, Copies available from AIAA library, 750 3rd Ave., New Yoke, NY 10017.
Doviak, R.J., and R.D. Palmer. 2014. Polarimetric doppler weather radar. In *Encyclopedia of atmospheric science*, ed. G. North. 2nd ed. London: Elsevier.
Doviak, R.J., and P.S. Ray. 1976. Error estimation in wind fields derived from dual-Doppler radar measurement. *Journal of Applied Meteorology* 15: 868–878.
Doviak, R.J., and C.M. Weil. 1972. Bistatic radar detection of the melting layer. *Journal of Applied Meteorology* 11: 1012–1016.
Doviak, R.J., and D.S. Zrnić. 1984. *Doppler radar and weather observations*, 458 pp. Orlando: Academic.
Doviak, R.J., and D.S. Zrnić. 2006. *Doppler radar and weather observations*, 562 pp. 2nd ed. Mineola: Dover.
Doviak, R.J., J. Goldhirsh, and A.R. Miller. 1972. Bistatic-radar detection of high-altitude clear-air atmospheric targets. *Radio Science* 7: 993–1003.
Doviak, R.J., R.J. Lataitis, and C.L. Holloway. 1996. Cross correlations and cross spectra for spaced antenna wind profilers 1. Theoretical analysis. *Radio Science* 31: 157–180.
Doviak, R.J., V. Bringi, A. Ryzhkov, A. Zahrai, and D. Zrnić. 2000. Considerations for polarimetric upgrades to operational WSR-88D radars. *Journal of Atmospheric and Oceanic Technology* 17: 257–278.
Droegemeier, K.K., and R.B. Wilhelmson. 1987. Numerical simulation of thunderstorm outflow dynamics, Part I: Outflow sensitivity experiments and turbulence dynamics. *Journal of the Atmospheric Sciences* 44: 1180–1210.

Easterbrook, C.C. 1974. Estimating horizontal wind fields by two-dimensional curve fitting of single Doppler radar measurements. In *16th Conference on radar meteorology (preprints)*, 214–219. Boston: American Meteorological Society.

Ecklund, W.L., D.A. Carter, and B.B. Balsley. 1988. A UHF wind profiler for the boundary layer: Brief description and initial results. *Journal of Atmospheric and Oceanic Technology* 5: 432–441.

Einaudi, F., D.P. Lalas, and G.E. Perona. 1978. The role of gravity waves in tropospheric processes. *Pure and Applied Geophysics* 117: 627–663.

Evans, J.V. 1969. Theory and practice of troposphere study by Thomson scatter radar. *Proceedings of the IEEE* 57: 496–500.

Fang, M., and R.J. Doviak. 2008. Coupled contributions in the Doppler radar spectrum width equation. *Journal of Atmospheric and Oceanic Technology* 25: 2245–2258.

Fang, M., R.J. Doviak, and V. Melnikov. 2004. Spectrum width measured by WSR-88D: Error sources and statistics of various weather phenomena. *Journal of Atmospheric and Oceanic Technology* 21: 888–904.

Fang, M., R.J. Doviak, and B.A. Albrecht. 2012. Analytical expressions for Doppler spectra of scatter from hydrometeors observed with a vertically directed radar beam. *Journal of Atmospheric and Oceanic Technology* 29: 500–509.

Farley, D., H. Ierkic, and B. Fejer. 1981. Radar interferometry: A new technique for studying plasma turbulence in the ionosphere. *Journal of Geophysical Research* 86: 1467–1472.

Franke, S.J. 1990. Pulse compression and frequency domain interferometry with a frequency-hopped MST radar. *Radio Science* 25: 565–574.

Fraser, G.J. 1968. Seasonal variation of southern hemisphere mid-latitude winds at altitudes of 70–100 km. *Journal of Atmospheric and Terrestrial Physics* 30: 707–720.

Friend, A.W. 1949. Theory and practice of tropospheric sounding by radar. *Proceedings of the IRE* 37: 116–138.

Fritts, D.C. 1984. Shear excitation of atmospheric gravity waves, 2. Nonlinear radiation from a free shear layer. *Journal of the Atmospheric Sciences* 41: 524–537.

Fritts, D.C., and Z. Luo. 1992. Gravity wave excitation by geostrophic adjustment of the jet stream, Part 1: Two-dimensional forcing. *Journal of the Atmospheric Sciences* 49: 681–697.

Fritts, D.C., and G.D. Nastrom. 1992. Sources of mesoscale variability of gravity waves, Part 2: Frontal, convective and jet stream excitation. *Journal of the Atmospheric Sciences* 49: 111–127.

Fujita, T.T. 1985. *The downburst: Microburst and macroburst, SMRP Res. Rep.* 210, 122 pp. Chicago: University of Chicago.

Fujita, T.T., and J. McCarthy. 1990. The application of weather radar to aviation meteorology. In *Radar in meteorology*, ed. D. Atlas, 657–681. Boston: American Meteorological Society.

Fujiwara, M., M.K. Yamamoto, H. Hashiguchi, T. Horinouchi, and S. Fukao. 2003. Turbulence at the tropopause due to breaking Kelvin waves observed by the Equatorial Atmosphere Radar. *Geophysical Research Letters* 30: 1171. doi:10.1029/2002GL01627.

Fujiyoshi, Y. 2001. Three dimensional radar echo structure of a vortex-like disturbance developed in a strong horizontal wind shear zone. *Tenki, Meteorological Society of Japan* 48: 3–4 (in Japanese).

Fujiyoshi, Y., and B. Geng. 1995. Dual Doppler radar observation of a tropical rainband developed from two convective clouds. *Journal of the Meteorological Society of Japan* 73: 471–490.

Fujiyoshi, Y., N. Yoshimoto, and T. Takeda. 1998. A dual-Doppler radar study of longitudinal-mode snowbands, Part I: A three dimensional kinematic structure of meso-γ-scale convective cloud systems within a longitudinal-mode snowband. *Monthly Weather Review* 126: 72–91.

Fukao, S. 2006. Coupling processes in the equatorial atmosphere (CPEA): A project overview. *Journal of the Meteorological Society of Japan* 84: 1–18.

Fukao, S., and R.D. Palmer. 1991. Spatial and frequency domain interferometry using the MU radar: A tutorial and recent developments. *Journal of Geomagnetism and Geoelectricity* 43: 645–666.

Fukao, S., T. Sato, S. Kato, R.M. Harper, R.F. Woodman, and W.E. Gordon. 1979. Mesospheric winds and waves over Jicamarca on May 23–24. *Journal of Geophysical Research* 84: 4379–4386.

Fukao, S., S. Kato, T. Aso, M. Sasada, and T. Makihira. 1980a: Middle and upper atmosphere radar (MUR) under design in Japan. *Radio Science* 15: 225–231.

Fukao, S., K. Wakasugi, and S. Kato. 1980b. Radar measurement of short-period atmospheric waves and related scattering properties at the altitude of 13–25 km over Jicamarca. *Radio Science* 15: 431–438.

Fukao, S., T. Sato, T. Tsuda, S. Kato, K. Wakasugi, and T. Makihira. 1985a. The MU radar with an active phased array system: 1. Antenna and power amplifiers. *Radio Science* 20: 1155–1168.

Fukao, S., T. Tsuda, T. Sato, S. Kato, K. Wakasugi, and T. Makihira. 1985b. The MU radar with an active phased array system: 2. In-house equipment. *Radio Science* 20: 1169–1176.

Fukao, S., T. Sato, and S. Kato. 1985c. Monitoring of the MU radar antenna pattern by satellite OHZORA (EXOS-C). *Journal of Geomagnetism and Geoelectricity* 37: 431–441.

Fukao, S., K. Wakasugi, T. Sato, T. Tsuda, I. Kimura, N. Takeuchi, M. Matsuo, and S. Kato. 1985d. Simultaneous observation of precipitating atmosphere by VHF band and C/Ku band radars. *Radio Science* 20: 622–630.

Fukao, S., K. Wakasugi, T. Sato, S. Morimoto, T. Tsuda, I. Hirota, I. Kimura, and S. Kato. 1985e. Direct measurement of air and precipitation particle motion by very high frequency Doppler radar. *Nature* 316: 712–714.

Fukao, S., T. Sato, H. Hojo, I. Kimura, and S. Kato. 1986. A numerical consideration on edge effect of planar dipole phased arrays. *Radio Science* 21: 1–12.

Fukao, S., T. Sato, T. Tsuda, S. Kato, M. Inaba, and I. Kimura. 1988a. VHF Doppler radar determination of the momentum flux in the upper troposphere and lower stratosphere: Comparison between the three- and four-beam methods. *Journal of Atmospheric and Oceanic Technology* 5: 57–69.

Fukao, S., M. Inaba, I. Kimura, T. Sato, T. Tsuda, and S. Kato. 1988b. A systematic error in MST/ST radar measurement induced due to finite range volume effect: 2. Numerical considerations. *Radio Science* 23: 74–82.

Fukao, S., M.D. Yamanaka, H. Matsumoto, T. Sato, T. Tsuda, and S. Kato. 1989. Wind fluctuations near a cold vortex-tropopause funnel system observed by the MU radar. *Pure and Applied Geophysics* 130: 463–479.

Fukao, S., T. Sato, T. Tsuda, M. Yamamoto, M.D. Yamanaka, and S. Kato. 1990. MU radar: New capabilities and system calibrations. *Radio Science* 25: 477–485.

Fukao, S., M.C. Kelley, T. Shirakawa, T. Takami, M. Yamamoto, T. Tsuda, and S. Kato. 1991. Turbulent upwelling of the mid-latitude ionosphere: 1. Observational results by the MU radar. *Journal of Geophysical Research* 96: 3725–3746.

Fukao, S., N. Ao, M.D. Yamanaka, W.K. Hocking, T. Sato, M. Yamamoto, T. Nakamura, T. Tsuda, and S. Kato. 1994. Seasonal variability of vertical eddy diffusivity in the middle atmosphere I: Three-year observations by the MU radar. *Journal of Geophysical Research* 99: 18973–18987.

Fukao, S., H. Hashiguchi, M. Yamamoto, T. Tsuda, T. Nakamura, M. K. Yamamoto, T. Sato, M. Hagio, and Y. Yabugaki. 2003. Equatorial Atmosphere Radar (EAR): System description and first results. *Radio Science* 38: 1053. doi:10.1029/2002RS002767.

Furumoto, J. 2002. *Observation of turbulence echo characteristics and humidity profiles with the MU radar-RASS*, 134 pp. Ph.D. dissertation. Kyoto University.

Furumoto, J., K. Kurimoto, and T. Tsuda. 2003. Continuous observations of humidity profiles with the MU radar-RASS combined with GPS and rawinsonde measurements. *Journal of Atmospheric and Oceanic Technology* 20: 23–41.

Furumoto, J., T. Tsuda, S, Iwai, T, Kozu. 2006. Continuous humidity monitoring in a tropical region with the equatorial atmosphere radar. *Journal of Atmospheric and Oceanic Technology* 23: 538–551.

Gage, K.S. 1990. Radar observations of the free atmosphere: Structure and dynamics. In *Radar in meteorology*, ed. D. Atlas, 534–565. Boston: American Meteorological Society.

Gage, K.S., and B.B. Balsley. 1978. Doppler radar probing of the clear atmosphere. *Bulletin of the American Meteorological Society* 59: 1074–1093.
Gage, K.S, and B.B. Balsley. 1980. On the scattering and reflection mechanisms contributing to clear air radar echoes from the troposphere, stratosphere, and mesosphere. *Radio Science* 15: 243–257.
Gage, K.S., and J.L. Green. 1978. Evidence for specular reflection from monostatic VHF radar observations of the atmosphere. *Radio Science* 13: 991–1001.
Gage, K.S., and J.L. Green. 1979. Tropopause detection by partial specular reflection with very-high-frequency radar. *Science* 203: 1238–1240.
Gage, K.S., and J.L. Green. 1982. An objective method for the determination of tropopause height from VHF radar observations. *Journal of Applied Meteorology* 21: 1150–1154.
Gage, K.S., B.B. Balsley, and J.L. Green. 1981. Fresnel scattering model for the specular echoes observed by VHF radar. *Radio Science* 16: 1447–1453.
Gage, K.S., W.L. Ecklund, and B.B. Balsley. 1985. A modified Fresnel scattering model for the parameterization of Fresnel returns. *Radio Science* 20: 1493–1501.
Gage, K.S., B.B. Balsley, W.L. Ecklund, D.A. Carter, and J.R. McAfee. 1991. Wind profiler-related research in the tropical Pacific. *Journal of Geophysical Research* 96: 3209–3220.
Gage, K.S., J.R. McAfee, and C.R. Williams. 1996. On the annual variation of tropospheric zonal winds observed above Christmas Island in the central equatorial Pacific. *Journal of Geophysical Research* 101: 15061–15070.
Gavrilov, N.M. 1992. Internal gravity waves in the mesopause region: Hydrodynamic sources and climatological patterns. *Advances in Space Research* 12: 10113–10121.
Gavrilov, N.M., S. Fukao, T. Nakamura, T. Tsuda, M.D. Yamanaka, and M. Yamamoto. 1996. Statistical analysis of gravity waves observed with the middle and upper atmosphere radar in the middle atmosphere, 1. Method and general characteristics. *Journal of Geophysical Research* 101: 29511–29521.
Gavrilov, N.M., S. Fukao, and H. Hashiguchi. 1999. Multi-beam MU radar measurements of advective accelerations in the atmosphere. *Geophysical Research Letters* 26: 315–318.
Geerts, B., and P.V. Hobbs. 1991. Organization and structure of clouds and precipitation on the Mid-Atlantic Coast of the United States of America. Part IV: Retrieval of the thermodynamic and cloud microphysical structure of a frontal rainband from Doppler radar data. *Journal of the Atmospheric Sciences* 48: 1287–1305.
Geller, M.A. 1983. Dynamics of the middle atmosphere (Tutorial lecture). *Space Science Reviews* 34: 359–375.
Gill, A.E. 1982. *Atmosphere-ocean dynamics*, 662 pp. London: Academic.
Gordon, W.E. 1958. Incoherent scattering of radio waves by free electrons with applications to space explorations by radar. *Proceedings of the IRE* 46: 1824–1829.
Gordon, W.E., and L.M. LaLonde. 1961. The design and capabilities of an ionospheric radar probe. *IRE Transactions on Antennas and Propagation* AP-9: 17–22.
Gorgucci, E., G. Scarchilli, and V. Chandrasekar. 1999. Specific differential phase estimation in the presence of nonuniform rainfall medium along path. *Journal of Atmospheric and Oceanic Technology* 16: 1690–1697.
Gossard, E.E., and R.G. Strauch. 1983. *Radar observation of clear air and clouds*, 280 pp. Amsterdam: Elsevier.
Gossard, E.E., J.H. Richter, and D. Atlas. 1970. Internal waves in the atmosphere from high-resolution radar measurements. *Journal of Geophysical Research* 75: 3523–3536.
Gossard, E.E., S. Gutman, B.B. Stankov, and D.E. Wolfe. 1999. Profile of radio refractive index and humidity derived from radar wind profilers and the Global Positioning System. *Radio Science* 34: 371–383.
Gunn, K.L.S., and T.W.R. East. 1954. The microwave properties of precipitation particles. *Quarterly Journal of the Royal Meteorological Society* 80: 522–545.
Gunn, R., and G.D. Kinzer. 1949. The terminal velocity of fall for water droplets in stagnant air. *Journal of Meteorology* 6: 243–248.

Gunn, K.L.S., and R.S. Marshall. 1958. The distribution of size of aggregate snowflakes. *Journal of Meteorology* 15: 452–466.
Hamazu, K. 2002. *Development of Doppler radars for studying aviation weather*, 207 pp. Ph.D. dissertation. Kyoto University.
Hamazu, K., K. Hata, M. Ishihara, H. Hashiguchi, and S. Fukao. 2000a. Development of a C-band Doppler radar for low-level wind shear detection. *IEICE, J83-B* 6: 894–909 (in Japanese).
Hamazu, K., M. Ishihara, K. Hata, H. Hashiguchi, and S. Fukao. 2000b. Development of a low-level wind shear detection algorithm for a Doppler weather radar. *IEICE, J83-B* 7: 1067–1080 (in Japanese).
Hamazu, K., K. Hemmi, K. Hayashi, H. Hashiguchi, and S. Fukao. 2002. Development of a 5.3-GHz klystron for a pulse Doppler radar. *IEICE* E85-B: 1152–1159.
Hamazu, K., H. Hashiguchi, T. Wakayama, T. Matsuda, R.J. Doviak, and S. Fukao. 2003. A 35-GHz scanning Doppler radar for fog observations. *Journal of Atmospheric and Oceanic Technology* 20: 972–986.
Hanle, E. 1986. Survey of bistatic and multistatic radar. *Proceedings of the IEE* 133: 587–595.
Hansen, R.C. 2009. *Phased array antennas*, 547 pp. 2nd ed. Hoboken: Wiley.
Hardy, K.R., and H. Ottersten. 1969. Radar investigations of convective patterns in the clear atmosphere. *Journal of the Atmospheric Sciences* 26: 666–672.
Harris, F.J. 1978. On the use of windows for harmonic analysis with the discrete Fourier transform. *Proceedings of the IEEE* 66: 51–83.
Hartree, D.R., J.G. Michel, and P. Nicolson. 1946. Practical methods for the solution of the equations of tropospheric refraction. In *Meteorological factors in radio wave propagation*, 127–168. London: Physical Society.
Hashiguchi, H., M. Yamamoto, S. Fukao, T. Tsuda, M.D. Yamanaka, T. Nakamura, T. Sato, S. Kato, T. Makihira, and K. Hamazu. 1992. Development of a boundary layer radar. In *International symposium on middle atmosphere science (preprints)*, 23–37 March, Kyoto, 46–47.
Hashiguchi, H., M.D. Yamanaka, T. Tsuda, M. Yamamoto, T. Nakamura, T. Adachi, S. Fukao, T. Sato, and D.L. Tobing. 1995a. Diurnal variations of the planetary boundary layer observed with an L-band clear-air Doppler radar. *Boundary-Layer Meteorology* 74: 419–424.
Hashiguchi, H., S. Fukao, T. Tsuda, M.D. Yamanaka, D.L. Tobing, T. Sribimawati, S.W.B. Harijono, and H. Wiryosumarto. 1995b. Observations of the planetary layer over equatorial Indonesia with an L-band clear-air Doppler radar: Initial results. *Radio Science* 30: 1043–1054.
Hashiguchi, H., S. Fukao, Y. Moritani, T. Wakayama, and S. Watanabe. 2004. A lower troposphere radar: 1.3-GHz active phased-array type wind profiler with RASS. *Journal of the Meteorological Society of Japan* 82: 915–931.
Hassenpflug, G., P.B. Rao, M. Yamamoto, and S. Fukao. 2003. MU radar spaced antenna observations with varying apertures: Scatterer and antenna contributions to the ground diffraction pattern. *Radio Science* 38: 1043. doi:10.1029/2002RS002751.
Hassenpflug, G., M. Yamamoto, H. Luce, and S. Fukao. 2008. Description and demonstration of the new middle and upper atmosphere radar imaging system: 1-D, 2-D, and 3-D imaging of troposphere and stratosphere. *Radio Science* 43: 2013. doi:10.1029/2006RS003603.
Hauser, D., and P. Amayenc. 1981. A new method for deducing hydrometeor-size distributions and vertical air motions from Doppler radar measurements at vertical incidence. *Journal of Applied Meteorology* 20: 547–555.
Hauser, D., F. Roux, and P. Amayenc. 1988. Comparison of two method for the retrieval of thermodynamic and microphysical variables from Doppler radar measurements: Application to the case of a tropical squall line. *Journal of the Atmospheric Sciences* 45: 1285–1303.
Heisenberg, W. 1948. On the theory of statistical and isotropic turbulence. *Proceedings of the Royal Society A* 195: 402–406.
Heiss, W.H., D.L. McGrew, and D.S. Sirmans. 1990. NEXRAD; Next generation weather radar (WSR-88D). *Microwave Journal* 33: 79–98.

Hélal D., M. Crochet, H. Luce, and E. Spano. 2001. Radar imaging and high-resolution array processing applied to a classical VHF-ST profiler. *Journal of Atmospheric and Solar-Terrestrial Physics* 63: 263–274.
Herman, B.M., and L.J. Battan. 1961. Calculations of mie back-scattering from melting ice spheres. *Journal of Meteorology* 18: 468–478.
Hill, R.J. 1978. Spectra of fluctuations in refractivity, temperature, humidity, and the temperature–humidity cospectrum in the inertial and dissipation range. *Radio Science* 13: 935–961.
Hill, R.J., and S.F. Clifford. 1978. Modified spectrum of atmospheric temperature fluctuations and its application to optical propagation. *Journal of the Optical Society of America* 68: 892–899.
Hines, C.O. 1968. A possible source of waves in noctilucent clouds. *Journal of the Atmospheric Sciences* 25: 937–942.
Hirono, M., H. Luce, M. Yamamoto, and S. Fukao. 2004. Horizontal maps of echo power in the lower stratosphere using the MU radar. *Annales Geophysicae* 22: 717–724.
Hirota, I., and T. Niki. 1986. Inertia-gravity waves in the troposphere and stratosphere observed by the MU radar. *Journal of the Meteorological Society of Japan* 64: 995–999.
Hitschfeld, W.F., and J. Bordan. 1954. Errors inherent in the radar measurement of rainfall at attenuating wavelengths. *Journal of Meteorology* 11: 58–67.
Hobbs, P.V. 1978. Organization and structure of clouds and precipitation on the mesoscale and microscale in cyclonic storms. *Reviews of Geophysics and Space Physics* 16: 741–755.
Hocking, W.K. 1983. On the extraction of atmospheric turbulence parameters from radar backscatter Doppler spectra—I. Theory. *Journal of Atmospheric and Terrestrial Physics* 45: 89–102.
Hocking, W.K. 1985. Measurement of turbulent energy dissipation rates in the middle atmosphere by radar techniques. A review. *Radio Science* 20: 1403–1422.
Hocking, W.K. 1988. Two years of continuous measurements of turbulence parameters in the upper mesosphere and lower thermosphere made with a 2-MHz radar. *Journal of Geophysical Research* 93: 2475–2491.
Hocking, W.K. 1997a. System design, signal processing procedures and preliminary results for the Canadian (London, Ontario) VHF atmospheric radar. *Radio Science* 32: 687–706.
Hocking, W.K. 1997b. Recent advances in radar instrumentation and techniques for studies of the mesosphere, stratosphere, and troposphere. *Radio Science* 32: 2241–2270.
Hocking, W.K., T. Thayaparan, and J. Jones. 1997. Meteor decay times and their use in determining a diagnostic mesospheric temperature–pressure parameter: Methodology and one year of data. *Geophysical Research Letters* 24: 2977–2980.
Holloway, C.L., R.J. Doviak, and S.A. Cohen. 1997a. Cross correlations of fields scattered by horizontally anisotropic refractive index irregularities. *Radio Science* 32: 1911–1920.
Holloway, C.L., R.J. Doviak, S.A. Cohn, R.J. Lataitis, and J.S. Van Baelen. 1997b. Cross correlations and cross spectra for spaced antenna wind profilers 2. Algorithms to estimate wind and turbulence. *Radio Science* 32: 967–982.
Houze, R.A. Jr. 1993. *Cloud dynamics*, 570 pp. San Diego: Academic.
Houze, R.A. Jr., J.D. Locatelli, and P.V. Hobbs. 1976. Dynamics and cloud microphysics of the rainbands in an occluded frontal system. *Journal of the Atmospheric Sciences* 33: 1921–1936.
Houze, R.A. Jr., S.A. Rutledge, M.I. Biggerstaff, and B.F. Smull. 1989. Interpretation of Doppler weather-radar displays in midlatitude mesoscale convective systems. *Bulletin of the American Meteorological Society* 70: 608–619.
Houze R.A. Jr., B.F. Smull, and P. Dodge. 1990. Mesoscale organization of springtime rainstorms in Oklahoma. *Monthly Weather Review* 117: 613–654.
Hubbert, J., and V.N. Bringi. 1995. An iterative filtering technique for the analysis of copolar differential phase and dual-frequency radar measurements. *Journal of Atmospheric and Oceanic Technology* 12: 643–648.
Hubbert, J., V. Chandrasekar, and V.N. Bringi. 1993. Processing and interpretation of coherent dual-polarized radar measurements. *Journal of Atmospheric and Oceanic Technology* 10: 155–164.
Hubbert, J.C., V.N. Bringi, and D. Brunkow. 2003. Studies of the polarimetric covariance matrix. Part I: Calibration methodology. *Journal of Atmospheric and Oceanic Technology* 20: 696–706.

Hubbert, J.C., S.M. Ellis, M. Dixon, and G. Meymaris. 2010. Modeling, error analysis, and evaluation of dual-polarization variables obtained from simultaneous horizontal and vertical polarization transmit radar, Part II: Experimental data. *Journal of Atmospheric and Oceanic Technology* 27: 1599–1607.

Ice, R.L., R.D. Rhoton, D.S. Saxion, N.K. Patel, D. Sirmans, D.A. Warde, D.L. Rachel, and R.G. Fehlen. 2004. Radar Operations Center (ROC) evaluation of the WSR-88D open radar data acquisition (ORDA) system signal processing. In *20th International conference on interactive information and processing systems for meteorology, oceanography, and hydrology*, Seattle, WA. Boston: American Meteorological Society, paper 5.5.

Ice, R.L., G.T. McGehee, R.D. Rhoton, D.S. Saxion, D.A. Warde, R.G. Guenther, D. Sirmans, and D.L. Rachel. 2005. Radar Operations Center (ROC) evaluation of new signal processing techniques for the WSR-88D. In *21th international conference on interactive information and processing systems for meteorology, oceanography, and hydrology*, San Diego, CA. Boston: American Meteorological Society, paper P1.4.

IEICE. 2008. *Antenna engineering handbook*, 1098 pp. 2nd ed. Tokyo: Ohmsha (in Japanese).

Ishihara, M., Z. Yanagisawa, H. Sakakibara, K. Matsuura and J. Aoyagi. 1986. Structure of a typhoon rainband observed by two Doppler radars. *Journal of the Meteorological Society of Japan* 64: 923–939.

Ishihara, M., Y. Kato, T. Abo, K. Kobayashi, and Y. Izumikawa. 2006. Characteristics and performance of the operational wind profiler network of the Japan Meteorological Agency. *Journal of the Meteorological Society of Japan* 84: 1085–1096.

ITU-R. 2001. Attenuation by atmospheric gases. In *ITU-R Recommendations* ITU-R P. 676-5. Geneva, Switzerland: International Telecommunications Union.

Iwanami, K., R. Misumi, M. Maki, T. Wakayama, K. Hata, and S. Watanabe. 2001. Development of a multiparameter radar system on mobile platform. In *30th Conference on radar meteorology (preprints)*, 104–106. Boston: American Meteorological Society.

Jackson, M.C. 1986. The geometry of bistatic radar systems. *Proceedings of the IEE* 133: 604–612.

Jameson, A.R. 1985. Microphysical interpretation of multiparameter radar measurements in rain, Part III: Interpretation and measurement of propagation differential phase shift between orthogonal linear polarizations. *Journal of the Atmospheric Sciences* 42: 607–614.

Jameson, A.R. 1992. The effect of temperature on attenuation correction schemes in rain using polarization propagation differential phase shift. *Journal of Applied Meteorology* 31: 1106–1118.

Jameson, A.R., and D.B. Johnson. 1990. Cloud microphysics and radar. In *Radar in meteorology*, ed. D. Atlas, 323–340. Boston: American Meteorological Society.

Janssen, L., and G. van der Spek. 1985. The shape of Doppler spectra from precipitation. *IEEE Trans. Aerosp. Electron. Syst.* 21: 208–219.

Johnson, R.H., and P.J. Hamilton. 1988. The relationship of surface pressure features to the precipitation and airflow structure of an intense midlatitude squall line. *Monthly Weather Review* 116: 1444–1472.

Jorgensen, D.P., and P.T. Willis. 1982. A Z-R relationship for hurricanes. *Journal of Applied Meteorology* 21: 356–366.

Joss, J., and A. Waldvogel. 1970. A method to improve the accuracy of radar-measured amounts of precipitation. In *14th Conference on radar meteorology (preprints)*, 237–238. Boston: American Meteorological Society.

Kato, S. 1980. *Dynamics of the upper atmosphere*, 233 pp. Dordrecht: D. Reidel Publishing.

Kato, S., K. Fukuyama, K. Wakasugi, T. Sato, and S. Fukao. 1982. Middle atmosphere observations with large scale MST radar. Research Note. *Journal of the Meteorological Society of Japan* **144**, 1–55 (in Japanese).

Kato, S., T. Ogawa, T. Tsuda, T. Sato, I. Kimura, and S. Fukao. 1984. The middle and upper atmosphere radar: First results using a partial system. *Radio Science* 19: 1475–1484.

Kawashima, M., K. Tsuboki, and T. Asai. 1995. Maintenance mechanism and thermodynamic structure of a Baiu frontal rainband retrieved from dual Doppler radar observations. *Journal of the Meteorological Society of Japan* 73: 717–735.

Keenan, T.D. 2003. Hydrometeor classification with a C-band polarimetric radar. *Australian Meteorological Magazine* 52: 23–31.

Keer, D.E. 1951. *Propagation of short radio waves*, 728 pp. New York: McGraw-Hill.

Kilburn, C., S. Fukao, and M. Yamamoto. 1995. Extended period frequency domain interferometry observations at stratospheric and tropospheric heights. *Radio Science* 30: 1099–1109.

Kingsmill, D.E., and R.M. Wakimoto. 1991. Kinematic, dynamic, and thermodynamic analysis of a weakly sheared thunderstorm over northern Alabama. *Monthly Weather Review* 119: 262–297.

Kobayashi, T., and A. Adachi. 2001. Measurements of rain-drop breakup by using UHF wind profilers. *Geophysical Research Letters* 28: 4071–4072.

Kodaira, N. 1990. History of radar meteorology in Japan. In *Radar in meteorology*, ed. D. Atlas, 69–76. Boston: American Meteorological Society.

Komabayashi, M., T. Gonda, and K. Isono. 1964. Lifetime of water drops before breaking and size distribution of fragment drops. *Journal of the Meteorological Society of Japan* 42: 330–340.

Konrad, T.G. 1970. The dynamics of the convective process in clear air as seen by radar. *Journal of the Atmospheric Sciences* 27: 1138–1147.

Koscielny, A.J., R.J. Doviak, and R. Rabin. 1982. Statistical considerations in the estimation of divergence from single-Doppler radar and application to prestorm boundary-layer observations. *Journal of Applied Meteorology* 21: 197–210.

Kozu, T., T. Kawanishi, H. Kuroiwa, M. Kojima, K. Oikawa, H. Kumagai, K. Okamoto, M. Okumura, H. Nakatsuka, and K. Nishikawa. 2001. Development of precipitation radar onboard the Tropical Rainfall Measuring Mission (TRMM) satellite. *IEEE Transactions on Geoscience and Remote Sensing* 39: 102–116.

Kozu, T., T. Shimomai, Z. Akramin, Marzuki, Y. Shibagaki, and H. Hashiguchi. 2005. Intraseasonal variation of raindrop size distribution at Koto Tabang, West Sumatra, Indonesia. *Geophysical Research Letters* 32: L07803. doi:10.1029/2004GL022340.

Kropfli, R.A., S.Y. Matrosov, T. Uttal, B.W. Orr, A.S. Frisch, K.A. Clark, B.W. Bartram, R.F. Reinking, J.B. Snider, and B.E. Martner. 1995. Cloud physics studies with 8 mm wavelength radar. *Atmospheric Research* 35: 299–313.

Kudeki, E., and G.R. Stitt. 1987. Frequency domain interferometry: A high resolution radar technique for studies of atmospheric turbulence. *Geophysical Research Letters* 14: 198–201.

Kudeki, E., and F. Sürücü. 1991. Radar interferometric imaging of field-aligned plasma irregularities in the equatorial electrojet. *Geophysical Research Letters* 18: 41–44.

Kudeki, E., and R.F. Woodman. 1990. A post-statistic steering technique for MST radar applications. *Radio Science* 25: 591–594.

Kurosaki, S., M.D. Yamanaka, H. Hashiguchi, T. Sato, and S. Fukao. 1996. Vertical eddy diffusivity in the lower and middle atmosphere: A climatology based on the MU radar observations during 1986–1992. *Journal of Atmospheric and Terrestrial Physics* 58: 727–734.

Laird, B.G. 1981. On ambiguity resolution by random phase processing. In *20th International conference on radar meteorology (preprints)*, 327–331. Boston: American Meteorological Society.

Laird, B.G., and J.E. Evans. 1982. FAA weather radar surveillance requirements in the context of NEXRAD. *MIT Lincoln Laboratory Project Rep.*, ATC-112, DOT/FAA-RD-81-111.

Lane, J.A., and J.A. Saxton. 1952. Dielectric dispersion in pure polar liquids at very high radio frequencies. *Proceedings of the Royal Society A* 213: 400–408.

Larsen, M.F., R.D. Palmer, S. Fukao, R.F. Woodman, M. Yamamoto, T. Tsuda, and S. Kato. 1992. An analysis technique for deriving vector winds and in-beam incidence angles from radar interferometer measurements. *Journal of Atmospheric and Oceanic Technology* 9: 3–14.

Law, D.C., K.P. Moran, R.G. Frehlich, and R.G. Strauch. 1994. Maximum likelihood estimation of spectral moments in the presence of clutter. In *Extended abstracts of the third international symposium on tropospheric profiling: needs and technologies*, August 30–September 2, Hamburg, Germany, 216–218.

Le, K.D., R.D. Palmer, B.L. Cheong, T.-Y. Yu, G. Zhang, and S.M. Torres. 2010. Reducing the effects of noise on atmospheric imaging radars using multilag correlation. *Radio Science* 45: 1008. doi:10.1029/2008RS003989.

Leary, C.A., and R.A. Houze, Jr. 1979. The structure and evolution of convection in a tropical cloud cluster. *Journal of the Atmospheric Sciences* 36: 437–457.

Lhermitte, R.M. 1968. Turbulent air motion as observed by Doppler radar. In *13th Conference on radar meteorology (preprints)*, 498–503. Boston: American Meteorological Society.

Lhermitte, R.M. 1970. Dual-Doppler radar observations of convective storm circulation. In *14th Conference on radar meteorology (preprints)*, 139–144. Boston: American Meteorological Society.

Lhermitte, R.M., and L.J. Miller. 1970. Doppler radar methodology for the observation of convective storms. In *14th Conference on radar meteorology (preprints)*, 133–138. Boston: American Meteorological Society.

Liebe, H.J. 1985. An updated model for millimeter wave propagation in moist air. *Radio Science* 20: 1069–1089.

Lilly, D.K., D.E. Waco, and S.I. Adelfang. 1974. Stratospheric mixing estimated from high-latitude turbulence measurements. *Journal of Applied Meteorology* 13: 488–493.

Lindzen, R.S. 1984. Gravity waves in the mesosphere. In *Dynamics of the middle atmosphere*, 3–18. Tokyo: Dordrecht Press.

Liu, H., and V. Chandrasekar. 2000. Classification of hydrometeors based on polarimetric radar measurements: Development of fuzzy logic and neuro-fuzzy systems, and in situ verification. *Journal of Atmospheric and Oceanic Technology* 17: 140–164.

Liu, C.H., J. Röttger, G. Dester, S.J. Franke, and C.-J. Pan. 1991. The oblique spaced antenna method for measuring the atmospheric wind field. *Journal of Atmospheric and Oceanic Technology* 8: 247–258.

Liu, S., Q. Xu, and P. Zhang. 2005. Identifying Doppler velocity contamination caused by mitigating birds, Part II: Bays identification and probability tests. *Journal of Atmospheric and Oceanic Technology* 22: 1114–1121.

Lopez-Dekker, P.L., and S.J. Frasier. 2004. Radio acoustic sounding with a UHF volume imaging radar. *Journal of Atmospheric and Oceanic Technology* 21: 766–776.

Lott, F., and H. Teitelbaum. 1993. Topographic waves generated by a transient wind. *Journal of the Atmospheric Sciences* 50: 2607–2654.

Luce, H., M. Crochet, F. Dalaudier, and C. Sidi. 1995. Interpretation of VHF ST radar vertical echoes from in-situ temperature sheet observations. *Radio Science* 30: 1002–1025.

Luce, H., S. Fukao, and M. Yamamoto. 2001a. Validation of winds measured by MU radar with GPS radiosondes during the MUTSI campaign. *Journal of Atmospheric and Oceanic Technology* 18: 817–829.

Luce, H., M. Yamamoto, S. Fukao, and M. Crochet. 2001b. Extended radar observations with the frequency radar domain interferometric imaging (FII) technique. *Journal of Atmospheric and Solar-Terrestrial Physics* 63: 221–234.

Luce, H., S. Fukao, F. Dalaudier, and M. Crochet. 2002. Strong mixing events observed near the tropopause with the MU radar and high-resolution balloon techniques. *Journal of the Atmospheric Sciences* 59: 2885–2896.

Luce, H., G. Hassenpflug, M. Yamamoto, and S. Fukao. 2007. Comparisons of refractive index gradient and stability profiles measured by balloons and the MU radar at a high vertical resolution in the lower stratosphere. *Annales Geophysicae* 25: 47–57.

Luce, H., G. Hassenpflug, M. Yamamoto, S. Fukao, and K. Sato. 2008. High-resolution observations with MU radar of a KH instability triggered by an inertia-gravity wave in the upper part of a jet-stream. *Journal of the Atmospheric Sciences* 65: 1711–1718.

Lyons, R.G. 2004. *Understanding Digital Signal Processing*, 665 pp. 2nd ed. Boston: Prentice Hall.

Maeda, K., H. Alvarez, J. Aparici, J. May, and P. Reich. 1999. A 45-MHz continuum survey of the northern hemisphere. *Astronomy and Astrophysics Supplement Series* 140: 145–154.

Maekawa, Y., S. Fukao, T. Sato, S. Kato, and R.F. Woodman. 1984. Internal inertia-gravity waves in the tropical lower stratosphere observed by the Arecibo radar. *Journal of the Atmospheric Sciences* 41: 2359–2367.

Maguire, W.B., and S.K. Avery. 1995. Retrieval of raindrop size distributions using two Doppler wind profilers: model sensitivity testing. *Journal of Applied Meteorology* 33: 1623–1635.

Mahapatra, P.R. 1999. *Aviation weather surveillance systems*, 453 pp. Reston: AIAA.

Maki, M., S.G. Park, and V.N. Bringi. 2005. Effect of natural variations in rain drop size distributions on rain rate estimation of 3 cm wavelength polarimetric radar. *Journal of the Meteorological Society of Japan* 83: 871–893.

Manzini, E., and K. Hamilton. 1993. Middle atmospheric travelling waves forced by latent and convective heating. *Journal of the Atmospheric Sciences* 50: 2180–2200.

Marcuvitz, N. 1965. *Waveguide handbook*, 428 pp. New York: Dover Publications.

Marks, F.D., Jr., and R.A. Houze, Jr. 1987. Inner-core structure of Hurricane Alicia from airborne Doppler-radar observations. *Journal of the Atmospheric Sciences* 44: 1296–1317.

Marshall, J.S., and W.M. Palmer. 1948. The distribution of raindrops with size. *Journal of Meteorology* 5: 165–166.

Marshall, J.S., W. Hitschfeld, and K.L.S. Gunn. 1955. Advances in radar weather. *Advances in Geophysics* 2: 1–56.

Marshall, J.M., A.M. Peterson, and A.A. Barnes, Jr. 1972. Combined radar-acoustic sounding system. *Applied Optics* 11: 108–112.

Masuda, Y. 1988. Influence of wind and temperature on the height limit of a radio acoustic sounding system. *Radio Science* 23: 647–654.

Masuda, Y., J. Awaka, K. Nakamura, T. Adachi, and T. Tsuda. 1992. Analysis of the radio sounding system using a chirped acoustic wave. *Radio Science* 27: 681–691.

Matrosov, S.Y., K.A. Clark, and B.E. Martner. 2002. X-band polarimetric radar measurements of rainfall. *Journal of Applied Meteorology* 41: 941–952.

Matson, R.J., and A.W. Huggins. 1980. The direct measurement of the size, shapes and kinematics of falling hailstones. *Journal of the Atmospheric Sciences* 37: 1107–1125.

Matuura, N., Y. Masuda, H. Inuki, S. Kato, S. Fukao, T. Sato, and T. Tsuda. 1986. Radio acoustic measurement of temperature profile in the troposphere and stratosphere. *Nature* 323: 426–428.

May, P.T., and T.D. Keenan. 2005. Evaluation of microphysical retrievals from polarimetric radar with wind profiler data. *Journal of Applied Meteorology* 44: 827–838.

May, P.T., S. Fukao, P.J. Neiman, M.W. Kozleski, M.D. Yamanaka, S. Kato, M. Yamamoto, T. Sato, and T. Tsuda. 1992. MU radar observations of the wind field in the vicinity of the Baiu front during early July, 1987. *Beitraege zur Physik der Atmosphaere (Contributions To Atmospheric Physics)* 65: 3–11.

May, P.T., G.T. Holland, and W.L. Ecklund. 1994. Wind profiler observation of tropical storm Flo at Saipan. *Weather and Forecasting* 9: 410–426.

McCarthy, J., and J.W. Wilson. 1985. The Classify, Locate, and Avoid Wind Shear (CLAWS) project at Denver's Stapleton International Airport: Operational testing of terminal weather hazard warnings with an emphasis on microburst wind shear. In *Second international conference on the aviation weather system*, Montreal, 247–256. Boston: American Meteorological Society.

McCarthy, J., W. Frost, B. Terkel, R.J. Doviak, D.W. Camp, E.F. Blick, and K.L. Elmore. 1980. An airport wind shear detection and warning system using Doppler radar. In *19th Conference on radar meteorology (preprints)*, Miami, FL, 135–142. Boston: American Meteorological Society.

McKinley, D.W.R. 1961. *Meteo science and engineering*, 309 pp. New York: McGraw-Hill.

Mead, J.B., G. Hopcraft, S.J. Frasier, B.D. Pollard, C.D. Cherry, D.H. Schaubert, and R.E. McIntoch. 1998. A volume-imaging radar wind profiler for atmospheric boundary layer turbulence studies. *Journal of Atmospheric and Oceanic Technology* 15: 849–859.

Melnikov, V.M. 2006. One-lag estimators for cross-polarization measurements. *Journal of Atmospheric and Oceanic Technology* 23: 915–926.

Melnikov, V.M., R.J. Doviak, D.S. Zrnić, and D.J. Stensrud. 2011. Mapping Bragg scatter with a polarimetric WSR-88D. *Journal of Atmospheric and Oceanic Technology* 28: 1273–1285.

Meymaris, G. 2007. The use of spectral processing to improve radar spectral moment. In *23rd Conference on interactive information and processing systems for meteorology, oceanography, and hydrology (preprints)*, San Antonio, TX. Boston: American Meteorological Society, 8A.4.

Michelson, M., W.W. Shrader, and J.G. Wieler. 1990. Terminal Doppler weather radar. *Microwave Journal* 33: 139–148.

Miyashita, H., H. Ohmine, K. Nishizawa, S. Makino, and S. Urasaki. 1999. Electromagnetically coupled coaxial dipole array antenna. *IEEE Transactions on Antennas and Propagation* 47: 1716–1726.

Moran, K.P., E.B. Martner, M.J. Post, R.A. Kropfli, D.C. Welsh, and K.B. Widener. 1998. An unattended cloud-profiling radar for use in climate research. *Bulletin of the American Meteorological Society* 79: 443–455.

Mori, S., J.-I. Hamada, M.D. Yamanaka, Y.-M. Kodama, M. Kawashima, T. Shimomai, Y. Shibagaki, H. Hashiguchi and T. Sribimawati. 2006. Vertical wind characteristics in precipitating cloud systems over West Sumatera, Indonesia, observed with Equatorial Atmosphere Radar: Case Study of 23–24 April 2004 during the first CPEA campaign period. *Journal of the Meteorological Society of Japan* 84: 113–131.

Muraoka, Y., K. Kawahira, T. Sato, T. Tsuda, S. Fukao, and S. Kato. 1987. Characteristics of inertial gravity waves in the mesosphere observed by the MU radar. *Geophysical Research Letters* 14: 1154–1157.

Murayama, Y., T. Tsuda, and S. Kato. 1994. Seasonal variation of gravity wave activity in the lower atmosphere observed with the MU radar. *Journal of Geophysical Research* 99: 23057–23069.

Muschinski, A., and C. Wode. 1998. First in situ evidence for coexisting submeter temperature and humidity sheets in the lower free troposphere. *Journal of the Atmospheric Sciences* 55: 2893–2906.

Nakamura, K., and Y. Masuda. 1992. Development of a lower troposphere wind profiler at Communications Research Laboratory. In *International symposium on middle atmosphere science (preprints)*, Kyoto, 23–37 March, 48–49.

Nastrom, G.D., and D.C. Fritts. 1992. Sources of mesoscale variability of gravity waves, Part 1: Topographic excitation. *Journal of the Atmospheric Sciences* 49: 101–110.

Nathanson, F.E. 1991. *Radar design principles*, 720 pp. 2nd ed. Mendham: SciTech.

Nathanson, F.E., and J.P. Reilly. 1968. Radar precipitation echoes. *IEEE Transactions on Aerospace and Electronic Systems* AES-4: 505–514.

Nathanson, F.E., and P.L. Smith. 1972. A modified coefficient for the weather radar equation. In *15th Conference radar meteorology (preprints)*, 228–230. Boston: American Meteorological Society.

Nelson, S.P., and R.A. Brown. 1987. Error sources and accuracy of vertical velocities computed from multiple-Doppler radar measurements in deep convective storms. *Journal of Atmospheric and Oceanic Technology* 4: 234–238.

Nickel, U. 1988. Algebraic formulation of Kumaresan-Tuffs superresolution method, showing relation to ME and MUSIC method. *IEE Proceedings* 135: 7–10.

Nishi, N., M.K. Yamamoto, T. Shimomai, A. Hamada, and S. Fukao. 2007. Fine structure of vertical motion in the stratiform precipitation region observed by a VHF Doppler radar installed in Sumatra, Indonesia. *Journal of Applied Meteorology and Climatology* 46: 522–537.

Ochs, G.R. 1965. The large 50 Mc/s dipole array at Jicamarca Radar Observatory. *NBS Rep.*, 8772, 61 pp. National Bureau of Standards, Boulder, CO.

Ogawa, T., and T. Shimazaki. 1975. Diurnal variations of odd nitrogen and ionic densities in the mesosphere and lower thermosphere: Simultaneous solution of photochemical-diffusive equations. *Journal of Geophysical Research* 80: 3945–3960.

Oguchi, T. 1983. Electromagnetic wave propagation and scattering in rain and other hydrometeors. *Proceedings of the IEEE* 71: 1029–1078.

Ogura, H. 1998. *Introduction to stochastic process*, 212 pp. Tokyo: Morikita-Suppan (in Japanese).

Ogura, Y., and N.A. Phillips. 1962. Scale analysis of deep and shallow convection in the atmosphere. *Journal of the Atmospheric Sciences* 19: 1458–1476.

Ogura, H., and Y. Yoshida. 1981. Spectral analysis and subtraction of noise in radar signals. *IEEE Transactions on Aerospace and Electronic Systems* AES-17: 62–71.

Orlansky, I. 1975. A rational subdivision of scales for atmospheric processes. *Bulletin of the American Meteorological Society* 56: 527–530.

O'Sullivan, D., and T.J. Dunkerton. 1995. Generation of inertia-gravity waves in a simulated life cycle of baroclinic instability. *Journal of the Atmospheric Sciences* 52: 3695 3716.

Ottersten, H. 1969a. Atmospheric structure and radar backscattering in clear air. *Radio Science* 4: 1179–1193.

Ottersten, H. 1969b. Radar backscattering from the turbulent clear atmosphere. *Radio Science* 4: 1251–1255.

Ottersten, H. 1969c. Mean vertical gradient of potential refractive index in turbulent mixing and radar detection of CAT. *Radio Science* 4: 1247–1249.

Palmer, R.D., R.F. Woodman, S. Fukao, T. Tsuda, and S. Kato. 1990. Three-antenna poststatistic steering using the MU radar. *Radio Science* 25: 1105–1110.

Palmer, R.D., S. Fukao, M.F. Larsen, R.F. Woodman, M. Yamamoto, T. Tsuda, and S. Kato. 1991. VHF radar interferometry measurements of vertical velocity and the effects of tilted refractivity surfaces on standard Doppler measurements. *Radio Science* 26: 417–427.

Palmer, R.D., M.F. Larsen, E.L. Sheppard, S. Fukao, M. Yamamoto, T. Tsuda, and S. Kato. 1993. Poststatistic steering wind estimation in the troposphere and lower stratosphere. *Radio Science* 28: 261–271.

Palmer, R.D., S. Gopalam, T.Y. Yu, and S. Fukao. 1998. Coherent radar imaging using the Capon's method. *Radio Science* 33: 1585–1598.

Palmer, R.D., T.Y. Yu, and P.B. Chilson. 1999. Range imaging using frequency diversity. *Radio Science* 34: 1485–1496.

Palmer, R.D., B.L. Cheong, M.W. Hoffman, S.J. Frasier, and F.J. López-Dekker. 2005. Observations of the small-scale variability of precipitation using an imaging radar. *Journal of Atmospheric and Oceanic Technology* 22: 1122–1137.

Papoulis, A. 1991. *Provability, random variables, and stochastic processes*, 666 pp. 3rd ed. Boston: WCB/McGraw-Hill.

Park, S.-G., V.N. Bringi, V. Chandrasekar, M. Maki, and K. Iwanami. 2005a. Correction of radar reflectivity and differential reflectivity for rain attenuation at X band, Part I: Technical and empirical basis. *Journal of Atmospheric and Oceanic Technology* 22: 1621–1632.

Park, S.-G., M. Maki, K. Iwanami, V.N. Bringi, and V. Chandrasekar. 2005b. Correction of radar reflectivity and differential reflectivity for rain attenuation at X band, Part II: Evaluation and application. *Journal of Atmospheric and Oceanic Technology* 22: 1633–1655.

Park, H., A.V. Ryzhkov, D.S. Zrnić, and K.E. Kim. 2009. The hydrometeor classification algorithm for the polarimetric WSR-88D: Description and application to an MCS. *Weather and Forecasting* 24: 730–748.

Pasqualucci, F. 1984. Drop size distribution measurements in convective storms with a vertically pointing 35-GHz Doppler radar. *Radio Science* 19: 177–183.

Pekour, M.S., and R.L. Coulter. 1999. A technique for removing the effect of migrating birds in 915-MHz wind profiler data. *Journal of Atmospheric and Oceanic Technology* 16: 1941–1948.

Pfeifer, M., G. C. Craig, M. Hagen, and C. Keil. 2008. A polarimetric radar forward operator for model evaluation. *Journal of Applied Meteorology and Climatology* 47: 3202–3220.

Pfister, L., K.R. Chan, T.P. Bui, S. Bowen, M. Legg, B. Gary, K. Kelly, M. Proffitt, and W. Starr. 1993. Gravity waves generated by a tropical cyclone during the STEP tropical field program: A case study. *Journal of Geophysical Research* 98: 8611–8638.

Prichard, I.T., L. Thomas, and R.M. Worthington. 1995. The characteristics of mountain waves observed by radar near the west coast of Wales. *Annales Geophysicae* 13: 757–767.

Probert-Jones, J.R. 1962. The radar equation in meteorology. *Quarterly Journal of the Royal Meteorological Society* 88: 485–495.

Protat, A., and I. Zawadzki. 1999. A variational method for real-time retrieval of three-dimensional wind multiple-Doppler bistatic radar network data. *Journal of Atmospheric and Oceanic Technology* 16: 432–449.

Pruppacher, H.R., and K.V. Beard. 1970. A wind tunnel investigation of the internal circulation and shape of water drops falling at terminal velocity in air. *Quarterly Journal of the Royal Meteorological Society* 96: 247–256.

Ramo, S., J.R. Whinnery, and T. Van Duzer. 1965. *Fields and waves in communication electronics*, 265 pp. New York: Wiley.

Rao, P.B., A.R. Jain, P. Kishore, P. Balamuralidhar, S.H. Damle, and G. Viswanathan. 1995. Indian MST radar, 1. System description and sample vector wind measurements in ST mode. *Radio Science* 30: 1125–1138.

Rao, Q., H. Hashiguchi, and S. Fukao. 2003. Study on ground clutter prevention fences for boundary layer radars. *Radio Science* 38: 1030. doi:10.1029/2001RS002489.

Ratcliffe, J.A. 1956. Some aspects of diffraction theory and their application to the ionosphere. *Reports on Progress in Physics* 19: 188–267.

Ray, P.S., R.J. Doviak, G.B. Walker, D. Sirmans, J. Carter, and B. Bumgarner. 1975. Dual-Doppler observation of tornadic storm. *Journal of Applied Meteorology* 17: 1201–1212.

Ray, P.S., C.L. Zeigler, R.J. Serafin, and W. Bumgarner. 1980. Single- and multiple-Doppler radar observations of tornadic storms. *Monthly Weather Review* 108: 1607–1625.

Reid, I.M., and R.A. Vincent. 1987. Measurements of mesospheric gravity wave momentum fluxes and mean flow acceleration at Adelaide, Australia. *Journal of Atmospheric and Terrestrial Physics* 49: 443–460.

Renggono, F., H. Hashiguchi, S. Fukao, M.D. Yamanaka, S.-Y. Ogino, N. Okamoto, F. Murata, B.P. Sitorus, M. Kudsy, M. Kartasasmita, and G. Ibrahim. 2001. Precipitating clouds observed by 1.3-GHz boundary layer radars in equatorial Indonesia. *Annales Geophysicae* 19: 889–897.

Renggono, F., M.K. Yamamoto, H. Hashiguchi, S. Fukao, T. Shimomai, M. Kawashima, and M. Kudsy. 2006. Raindrop size distribution observed with the Equatorial Atmosphere Radar (EAR) during the CPEA-I observation campaign. *Radio Science* 41: RS5002. doi:10.1029/2005RS003333.

Rodi, A.R., K.L. Elmore, and W.P. Mahoney. 1983. Aircraft and Doppler air motion comparisons in a JAWS microburst. *21st Conference on Radar Meteorology (preprints)*, Edmonton, 624–629. Boston: American Meteorological Society.

Röttger, J. 1979. VHF radar observations of a frontal passage. *Journal of Applied Meteorology* 18: 85–91.

Röttger, J. 1980a. Reflection and scattering of VHF radar signals from atmospheric refractivity structures. *Radio Science* 15: 259–276.

Röttger, J. 1980b. Structure and dynamics of the stratosphere and mesosphere revealed by VHF radar investigations. *Pure and Applied Geophysics* 118: 494–527.

Röttger, J. 1981. Investigations of lower and middle atmosphere dynamics with spaced antenna drifts radars. *Journal of Atmospheric and Terrestrial Physics* 43: 277–292.

Röttger, J., and H.M. Ierkic. 1985. Postbeam steering and interferometry applications of VHF radars to study winds, waves, and turbulence in the lower and middle atmosphere. *Radio Science* 20: 1461–1480.

Röttger, J., and M.F. Larsen. 1990. UHF/VHF radar techniques for atmospheric research and wind profiler applications.In *Radar in meteorology*, ed. D. Atlas, 235–281. Boston: American Meteorological Society.

Röttger, J., and C.H. Liu. 1978. Partial reflection and scattering of VHF radar signals from the clear atmosphere. *Geophysical Research Letters* 5: 357–360.

Röttger, J., and R.A. Vincent. 1978. VHF radar studies of tropospheric velocities and irregularities using spaced antenna techniques. *Geophysical Research Letters* 5: 917–920.

Röttger, J., J. Klostermeyer, P. Czechowsky, R. Rüster, and G. Schmidt. 1978. Remote sensing of the atmosphere by VHF radar experiment. *Naturwissenschaften* 65: 285–296.

Röttger, J., C.-H. Liu, J.K. Chao, A.J. Chen, Y.H. Chu, I.-J. Fu, C.M. Huang, Y.W. Kiang, F.S. Kuo, and C.J. Pan. 1990a. The Chung-Li VHF radar: Technical layout and a summary of initial results. *Radio Science* 25: 478–502.

Röttger, J., C.-H. Liu, C.J. Pan, and I.-J. Fu. 1990b. Spatial interferometry measurements with the Chung-Li VHF radar. *Radio Science* 25: 503–515.

Ryde, J.W. 1946. The attenuation and radar echoes produced at centimetre wavelengths by various meteorological phenomena. In *Meteorological factors in radio wave propagation*, 169–188. London: Physical Society.

Ryzhkov, A.V., and D.S. Zrnić. 1998. Polarimetric rainfall estimation in the presence of anomalous propagation. *Journal of Atmospheric and Oceanic Technology* 15: 1320–1330.

Ryzhkov, A.V., S.E. Giangrande, and T.J. Schuur. 2005. Rainfall estimation with a polarimetric prototype of WSR-88D. *Journal of Applied Meteorology* 44: 502–515.

Saad, Y. 2003. *Iterative methods for space linear systems*, 528 pp. Philadelphia: Siam.

Sachindananda, M., and D.S. Zrnić. 1985. Z_{DR} measurement considerations for a fast scan capability radar. *Radio Science* 20: 907–922.

Sachidananda, M., and D.S. Zrnić. 1986. Recovery of spectral moments from overlaid echoes in a Doppler weather radar. *IEEE Transactions on Geoscience and Remote Sensing* 24: 751–764.

Sachidananda, M., and D.S. Zrnić. 1999. Systematic phase codes for resolving range overlaid signals in a Doppler weather radar. *Journal of Atmospheric and Oceanic Technology* 16: 1351–1363.

Sakakibara, H., M. Ishihara, A. Tabata, K. Akaeda, and T. Yokoyama. 1991. Evolution and structure of a cold-frontal precipitation system over the subtropical ocean. In *International conference on mesoscale meteorology and TAMEX (preprints)*, Taipei, R. O. C., 173–181. Boston: American Meteorological Society.

Sasaoka, M. 2003. Improvement of wind boundary layer radar using a grouping algorithm. *Tenki, Journal of the Meteorological Society* 50: 161–174 (in Japanese).

Sato, K. 1989. An inertia gravity wave associated with a synoptic-scale pressure trough observed by the MU radar. *Journal of the Meteorological Society of Japan* 67: 325–333.

Sato, K. 1990. Vertical wind disturbances in the troposphere and lower stratosphere observed by the MU radar. *Journal of the Atmospheric Sciences* 47: 2803–2817.

Sato, K. 1993. Small-scale wind disturbances observed by the MU radar during the passage of Typhoon Kelly. *Journal of the Atmospheric Sciences* 50: 519–537.

Sato, T. 1988. Radar principles. In *Lecture notes of International School on Atmospheric Radar (ISAR)*, ed. S. Fukao, 19–53. Kyoto: Kyoto University.

Sato, T., and R.F. Woodman. 1982. Spectral parameter estimation of CAT radar echoes in the presence of fading clutter. *Radio Science* 17: 817–826.

Sato, K., H. Hashiguchi, and S. Fukao. 1995. Gravity waves and turbulence associated with cumulus convection observed with the UHF/VHF clear-air Doppler radars. *Journal of Geophysical Research* 100: 7111–7119.

Sato, K., M. Tsutsumi, T. Sato, A. Saito, Y. Tomikawa, T. Aso, T. Yamauchi, and M. Ejiri. 2006. Program of the Antarctic Syowa MST/IS radar (Pansy). *European Geosciences Union* 8: 1607-796/gra/EGU06-A-05594.

Sato, T., H. Iwai, I. Kimura, S. Fukao, M. Yamamoto, T. Tsuda, and S. Kato. 1990. Computer processing for deriving drop-size distributions and vertical air velocities from VHF Doppler radar spectra. *Radio Science* 25: 961–973.

Sato, T., N. Ao, M. Yamamoto, S. Fukao, T. Tsuda, and S. Kato. 1991. A typhoon observed with the MU Radar. *Monthly Weather Review* 119: 755–768.

Satoh, S., and J. Wurman. 2003. Accuracy of wind fields observed by a bistatic Doppler radar network. *Journal of Atmospheric and Oceanic Technology* 20: 1077–1091.

Sauvageot, H. 1992. *Radar meteorology*, 366 pp. Boston: Artech House.

Schmidt, R.O. 1986. Multiple emitter location and signal parameter estimation. *IEEE Transactions on Antennas and Propagation* AP-34: 276–280.

Sekhon, R.S., and R.C. Srivastava. 1970. Snow-size spectra and radar reflectivity. *Journal of the Atmospheric Sciences* 27: 299–307.

Sekhon, R.S., and R.C. Srivastava. 1971. Doppler radar observations of drop-size distributions in a thunderstorm. *Journal of the Atmospheric Sciences* 28: 983–994.

Seliga, T.A., and V.N. Bringi. 1976. Potential use of radar differential reflectivity measurements at orthogonal polarizations for measuring precipitations. *Journal of Applied Meteorology* 15: 69–76.

Sempre-Torres D., J.M. Porra, and J.D. Creutin. 1994. A general formulation for raindrop size distribution. *Journal of Applied Meteorology* 33: 1494–1502.

Shibagaki, Y., M.D. Yamanaka, H. Hashiguchi, A. Watanabe, H. Uyeda, Y. Maekawa, and S. Fukao. 1997. Hierarchical structures of vertical velocity variations and precipitating clouds near the Baiu frontal cyclone center observed by the MU and meteorological radars. *Journal of the Meteorological Society of Japan* 75: 569–596.

Shibagaki, Y., M.D. Yamanaka, S. Shimizu, H. Uyeda, A. Watanabe, Y. Maekawa, and S. Fukao. 2000. Meso-β to -γ-scale wind circulations associated with precipitating clouds near Baiu front observed by the MU and meteorological radars. *Journal of the Meteorological Society of Japan* 78: 69–91.

Shibagaki, Y., M.D. Yamanaka, M. Kita-Fukase, H. Hashiguchi, Y. Maekawa, and S. Fukao. 2003. Meso-α-scale wind field and precipitating clouds in Typhoon 9426 (Orchid) observed by the MU radar. *Journal of the Meteorological Society of Japan* 81: 211–228.

Shusse, Y., K. Nakagawa, N. Takahashi, S. Satoh, and T. Iguchi. 2009. Characteristics of polarimetric radar variables in three types of rainfalls in a baiu front event over the East China Sea. *Journal of the Meteorological Society of Japan* 87: 865–875.

Siggia, A. 1983. Processing phase codes radar signals with adaptive digital filters. In *21st International conference on radar meteorology (preprints)*, Edmonton, AB, Canada, 167–172. Boston: American Meteorological Society.

Siggia, A.D., and R.E. Passarelli, Jr. 2004. Gaussian model adaptive processor (GMAP) for improved ground clutter cancellation and moment calculation. In *Proceedings. Third European Conference on Radar Meteorology (ERAD)*, Visby, Island of Gotland, Sweden, 67–73.

Silver, S. 1949. *Microwave antenna theory and design*, 623 pp. New York: McGraw-Hill.

Silverman, R.A. 1956. Turbulent mixing theory applied to radio scattering. *Journal of Applied Physics* 27: 690–705.

Skolnik, M.I. ed. 1970. Radar handbook, 1536 pp. New York: McGraw-Hill.

Skolnik, M.I. ed. 1981. *Introduction to radar systems*, 581 pp. 2nd ed. New York: McGraw-Hill.

Skolnik, M.I. ed. 1990. *Radar handbook*, 1200 pp. 2nd ed. New York: McGraw-Hill.

Skolnik, M.I. ed. 2001. *Introduction to radar systems*, 772 pp. 3rd ed. New York: McGraw-Hill.

Smith, P.L. 1984. Equivalent radar reflectivity factor for snow and ice particles. *Journal of Climate and Applied Meteorology* 23: 1258–1260.

Smith, P.L. 1986. On the sensitivity of weather radar. *Journal of Atmospheric and Oceanic Technology* 3: 704–713.

Smith, S.A., D.C. Fritts, and T.E. VanZandt. 1987. Evidence of a saturation spectrum of atmospheric waves. *Journal of the Atmospheric Sciences* 44: 1404–1410.

Spano, E., and O. Ghebrebrhan. 1996a. Pulse coding techniques for ST/MST radar systems: A general approach based on a matrix formulation. *IEEE Transactions on Geoscience and Remote Sensing* GE-34: 304–316.

Spano, E., and O. Ghebrebrhan. 1996b. Sequences of complementary codes for the optimum decoding of truncated ranges and high sidelobe suppression factors for ST/MST radar systems. *IEEE Transactions on Geoscience and Remote Sensing* GE-34: 330–345.

Srivastava, R.C. 1967. On the role of coalescene between raindrops in shaping their size distribution. *Journal of the Atmospheric Sciences* 24: 287–292.

Srivastava, R.C. 1971. Size distribution of raindrops generated by their breakup and coalescence. *Journal of the Atmospheric Sciences* 28: 410–415.

Srivastava, R.C., A. R. Jameson, and P. H. Hildebrand. 1979. Time-domain computation of mean and variance of Doppler spectra. *Journal of Applied Meteorology* 18: 189–194.

Straka, J.M., D. Zrnić, A.V. Ryzhkov. 2000. Bulk hydrometeor classification and quantification using polarimetric radar data: Synthesis of relations. *Journal of Applied Meteorology* 39: 1341–1372.

Stratton, J.A. 2007. *Electromagnetic theory*, 615 pp. Piscataway: IEEE Press.

Stutzman, W.L. 1998. Estimating directivity and gain of antennas. *IEEE Antennas and Propagation Magazine* 40: 7–11.

Sumi, A. 1989. Short-period fluctuation of the lower tropospheric winds observed by the MU radar. *Journal of the Meteorological Society of Japan* 67: 167–175.

Tabata, A., H. Sakakibara, M. Ishihara, K. Matsuura, and Z. Yanagisawa. 1992. A general view of the structure of Typhoon 8514 observed by dual-Doppler radar—From outer rainbands to eyewall clouds. *Journal of the Meteorological Society of Japan* 70: 897–917.

Takeda, S., T. Nakamura, and T. Tsuda. 2001. An improvement of wind velocity estimation from radar Doppler spectra in the upper mesosphere. *Annales Geophysicae* 19: 837–843.

Tatarskii, V.I. 1971. *The effects of the turbulent atmosphere on wave propagation*, 472 pp. Translated from Russian, Jerusalem.

Teschl, T., W.L. Randeu, M. Schönhuber, and R. Teschl. 2008. Simulation of polarimetric radar variables in rain at S-, C- and X-band wavelengths. *Advances in Geosciences* 16: 27–32

Teshiba, M., H. Hashiguchi, S. Fukao, and Y. Shibagaki. 2001. Typhoon 9707 observations with the MU radar and L-band boundary layer radar. *Annales Geophysicae* 19: 925–931.

Teshiba, M., H. Fujita, H. Hashiguchi, Y. Shibagaki, M.D. Yamanaka, and S. Fukao. 2005. Detailed structure within a tropical cyclone "eye." *Geophysical Research Letters* 32: L24805. doi:10.1029/2005GL023242.

Testud, J. 1982. Three-dimensional wind field analysis from Doppler radar data. In *Mesoscale meteorology—theories, observations and models*, eds. D.L. Lilly, and T. Gal-Chen, 711–754. Dordrecht: D. Reidel Publishing.

Torlaschi, E., and A.R. Holt. 1993. Separation of propagation and backscattering effects in rain for circular polarization diversity S-band radar. *Journal of Atmospheric and Oceanic Technology* 10: 465–477.

Townsend, A.A. 1965. Excitation of internal waves by a turbulent boundary layer. *Journal of Fluid Mechanics* 22: 241–252.

Townsend, A.A. 1966. Internal waves produced by a convective layer. *Journal of Fluid Mechanics* 24: 307–319.

Tsuda, T. 2001. Measurements of atmospheric temperature with the RASS. *Choonpa Techno* 13: 18–22 (in Japanese).

Tsuda, T., T. Sato, K. Hirose, S. Fukao, and S. Kato. 1986. MU radar observations of the aspect sensitivity of backscattered VHF echo power in the troposphere and lower stratosphere. *Radio Science* 21: 971–980.

Tsuda, T., Y. Masuda, H. Inuki, K. Takahashi, T. Takami, T. Sato, S. Fukao, and S. Kato. 1989a. High time resolution monitoring of tropospheric temperature with a Radio Acoustic Sounding System (RASS). *Pure and Applied Geophysics* 130: 497–507.

Tsuda, T., T. Inoue, D.C. Fritts, T.E. VanZandt, S. Kato, T. Sato, and S. Fukao. 1989b. MST radar observations of a saturated gravity wave spectrum. *Journal of the Atmospheric Sciences* 46: 2440–2447.

Tsuda, T., S. Kato, T. Yokoi, T. Inoue, M. Yamamoto, T.E. VanZandt, S. Fukao, and T. Sato. 1990a. Gravity waves in the mesosphere observed with the middle and upper atmosphere radar. *Radio Science* 26: 1005–1018.

Tsuda, T., Y. Murayama, M. Yamamoto, S. Kato, and S. Fukao. 1990b. Seasonal variation of momentum flux in the mesosphere observed with the MU radar. *Geophysical Research Letters* 17: 725–728.

Tsuda, T., T. Adachi, Y. Masuda, S. Fukao, and S. Kato. 1994. Observations of tropospheric temperature fluctuations with the MU radar-RASS. *Journal of Atmospheric and Oceanic Technology* 11: 50–62.

Tsuda, T., T.E. VanZandt, and H. Saito. 1997. Zenith-angle dependence of VHF specular reflection echoes in the lower atmosphere. *Journal of Atmospheric and Terrestrial Physics* 59: 766–776.

Tsuda, T., M. Miyamoto, and J. Furumoto. 2001. Estimation of a humidity profile using turbulence echo characteristics. *Journal of Atmospheric and Oceanic Technology* 18: 1214–1222.

Tsutsumi, M., T. Tsuda, and T. Nakamura. 1994. Temperature fluctuations near the mesopause inferred from meteor observations with the middle and upper atmosphere radar. *Radio Science* 29: 599–610.

Ulaby, F.T., R.K. Moore, and A.K. Fung. 1981. *Microwave remote sensing I, microwave remote sensing fundamentals and radiometry*, 456 pp. Norwood: Artech House.

Ulaby, F.T., R.K. Moore, and A.K. Fung. 1986. *Microwave remote sensing III, from theory to application*, 2162 pp. Norwood: Artech House.

Ulbrich, C.W. 1983. Natural variations in the analytical form of the raindrop-size distribution. *Journal of Climate and Applied Meteorology* 22: 1764–1775.

Umemoto, Y., M. Teshiba, Y. Shibagaki, H. Hashiguchi, M.D. Yamanaka, S. Fukao, and X-BAIU-99 and X-BAIU-02 Observational Groups. 2004. Combined wind profiler-weather radar observations of orographic rainband around Kyushu, Japan in the Baiu season. *Annales Geophysicae* 22: 3971–3982.

Ushiyama, T., M. Kawashima, and Y. Fujiyoshi. 2003. Heating distribution by cloud systems derived from Doppler radar observation in TOGA-COARE. *Journal of the Meteorological Society of Japan* 81: 1407–1434.

Ushiyama, T., M. Katsumata, and R. Shiooka. 2005. Idealized simulation of dual Doppler radar observation using numerically simulated clouds. *JAMSTEC Report of Research and Development* 1: 37–43.

Uyeda, H., and D.S. Zrnić. 1986. Automatic detection of gust fronts. *Journal of Atmospheric and Oceanic Technology* 3: 36–50.

Van Baelen, J.S., A.D. Richmond, T. Tsuda, S.K. Avery, S. Kato, S. Fukao, and M. Yamamoto. 1991. Radar interferometry technique and anisotropy of the echo power distribution: First results. *Radio Science* 26: 1315–1326.

Van de Hulst, H.D. 1957. *Light scattering by small particles*, 470 pp. New York: Wiley.

Van Vleck, J.H. 1947a. Absorption of microwaves by oxygen. *Physical Review* 71: 413–424.

Van Vleck, J.H. 1947b. The absorption of microwaves by uncondensed water vapor. *Physical Review* 71: 425–433.

VanZandt, T.E., and R.A. Vincent. 1983. Is VHF Fresnel reflectivity due to low-frequency waves. In *Handbook for MAP*, vol. 9, 78–80. Urbana: ICSU Scientific Committee on Solar-Terrestrial Physics (SCOSTEP).

Vincent, R.A., and I.M. Reid. 1983. HF Doppler measurements of mesospheric gravity wave momentum fluxes. *Journal of the Atmospheric Sciences* 40: 1321–1333.

Vincent, R.A., S. Dullaway, A. MacKinnon, I.M. Reid, F. Zink, P.T. May, and B.H. Johnson. 1998. A VHF boundary layer radar: first results. *Radio Science* 33: 845–860.

Vivekanandan, J., G. Zhang, and E. Brandes. 2004. Polarimetric radar estimators based on a constrained gamma drop size distribution model. *Journal of Applied Meteorology* 43: 217–230.

Wada, M., and H. Uyeda. 2011. Solid-state weather radar which reached the practical use stage. In *35th Conference on radar meteorology (preprints)*. Pittsburgh: American Meteorological Society, p12.163.

Wait, J.R. 1962. *Electromagnetic waves in stratified media*, 372 pp. Oxford: Pergamon.

Wakasugi, K., A. Mizutani, M. Matsuo, S. Fukao, and S. Kato. 1986. A direct method for deriving drop-size distribution and vertical air velocities from VHF Doppler radar spectra. *Journal of Atmospheric and Oceanic Technology* 3: 623–629.

Wakimoto, R.M. 1985. Forecasting dry microburst activity over the high plains. *Monthly Weather Review* 113: 1131–1143.

Waldteufel, P., and H. Corbin. 1979. On the analysis of single Doppler data. *Journal of Applied Meteorology* 18: 532–542.

Wang, Y., V. Chandrasekar, and B. Dolan. 2008. Development of scan strategy for dual Doppler retrieval in a networked radar system. In *IEEE international of geoscience and remote sensing symposium, 2008*, vol. 5, V-322–V-325.

Watanabe, A., S. Fukao, M.D. Yamanaka, A. Sumi, and H. Uyeda. 1994. A rotor circulation near the Baiu front observed by the MU radar. *Journal of the Meteorological Society of Japan* 72: 91–105.

Weinman, J.A., R. Meneghini, and K. Nakamura. 1990. Retrieval of precipitation profiles from airborne radar and passive radiometer measurements: Comparison with dual-frequency radar measurements. *Journal of Applied Meteorology* 29: 981–993.

Weinstock, J. 1978. Vertical turbulent diffusion in a stably stratified fluid. *Journal of the Atmospheric Sciences* 35: 1022–1027.

Weinstock, J. 1981. Energy dissipation rates of turbulence in the stable free atmosphere. *Journal of the Atmospheric Sciences* 38: 880–883.

Wexler, R. and D. Atlas. 1963. Radar reflectivity and attenuation of rain. *Journal of Applied Meteorology* 2: 276–280.

Wilczak, J.M., R.G. Strauch, F.M. Ralph, B.L. Weber, D.A. Merritt, J.R. Jordan, D.E. Wolfe, L.K. Lewis, D.B. Wuertz, J.E. Gaynor, S.A. McLaughlin, R.R. Rogers, A.C. Riddle, and T.S. Dye. 1995. Contamination of wind profiler data by migrating birds: Characteristics of corrupted data and potential solutions. *Journal of Atmospheric and Oceanic Technology* 12: 449–467.

Williams, C.R., W.L. Ecklund, and K.S. Gage. 1995. Classification of precipitating clouds in the tropics using 915-MHz wind profilers. *Journal of Atmospheric and Oceanic Technology* 12: 996–1012.

Willis, P.T. 1984. Functional fits to some observed drop size distributions and parameterization of rain. *Journal of the Atmospheric Sciences* 41: 1648–1661.

Wilson, R. 2004. Turbulent diffusivity in the free atmosphere inferred from MST radar measurements: A review. *Annales Geophysicae* 22, 3869–3887.

Wolfson, M. M. 1983. Doppler radar observations of an Oklahoma downburst. In *21st Conference on radar meteorology (preprints)*, 590–595. Edmonton: American Meteorological Society.

Wood, V.T., and R.A. Brown. 1983. Single Doppler velocity signatures: An atlas of patterns in clear air/widespread precipitation and convective storms. In *NOAA technical memorandum ERL NSSL-95*, 71 pp. Norman: NOAA Environmental Research Laboratories.

Woodman R.F. 1980. High altitude-resolution stratospheric measurements with the Arecibo 2380-MHz radar. *Radio Science* 15: 423–430.

Woodman, R.F., and A. Guillén. 1974. Radar observations of winds and turbulence in the stratosphere and mesosphere. *Journal of the Atmospheric Sciences* 31: 493–505.

Worthington, R.M., R.D. Palmer, and S. Fukao. 1999. Complete maps of the aspect sensitivity of VHF atmospheric radar echoes. *Annales Geophysicae* 17: 1116–1119.

Wurman, J., S. Heckman, and D. Boccipio. 1993. A bistatic multiple-Doppler radar network. *Journal of Applied Meteorology* 32: 1802–1814.

Wurman, J., J. Straka, and E. Rasmussen. 1996. Fine scale Doppler radar observation of tornadoes. *Science* 272: 1774–1777.

Yamada, Y. 1997. Numerical estimation of error variance in horizontal divergence for the adjustment of vertical winds derived from conical-scan-based dual-Doppler radar data based on the "floating boundary condition" concept. *Papers in Meteorology and Geophysics* 48: 49–65.

Yamamoto, M., T. Tsuda, S. Kato, T. Sato, and S. Fukao. 1987. A saturated inertia gravity wave in the mesosphere observed by the middle and upper atmosphere radar. *Journal of Geophysical Research* 92: 11993–11999.

Yamamoto, M., T. Tsuda, S. Kato, T. Sato, and S. Fukao, 1988. Interpretation of the structure of mesospheric turbulence layers in terms of inertia gravity waves. *Physica Scripta* 37: 645–650.

Yamamoto, M., S. Fukao, R.F. Woodman, T. Ogawa, T. Tsuda, and S. Kato. 1991. Mid-latitude E-region field-aligned irregularities observed with the MU radar. *Journal of Geophysical Research* 96: 15943–15949.

Yamamoto, M.K., H. Hashiguchi, S. Fukao, Y. Shibano, and K. Imai. 2002. Development of a transportable 3-GHz wind profiler for wind and precipitation studies. *Journal of the Meteorological Society of Japan* 80: 273–283.

Yamamoto, M.K., M. Oyamatsu, T. Horinouchi, H. Hashiguchi, and S. Fukao. 2003a. High time resolution determination of the tropical tropopause by the equatorial atmosphere radar. *Geophysical Research Letters* 30: 2094. doi:10.1029/2003GL018072.

Yamamoto, M.K., M. Fujiwara, T. Horinouchi, H. Hashiguchi, and S. Fukao. 2003b. Kelvin-Helmholtz instability around the tropical tropopause observed with the equatorial atmosphere radar. *Geophysical Research Letters* 30: 1476. doi:10.1029/2002GL016685.

Yamanaka, M.D., S. Fukao, H. Matsumoto, T. Sato, T. Tsuda, and S. Kato. 1989. Internal gravity wave selection in the upper troposphere and lower stratosphere observed by the MU radar: Preliminary results. *Pure and Applied Geophysics* 130: 481–495.

Yamauchi, H., O. Suzuki, and K. Akaeda. 2006. A hybrid multi-PRI method to dealias Doppler velocities. *Scientific Online Letters on the Atmosphere* 2: 92–95.

Yamauchi, H., O. Suzuki, and K. Akaeda. 2007. Range extension of Doppler radar by combined use of low-PRF and phase diversity processed dual-PRF observations. In *33rd Conference on radar meteorology (preprints)*. Cairns: American Meteorological Society, p7.5.

Yeh, K.C., and C.H. Liu. 1972. *Theory of ionospheric waves*, 464 pp. New York: Academic.

Yoshizaki, M., and H. Seko. 1994. A retrieval of thermodynamic and microphysical variables by using wind data in simulated multi-cellular convective storms. *Journal of the Meteorological Society of Japan* 72: 31–42.

Yu, T.-Y., and R.D. Palmer. 2001. Atmospheric radar imaging using multiple-receiver and multiple-frequency techniques. *Radio Science* 36: 1493–1503. doi:10.1029/2000RS002622.

Yu, T.-Y., and W.O.J. Brown. 2004. High-resolution atmospheric profiling using combinated spaced antenna and range imaging techniques. *Radio Science* 39: 1011. doi:10.1029/2003RS002907.

Yu, T.-Y., J. Furumoto, and M. Yamamoto. 2010. Clutter suppression for high-resolution atmospheric observations using multiple receivers and multiple frequencies. *Radio Science* 45: RS4011. doi:10.1029/2009RS004330.

Zhang, P., S. Liu, and Q. Xu. 2005. Identifying Doppler velocity contamination caused by mitigating birds, Part I: Feature extraction and quantification. *Journal of Atmospheric and Oceanic Technology* 22: 1105–1113.

Zhong, S., J.D. Fast, and X. Bian. 1996. A case study of the Great Plains low-level jet using wind profiler network data and a high-resolution mesoscale model. *Monthly Weather Review* 124: 785–806.

Ziemer, R.E., W.H. Tranter, and D.R. Fannin. 1998. *Signals and systems: Continuous and discrete*. 4th ed, 622 pp. New Jersey: Prentice Hall.

Zrnić, D.S. 1979. Estimation of spectral moments for weather echoes. *IEEE Transactions on Geoscience Electronics* GE-17: 113–128.

Zrnić, D.S., and P. Mahapatra. 1985. Two methods of ambiguity resolution in pulsed Doppler weather radars. *IEEE Transactions on Aerospace and Electronic Systems* 21: 470–483.

Zrnić, D.S., A. Ryzhkov, J. Straka, Y. Liu, and J. Vivekanandan. 2001. Testing a procedure for automatic classification of hydrometeor types. *Journal of Atmospheric and Oceanic Technology* 18: 892–913.

Zrnić, D.S., V.M. Melnikov, and J.K. Carter. 2006. Calibrating differential reflectivity on the WSR-88D. *Journal of Atmospheric and Oceanic Technology* 23: 892–913.

Zrnić, D.S., V.M. Melnikov, and R.J. Doviak. 2012. A draft report on issues and challenges for polarimetric measurement of weather with an agile-beam phased array radar. http://publications.nssl.noaa.gov/mpar_reports/MPAR-WEB_RPT.pdf, 132pp.

Index

S. Fukao and K. Hamazu, *Radar for Meteorological and Atmospheric Observations*,
DOI 10.1007/978-4-431-54334-3, © Springer Japan 2014

MIX
Papier aus verantwortungsvollen Quellen
Paper from responsible sources
FSC® C105338

If you have any concerns about our products,
you can contact us on
ProductSafety@springernature.com

In case Publisher is established outside the EU,
the EU authorized representative is:
Springer Nature Customer Service Center GmbH
Europaplatz 3, 69115 Heidelberg, Germany

Printed by Libri Plureos GmbH
in Hamburg, Germany